DIGITAL SYSTEM DESIGNS AND PRACTICES

Using Verilog HDL and FPGAs

DIGITAL SYSTEM DESIGNS AND PRACTICES
Using Verilog HDL and FPGAs

Ming-Bo Lin

Department of Electronic Engineering
National Taiwan University of Science and Technology
Taipei, Taiwan

John Wiley & Sons (Asia) Pte Ltd

Other Wiley Editorial Offices

John Wiley & Sons, Ltd, The Atrium, Southern Gate, Chichester, West Sussex, PO19 8SQ, UK

John Wiley & Sons Inc., 111 River Street, Hoboken, NJ 07030, USA

Jossey-Bass, 989 Market Street, San Francisco, CA 94103-1741, USA

Wiley-VCH Verlag GmbH, Boschstr. 12, D-69469 Weinheim, Germany

John Wiley & Sons Australia Ltd, 42 McDougall Street, Milton, Queensland 4064, Australia

John Wiley & Sons Canada Ltd, 6045 Freemont Blvd, Mississauga, ONT, L5R 4J3, Canada

Wiley also publishes its books in a variety of electronic formats. Some content that appears in print may not be available in electronic books.

Library of Congress Cataloging-in-Publication Data

Lin, Ming-Bo.
    Digital system designs and practices: using Verilog HDL and FPGAs / Ming-Bo Lin.
        p. cm.
    Includes bibliographical references and index.
    ISBN 978-0-470-82323-1 (cloth)
    1. Digital electronics. 2. Field programmable gate arrays. 3. Verilog (Computer hardware description
        language) I. Title.
    TK7868.D5L535 2008
    621.381–dc22   2008002506

ISBN 978-0-470-82323-1 (HB)

Typeset in 10.5/12pt Times by Thomson Digital, Noida, India.
Printed and bound in Singapore by Markono Print Media Pte Ltd, Singapore.
This book is printed on acid-free paper responsibly manufactured from sustainable forestry in which at least two trees are planted for each one used for paper production.

To Alfred, Fanny, Alice and Frank
and in memory of my parents

CONTENTS

PREFACE

With the advance of the semiconductor and communication industries, the use of system-on-a-chip (SoC) has become an essential technique to decrease product costs. It has become increasingly important for electrical engineers to develop a good understanding of the key stages of hardware description language (HDL) design flow based on cell-based libraries or field-programmable gate array (FPGA) devices. This book addresses the need for teaching such a topic based on Verilog HDL and FPGAs.

The objective of this book, *Digital System Designs and Practices: Using Verilog HDL and FPGAs*, is intended to be useful both as a text for students and as a reference book for practicing engineers or a self-study book for readers. For class use, each chapter includes many worked problems and review questions for helping readers test their understanding of the context. In addition, throughout the book, an abundance of examples are provided for helping readers realize the basic features of Verilog HDL and grasp the essentials of digital system designs as well.

The contents of this book stem largely from the course *FPGA System Designs and Practices*, given at our campus over the past few years. This course is an 'undergraduate-elective' and first-year graduate course. This book is structured so that it can be used as a sequence of courses, such as *Hardware Description Language*, *FPGA System Designs and Practices*, *Digital System Designs*, *Advanced Digital System Designs*, and others.

CONTENTS OF THIS BOOK

The contents of this book can be roughly divided into four parts. The first part includes Chapters 1 to 7 and introduces the basic features and capabilities of Verilog HDL. This part can also be used as a reference for Verilog HDL. The second part covers Chapters 8 to 10 and contains basic combinational and sequential modules. In addition, the various options for implementing a digital system are discussed in detail. The third part consists of Chapters 11 to 13 and examines the three closely related topics: design, synthesis and verification. The last part considers the register-transfer level (RTL) and system level design examples and the techniques of testing and testable design. This part is composed of Chapters 14 to 16.

Chapter 1 introduces the features and capabilities of Verilog HDL and gives a tutorial example to illustrate how to use it to model a design at various levels of abstraction and in various modeling styles. In addition, we also demonstrate the use of Verilog HDL to verify a design after the description of a design is completed.

Chapter 2 deals with how to model a design in structural style. In this style, a module is described as a set of interconnected components, which include modules, user-defined primitives (UDPs), gate primitives and switch primitives. In this chapter, we introduce the structural modeling at the gate and switch levels. The UDPs and modules are dealt with separately in Chapters 5 and 6.

Chapter 3 describes the essentials of dataflow modeling style. In this modeling style, the most basic statement is the continuous assignment, which consists of operators and operands in turn. The continuous assignment continuously drives a value onto a net and is usually used to model a combinational logic.

Chapter 4 is concerned with the behavioral modeling style, which provides users with the capability of modeling a design in a way like that of most high-level programming languages. In this modeling style, the most common statements include procedural assignments, selection statements and iterative (loop) statements. In addition, timing controls are also dealt with in detail.

Chapter 5 describes three additional behavioral ways provided by Verilog HDL which are widely used to model designs. These include tasks, functions and user-defined primitives (UDPs). Tasks and functions provide the ability to re-use the same piece of code from many places in a design and UDPs provide a means to model a design with a truth table. In addition, the predefined system tasks and functions are introduced. These system tasks and functions are useful when modeling, abstractly, a design in behavioral style or writing test benches for designs.

Chapter 6 discusses three closely related issues of hierarchical structural modeling. These include instantiations, generate statements and configurations. The instantiation is the mechanism through which the hierarchical structure is formed by modules being embedded into other modules. The generate statements can conditionally generate declarations and instantiations into a design. By using configurations, we may specify a new set of target libraries so as to change the mapping of a design without having to change the source description.

Chapter 7 deals with the additional features of Verilog HDL. These features include block constructs, procedural continuous assignments, specify blocks, timing checks and compiler directives.

Chapters 8 and 9 examine some basic combinational and sequential modules that are often used as basic building blocks to construct a complex design. In particular, these modules are the basic building blocks of a datapath when using the datapath and controller approach in a complex design.

Chapter 8 is concerned with the most commonly used combinational logic modules, which include encoders and decoders, multiplexers and demultiplexers, and magnitude comparators. In addition, a multiplexing-driven seven-segment light-emitting diode (LED) display system which combines the use of a decoder, as well as a multiplexer, is discussed in detail.

Chapter 9 examines several basic sequential modules that are widely used in digital systems. These include flip-flops, synchronizers, a switch-debouncing circuit, registers, data registers, register files, shift registers, counters (binary, BCD, Johnson), CRC generators and detectors, clock generators and pulse generators, as well as timing generators.

Chapter 10 describes various design options of digital systems. These options include application-specific integrated circuits (ASICs) and field-programmable devices. ASICs are devices that must be fabricated in IC foundries and can be designed with one of the following: full-custom, cell-based and gate-array-based approaches.

Field-programmable devices are the ones that can be personalized in laboratories and include programmable logic devices (PLDs), complex PLDs (CPLDs) and field-programmable gate arrays (FPGAs). In addition, the issues of interfacing two logic modules or devices with different logic levels and power-supply voltages are also dealt with in detail in this chapter.

The next three chapters consider three closely related issues: design, synthesis and verification. Chapter 11 introduces two useful techniques by which a system can be designed. These techniques include the finite-state machine (FSM) and register-transfer level (RTL) design approaches. The former may be described by using a state diagram or an algorithmic state machine (ASM) chart; the latter may be described by an ASM chart or by using the datapath and controller (DP+CU) paradigm. For a simple system, a three-step paradigm introduced in the chapter may be used to derive the datapath and controller of a design from its ASM chart. For complex systems, their datapaths and controllers are often derived from specifications in a state-of-the-art manner. An example of displaying four-digit data on a commercial dot-matrix liquid-crystal display (LCD) module is used to illustrate this approach. In this chapter, we also emphasize the concept that a hardware algorithm can usually be realized by using either a multiple-cycle or a single-cycle structure. The choice is based on the tradeoff among area (hardware cost), performance (operating frequency or propagation delay) and power consumption.

Chapter 12 is concerned with the principles of logic synthesis and the general architecture of synthesis tools. The function of logic synthesis is to transform an RTL representation into gate-level netlists. In order to make good use of synthesis tools, we need to provide the design environment and design constraints along with an RTL code and technology library. Moreover, we give some guidelines about how to write a good Verilog HDL code such that it can be accepted by most logic synthesis tools and can achieve the best compile times and synthesis results. These guidelines also include clock signals, reset signals and how to partition a design.

Verification is a necessary process that makes sure a design can meet its specifications both in function and timing. Chapter 13 deals with this issue in more detail and gives a comprehensive example based on FPGA design flow to illustrate how to enter, synthesize, implement and configure the underlying FPGA device of a design. Along with the design flow, static timing analyses are also given and explained. In addition, design verification through dynamic timing simulations, incorporating the delays of logic elements and interconnect, is introduced.

The next two chapters are concerned with more complex modules. Chapter 14 examines many frequently used arithmetic modules, including addition, multiplication, division, ALU, shift and two digital-signal processing (DSP) filters as well. Along with these arithmetic operations and their algorithms, we also re-emphasize the concept that a hardware algorithm can often be realized by using either a multiple-cycle or a single-cycle structure.

Chapter 15 describes the design of a small μC system, which is the most complex design example in the book. This system includes a general-purpose input and output (GPIO), timers and a universal asynchronous receiver and transmitter (UART) being connected by a system bus composed of an address bus and a data bus, as well as a control bus. The 16-bit CPU provides 27 instructions and 7 addressing modes.

The final chapter is concerned with the topic of testability and testable design. Testing is the only way to ensure that a system or a circuit may function properly. The goal of testing is to find any existing faults in a system or a circuit. In this chapter, we examine fault models, test vector generations, testable circuit design or design for testability. In addition, system-level testing, such as SRAM, a core-based system and system-on-a-chip (SoC), are also briefly dealt with.

Appendix A contains a complete syntax reference of Verilog HDL, including the keywords and formal definition of the Verilog-2001 standard in Backus-Naur Form (BNF).

SUPPLEMENTS

Two important and useful supplements are available for this book at the following URL: www.wiley.com/go/mblin. The first is student supplements, including source files of Verilog HDL examples in the book and the pdf files of lecture notes. The second is the instructor's supplements, containing figures, a solution manual and lecture notes in power-point files, in addition to the student supplements.

STUDENT PROJECTS

Many end-of-chapter problems may be assigned as student projects, in particular, the problems of Chapters 11, 14 and 15. Of course, many other chapters may also contain problems that may be used for the same purpose.

ACKNOWLEDGMENTS

Most material of this book has been taken from the course ET5009 offered at the National Taiwan University of Science and Technology over the past few years. My thanks go to the students of this course, who suffered through many of the experimental class offerings based on the draft of this book. Valuable comments from the participants of the course have helped in evolving the contents of this book and are greatly appreciated. Thanks to my mentor, Ben Chen, who is also a cofounder of the Chuan Hwa Book Company, who brought me into this colorful digital world about thirty years ago. In addition, he has also kindly allowed me to freely use figures, tables and even parts of material from my earlier Chinese Books, published by the Chuan Hwa Book Company, in this current book. Without this permission, it would have needed much more time to prepare the manuscript of this book. Finally but not least, I would like to thank my children, Alice and Frank, and my wife, Fanny, for their patience in enduring my absence from them during the writing of this book. I am also grateful to the publisher's staff for their support, encouragement and willingness to give prompt assistance during this book project.

Ming-Bo Lin
Taipei, Taiwan

## CHAPTER 1

INTRODUCTION

USING HARDWARE description languages (HDLs) to design digital systems has become an essential way in modern electronic engineering. A major feature inherent to HDLs is that it has the capability of modeling a digital system at many levels of abstraction, ranging from the algorithm-level to the gate-level, and even to switch-level.

Just like C programming language, Verilog HDL is very easy to learn and is much closer to the behavior of hardware modules. In this introductory chapter, we address the basic concepts of a Verilog HDL program (usually called a *module*), such as how to describe a design in various modeling styles, including structural, dataflow and behavioral. In addition, a tutorial example is given in the final section to illuminate how a verification process can be done after the description of a design is completed.

1.1 INTRODUCTION

HDL is an acronym of hardware description language. The two most commonly used HDLs in industry are Verilog HDL (some texts call it Verilog for short) and very high-speed integrated circuit (VHSIC) hardware description language (VHDL). Designs can be described at a very abstract level by using HDLs. By describing designs in HDLs, functional verification of the design can be done early in the design cycle. Designing with HDLs is analogous to computer programming.

1.1.1 Popularity of Verilog HDL

Verilog HDL originated in 1983 at Gateway Design Automation,[1] as a hardware modeling language associated with their simulator products. Since then, Verilog HDL has gradually gained popularity in industry. In order to increase its popularity, Verilog HDL was placed in the public domain in 1990. In 1995, it was standardized by IEEE as IEEE

---

[1] Gateway Design Automation has been acquired by Cadence Design Systems.

Digital System Designs and Practices   Ming-Bo Lin
© 2008 John Wiley & Sons (Asia) Pte Ltd

Std 1364-1995, which has been promoted by Open Verilog International (OVI) since 1992. An updated version of the language was standardized by IEEE in 2001 as IEEE Std 1364-2001. The new version included many new features, including alternate port declaration styles, configurations and generate statements.

The popularity of Verilog HDL manifests itself on the following features:

- Verilog HDL is a general-purpose, easy to learn and easy to use HDL.
- Verilog HDL allows different levels of abstraction to be mixed in the same module.
- Verilog HDL allows modeling a design entirely at switch level by only using its built-in switch-level primitives.
- Most popular logic synthesis tools support Verilog HDL.
- All fabrication vendors provide Verilog HDL libraries for post-logic synthesis simulation.

In addition, a powerful programming language interface (PLI) is provided by Verilog HDL so as to allow users to develop their own computer-aided design (CAD) tools such as delay calculator.

1.1.2 Simple Examples of Verilog HDL

A simple Verilog HDL module is shown in Example 1.1, which displays a line of text on screen.

EXAMPLE 1.1 *A Simple Example to Display a Line of Text*

```
module simple_example_1;
// a simple example to display a line of text.
initial begin
    $display("Welcome to Verilog HDL World!\n");
end
endmodule
```

From this example, you can see several syntactic features of a Verilog HDL module. The function of the $display system task is similar to the printf library function in C programming language, being used to display something on basic I/O devices. In addition, the Verilog HDL module begins with the keyword module followed by module name, an initial statement, and ends with the keyword endmodule.

The texts after // are ignored by Verilog HDL compilers and simulators. They are only used for illustration purposes. That is, they are comments.

The following example is another simple Verilog HDL module, which displays the sum of two integers.

EXAMPLE 1.2 *A Simple Example to Display the Sum of Two Numbers*

Like in other high-level programming languages, a variable in Verilog HDL must be declared before it can be used. Hence, we declare variables a and b as integer type.

To print out their values, we have to declare their output format. For simplicity, both variables a and b and their sum are assumed to be of two decimal digits. Therefore, output format %2d is used to print out their values.

```
module simple_example_2;
// a simple example to display the sum of two numbers.
integer a, b;
initial begin
    a = 5; b = 8;
    $display("The sum of %2d and %2d is: %2d\n",
            a, b, a + b);
end
endmodule
```

Roughly, a Verilog HDL module is much like a high-level language program except that it is used to describe a hardware module rather than a software function. One of the major differences between a software program and a hardware module is the timing feature inherent to the hardware module. Hence, a practical HDL such as Verilog HDL must provide some mechanisms to describe the timing feature of hardware. The following example illustrates this idea.

EXAMPLE 1.3 *A Simple Verilog HDL Module*

In this example, a module is designed to add two *n*-bit operands and produce a result of $n + 1$ bits. In order to exactly model the operations of actual hardware, we need to include the timing characteristics of hardware. As a consequence, a delay #5 is added to the assign statement. The function of the entire assign statement is to add two input operands along with the input carry and produces a carry out as well as a sum in 5 time units.

```
module nbit_adder(x, y, c_in, sum, c_out);
/* I/O port declarations */
parameter N = 4;  // define the default value of N.
input   [N-1:0] x, y;
input   c_in;
output [N-1:0] sum;
output c_out;
// specify the function of an n-bit adder using assign
// statement.
    assign #5 {c_out, sum} = x + y + c_in;
endmodule
```

In summary, from the above three examples we can see that a more complete Verilog HDL module usually begins with the keyword module followed by the module name and a port list (if any), declarations of each item in the port list (if any), an assign statement or other statements and ends with the keyword endmodule.

1.1.3 HDL-Based Design

The HDL-based design paradigm has become dramatically popular recently due to the fact that the traditional schematic-based design has faced a bottleneck in that it cannot provide the capability of design re-use and easily handle designs with large numbers of gate counts. Another reason is that the computation power of modern computers is now good enough to provide the required computation of CAD tools.

An HDL-based design flow is depicted in Figure 1.1, which can be divided into target-independent and target-dependent parts. The target-independent part begins with design specification and ends with functional verification. The design entry includes many available tools such as schematic, ABEL program, VHDL or Verilog HDL modules. Various design entries are then represented as a unified data structure for use in the next stage – functional verification. The functional verification assures that the function of design entry is correct and conforms to the specification, in addition to

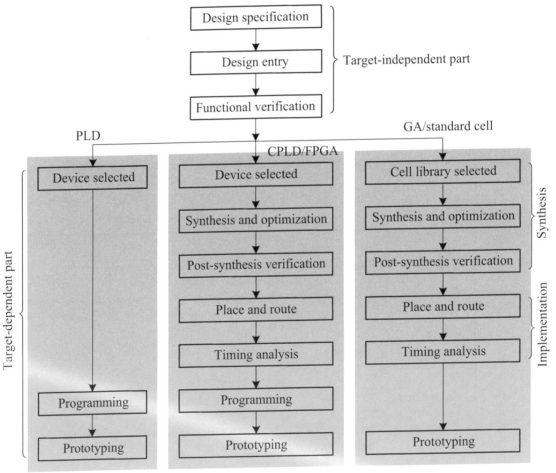

FIGURE 1.1 A design flow of the HDL-based paradigm

performing some basic checks such as syntax error in HDL. After this, the work of the target-independent part is completed.

In the target-dependent part, we have to choose a target, such as a programmable logic device (PLD), complex PLD (CPLD), field-programmable gate array (FPGA) or cell library. Most of the work of this part is carried out automatically by the CAD tools. The simplest devices are PLDs, which are usually used to implement designs with gate counts ranging from a few gates to several hundred gates. The target-dependent part of using a PLD only consists of two steps: selecting device and programming.

For CPLD/FPGA design flow, due to the complicated structure of the devices the target-dependent part is further divided into many sub-steps, as shown in Figure 1.1. These sub-steps can be roughly classified as three major steps: *synthesis*, *implementation* and *programming*. The synthesis step is composed of the device selected, synthesis and optimization, as well as post-synthesis verification. The implementation step consists of place and route, and timing analysis. The function of the synthesis step is to optimize the switching functions designed by the designers, based on the device selected, and to map the abstract logic elements to actual logic blocks provided by the device. After completion of the synthesis and optimization step, it is usually necessary to verify the synthesized result through post-synthesis simulation or formal verification.

The implementation step starts with the place and route operations, which place the logic blocks into the actual logic elements and then set up the related interconnect. Due to the inherent delays associated with logic elements and interconnect, it needs to perform a timing analysis to verify whether the timing constraints meet the specification. The widely used timing analysis can be either *dynamic* or *static*. The dynamic timing analysis (DTA) is performed by simulation whereas the static timing analysis (STA) is by analyzing the timing paths of the design without doing any actual simulation.

Once the implementation step is finished, the next step is to generate the programming file in order to program the device. Of course, after programming a device, we often use it in a real-world test to see if it indeed works as we might have expected.

The design flows of cell-based (standard cell) and gate-array based designs are almost the same as that of CPLD/FPGA, with only one exception that they do not have the programming step. Actually, they must be "programmed" in an IC (integrated circuit) foundry; namely, they must be programmed by masks.

A final comment on the above mentioned design flow is that the target-independent part plus synthesis is often referred as *front-end design* and the remaining part as *back-end design* in industry. We will return to this topic in a dedicated chapter later.

Review Questions

Q1.1   Why is HDL so popular in modern digital system design?

Q1.2   Is Verilog HDL a standard of IEEE?

Q1.3   What are the major differences between a software program and a hardware module?

Q1.4   What can a Verilog HDL module be?

Q1.5   How do we introduce a comment into a Verilog HDL module?

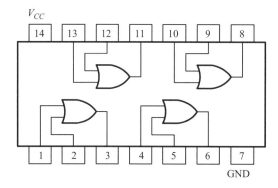

FIGURE 1.2 A simple example of a hardware module

1.2 INTRODUCTION TO VERILOG

In this section, we introduce the basic syntax and some of the most basic components of Verilog HDL.

1.2.1 Module Concept

The basic unit of a digital system is a *module*, which consists of a core circuit (called *internal* or *body*) and an *interface* (called *ports*). The body performs the required function of the module while the interface carries out the required communication between the core circuit and the outside world. In addition, power and ground need to be provided by the interface. A simple example of a hardware module is shown in Figure 1.2. The module embodies four 2-input OR gates along with their power (V_{CC}) and ground (GND).

To be consistent with the concepts of hardware modules, the basic building block in Verilog HDL is called a module, which describes the corresponding hardware module.

Like a hardware module, a Verilog HDL module can be an element or a collection of lower-level design blocks (modules). A basic module structure is shown in Figure 1.3, which includes many different declarations, statements, tasks and functions, among others. We will detail each of these in the appropriate sections of the book.

---

module Module name

Port List, Port declarations (if any)

Parameters (if any)

Declarations of *wires*, *regs* and other variables

Instantiation of lower-level modules or primitives

Data flow statements (*assign*)

always and *initial* blocks (all behavioral statements go into these blocks)

Tasks and functions

endmodule Statement

FIGURE 1.3 The basic structure of a Verilog module

1.2.2 Lexical Conventions

Like any other high-level programming languages, Verilog HDL uses almost the same lexical conventions except that it needs to describe the unique features of a hardware module.

1.2.2.1 Identifiers Any identifier is a sequence of alphanumeric characters, _ and $, with the restriction that the first character must be a letter, _ or $. The lengths of identifiers are limited by implementations but they should be at least 1024 characters. If an identifier exceeds the implementation-specified length limit, an error will be reported. The identifiers beginning with $ are reserved for system tasks and functions as well as timing checks, which will be described in Sections 5.3 and 7.3.3, respectively. Here are some examples of identifiers:

```
counter
four_bit_adder
a
b
_4b_adder
```

Note that Verilog HDL is a case-sensitive language just like C. Hence, `adder` and `Adder` are two different identifiers.

An escaped identifier begins with a \ (back-slash) character and ends with a white space, which includes blank space (\b), tab (\t), and new line (\n).

Like most high-level programming languages such as C, there is a list of reserved identifiers in Verilog HDL called *keywords*, which can only be used in some contexts. Here are some examples of keywords:

```
and        assign      begin       case        else
end        endcase     endmodule   for         if
initial    inout       input       integer     module
nand       negedge     nor         not         or
output     parameter   posedge     real        realtime
reg        repeat      signed      unsigned    xnor
xor
```

A keyword can also be escaped and treated as not having the same meaning as its original keyword. For example, `\always` will have a different meaning to the keyword `always`. However, in order to make the module clear and readable, it should avoid using keywords with a different case or escaped keywords as identifiers.

1.2.2.2 Comments The Verilog HDL has two ways to introduce comments. A one-line comment starts with the two characters // and ends with a new line. A block comment begins with /* and ends with */. Block comments cannot be nested and the one-line comment token // has no special meaning in a block comment.

Coding Styles

1. It is good practice to:

- *use lowercase letters for all signal names, variable names, and port names;*
- *use uppercase letters for names of constants and user-defined types;*
- *use meaningful names for signals, ports, functions and parameters;*
- *use the same names for both module and file.*

1.2.3 Value Set

Usually, the logic value has only two values: 0 and 1. However, in actual hardware circuits, it is common to include a tristate (or three state) z to indicate a high-impedance condition of a node or net. In addition, some circuit nodes such as the outputs of *D*-type flip-flops, will have an unknown logic value when the power is just on and before it samples any input data. In order to model these conditions, another logic value called x in Verilog HDL is used to specify an unknown logic value of a net or node. The resulting system is called a *four-value logic* system.

- 0: logic 0, false condition
- 1: logic 1, true condition
- x: unknown logic value
- z: high-impedance.

1.2.4 Constants

There are three types of constants provided in Verilog HDL. These are *integer*, *real* and *string*. Integer constants can be specified in decimal, hexadecimal, octal or binary formats. There are two forms to express integer constants: *simple decimal form* and *base format notation*. In the simple decimal form, a number is specified as a sequence of digits 0 through 9, optionally beginning with a plus (+) or minus (−) unary operator. For example:

```
-123    // is decimal -123
12345   // is decimal 12345
```

An integer value in this form represents a signed number. A negative number is represented in two's complement form. When the base format is used, a number is composed of up to three parts: an optional size constant, a single quote followed by a base format character and the digits representing the value of the number. For example:

```
4'b1011    // a 4-bit binary number
8'habc     // an 8-bit hexadecimal number
   2006    // unsized number - a 32-bit decimal
           // number by default
```

Real numbers can be specified in either decimal notation or scientific notation. When a real number is represented in decimal notation, there must be at least one digit on each side of the decimal point. The following are some examples of real numbers with legal and illegal representations.

```
3.4          //
 .2          // illegal ---
294.872      //
1.44E9       // the exponent symbol can be e or E
```

A string is a sequence of characters enclosed by double quotes (" ") and cannot be split into multiple lines. One character is represented as an 8-bit ASCII code. Hence, a string is considered as an unsigned integer constant represented by a sequence of 8-bit ASCII codes.

String variables are variables of `reg` type with widths equal to eight times the number of characters in the string. For example, to store the string `"Verilog HDL"` requires a `reg of 8*11`, or 88 bits.

```
module string_test;
reg [8*11:1] str1;
initial begin
   str1 = "Verilog HDL";
   $display("%s\n", str1);
end
endmodule
```

The output is:

```
Verilog HDL
```

Definition of Constants There are three ways in Verilog HDL that may be used to define a constant in the source description of a design. These are compiler directive, `define`, and module parameters, `parameter` and `localparam`.

The `define` compiler directive is used to create a macro for text substitution. For instance:

```
`define BUS_WIDTH 8
```

Any place in the source description with the occurrence of `BUS_WIDTH` will be replaced with 8. The `define` compiler directive is usually placed at the head of a file or a separated file and can be used both inside and outside of the module definitions.

The module `parameter` is the most common approach used to define parameters that can be overridden by `defparam` or module instance parameter value assignment. The issues of parameter override will be discussed in Section 6.1.4 in more detail. The following is an example of using `parameter` to define a constant.

```
parameter BUS_WIDTH = 8;
```

Another module parameter `localparam` is identical to `parameter` except that it cannot be modified with the `defparam` statement or by module instance parameter

TABLE 1.1 Data types of Verilog HDL

Nets		Variables
wire	supply0	reg
tri	supply1	integer
wand	tri0	real
wor	tril1	time
triand	trireg	realtime
trior		

value assignment.

```
localparam BUS_WIDTH = 8;
```

Hence, the `localparam` is often used to define constants locally used in the module only.

1.2.5 Data Types

Verilog HDL has two classes of data types: *nets* and *variables*. Nets mean any hardware connection points, and variables represent any data storage elements. Note that net is not a reserved word but represents a class of data types listed in Table 1.1. A variable represents a data storage element. It stores a value from one assignment to the next. The variable data types include five different kinds: `reg`, `integer`, `time`, `real` and `realtime`. In this sub-section, we only deal with nets and `reg` variable data type. The details of net and various variable data types will be dealt with in Chapter 2 and Section 3.2.2, respectively.

1.2.5.1 Net Data Types
The nets represent physical connections between structural entities, such as gates and modules. A net does not store a value (except for the `trireg` net). A net can be referenced anywhere in a module and must be driven by a primitive (include gates, switches, and UDPs), continuous assignment `assign`, `force`/`release` statement or module ports. If no driver is connected to a net, its value is a high-impedance (z) unless the net is a `trireg`, in which case it holds the previously driven value.

A form of net declaration is as follows:

```
net_type [signed] [[msb:lsb]] net1, net2, ..., netn;
```

where `net_type` is any type in the column nets of Table 1.1. A net data type can be declared to store a signed value by using the keyword `signed` following `net_type`. The default value of a net is unsigned. The range [msb:lsb] is an optional part and is used to specify the range of the net. Both msb (most significant bit) and lsb (least significant bit) are constant expressions. The msb is always at the left-most bit and the lsb is always at the right-most bit despite what sequence, i.e. [high:low] or [low:high] is used to specify the range. If no range is specified, it defaults to a single 1-bit net and is called *scalar*. When a range of greater than 1 bit is specified, the net is called a *vector*. Among net data types, `wire` is the most frequently used. The following examples demonstrate how `wire` net data types are declared:

```
wire a, b, c_in;    // nets a, b, and c_in are
                    // 1-bit wires
wire [7:0] data_a;  // data_a is an 8-bit wire, the
                    // msb is bit 7
wire [0:7] data_b;  // data_b is an 8-bit wire,
                    // the msb is bit 0
wire signed [7:0] d;// d is an 8-bit signed wire
```

The other net data types will be described in the next chapter.

1.2.5.2 `reg` *Variable Data Type*   A `reg` variable holds a value between assignments; it can be used to model hardware registers such as edge-sensitive (i.e. flip-flops) and level sensitive (i.e. latches) storage elements. A form of `reg` declaration is as follows:

```
reg [signed] [[msb:lsb]] reg1, reg2, ..., regn;
```

where the keyword `signed` is employed to declare the storage of a signed value. The default value of a `reg` variable is unsigned. The range [msb:lsb] is an optional part and is used to specify the range of the `reg` variable, i.e. the width of the vector. The default size of a `reg` variable is a single bit, i.e. a scalar. Some examples of `reg` declarations are as follows:

```
reg a, b, c_in;    // reg a, b, and c_in are 1-bit
                   // reg variables
reg [7:0] data_a;  // data_a is an 8-bit reg, the msb
                   // is bit 7
reg [0:7] data_b;  // data_b is an 8-bit reg, the msb
                   // is bit 0
reg signed [7:0] d;// d is an 8-bit signed reg
```

1.2.6 Primitives

Verilog HDL has two types of primitives: built-in primitives, which include 12 gate primitives and 16 switch primitives, and user-defined primitives (UDPs), which can be used to define a new combinational or sequential logic module. The gate primitives support gate-level modeling while switch primitives support switch-level modeling. User-defined primitives (UDPs) provide users with a convenient way to define their own primitive functions. Two major classes of user-defined primitives are combinational UDPs and sequential UDPs. The instantiation of primitives are as follows:

```
primitive_name [instance_name]
       (output, input1, input2,...,inputn);
```

For both built-in and user-defined primitives, `instance_name` is optional. Primitives can be instantiated only within modules. The first port of an instantiated primitive is always an output and the other ports are all inputs. The details of UDPs will be dealt with in Chapter 5.

1.2.7 Attributes

An attribute is a mechanism provided by Verilog HDL for annotating information about objects, statements and groups of statements in the source description to be used by various related tools. Almost all statements can be attributed in Verilog HDL. There are no standardized attributes. An attribute can be attached as a prefix to a declaration, a module item, a statement or a port connection and as a suffix to an operator or a function name in an expression. Attributes have the following syntax:

```
(*attr_spec {, attr_spec}*)
```

where `attr_spec` can be either `attr_name` or `attr_name = const_expr`. If an attribute is not assigned to a value, its value defaults to 1. The following are some examples:

```
// example 1: attach full_case attribute only
(* full_case=1, parallel_case = 0 *)
case (selection)
  <rest_of_case_statement>
endcase

// example 2: attach an attribute to a module
(* dont_touch *) module array_multiplier
    (x, y, product);
  <the body of array_multiplier>
endmodule

// example 3: attach an attribute to a reg variable
(* fsm_state *)   reg [1:0] state1;
(* fsm_state=1 *) reg [1:0] state2, state3;

// example 4: attach an attribute to a function call
x = add(* mode = "cla" *)(y, z);

// example 5: attach an attribute to an operator
x = y + (* mode = "cla" *) z;
```

Review Questions

Q1.6 How would you describe a bundle of signals in Verilog HDL when it is used to describe hardware modules?

Q1.7 How many possible values may a net or `reg` variable have?

Q1.8 How many data types are included in net data types? What are these?

Q1.9 How many data types are included in variable data types? What are these?

Q1.10 Define the following two terms: *scalar* and *vector*, in the sense of Verilog HDL.

Q1.11 What kinds of primitives are provided in Verilog HDL?

1.3 MODULE MODELING STYLES

In order to understand the design methodology of modern digital systems, it is instructive to distinguish among *design*, *model*, *synthesis*, *implementation* or *realization*. Design is a series of transformations from one representation of a system to another until a representation that can be fabricated exists; that is, it can be created, fashioned, executed or constructed according to a plan. Model is a process that converts a specification document into a HDL module. Model is a system of postulates, data and inferences presented as a mathematical description of an entity or the state of affair. Synthesis is a process that converts HDL modules into a structural representation. Synthesis can be divided into logic synthesis and high-level synthesis. Logic synthesis is a process that converts an RTL description into a gate-level netlist, while high-level synthesis is a process that converts a high-level description (i.e. specification) into RTL results. Implementation (or realization) is the process of transforming design abstraction into physical hardware components such as FPGAs or cell-based ICs (integrated circuits).

1.3.1 Modules

As mentioned previously, a Verilog HDL module consists of two major parts: the interface and the internal.

1.3.1.1 *Modeling the Internal of a Module*   For each module in Verilog HDL, the internal (or body) can be modeled as one of the following styles:

1. *Structural style*. A design is described as a set of interconnected components. The components can be modules, UDPs, primitive gates and/or primitive switches.

   (a) Gate level. A design is said to be modeled at gate-level when it only comprises a set of interconnected gate primitives.

   (b) Switch level. A design is said to be modeled at switch-level when it only consists of a set of interconnected switch primitives.

   The structural style is a mechanism that can be used to construct a hierarchical design for a large digital system.

2. *Dataflow style*. The module is described by specifying the dataflow (i.e. data dependence) between registers and how the data is processed.

   (a) A module is specified as a set of continuous assignment statements.

3. *Behavioral* or *algorithmic style*. The design is described in terms of the desired design algorithm without concerning the hardware implementation details.

   (a) Designs can be described in any high-level programming languages.

4. *Mixed style*. The design is described in terms of the mixing use of above three modeling styles.

   (a) Mixed styles are most commonly used in modeling large designs.

In industry, the term *register-transfer level* (RTL) is often used to mean a structure that combines both behavioral and dataflow constructs and can be acceptable by logic synthesis tools.

1.3.1.2   Port Declaration   The interface signals (including supply and ground) of any Verilog HDL module can be cast into one of the following three types:

(1) input: Declare a group of signals as input ports.

(2) output: Declare a group of signals as output ports.

(3) inout: Declare a group of signals as bidirectional ports, that is, they can be used as input or output ports but not at the same time.

The simplest way to describe the complete interface of a module is to divide it into three parts: port list, port declaration and data type declaration of each port. Usually, an output port, except that it is a net data type, must be declared a data type associated with it. However, input ports are often left with their data types undeclared. For example:

```
// port list style
module adder(x, y, c_in, sum, c_out);
input [3:0] x, y;
input c_in;
output [3:0] sum;
output c_out;
reg [3:0] sum;
reg c_out;
```

This style of port declaration is known as *port list style*. The declaration of a port and its associated data type can be combined into a single line. Based on this idea, the above interface portion of the module can be rewritten as follows:

```
// port list style
module adder(x, y, c_in, sum, c_out);
input [3:0] x, y;
input c_in;
output reg [3:0] sum;
output reg c_out;
```

Of course, port list, port declarations and their associated data types can also be put together into a single list. This style is often called *port list declaration style*. Hence, the above module interface can be rewritten as follows:

```
// port list declaration style
module adder(input [3:0] x, y,
          input c_in,
          output reg [3:0] sum,
          output reg c_out
); // sometimes called ANSI C style
```

This style is the same as that of ANSI C programming language so that we often name it as the ANSI C style. Note that all of the above interface styles are valid in Verilog HDL.

1.3.1.3 Port Connection Rules
The port connection (also called port association) rules of Verilog HDL modules are consistent with those of actual hardware modules. That is, Verilog HDL allows ports to remain unconnected and with different sizes. In addition, unconnected inputs are driven to the "z" state; unconnected outputs are not used. Connecting ports to external signals can be done by one of the following two methods:

- *Named association.* The ports to be connected to external signals are specified by listing their names. The port order is not important.
- *Positional association.* The ports are connected to external signals by an ordered list. The signals to be connected must have the same order as the ports in the port list, leaving the unconnected port blank.

However, these two methods cannot be mixed in the same module. Moreover, Verilog HDL primitives (including both built-in and user-defined) can only be connected by positional association.

The operation to "call" (much like the *macro expansion* in assembly language) a built-in primitive, a user-defined primitive or the other module is called *instantiation* and each copy of the called primitive or module is called an *instance*. Figure 1.4 shows how to instantiate gate primitives and user-defined modules, as well as how to connect their ports through nets and input/output ports. The built-in primitives can only be connected by using positional association. They cannot be connected through named association. In addition, as mentioned before, the instance names of these primitives are optional.

The module full_adder depicted in Figure 1.4 shows how to instantiate an already defined module and how to connect their ports through nets. The user-defined modules can be connected by using either named association or positional association. In addition, the instance names of these instantiations are necessary.

Coding Styles

1. A module cannot be declared within another module.

2. A module can instantiate other modules.

3. A module instantiation must have a module identifier (instance name) except for built-in primitives, gate and switch primitives and user-defined primitives (UDPs).

```
module half_adder (x, y, s, c);
input  x, y;
output s, c;
// -- half adder body-- //
// instantiate primitive gates
   xor xor1 (s, x, y);    Can only be connected by using positional association
   and and1 (c, x, y);
endmodule          ──────── Instance name is optional.

module full_adder (x, y, cin, s, cout);
input  x, y, cin;
output s, cout;
wire   s1,c1,c2;  // outputs of both half adders
// -- full adder body-- //          ──── Connecting by using positional association
// instantiate the half adder
   half_adder ha_1 (x, y, s1, c1);       ── Connecting by using named association
   half_adder ha_2 (.x(cin), .y(s1), .s(s), .c(c2));
   or (cout, c1, c2);   ── Instance name is necessary.
endmodule
```

FIGURE 1.4 Port connection rules

4. It should use named association at the top-level modules to avoid confusion that may arise from synthesis tools.

1.3.2 Structural Modeling

As mentioned previously, the structural modeling of a design is by connecting required instantiations of built-in primitives, user-defined primitives or other (user-defined) modules through nets. The following example shows several examples of structural modeling:

EXAMPLE 1.4 *An Example of Structural Modeling at Gate Level*

The half_adder instantiates two gate primitives and full_adder instantiates two half_adder modules and one gate primitive. Finally, four_bit_adder is constructed by four full_adder instances in turn.

```
// gate-level hierarchical description of 4-bit adder
// gate-level description of half adder
module half_adder (x, y, s, c);
input  x, y;
output s, c;
// half adder body
// instantiate primitive gates
   xor (s,x,y);
   and (c,x,y);
endmodule
```

```
// gate-level description of full adder
module full_adder (x, y, cin, s, cout);
input  x, y, cin;
output s, cout;
wire   s1, c1, c2;  // outputs of both half adders
// full adder body
// instantiate the half adder
   half_adder ha_1 (x, y, s1, c1);
   half_adder ha_2 (cin, s1, s, c2);
   or (cout, c1, c2);
endmodule

// gate-level description of 4-bit adder
module four_bit_adder (x, y, c_in, sum, c_out);
input  [3:0] x, y;
input  c_in;
output [3:0] sum;
output c_out;
wire   c1, c2, c3; // intermediate carries
// four_bit adder body
// instantiate the full adder
   full_adder fa_1 (x[0], y[0], c_in, sum[0], c1);
   full_adder fa_2 (x[1], y[1], c1, sum[1], c2);
   full_adder fa_3 (x[2], y[2], c2, sum[2], c3);
   full_adder fa_4 (x[3], y[3], c3, sum[3], c_out);
endmodule
```

In fact, the structural style is one way to model a complex digital system in a hier-archical manner. An example of a 4-bit adder constructed in a hierarchical manner is depicted in Figure 1.5. Here, the 4-bit adder is composed of four full-adders and then each full-adder is built by basic logic gates in turn. Although this example is quite simple, it manifests several important features when designing a large digital system.

1.3.3 Dataflow Modeling

The essential structure used to model a design in the dataflow style is the continuous assignment. In a continuous assignment, a value is assigned onto a net. It must be a net because continuous assignments are used to model the behavior of combinational logic circuits. A continuous assignment starts with the keyword `assign` and has the syntax:

```
assign [delay] l_value = expression;
```

Anytime the value of an operand used in the `expression` changes, the `expression` is evaluated and the result is assigned to `l_value` after the specified `delay`. The

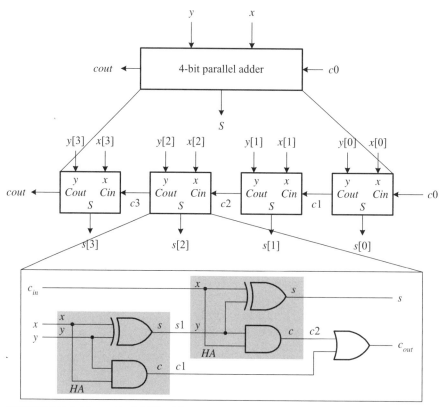

FIGURE 1.5 A hierarchical 4-bit adder

`delay` specifies the amount of time between a change of operand used in the expression and the assignment to `l_value`. If no `delay` is specified, the default is zero delay. Continuous assignments in a module execute concurrently regardless of the order they appear.

The following example illustrates how a continuous assignment is used to describe the 1-bit adder (i.e. full adder) depicted in Figure 1.6.

EXAMPLE 1.5   *A Full Adder Modeling in Dataflow Style*

In this example, we assume that the adder requires 5 time units to complete its operations. The delay will be ignored by the synthesis tools when the module is

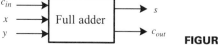

FIGURE 1.6 The block diagram of a full adder

synthesized because the delay will be replaced by the actual delays of the gates used to realize the adder.

```
module full_adder_dataflow(x, y, c_in, sum, c_out);
// I/O port declarations
input  x, y, c_in;
output sum, c_out;

// specify the function of a full adder
   assign #5 {c_out, sum} = x + y + c_in;
endmodule
```
∎

1.3.4 Behavioral Modeling

The behavioral style uses the following two procedural constructs: `initial` and `always`. The `initial` statement can only be executed once and therefore is usually used to set up initial values of variable data types whereas the `always` statement, as the name implies, is executed repeatedly. The `always` statements are used to model combinational or sequential logic. Each `always` corresponds to a piece of logic.

The `l_value` used in an expression within an `initial` or `always` statement must be a variable data type, which retains its value until a new value is assigned. All `initial` statements and `always` statements begin their execution at simulation time 0 concurrently.

The following example illustrates how a procedural construct is used to describe the 1-bit adder (i.e. full adder) depicted in Figure 1.6.

EXAMPLE 1.6 *A Full Adder Modeling in Behavioral Style*

Basically, the expression used to describe the operations of a 1-bit full adder in behavioral style is the same as that of the dataflow style except that it needs to be put inside an `always` statement. In addition, the `@(x, y, c_in)` is used to sensitize the changes of input signals.

```
module full_adder_behavioral(x, y, c_in, sum, c_out);
// I/O port declarations
input  x, y, c_in;
output sum, c_out;
reg    sum, c_out;  // sum and c_out need to be declared
                    // as reg types.

// specify the function of a full adder
always @(x, y, c_in)// can also use always @(*) or
                   // always@(x or y or c_in)
   #5 {c_out, sum} = x + y + c_in;
endmodule
```
∎

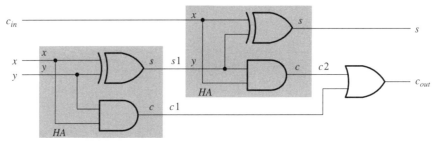

FIGURE 1.7 A full adder constructed with basic logic gates

1.3.5 Mixed-Style Modeling

As mentioned above, the mixed-style modeling is usually used to construct a hierarchical design in large systems. However, we are still able to model a simple design in mixed style. For example, as depicted in Figure 1.7 the full adder is constructed with two half adders and an OR gate. An example of the full adder being modeled in mixed style is shown in the following example.

EXAMPLE 1.7 *A Full Adder Module in Mixed Style*

The first half adder is modeled in structural style, the second half adder in dataflow style and the OR gate in behavioral style.

```
module full_adder_mixed_style(x, y, c_in, s, c_out);
// I/O port declarations
input   x, y, c_in;
output  s, c_out;
reg     c_out;
wire    s1, c1, c2;
// structural modeling of HA 1.
    xor xor_ha1 (s1, x, y);
    and and_ha1(c1, x, y);
// dataflow modeling of HA 2.
    assign s = c_in ^ s1;
    assign c2 = c_in & s1;
// behavioral modeling of output OR gate.
always @(c1, c2) // can also use always @(*)
    c_out = c1 | c2;
endmodule
```

Review Questions

Q1.12 What are the differences of design, model, synthesis and implementation (also called realization)?

Q1.13   Describe the features of structural style.

Q1.14   What is the basic statement used in dataflow style?

Q1.15   What are the basic statements used in behavioral style?

Q1.16   Can we write a module by mixing the use of various modeling styles?

1.4 SIMULATION

For a design to be useful, it must be verified so as to make sure that it can correctly operate according to the requirement. Verilog HDL not only provides capabilities to model a design, but also provides facilities to generate and control stimuli, monitor and store responses, and check the results. In this section, we use a 4-bit adder as an example to illustrate how to verify a design entirely through the mechanism provided by Verilog HDL.

1.4.1 Basic Simulation Constructs

Two basic simulation structures in Verilog HDL are shown in Figure 1.8. The first structure is to take the unit under test (UUT) as an instantiated module in the stimulus module. This is often used in simple or small projects since it is intuitively simple to write. The second construct considers both stimulus block and UUT as the separate instantiated module at the top-level module. This is suitable for large projects. In this book, we will use the first structure when writing a test bench.

In general, a test bench comprises several basic parts: an instantiation of the UUT, stimulus generation and control, response monitoring and storing, and result checking. Except for the instantiation of the UUT, the rest of the parts are usually modeled in behavioral style. In the rest of this section, we deal with these features of a typical test

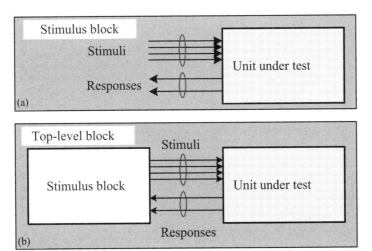

FIGURE 1.8 Two basic constructs for a simulation: (a) stimulus block at the top-level module; (b) stimulus block considered as a separate module

bench by using an example. The details of how to write test benches along with other insights of verification will be considered in a dedicated chapter.

1.4.2 Related Compiler Directive and System Tasks

In this sub-section, we introduce the most basic and widely used compiler directive, `timescale`, display system tasks, `$display` and `$monitor`, simulation time system functions, `$time` and `$realtime`, as well as simulation control system tasks, `$finish` and `$stop`.

1.4.2.1 **`timescale` *Compiler Directive***   In carrying out simulations, we need to specify the physical unit of measure, or time scale of a numerical time delay value. It also needs to specify the resolution of the time scale, i.e. the minimum step size of the scale during simulation. In Verilog HDL, this is accomplished by using a compiler directive:

```
`timescale time_unit / time_precision
```

where " ` " is the back quote. The `time_unit` specifies the unit of measure for times and delays; the `time_precision` specifies how delay values are rounded during simulation. Only the integers 1, 10 and 100 may be used to specify `time_unit` and `time_precision`; the valid units are s, ms, us, ns, ps and fs. For instance:

```
`timescale 10 ns /1 ns

#3.55 a = b + 1; // corresponds to 36 ns
#3.54 b = c + 1; // corresponds to 35 ns
```

The `timescale 10 ns / 1 ns` specifies that the time unit is 10 ns. As a result, the time values in the module are multiples of 10 ns, rounded to the nearest 1 ns. Therefore, the delay 3.55 and 3.54 are scaled rounded to 36 ns and 35 ns, respectively. As another example:

```
`timescale 1 ns /10 ps

#3.55 a = b + 1; // corresponds to 3.55 ns
#3.54 b = c + 1; // corresponds to 3.54 ns
```

Here, all time values are multiples of 1 ns because the `time_unit` is 1 ns. Delays are rounded to real numbers with two decimal places because the `time_precision` is 10 ps.

The `time_precision` must not exceed the `time_unit`. Hence, `timescale 1 ns /10 ns` is illegal.

It should be careful to use the same time unit in both behavioral and gate-level modeling; otherwise, the incorrect results might be obtained. As a rule of thumb, for FPGA designs it is suggested to use ns as the first attempt time unit in the test bench since the propagation delay of gate-level for the current devices is of the order of ns. In addition, the effect of timescale lasts for all modules that follow this compiler directive until another timescale compiler directive is read.

1.4.2.2 Display System Tasks

During simulation, we need to display information about the design for debugging or other useful purposes. Verilog HDL provides two widely used system tasks: $display and $monitor, for displaying information to standard output. The $display system task displays information only when it is called but the $monitor system task continuously monitors and displays the values of any variables or expressions. They have the same form:

```
task_name[(arguments)];
```

where task_name is either $display or $monitor. The arguments are used to specify strings to be displayed literally. To print information with a specified format, an escape sequence is used, much like that of C programming language. For example:

```
$display($time,"ns %d %d %h %h", x, y, c_in,
        {c_out, sum});
$monitor($realtime,"ns %d %d %h %h", x, y, c_in,
        {c_out, sum});
```

where %d and %h are used to specify that variable values will be displayed in decimal and hexadecimal, respectively.

Only one $monitor display list can be active at any time; however, a new $monitor system task with a new display list can be issued any number of times during simulation. In addition to the $monitor system task, $monitoron and $monitoroff system tasks are two related tasks widely used to control the monitoring operations. The $monitoron and $monitoroff system tasks enable and disable the monitoring operations, respectively.

1.4.2.3 Simulation Time System Functions

There are two system functions that provide access to current simulation time: $time and $realtime. The $time system function returns a 64-bit integer of time and the $realtime system function returns a real number of time. All of the returned values from these two system functions are scaled to the time unit of the module that calls them. For example:

```
`timescale 10 ns/ 1 ns
module time_usage;
reg a;
initial begin
   $monitor($time,, "a = ", a);
   #3.55 a = 0;
   #3.55 a = 1;
end
endmodule
```

The result is:

```
#0 a = x
#4 a = 0
#7 a = 1
```

The simulation times are 35 and 70 ns. They are scaled to 4 time units and 7 time units, respectively, because the time unit of the module is 10 ns.

The $realtime system function returns a real number. Hence, if we replace the first argument of the $monitor system task in the above module, $time, with $realtime, the simulation results would be as follows:

```
#0    a = x
#3.6 a = 0
#7.2 a = 1
```

1.4.2.4 *Simulation Control System Tasks*   There are two simulation control system tasks: $finish and $stop. The $stop system task suspends the simulation whereas the $finish system task terminates the simulation.

1.4.3 A Tutorial Example

In this sub-section, we use a simple synthesizable example to demonstrate how to verify a design through a test bench.

EXAMPLE 1.8   *A 4-Bit Adder Module*

In this example, four full-adder instances are instantiated to construct a 4-bit adder. The full-adder module is described in the preceding section.

```
// gate-level description of 4-bit adder
module four_bit_adder (x, y, c_in, sum, c_out);
input  [3:0] x, y;
input  c_in;
output [3:0] sum;
output c_out;
wire   c1, c2, c3; // intermediate carries
// -- four_bit adder body-- //
// instantiate the full adder
   full_adder fa_1 (x[0], y[0], c_in, sum[0], c1);
   full_adder fa_2 (x[1], y[1], c1, sum[1], c2);
   full_adder fa_3 (x[2], y[2], c2, sum[2], c3);
   full_adder fa_4 (x[3], y[3], c3, sum[3], c_out);
endmodule
```

■

Usually, it is instructive to use the result from synthesis tools as an aid to help verify the design because we may discern the differences between the results from our 'mind' and those from the synthesis tools. In addition, we might also be able to control the

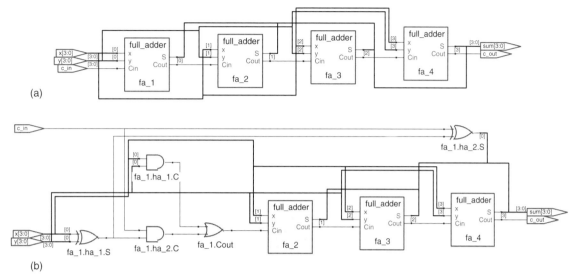

(a)

(b)

FIGURE 1.9 Synthesized results of the 4-bit adder module: (a) the original synthesized result; (b) the result after dissolving the first full-adder module

synthesized structure from the synthesis tools by structuring the source description. The synthesized results of the preceding example from a synthesis tool are shown in Figure 1.9. Figure 1.9(a) shows the original synthesized result and Figure 1.9(b) is the result after dissolving the first full-adder module.

The following example describes how a typical test bench is organized.

EXAMPLE 1.9 *The Test Bench for the 4-Bit Adder Module*

In general, a test bench consists of three major parts, as depicted in Figure 1.8, an instantiation of the UUT, the stimulus generation block and the response monitoring and checking block. The result checking may be done by comparing the text outputs or viewing the waveform generated by a waveform viewer through reinterpreting the text outputs.

```
`timescale 1 ns / 100 ps   // time unit is ns.
module four_bit_adder_tb;
// internal signals declarations:
reg [3:0] x;
reg [3:0] y;
reg c_in;
wire [3:0] sum;
wire c_out;
// Unit Under Test instantiation and port map
    four_bit_adder UUT (.x(x), .y(y), .c_in(c_in),
                        .sum(sum), .c_out(c_out));
reg [7:0] i;
```

```
initial begin // for use in post-map and post-par
              // simulations.
// $sdf_annotate ("four_bit_adder_map.sdf", four_bit_
// adder);
// $sdf_annotate ("four_bit_adder_timesim.sdf", four_
// bit_adder);
end
initial       // stimulus generation block
   for  (i = 0; i <= 255; i = i + 1) begin
        x[3:0] = i[7:4]; y[3:0] = i[3:0]; c_in =1'b0;
   #20;   end
initial #6000 $finish;
initial       // response monitoring block
   $monitor($realtime,"ns %h %h %h %h", x, y, c_in,
          {c_out, sum});
endmodule
```

The simulation results are as follows:

```
     0ns 0 0 0 00
#   20ns 0 1 0 01
#   40ns 0 2 0 02
#   60ns 0 3 0 03
#   80ns 0 4 0 04
# 100ns 0 5 0 05
# 120ns 0 6 0 06
# 140ns 0 7 0 07
# 160ns 0 8 0 08
# 180ns 0 9 0 09
# 200ns 0 a 0 0a
# 220ns 0 b 0 0b
# 240ns 0 c 0 0c
# 260ns 0 d 0 0d
# 280ns 0 e 0 0e
# 300ns 0 f 0 0f
# 320ns 1 0 0 01
# 340ns 1 1 0 02
# 360ns 1 2 0 03
# 380ns 1 3 0 04
# 400ns 1 4 0 05
# 420ns 1 5 0 06
# 440ns 1 6 0 07
# 460ns 1 7 0 08
# 480ns 1 8 0 09
# 500ns 1 9 0 0a
# 520ns 1 a 0 0b
# 540ns 1 b 0 0c
```

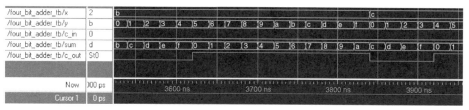

FIGURE 1.10 The waveform of the simulation results

The above results can also be viewed through a waveform viewer as shown in Figure 1.10.

Review Questions

Q1.17 What are the major components of a test bench?

Q1.18 Why is the `timescale` compiler directive required during simulation?

Q1.19 How would you monitor the response of a design during simulation?

Q1.20 Describe the differences between the $display and $monitor system tasks.

Q1.21 Describe the functions of the $time and $realtime system functions.

Q1.22 Describe the functions of the $stop and $finish system tasks.

SUMMARY

Using HDLs to design digital systems has become an essential procedure in modern electronic engineering. The major feature inherent to HDLs is that they have the capability of modeling a digital system at many levels of abstraction, ranging from the algorithm-level to the gate-level, and even to switch-level. HDL is an acronym for hardware description language. Designing with HDLs is analogous to computer programming.

The basic program structure in Verilog HDL is called a *module*. A Verilog HDL module is much like a high-level language program except that it is used to describe a hardware module rather than a software function. One of the major differences between a software program and a hardware module is the timing feature inherent to the hardware module. Hence, a practical HDL, such as Verilog HDL, must provide some mechanisms to describe the timing feature of the hardware.

A Verilog HDL module consists of two major parts: the interface and the internal. The internal (or body) of a module can be modeled as structural style, dataflow style, behavioral or algorithmic style, or a mix of them. In structural style, a design is described as a set of interconnected components. The components can be modules, UDPs, primitive gates and/or primitive switches. A design is said to be modeled at gate-level when it only consists of a set of interconnected gate primitives and at switch-level when it comprises a set of interconnected switch primitives. The structural style is a mechanism used to construct a hierarchical design for a large digital system.

In dataflow style, the module is described by specifying the dataflow (i.e. data dependence) between registers and how the data are processed. A module is specified as a set of continuous assignment statements. In behavioral or algorithmic style, the design is described in terms of the desired design algorithm without concerning the hardware implementation details. Designs can be described in any high-level programming languages. In mixed style, the design is described in terms of the mixing use of the above three styles. Mixed styles are most commonly used in specifying large designs. In industry, the term RTL is often used to mean a structure that combines both behavioral and dataflow constructs and is acceptable to logic synthesis tools.

For a design to be useful, it must be verified so as to make sure that it can function correctly. Verilog HDL not only provides capabilities to model a design but also provides facilities to generate and control stimuli, store responses, and check the results. The basic component of simulation-based verification is a test bench. In general, a test bench comprises three basic parts: an instantiation of the UUT, the stimulus generation block and the response monitoring and checking block. Both stimulus generation and response monitoring and checking blocks are usually modeled in behavioral style. The result checking may be done by comparing the text outputs or viewing the waveform generated by a waveform viewer through reinterpreting the text outputs.

REFERENCES

1. J. Bhasker, *A Verilog HDL Primer,* 3rd Edn, Star Galaxy Publishing, 2005.
2. M.D. Ciletti, *Modeling, Synthesis and Rapid Prototyping with the Verilog HDL,* Prentice-Hall, Upper Saddle River, NJ, USA, 1999.
3. IEEE 1364-2001 Standard, *IEEE Standard Verilog Hardware Description Language,* 2001.
4. M.-B. Lin, *Digital System Design: Principles, Practices and ASIC Realization,* 3rd Edn, Chuan Hwa Book Company, Taipei, Taiwan, 2002.
5. S. Palnitkar, *Verilog HDL: A Guide to Digital Design and Synthesis,* 2nd Edn, SunSoft Press, 2003.
6. J.P. Uyemura, *Introduction to VLSI Circuits and Systems,* John Wiley & Sons, Inc., Hoboken, NJ, USA, 2002.

PROBLEMS

1.1 A half subtractor (also subtracter) is a device that accepts two inputs, x and y, and produces two outputs, b and d. The full subtractor is a device that accepts three inputs, x, y and b_in and produces two outputs, b_out and d according to the truth table shown in Table 1.2.

(a) Derive the minimal expressions of both b_out and d of the full subtractor in terms of two half subtractors and one two-input OR gate.

(b) Draw the logic diagram of the switching expressions: b_out and d.

1.2 (structural style)

(a) Write a Verilog HDL module to describe the full subtractor obtained from Problem 1.1 in structural style using instantiations of gate primitives.

(b) Write a test bench and verify if the module behaves as a full subtractor.

TABLE 1.2 The truth table of a full subtractor

x	y	b_in	b_out	d
0	0	0	0	0
0	0	1	1	1
0	1	0	1	1
0	1	1	1	0
1	0	0	0	1
1	0	1	0	0
1	1	0	0	0
1	1	1	1	1

1.3 (dataflow style)

(a) Write a Verilog HDL module to describe the full subtractor obtained from Problem 1.1 in dataflow style using bit-wise operators.

(b) Write a test bench and verify if the module behaves as a full subtractor.

1.4 (behavioral style)

(a) Write a Verilog HDL module to describe the full subtractor obtained from Problem 1.1 in behavioral style.

(b) Write a test bench and verify if the module behaves as a full subtractor.

1.5 (mixed-style modeling)

(a) Model the first half subtractor obtained from Problem 1.1 in structural style and the second one in dataflow style.

(b) Model the switching expression b_out of the full subtractor obtained from Problem 1.1 in behavioral style.

(c) Write a test bench and verify if the module behaves as a full subtractor.

## CHAPTER 2

STRUCTURAL MODELING

IN **STRUCTURAL** style, a module is described as a set of interconnected components. The components can be modules, user-defined primitives (UDPs), gate primitives and switch primitives.

1. *Modules*. A module can instantiate any number of other modules. Whether a module is synthesizable or not depends on the contents of the module. A gate-level module is usually synthesizable.

2. *Gate primitives*. In Verilog HDL, there are 12 gate primitives. These primitives may be used in any modules at any time whenever they are appropriate. Gate primitives are synthesizable.

3. *Switch primitives*. In Verilog HDL, there are 16 switch primitives. These primitives may be used in any modules at any time whenever they are appropriate. Switch primitives are usually used to model a new logic gate at switch level and are not synthesizable, in general.

4. *UDPs*. A UDP is like a module but it cannot instantiate any other UDPs or modules. In addition, it is generally not supported by synthesizers.

In this chapter, we introduce the structural modeling at the gate and switch levels. The UDPs and modules are dealt with separately in Chapters 5 and 6.

2.1 GATE-LEVEL MODELING

As mentioned before, gate-level modeling describes a design which only uses gate primitives. In this section, we will describe this modeling style in detail and give some more insight about how to use gate primitives to model an actual logic circuit.

Digital System Designs and Practices   Ming-Bo Lin
© 2008 John Wiley & Sons (Asia) Pte Ltd

2.1.1 Gate Primitives

In Verilog HDL, there are 12 predefined gate primitives. These primitives can be cast into two groups: and/or gates and buf/not gates. The group of and/or gates includes and, nand, or, nor, xor and xnor. The group of buf/not gates includes buf, not, bufif0, bufif1, notif0 and notif1. Two from the group of buf/not gates: buf and not, are also described in this sub-section because they are usually combined with gates from the and/or group to form useful logic circuits.

2.1.1.1 The Group of and/or Gates   All gates of the and/or group have one scalar output and multiple scalar inputs and are used to realize the basic logic

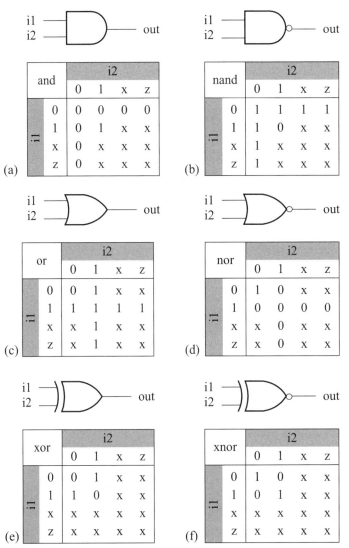

FIGURE 2.1 The group of and/or gates: (a) and gate; (b) nand gate; (c) or gate; (d) nor gate; (e) xor gate; (f) xnor gate

operations such as AND, OR or XOR. In addition, the complementary parts are provided. The symbols and truth tables of these gates are depicted in Figures 2.1(a) to 2.1(f), respectively.

To instantiate a gate from the group of and/or gates, the following syntax is used:

```
gate_name [instance_name] (output, input1, input2,
    ..., inputn);
```

where `gate_name` can be any of the set {and, nand, or, nor, xor, xnor} and the [instance_name] is optional. The first port is always the output and the other ports are inputs. It is also possible to instantiate multiple instances of the same gate in one statement by using the following syntax:

```
gate_name
    [instance_name1] (output, input1, input2, ..., inputn),
    [instance_name2] (output, input1, input2, ..., inputn),
    ...
    [instance_namen] (output, input1, input2, ..., inputn);
```

2.1.1.2 The `buf` and `not` Gates

Both `buf` and `not` gates have one scalar input and one or multiple scalar outputs. They are used to realize the NOT operation or to buffer the output of an AND/OR gate. The symbols and truth tables of these gates are depicted in Figures 2.2.

To instantiate a `buf` or `not`, the following syntax is used:

```
buf_not [instance_name] (output1, output2, ..., outputn,
    input);
```

where the [instance_name] is optional. The last port is always the input and the other ports are outputs. It is also possible to instantiate multiple instances of the same gate in one statement by using the following syntax:

```
buf_not
    [instance_name1](output1, output2, ..., outputn, input),
    [instance_name2](output1, output2, ..., outputn, input),
    ...
    [instance_namen](output1, output2, ..., outputn, input);
```

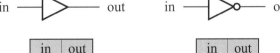

in	out
0	0
1	1
x	x
z	x

(a)

in	out
0	1
1	0
x	x
z	x

(b)

FIGURE 2.2 Buffer (a) and not (b) gates

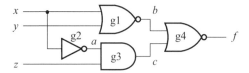

FIGURE 2.3 A simple application of basic gates

The following example is a simple application of basic gates to implement the logic circuit shown in Figure 2.3.

EXAMPLE 2.1 *A Simple Application of Basic Gates*

In this example, we need to declare three internal nets, a, b and c, to connect the output port from one gate to the input port of another.

```
module basic_gates (x, y, z, f) ;
input   x, y, z;
output f ;
wire    a, b, c;  // internal nets
// structural modeling using basic gates.
   nor g1 (b, x, y);
   not g2 (a, x);
   and g3 (c, a, z);
   nor g4 (f, b, c);
endmodule
```

It is possible to instantiate an array of gates from the group of and/or gates by using the following syntax:

```
gate_name [instance_name[range]] (output, input1, ...,
inputn);
```

However, the array instantiations of a primitive gate might be a synthesizer-dependent! So, the suggestion is that you had better check this feature from your synthesizer before using array instantiations.

EXAMPLE 2.2 *An Example of Array of Instances*

The following module demonstrates how to instantiate an array of instantiations of the nand gate. Here, we also show that the effect of an array of instantiations is actually a concise representation of multiple instantiations using the same vector arguments.

```
wire [3:0] out, in1, in2;
// an array instantiations of nand gate.
   nand n_gate[3:0] (out, in1, in2);

   // this is equivalent to the following:
   nand n_gate0 (out[0], in1[0], in2[0]);
   nand n_gate1 (out[1], in1[1], in2[1]);
   nand n_gate2 (out[2], in1[2], in2[2]);
   nand n_gate3 (out[3], in1[3], in2[3]);
```

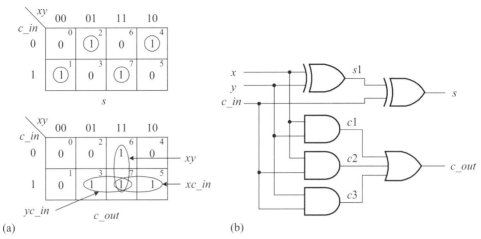

(a)   (b)

FIGURE 2.4 A 1-bit full adder: (a) Karnaugh maps for s and c-out; (b) logic diagram

Gate-level style is usually used to model a logic circuit at gate level. Figure 2.4 shows a full adder constructed from basic gates along with the Karnaugh maps used to reduce both sum and carry out functions.

The full adder can be modeled in a straightforward manner using basic gates, as in the following example.

EXAMPLE 2.3 *An Example of 1-Bit Full Adder*

From the Karnaugh maps shown in Figure 2.4(a), we obtain the logic diagram shown in Figure 2.4(b). Hence, by instantiating the basic primitive gates and connecting them together according to the logic diagram, the following module results.

```
module full_adder_structural(x, y, c_in, s, c_out);
// I/O port declarations
input   x, y, c_in;
output  s, c_out;
wire    s1, c1, c2, c3;
// structural modeling of the 1-bit full adder.
   xor  xor_s1(s1, x, y);      // compute sum.
   xor  xor_s2(s, s1, c_in);
   and  and_c1(c1, x, y);      // compute carry out.
   and  and_c2(c2, x, c_in);
   and  and_c3(c3, y, c_in);
   or   or_cout(c_out, c1, c2, c3);
endmodule
```

A 4-to-1 multiplexer is a device used to route the data from one of the four inputs to its output end. The logic symbol and function table are shown in Figures 2.5(a) and 2.5 (b), respectively. From the function table, we can obtain the following switching

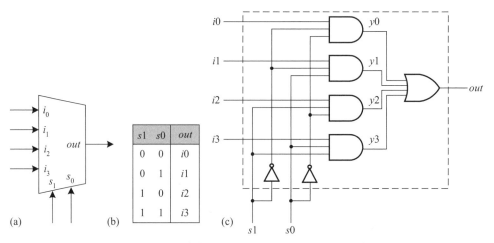

FIGURE 2.5 A 4-to-1 multiplexer: (a) logic symbol; (b) function table; (c) logic circuit

expression:

$$out = i0 \cdot s1' \cdot s0' + i1 \cdot s1' \cdot s0 + i2 \cdot s1 \cdot s0' + i3 \cdot s1 \cdot s0$$

The logic circuit that corresponds to this expression is depicted in Figure 2.5(c). The 4-to-1 multiplexer can be modeled structurally using basic gates as in the following example.

EXAMPLE 2.4   *An Example of 4-to-1 Multiplexer*

In this example, two not gates and four and gates are required to uniquely select an input out of the four inputs. One additional or gate is required to join together the four possible chosen routes. The resulting module immediately follows.

```
module mux4_to_1_structural (i0, i1, i2, i3, s1, s0, out);
input   i0, i1, i2, i3, s1, s0;
output out;
wire    s1n, s0n;       // internal wire declarations
wire    y0, y1, y2, y3;
// gate instantiations
   not (s1n, s1);       // create s1n and s0n signals.
   not (s0n, s0);
   and (y0, i0, s1n, s0n);     // 3-input and gates
                               // instantiated
   and (y1, i1, s1n, s0);
   and (y2, i2, s1, s0n);
   and (y3, i3, s1, s0);
    or (out, y0, y1, y2, y3); // 4-input or gate
                               // instantiated
endmodule
```

The final example of this subsection is a 9-bit parity generator, which can be described by the following switching expression:

$$ep = x[0] \oplus x[1] \oplus x[2] \oplus x[3] \oplus x[4] \oplus x[5] \oplus x[6] \oplus x[7] \oplus x[8]$$

$$op = x[0] \oplus x[1] \oplus x[2] \oplus x[3] \oplus x[4] \oplus x[5] \oplus x[6] \oplus x[7] \odot x[8]$$

The above two expressions can be realized using XOR gates according to the sequence of the inputs $x[0], x[1], \cdots, x[8]$, i.e. $(\cdots((x[0] \oplus x[1]) \oplus x[2]) \cdots x[8])$. However, this kind of linear implementation takes $(n-1)t_{g,xor}$, where $t_{g,xor}$ is the propagation delay of one XOR gate. A much faster realization at no extra cost is to rearrange the XOR gates as a binary tree, as shown in the following two expressions:

$$ep = \{[(x[0] \oplus x[1]) \oplus (x[2] \oplus x[3])] \oplus [(x[4] \oplus x[5]) \oplus (x[6] \oplus x[7])]\} \oplus x[8]$$

$$op = \{[(x[0] \oplus x[1]) \oplus (x[2] \oplus x[3])] \oplus [(x[4] \oplus x[5]) \oplus (x[6] \oplus x[7])]\} \odot x[8]$$

Using a binary tree structure, the total propagation delay is reduced from $(n-1)t_{g,xor}$ to $\lceil \log_2 n \rceil t_{g,xor}$. Hence, the total propagation delay of the 9-bit parity generator is only $4t_{g,xor}$ when using a binary tree structure and is $8t_{g,xor}$ when using a linear structure.

The logic circuit that corresponds to both expressions ep and op is depicted in Figure 2.6. The 9-bit parity generator can be modeled structurally using basic gates, as in the following example.

EXAMPLE 2.5 *An Example of 9-Bit Parity Generator*

In this example, eight `xor` gates and one `xnor` gate are required. By connecting together these nine primitive gates in accordance with the logic diagram shown in Figure 2.6, the module immediately follows.

```
module parity_gen_9b_structural(x, ep, op);
// I/O port declarations
input  [8:0] x;
output ep, op;
```

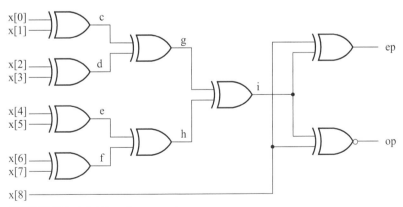

FIGURE 2.6 A 9-bit parity generator

```
wire   c, d, e, f, g, h, j;
   xor xor_11(c, x[0], x[1]);   // first level
   xor xor_12(d, x[2], x[3]);
   xor xor_13(e, x[4], x[5]);
   xor xor_14(f, x[6], x[7]);
   xor xor_21(g, c, d);          // second level
   xor xor_22(h, e, f);
   xor xor_31(i, g, h);          // third level
   xor xor_ep(ep, i, x[8]);      // fourth level
   xnor xnor_op(op, i, x[8]);
endmodule
```

2.1.1.3 *Concepts of Controlled Gates*

In designing some digital systems, it is quite useful to consider the AND/OR/XOR gates as controlled gates, as shown in Figure 2.7.

The first group of controlled gates is AND/NAND gates, as shown in Figure 2.7(a). From their truth tables, shown in Figures 2.1(a) and 2.1(b), we know for this group of gates that the output is a constant 0 or 1 while one of the inputs is fixed to 0, depending on whether the underlying gate is AND or NAND. To pass the input data through the gates to their outputs, all inputs except the one of interest must be fixed to 1. The data being passed to the output will be in its true or complement form, depending on whether the underlying gate is the AND or NAND type.

The second group of controlled gates is OR/NOR gates, as shown in Figure 2.7(b). From their truth tables, shown in Figures 2.1(c) and 2.1(d), we know for this group of gates that the output is a constant 1 or 0 while one of the inputs is fixed to 1, depending on whether the underlying gate is OR or NOR. To pass the input data through the gates

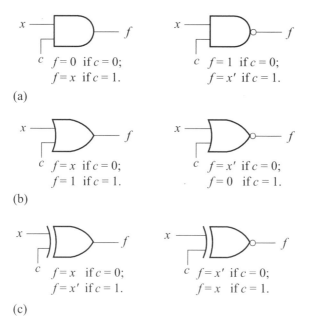

(a) $f = 0$ if $c = 0$;
$f = x$ if $c = 1$.
$f = 1$ if $c = 0$;
$f = x'$ if $c = 1$.

(b) $f = x$ if $c = 0$;
$f = 1$ if $c = 1$.
$f = x'$ if $c = 0$;
$f = 0$ if $c = 1$.

(c) $f = x$ if $c = 0$;
$f = x'$ if $c = 1$.
$f = x'$ if $c = 0$;
$f = x$ if $c = 1$.

FIGURE 2.7 Dealing with AND/OR/XOR gates as controlled gates: (a) AND and NAND gates; (b) OR and NOR gates; XOR and XNOR gates

to their outputs, all inputs except the one of interest must be fixed to 0. The data being passed to the output will be in its true or complement form, depending on whether the underlying gate is the OR or NOR type.

The third group of controlled gates is XOR/XNOR gates, as shown in Figure 2.7(c). From their truth tables, shown in Figures 2.1(e) and 2.1(f), we know for this group of gates that the output is a true or complement form of its input, depending on the status of the other input and the gate type. For a two-input XOR gate, the output takes the true form of one input while the other input is set to 0; otherwise, it takes the complement form. For the case of the XNOR gate, the polarity of output is reversed.

Review Questions

Q2.1 Describe the meaning of structural style.

Q2.2 Define the gate-level modeling.

Q2.3 What types of gates do the group of and/or gates contain?

Q2.4 What types of gates do the group of buf/not gates contain?

Q2.5 What is the meaning of array instantiations?

2.1.2 Tristate Buffers

When an output of a logic circuit can also be set to an extra state, high-impedance, in addition to its two normal states, 0 and 1, the logic circuit is said to be a tristate (three-state) logic circuit. The tristate output is usually controlled by an enable input. When the enable signal is activated, the logic circuit operates in its two normal output states, 0 and 1, but when the enable signal is disabled, the output of the logic circuit is set to a high-impedance.

Since the function of a tristate may be associated with buffers or inverters and the enable signal can be activated at low-logic level or high-logic level, there are four kinds of tristate buffers: active-low buffer (bufif0), active-low inverter (notif0), active-high buffer (bufif1) and active-high inverter (notif1), as shown in Figure 2.8. It is worth noting that the keyword is ended with "if0" for an active-low buffer/inverter and ended with "if1" for an active-high buffer/inverter.

Each tristate buf/not gate has three ports: one scalar output, one scalar input and one control. Tristate buf/not gates are usually used as controlled buffers/inverters. The symbols truth tables of these tristate buffers and inverters are depicted in Figures 2.8(a)–(d), respectively. The symbol L in Figure 2.8 represents 0 or z while symbol H denotes 1 or z.

To instantiate a buffer from this group, the following syntax is used:

```
buf_name [instance_name] (output, input, control);
```

where buf_name can be any one from the set {bufif0, bufif1, notif0, notif1} and the [instance_name] is optional. The first port is always an output port, the second port is an input port and the last port is control. It is also possible to instantiate multiple instances of the same tristate buffer in one statement by using the

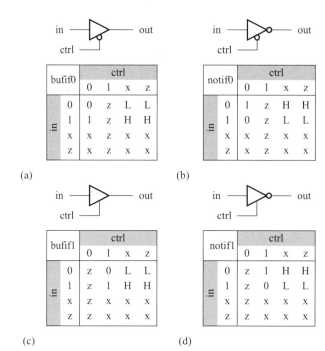

FIGURE 2.8 Tristate buffers and inverters: (a) bufif0; (b) notif0; (c) bufif1; (d) notif1

following syntax:

```
buf_name
    [instance_name1] (output, input, control),
    [instance_name2] (output, input, control),
    ...
    [instance_namen] (output, input, control);
```

2.1.2.1 wire *and* tri *Nets* Both wire and tri nets connect elements and have the same syntax and function except that a wire net is driven by a single driver, a gate or continuous assignment, but the tri net may be driven by multiple drivers. When multiple drivers drive a wire or a tri net, the effective value of the net is determined by the truth table shown in Table 2.1.

To use wire and tri nets, the following syntax can be used:

```
net_name [signed][[msb:lsb]] net1, net2, ..., netn;
```

where net_name can be either a wire or tri net.

TABLE 2.1 The truth table of wire and tri nets

wire/tri	0	1	x	z
0	0	x	x	0
1	x	1	x	1
x	x	x	x	x
z	0	1	x	z

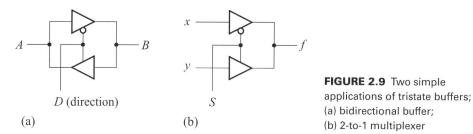

FIGURE 2.9 Two simple applications of tristate buffers; (a) bidirectional buffer; (b) 2-to-1 multiplexer

Figure 2.9 shows two simple applications of tristate buffers. By appropriately combining `bufif0` and `bufif1` gates, two useful circuits are produced. Figure 2.9(a) is a 1-bit bidirectional buffer and Figure 2.9(b) is a 2-to-1 multiplexer. The following example shows how to model the 2-to-1 multiplexer shown in Figure 2.9(b).

EXAMPLE 2.6   *A 2-to-1 Multiplexer Constructed from Tristate Buffers*

Because the output f is driven by two tristate buffers, it needs to be declared as a `tri` net. The tristate buffer `bufif0` is enabled if s is 0; the tristate buffer `bufif1` is enabled if s is 1. Hence, it is a 2-to-1 multiplexer.

```
// Data selector: 2-to-1 mux
module two_to_one_mux_tristate (x, y, s, f);
input  x, y, s;
output f;
// internal declaration
tri f;
// data selector body
   bufif0 b1 (f, x, s);  // enable if s = 0
   bufif1 b2 (f, y, s);  // enable if s = 1
endmodule
```

The bidirectional buffer shown in Figure 2.9(a) can also be easily modeled. Hence, it is left to the reader as an exercise.

Review Questions

Q2.6 What kinds of gates belong to the group of tristate gates?

Q2.7 Explain the operation of active-low inverters.

Q2.8 Explain the operation of active-high buffers.

Q2.9 Explain the meaning of `tri` net data type.

Q2.10 What is the difference between `wire` and `tri` net data types?

2.1.3 Wired Logic

In some applications, a logic circuit with an open-collector (for TTL) or open-drain (for CMOS) output stage is preferred because fewer logic gates or less area are required.

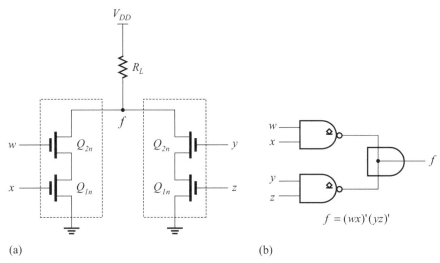

(a)   (b)

FIGURE 2.10 A wired NAND–AND logic circuit: (a) circuit; (b) logic symbol

For instance, the circuit portrayed in Figure 2.10 is a wired NAND–AND logic circuit. From this figure, we can easily derive the logic function of this circuit as follows:

$$f(w, x, y, z) = (w \cdot x)' \cdot (y \cdot z)'$$

which requires two NAND gates and one AND gate when realized directly from the above expression. However, it only requires two wired NAND gates in reality.

2.1.3.1 wor/trior *and* wand/triand *Nets*

In Verilog HDL, there are four net types, wor, wand, trior and triand, that can be used to model wired logic. The truth tables of these net types are shown in Table 2.2. The syntax and functionality of wor and trior nets are identical and the same situation applies for wand and triand nets. To use these nets, the following syntax can be used:

```
net_name [signed][[msb:lsb]] net1, net2, ..., netn;
```

where net_name can be any of the set {wor, wand, trior, triand}.

Because a wired net configuration may be driven by multiple drivers, a confliction resolution rule is required and set as follows: the value of a wor or trior net is 1

TABLE 2.2 The truth tables of triand/wand and trior/wor nets

triand/ wand	0	1	x	z	trior/ wor	0	1	x	z
0	0	0	0	0	0	0	1	x	0
1	0	1	x	1	1	1	1	1	1
x	0	x	x	x	x	x	1	x	x
z	0	1	x	z	z	0	1	x	z

when any of the drivers is 1; the value of a wand and triand net is 0 when any of drivers is 0.

With the aid of the wand net data type, Figure 2.10(a) may be modeled as in the following example.

EXAMPLE 2.7 *A Wired Nand-and Logic Circuit*

To model the wired NAND–AND logic circuit, the output port is declared as a wand net so that it may be driven by two nand gates.

```
module open_drain (w, x, y, z, f);
input   w, x, y, z;
output f;
wand    f;   // internal declaration
// wired AND logic gate
   nand n1 (f, w, x);
   nand n2 (f, y, z);
endmodule
```

2.1.3.2 tri0 *and* tri1 *Nets*
The tri0 and tri1 nets are also used to model wired-logic nets, i.e. a net with more than one driver. The salient features of these nets are as follows. For a tri0 net, its value is 0 when no driver is driving it; for a tri1 net, its value is 1 when no driver is driving it. The effective values of a tri0 and a tri1 net are determined by the truth tables shown in Table 2.3, respectively, when more than one drivers drive the net.

To use tri0 and tri1 nets, the following syntax can be used:

```
net_name [signed][[msb:lsb]] net1, net2, ..., netn;
```

where net_name can be either a tri0 or tri1 net.

Review Questions

Q2.11 Explain the meaning of the wor net data type.

Q2.12 Explain the meaning of the wand net data type.

Q2.13 Explain the meaning of a wired-and logic circuit.

Q2.14 Explain the meaning of a wired-or logic circuit.

Q2.15 Explain the meanings of tri0 and tri1 net data types.

TABLE 2.3 The truth tables of tri0 and tri1 nets

tri0	0	1	x	z		tri1	0	1	x	z
0	0	x	x	0		0	0	x	x	0
1	x	1	x	1		1	x	1	x	1
x	x	x	x	x		x	x	x	x	x
z	0	1	x	0		z	0	1	x	1

2.2 GATE DELAYS

Due to the existence of resistance (R) and capacitance (C) in logic circuits, all of the actual logic gates have finite *propagation delays*. In addition, in every interconnection between any two circuits there always exists an RC delay. This RC delay is called a *transport delay*. For these reasons, HDLs usually employ *inertial* and *transport* delay models to model the propagation delay of actual logic circuits and interconnection wire delay, respectively. In this section, we first describe both delay models in greater detail and then introduce the delay specifications for gate primitives.

2.2.1 Delay Models

As mentioned above, both the inertial delay model and transport delay model are used in Verilog HDL to separately capture the propagation delay of logic gates and the transport delay of interconnection wires. In the following, we describe each of these in more detail.

2.2.1.1 Inertial Delay Model   Due to the inherent resistances (R) and capacitances (C) in circuits, all digital circuits have a certain amount of inertia (analogous to inertia in physics); namely, it takes a finite amount of time and a certain amount of energy for the output of a gate to respond to a change on the input. This implies that the signal events do not persist long enough, will be filtered out and not propagated to the output of the gates. In other words, inertial delay has the effect of suppressing input pulses whose duration is not longer than the inertial delay of the logic gate. This propagation delay model is often referred to as the *inertial delay model*. It is usually used to model gate delays in HDLs, such as Verilog HDL and VHDL.

For example, as shown in Figure 2.11, the input pulse which occurred between time units 2 and 3 is filtered out because it is shorter than the propagation delay of the `not` gate, which is 3 time units. The other pulse occurring between time units 5 and 9 will be present at the output after a propagation delay of 3 time units.

EXAMPLE 2.8   *A More Complicated Example of Inertial Delay*

Another more complicated example is shown in Figure 2.12, where two inputs x and y are applied to the inputs of both AND gates. The output f is obtained by passing the output a through an inverter. Assume that the propagation delays of both AND gates are 4 time units and the propagation delay of the inverter is 1 time unit.

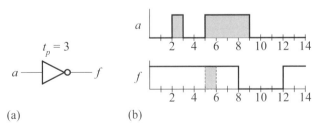

(a)   (b)

FIGURE 2.11 An example of the effects of inertial delay: (a) not gate; (b) timing diagram

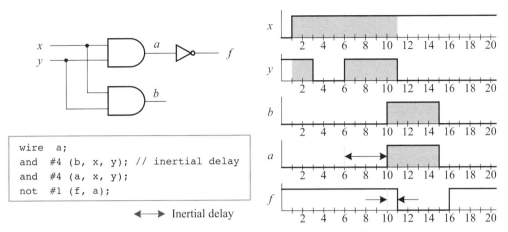

```
wire   a;
and    #4 (b, x, y);  // inertial delay
and    #4 (a, x, y);
not    #1 (f, a);
```

←——→ Inertial delay

FIGURE 2.12 An example of the effects of inertial delay

From the definition of the inertial delay model, the pulse width appearing in the input of a gate less than the specified inertial delay will not appear in the output of the gate. Hence, although a pulse of width 2 time units (from time units 1 to 3) results from the AND operation of both inputs x and y, it will not appear at the output end a. On the other hand, the pulse of width 5 time units (from time units 6 to 11) will propagate to the output end f after 5 time units. ■

2.2.1.2 Transport Delay Model   The inherent feature of an interconnection wire is that any signal events appearing at the input will be propagated to the output. This is also the essential feature of the *transport delay model*. The transport delay model is usually used to model net (i.e. wire) delays, that is, the time of flight of a signal passing through a wire. The default delay of a net is zero. For example, as shown in Figure 2.13, both input pulses are presented at the output after a propagation delay of 3 time units regardless of their pulse widths.

A more complicated example showing both the effects of inertial and transport delays is as follows.

EXAMPLE 2.9   *An Example Showing Both Effects of Inertial and Transport Delays*

An shown in Figure 2.14, the logic circuit of this example is the same as that of Figure 2.12 except that wire a has a transport delay of 2 time units. From the definition of the transport delay model, any signal events appearing at the input will be propagated

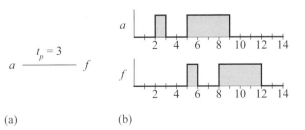

$$t_p = 3$$

a ———————— f

(a)   (b)

FIGURE 2.13 An example of the effects of transport delay: (a) a wire; (b) timing diagram

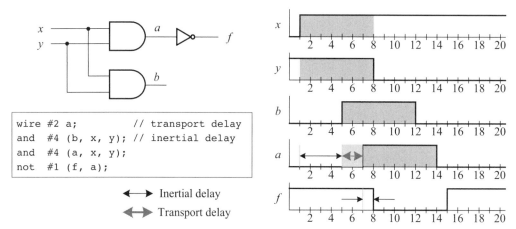

```
wire #2 a;         // transport delay
and  #4 (b, x, y); // inertial delay
and  #4 (a, x, y);
not  #1 (f, a);
```

◄──────► Inertial delay

◄──► Transport delay

FIGURE 2.14 An example of the effects of both inertial and transport delays

to the output after the specified delay. Hence, the pulse of width 7 time units (from time units 1 to 8) resulted from the AND operation of both inputs x and y will appear at the output end a after 6 time units, starting from time unit 7, where 6 time units are a combination of inertial delay (4 time units) of AND gate and transport delay of wire a (2 time units). This signal will appear at the output f one time unit later due to the delay of the inverter. ■

2.2.2 Delay Specifications

Now that we have the delay models of both gates and nets, the propagation delay of a gate or the transport delay of a net can then be specified by a delay specification, which consists of zero up to three *delay specifiers*. In the following, we deal with these two issues in order.

The signal propagation delay from any gate input to the gate output can be specified by *gate delays*. Propagation delays of gate primitives may be specified in one of the following ways:

1. Specify no delay:
   ```
   gatename [instance_name](output,in1,in2, ...);
   ```
2. Specify propagation delay only:
   ```
   gatename #(prop_delay) [instance_name](output,in1,in2,
       ...);
   ```
3. Specify both rise and fall times:
   ```
   gatename #(t_rise,t_fall) [instance_name](output,in1,
       in2, ...);
   ```
4. Specify rise, fall and turn-off times:
   ```
   gatename #(t_rise,t_fall,t_off) [instance_name](output,
       in1,in2,...);
   ```

where t_rise refers to the transition to the 1 value, t_fall refers to the transition to the 0 value and t_off refers to the transition to the high-impedance value.

The signal transports from any net input to its output end may be specified by *net delays*. To specify the delay value of specified nets, the following syntax can be used:

```
net_name [#delay][signed][[msb:lsb]] net1, net2, ...,
   netn;
```

Like gate delays, up to three delay values per net may be specified. For both gates and nets, the default delay is zero when no delay specification is given.

Each delay value within a delay specification may be specified by a delay specifier, which has the following format:

```
minimum:typical:maximum (min:typ:max)
```

where the minimum, typical and maximum values for each delay are specified as constant expressions separated by colons. For example, the following example shows `min:typ:max` values for rising, falling and turn-off delays:

```
// only specify one delay
and  #(5)  and1 (b, x, y);

// only specify one delay using min:typ:max
not  #(10:12:15) not1 (a, x);

// specify two delays using min:typ:max
and  #(10:12:15, 12:15:20) and2 (c, a, z);

// specify three delays using min:typ:max
or  #(10:12:15, 12:15:20, 12:13:16) or2 (f, b, c);
```

Review Questions

Q2.16 What are the two delay models used in Verilog HDL?

Q2.17 Define the inertial delay model. When would you use it?

Q2.18 Define the transport delay model. When would you use it?

Q2.19 Describe the ingredients of a delay specifier.

Q2.20 What does a one-value delay specification specify?

Q2.21 What does a two-value delay specification specify?

Q2.22 What does a three-value delay specification specify?

Q2.23 How would you specify a net delay value?

2.3 HAZARDS

In general, the output signal of a combinational logic is a combination of many signals propagated from different paths. Due to the different propagation delays of these paths,

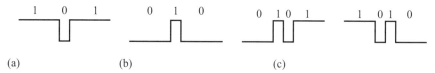

FIGURE 2.15 Static and dynamic hazards: (a) static-1 hazard; (b) static-0 hazard; (c) dynamic hazard

the output signal before it is stable must experience an amount of time during which fluctuations occur. This duration is called a *transient time* of the output signal and may result in several undesired short pulses called *glitches*. A *hazard* (or *timing hazard*) is said to be raised when fluctuation occurred during the transient time. Hazards can be divided into a *static hazard* and a *dynamic hazard*, as shown in Figure 2.15.

2.3.1 Static Hazards

A static hazard represents the situation where a circuit output may momentarily go to 0 (or 1) when it should remain at a constant 1 (or 0) during the transient time. In other words, a static hazard is a situation when the output produces a "0" glitch when its stable value is 1 and a "1" glitch when its stable value is 0. Static hazards can be further divided into a *static-0 hazard* and a *static-1 hazard*, as shown in Figures 2.15(a) and 2.15(b), respectively.

   In order to gain more insight into static hazards, consider the logic circuit shown in Figure 2.16(a). There are four paths from inputs, x, y and z, that can reach the output f. For simplicity, assume that all gates have the same propagation delay, t_{pd}. As shown in Figure 2.16(b), the output f will momentarily go to 0 for an amount of time t_{pd} when both inputs y and z are 1, and x changes from 1 to 0.

   What follows is an example of modeling the logic circuit depicted in Figure 2.16(a).

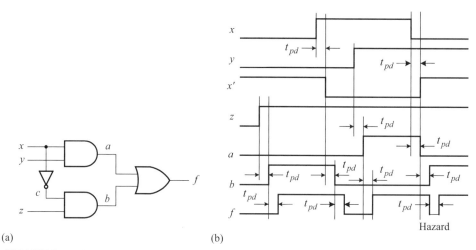

FIGURE 2.16 An example of static hazard: (a) logic circuit; (b) timing diagram

EXAMPLE 2.10   *An Example of Static Hazard and Its Effects*

This example simply models the logic circuit shown in Figure 2.16(a) in structural style. All gates are assumed to have the same propagation delays of 5 time units. By appropriately applying stimuli to the module, it can be seen that the static hazard is exactly the same as that described previously.

```
module hazard_static (x, y, z, f);
input  x, y, z;
output f;
// internal declaration
wire   a, b, c;  // internal net
// logic circuit body
   and #5 a1 (a, x, y);
   not #5 n1 (c, x);
   and #5 a2 (b, c, z);
   or  #5 o2 (f, b, a);
endmodule
```
■

To see the reason why the glitches may occur in combinational logic circuits, consider the Karnaugh map shown in Figure 2.17(a) of the logic circuit depicted in Figure 2.16(a). As mentioned previously, the "0" glitch occurs when the variables $y = z = 1$ and x changes from 1 to 0. This corresponds to the case that the function of output f switches from the product term xy to $x'z$. From the Karnaugh map, we can see that when the input combination switches from a minterm to another within the same product term, there is no glitch; otherwise, when the input combination switches from a minterm to another with two different product terms, there may be a glitch. To remove this glitch, we need to add a redundant product term to cover the gap between two prime implicants, for instance, the product term yz in the preceding example, so that the output f still remains at 1 between the switching time of two prime implicants.

From a hardware viewpoint, each product term is realized by an AND gate and each gate has a finite amount of propagation delay. The glitch is caused by the switching of AND gates and the propagation delays of the gates. The outputs of two AND gates may be at 1 or 0 at the same time for a finite amount of time, leaving a gap of signal

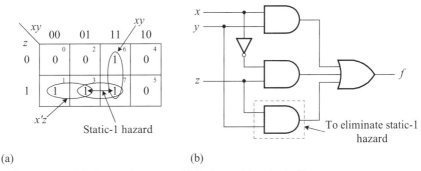

(a)   (b)

FIGURE 2.17 (a) Karnaugh map and (b) a hazard-free logic diagram

between them. To remedy this, we need to add a redundant gate to realize the redundant product term, i.e. yz, to sustain the signal level between the switching period of two AND gates so that the output f still remains at its signal level.

2.3.2 Dynamic Hazards

A dynamic hazard is a situation where the output of a combinational circuit changes from 0 to 1 and then to 0 (or 1 to 0 and then to 1); in other words, the output changes three or more times, as shown in Figure 2.15(c). Because three or more signal changes are required to have a dynamic hazard, a signal must arrive at the output at three different times. In other words, there must exist at least three paths from the signal input to the output of the underlying combinational circuit.

An example of a logic circuit having a dynamic hazard is shown in Figure 2.18 along with its associated timing diagram. It is easy to see from Figure 2.18(a) that the input signal w goes to the output f by way of three separate paths with different numbers of gates. As a consequence, the input signal w may reach the output f at three different times and it may cause a dynamic hazard. This is indeed the case. As we can see from the timing diagram depicted in Figure 2.18(b), the dynamic hazard indeed occurs when $x = y = z = 1$ and w changes from 1 to 0.

What follows is an example of modeling the logic circuit depicted in Figure 2.18(a).

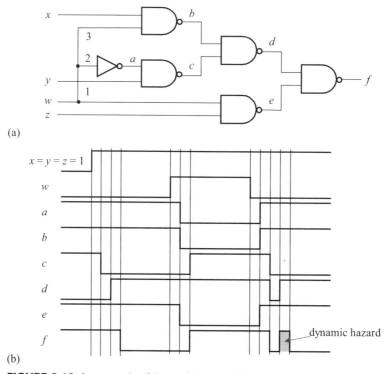

(a)

(b)

FIGURE 2.18 An example of dynamic hazard: (a) logic circuit; (b) timing diagram

EXAMPLE 2.11 *An Example of Dynamic Hazard and Its Effects*

This example simply models the logic circuit shown in Figure 2.18(a) in structural style. All gates are assumed to have the same propagation delays of 5 time units. By appropriately applying stimuli to the module, it can be seen that the dynamic hazard is exactly the same as that described previously.

```
// dynamic hazard example
module hazard_dynamic(w, x, y, z, f);
input   w, x, y, z;
output f;
// internal declaration
wire    a, b, c, d, e;   // internal net
// logic circuit body
   nand #5 nand1 (b, x, w);
   not  #5 n1    (a, w);
   nand #5 nand2 (c, a, y);
   nand #5 nand3 (d, b, c);
   nand #5 nand4 (e, w, z);
   nand #5 nand5 (f, d, e);
endmodule
```

A test bench used to drive the preceding module is as described in the following example.

EXAMPLE 2.12 *A Test Bench for the* `hazard_dynamic` *Module*

From the analysis of Figure 2.18(b), we know that in order to observe the dynamic hazard, the signals x, y and z must be set to the value 1. Then the input signal w is changed from 1 to 0. The resulting test bench module is as follows:

```
`timescale 1ns / 1ns
module hazard_dynamic_tb;
reg   w, x, y, z; //Internal signals declarations:
wire f;
// Unit Under Test port map
   hazard_dynamic UUT (
                .w(w),.x(x),.y(y),.z(z),.f(f));
initial
   begin
           w = 1'b0; x = 1'b0; y = 1'b0; z = 1'b0;
      #5   x = 1'b1; y = 1'b1; z = 1'b1;
      #30  w = 1'b1;
      #20  w = 1'b0;
      #190 $finish;  // terminate the simulation
   end
initial
   $monitor($realtime,,"ns %h %h %h %h %h ",w,x,y,z,f);
endmodule
```

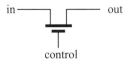

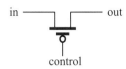

nmos	control			
	0	1	x	z
0	z	0	L	L
1	z	1	H	H
x	z	x	x	x
z	z	z	z	z

(a)

pmos	control			
	0	1	x	z
0	0	z	L	L
1	1	z	H	H
x	x	z	x	x
z	z	z	z	z

(b)

FIGURE 2.19 MOS switches: (a) nMOS; (b) pMOS

Review Questions

Q2.24 Define static hazard.

Q2.25 Define static-0 and static-1 hazards.

Q2.26 Define dynamic hazard.

2.4 SWITCH-LEVEL MODELING

When a logic circuit is modeled at the switch level,[1] all MOS transistors of it are regarded as ideal or nonideal switches. For ideal switches, there are no signal degradations when they are turned on, whereas for nonideal (resistive) switches, there have a certain finite amount of signal degradation. In Verilog HDL, all nonideal switches are prefixed with the letter "r" but all ideal switches are not.

2.4.1 MOS Switches

There are two MOS switches: `nmos`/`rnmos` and `pmos`/`rpmos`. These switches are usually used to model unidirectional switches through which data can be allowed to pass from input to output or be blocked by appropriately setting the control input(s). The `nmos` and `pmos` switches pass signals from their inputs to their outputs without degradation, whereas the `rnmos` and `rpmos` switches reduce the strength of the signals that propagate through them.

Figure 2.19 shows both `nmos` and `pmos` switches along with their truth tables. Since some combinations of input values and control values may cause these switches to output either of two values, without a preference for either value, the symbols L and H are used to represent the results that have a value 0 or z and a value 1 or z, respectively.

To instantiate a MOS switch element, the following syntax can be used:

```
switch_name[instance_name](output, input, control);
```

---

[1] This section may be omitted without loss of continuity.

where `instance_name` is optional. The first port is always the output, the second is the input port and the last is the control signal.

2.4.1.1 `supply0` *and* `supply1` *Nets*   The `supply0` and `supply1` nets are used to model *ground* and *power* nets, respectively. The logic value of `supply0` is 0 and the logic value of `supply1` is 1. We can declare these two nets using the following syntax:

```
supply0|supply1 [[msb:lsb]] net1, net2, ..., netn;
```

A simple application of MOS switches for constructing the CMOS inverter shown in Figure 2.20 is given in the following example.

EXAMPLE 2.13 *An Example of a CMOS Inverter*

The inverter consists of `nmos` and `pmos` switches with the same input and output nets. To properly model this circuit, we need another two nets, `supply1` and `supply0`, to provide the required `vdd` and `gnd` signals.

```
module mynot(input x, output f);
// internal declaration
supply1 vdd;
supply0 gnd;
// NOT gate body
   pmos p1 (f, vdd, x);  // source connected to vdd
   nmos n1 (f, gnd, x);  // source connected to ground
endmodule
```

A more complex application of MOS switches for constructing the CMOS NAND gate shown in Figure 2.21 is given in the following example.

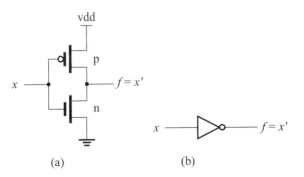

(a)   (b)

FIGURE 2.20 A CMOS inverter: (a) circuit; (b) logic symbol

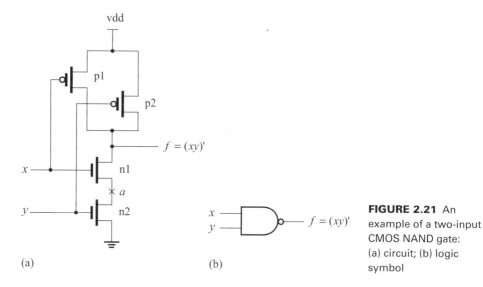

FIGURE 2.21 An example of a two-input CMOS NAND gate: (a) circuit; (b) logic symbol

EXAMPLE 2.14 *An Example of a Two-input CMOS NAND Gate*

To model this circuit, we need another two nets: `supply1` and `supply0`, to provide the required `vdd` and `gnd` signals. The two `nmos` switches are connected in series and the two `pmos` switches are connected in parallel. One end of the series-connected `nmos` switches, is also the output end. The other end of the series-connected `nmos` switches is connected to `gnd` signal. The controls of `pmos` and `nmos` switches are connected together in pair to form inputs x and y, respectively.

```
module my_nand (input x, y, output f);
// internal declaration
supply1   vdd;
supply0   gnd;
wire      a;        // terminal between two nMOS
// NAND gate body
   pmos p1 (f, vdd, x);   // source connected to vdd
   pmos p2 (f, vdd, y);   // parallel connection
   nmos n1 (f, a, x);     // serial connection
   nmos n2 (a, gnd, y);   // source connected to ground
endmodule
```

The following example applies MOS switches to construct the CMOS NOR gate shown in Figure 2.22.

EXAMPLE 2.15 *An Example of a Two-input CMOS NOR Gate*

Much the same as a NAND circuit, we need two nets: `supply1` and `supply0`, to provide the required `vdd` and `gnd` signals. The two `nmos` switches are connected in parallel and the two `pmos` switches are connected in series. Then one end of the

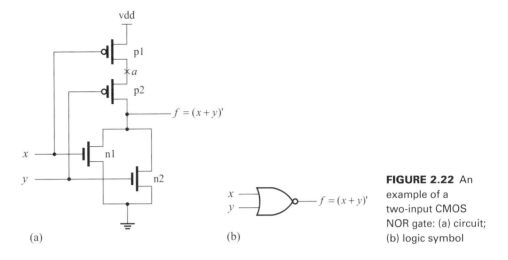

FIGURE 2.22 An example of a two-input CMOS NOR gate: (a) circuit; (b) logic symbol

series-connected pmos switches is connected to the vdd signal and the other to the nmos switches, which is also the output end. The other end of the parallel-connected nmos switches is connected to the gnd signal. The controls of pmos and nmos switches are connected together in a pair to form inputs x and y, respectively.

```
module my_nor(input x, y, output f);
// internal declaration
supply1  vdd;
supply0  gnd;
wire a;    // terminal between two pMOS
// NOR gate body
   pmos p1 (a, vdd, x);  // source connected to vdd
   pmos p2 (f, a, y);    // serial connection
   nmos n1 (f, gnd, x);  // parallel connection
   nmos n2 (f, gnd, y);  // source connected to ground
endmodule
```

2.4.1.2 `pullup` **and** `pulldown` ***Sources*** The pullup and pulldown sources separately place logic values 1 and 0 on the nets connected in its terminal list. No delay specification can be applied to these sources. The pullup and pulldown sources have the following form:

   pullup|pulldown [(strength)] [instance_name](net_name);

For example, the following statement declares two pullup instances:

   pullup (strong1) p1 (neta), p2 (netb);

In this example, the p1 instance drives neta and the p2 instance drives netb.

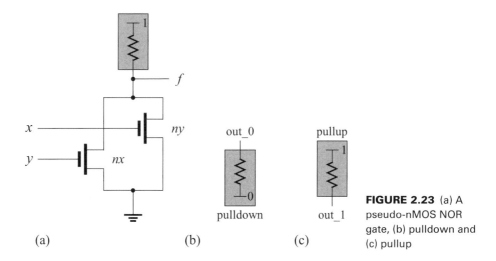

(a)   (b)   (c)

FIGURE 2.23 (a) A pseudo-nMOS NOR gate, (b) pulldown and (c) pullup

Figure 2.23 shows a pseudo-nMOS NOR gate consisting of two nmos switches and one pullup source. A simple application of a pullup net for constructing a two-input pseudo-nMOS nor gate is shown in the following example.

EXAMPLE 2.16 *An Example of a Two-input Pseudo-nMOS NOR Gate*

To model a pseudo-nMOS NOR gate, we need two nets: pullup and supply0, to provide the required pull-up and gnd signals. The two nmos switches are connected in parallel. One end of the parallel-connected nmos switches is connected to the pullup signal, which is also the output end, and the other end connected to gnd signal. The controls of the nmos switches are inputs x and y, respectively.

```
module my_pseudo_nor(input x, y, output f);
supply0   gnd;
// pseudo nMOS nor gate body
    nmos    nx (f, gnd, x);  // parallel connection
    nmos    ny (f, gnd, y);  // source connected to ground
    pullup  a (f);           // pull up output f
endmodule
```
∎

2.4.2 CMOS Switch

The cmos and rcmos switches have a data input, a data output and two control inputs. Figure 2.24 shows the cmos switch and its truth table. Like the nmos or pmos switch, some combinations of input values and control values may cause this switch to output either of two values, without a preference for either value. Hence, both the L and H symbols are also present in the truth table.

The cmos gate passes signals without reduction whereas the rcmos gate reduces the strength of signals passing through it. The cmos switch is virtually a combination of a pmos switch and an nmos switch. The rcmos switch is a combination of an

control		data			
n	p	0	1	x	z
0	0	0	1	x	z
0	1	z	z	z	z
0	x	L	H	x	z
0	z	L	H	x	z
1	0	0	1	x	z
1	1	0	1	x	z
1	x	0	1	x	z
1	z	0	1	x	z
x	0	0	1	x	z
x	1	L	H	x	z
x	x	L	H	x	z
x	z	L	H	x	z
z	0	0	1	x	z
z	1	L	H	x	z
z	x	L	H	x	z
z	z	L	H	x	z

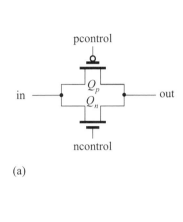

(a)

(b)

FIGURE 2.24 The CMOS switch (a) and its truth table (b)

rpmos switch and an rnmos switch. Hence:

```
cmos(out, in, ncontrol, pcontrol);
```

is equivalent to:

```
nmos(out, in, ncontrol);
pmos(out, in, pcontrol);
```

To instantiate a CMOS switch element, the following syntax can be used:

```
cmos|rcmos [instance_name](output, input, ncontrol,
    pcontrol);
```

where instance_name is optional. The first port is the output, the second is the input and the other two are ncontrol and pcontrol, which are connected to the *n*-channel control input and *p*-channel control input, respectively.

A simple application of cmos switches for constructing the 2-to-1 multiplexer depicted in Figure 2.25 is given in the following example.

EXAMPLE 2.17 *A 2-to-1 Multiplexer Constructed by* cmos *Switches*

In this example, we need a not gate and two cmos switches. These three elements are then conected in the way illustrated in the following module. The switch cmos_a turns on when the selection input s is 0; the switch cmos_b turns on when the selection input s is 1. Therefore, the circuit is a 2-to-1 multiplexer.

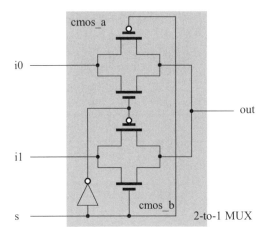

FIGURE 2.25 A 2-to-1 multiplexer constructed by CMOS switches

```
module my_mux(out, s, i0, i1);
output out;
input   s, i0, i1;
//internal wire
wire sbar; //complement of s
   not (sbar, s);

//instantiate cmos switches
   cmos (out, i0, sbar, s);
   cmos (out, i1, s, sbar);
endmodule
```

2.4.3 Bidirectional Switches

There are six bidirectional switches: tran, tranif0, tranif1, rtran, rtranif0 and rtranif1, as shown in Figure 2.26. These switches are usually used to model bidirectional switches through which data can be allowed to flow both ways by appropriately setting the control input. As in the cases of tristate buffers, for

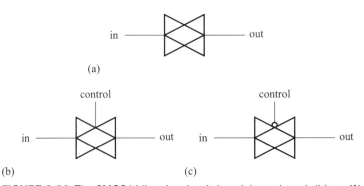

FIGURE 2.26 The CMOS bidirectional switches: (a) tran (rtran); (b) tranif1 (rtranif1); (c) tranif0 (rtranif0)

active-low bidirectional switches, the keyword is ended with "if0", such as `tranif0` and `rtranif0`; for active-high bidirectional switches, the keyword is ended with "if1", such as `tranif1` and `rtranif1`. The `tran`, `tranif0` and `tranif1` switches pass signals without reduction whereas the `rtran` `rtranif0` and `rtranif1` switches reduce the signals passing through them.

To instantiate bidirectional switches, the following syntax can be used:

```
tran/rtran         [instance_name](in, out);
tranif0/rtranif0 [instance_name](in, out, control);
tranif1/rtranif1 [instance_name](in, out, control);
```

where `instance_name` is optional. For `tran` and `rtran` switches, there are only two bidirectional data ports. For `tranif1`, `tranif0`, `rtranif1` and `rtranif0` switches, the first two are bidirectional ports that propagate signals to and from the switches, and the third port is a control input. As a consequence, `tran` and `rtran` switches cannot be turned off but the other four switches can be turned off through appropriately setting the values of control inputs.

2.4.4 Delay Specifications

Like gate primitives, all switch primitives, except `tran` and `rtran`, may also specify delays.

2.4.4.1 *MOS and CMOS Switches* The delay of ideal and resistive MOS and CMOS switches can be specified in one of the following ways:

1. Specify no delay:

```
mos_sw [instance_name](output, input, control);
cmos    [instance_name](output, input, ncontrol,
    pcontrol);
```

2. Specify propagation delay only:

```
mos_sw #(prop_delay)[instance_name](output, input,
    control);
cmos #(prop_delay)[instance_name]
                  (output, input, ncontrol, pcontrol);
```

3. Specify both rise and fall times:

```
mos_sw #(t_rise, t_fall)[instance_name]
                        (output, input, control);
cmos    #(t_rise, t_fall)[instance_name]
                        (output, input, ncontrol, pcontrol);
```

4. Specify rise, fall and turn-off times:

```
mos_sw #(t_rise, t_fall, t_off)[instance_name]
                          (output, input, control);
cmos    #(t_rise, t_fall, t_off)[instance_name]
                          (output, input, ncontrol, pcontrol);
```

where t_rise refers to the transition to the 1 value, t_fall refers to the transition to the 0 value and t_off refers to the transition to a high-impedance value.

2.4.4.2 Bidirectional Switches The bidirectional switches tran and rtran cannot specify delay since they are always on. The delay of ideal and resistive bidirectional switches, tranif1, tranif0, rtranif1 and rtranif0, can be specified in one of the following ways:

1. Specify no delay:
   bdsw_name [instance_name](in, out, control);
2. Specify turn-on and turn-off delay:
   bdsw_name #(t_on_off)[instance_name](in, out, control);
3. Specify separately turn-on and turn-off delays:
   bdsw_name #(t_on, t_off)[instance_name](in, out, control);

When two delays are specified, the first one is the turn-on delay and the second is the turn-off delay. If only one delay is specified, it specifies both the turn-on and the turn-off delays. If no delay is specified, then both turn-on and turn-off delays are zero.

2.4.5 Signal Strength

In Verilog HDL, in addition to four basic values, 0, 1, x and z, there are two kinds of signal strengths that can be specified to a scalar net: *driving strengths* and *charge storage strengths*. Signals with driving strengths propagate from gate outputs and continuous assignment outputs. There are four driving strengths, supply, strong, pull and weak. Signals with the charge storage strengths originate in the trireg net type. There are three charge storage strengths: large, medium and small.

Signal strength represents the ability of the source device to supply energy to drive the signal. The signal strengths defined in Verilog HDL are shown in Table 2.4.

A strength specification has two components, (strength1, strength0) or (strength0, strength1), where strength0 can be one of the following, supply0, strong0, pull0, weak0 and highz0, and strength1 can be one of the following: supply1, strong1, pull1, weak1 and highz1. The combinations (highz0, highz1) and (highz1, highz0) are not allowed. The default strength specification is (strong0, strong1).

TABLE 2.4 Strength levels for scalar net signal values

Strength	Strength0	Strength1	Type	Degree
supply	supply0	supply1	driving	strongest
strong	strong0	strong1	driving	↑
pull	pull0	pull1	driving	
large	large0	large1	storage	
weak	weak0	weak1	driving	
medium	medium0	medium1	storage	
small	small0	small1	storage	
highz	highz0	highz1	high Z	weakest

A drive strength can be specified for any of the following: a net in a net declaration assignment, the output port of a primitive gate instance and in a continuous assignment. A scalar net with drive strength has the following form:

```
net_name (strength1, strength0) [delay] net_id =
    expression;
```

where `net_name` can be any of the following: `wire`, `wand`, `wor`, `tri`, `triand`, `trior`, `trireg`, `tri0` and `tri1`.

An output port of a gate primitive with drive strength has the following form:

```
gate_name (strength1, strength0) [delay] [instancs_name]
    (port_list);
```

where `gate_name` can be any gate primitive or `pullup` or `pulldown`. A continuous assignment can also have drive strength associated with it. It has the following form:

```
assign (strength1, strength0) [delay] net_id =
    expression;
```

The signal strength can be printed using the `%v` format specification in the `$display`, `$strobe` or `$monitor` system tasks.

The signal strength can be weakened or attenuated by the resistance of the wires giving rise to signals of different strengths. The signal strength reduction rules when passing through resistive switches are shown in Figure 2.27.

2.4.5.1 Signal Contention

When multiple drivers drive a net at the same time, a contention occurs on the net. There are many rules applicable to resolve the contention. In what follows, we only consider the two most widely used cases:

1. *Combined signals with the same value and unequal strength.* If two signals with the same known values but different strengths drive the same net, the stronger signal dominates.

2. *Combined signals with an opposite value and equal strength.* If two signals with the opposite known values but equal strengths drive the same net, the result is an unknown value, `x`.

The details of the rest rules are referred to as LRM [5].

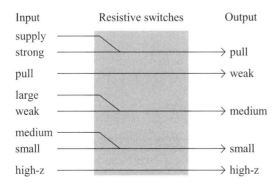

FIGURE 2.27 The signal strength reduction rules when passing through resistive switches

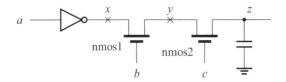

FIGURE 2.28 A circuit illustrating the relationship between a `trireg` net and its driver

2.4.6 `trireg` Net

The `trireg` net stores a value and is used to model charge storage nodes. It can be in one of two states:

1. *Driven state*. At least one driver drives a value of 1, 0 or x on the net. The value retains on the net. It takes the strength of the driver. The strength can be `supply`, `strong`, `pull` or `weak`.

2. *Capacitive state*. When all drivers to a `trireg` net are at the high-impedance (z), the net retains its last driven value. The strength can be `small`, `medium` (default) or `large`.

The following example employs the circuit depicted in Figure 2.28 to demonstrate the use of a `trireg` net in modeling a pass-element logic.

EXAMPLE 2.18 *An Example of a `trireg` Net*

In simulation time 0, control signals a, b and c are set to 1 so that the output x of the inverter is 0, which causes the net y to change its value to 0. The `trireg` net z then enters the driven state and discharges to `strong0`. At simulation time 10, the control signal b is cleared to 0, which causes net y to change to a high-impedance. The net z enters the capacitive state and stores its last driven value 0 with `medium` strength.

```
module triregexample;
reg   a, b, c;
wire x, y;
trireg (medium) z; // trireg declaration with medium
strength

not  not1  (x, a);
nmos nmos1 (y, x, b);
nmos nmos2 (z, y, c); // nmos that drives the trireg
initial begin
    $monitor("%0d a=%v b=%v c=%v x=%v y=%v z=%v ",
             $time, a , b, c, x, y, z);
    a = 1;
    b = 1;
    c = 1;
    // Toggle the control input b
    #10 b = 0;
```

```
    #30 b = 1;
    #10 b = 0;
    #100 $finish;
end
endmodule
```

The simulation result is as follows:

```
# 0   a=St1 b=St1 c=St1 x=St0 y=St0 z=St0
# 10  a=St1 b=St0 c=St1 x=St0 y=HiZ z=Me0
# 40  a=St1 b=St1 c=St1 x=St0 y=St0 z=St0
# 50  a=St1 b=St0 c=St1 x=St0 y=HiZ z=Me0
```

The problem of charge sharing is often encountered in practical circuits. Hence, we use a simple circuit, as shown in Figure 2.29, to explore this problem and give an example to illustrate how to model it.

EXAMPLE 2.19   *An Example Illustrating the Charge Sharing Problems*

At simulation time 0, the control signal a is 0 and b and c are 1. Hence, nodes x, y and z are driven to `strong1`. At simulation time 10, the control signal b changes to 0. Net y enters into the capacitive state and stores its last driven values 1 with `large` strength. The net z is still in the driven state and is driven to value 1 with `large` strength. At simulation time 20, the control signal c changes to 0, causing net z to enter into the capacitive state and store a value 1 of `small` strength. At simulation time 30, the control signal c changes to 1 again, connecting the two `trireg` nets y and z. These two nets now share the same charge. At simulation time 40, the control signal c changes to 0 one more time, causing net z to enter into the capacitive state and store a value 1 of `small` strength.

```
module triregChargeSharing;
reg   a, b, c;
wire x;
trireg (large) y; // declaration with large strength
trireg (small) z; // declaration with small strength
not   not1  (x, a);
nmos nmos1 (y, x, b);
nmos nmos2 (z, y, c); // nmos that drives the trireg
```

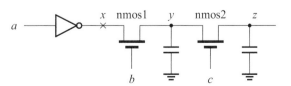

FIGURE 2.29 A circuit illustrating the charge sharing problem

```
initial begin
   $monitor("%0d a=%v b=%v c=%v x=%v y=%v z=%v ",
            $time, a , b, c, x, y, z);
   a = 0; b = 1; c = 1;
 // Toggle the control input c
   #10 b = 0;
   #10 c = 0;
   #10 c = 1;
   #10 c = 0;
   #100 $finish;
end
endmodule
```

The simulation results are listed as follows:

```
# 0   a=St0 b=St1 c=St1 x=St1 y=St1 z=St1
# 10  a=St0 b=St0 c=St1 x=St1 y=La1 z=La1
# 20  a=St0 b=St0 c=St0 x=St1 y=La1 z=Sm1
# 30  a=St0 b=St0 c=St1 x=St1 y=La1 z=La1
# 40  a=St0 b=St0 c=St0 x=St1 y=La1 z=Sm1
```

2.4.6.1 `trireg` *Net Charge Decay* Like all nets, a `trireg` net declaration can have up to three delays and has the following form:

```
#(t_rise, t_fall, t_decay)
```

The `t_decay` specifies the charge-decay time of a `trireg` net, which is the time duration between when its drivers turn off and the point that its stored charge can no longer be determined.

The charge decay process begins when the drivers turn off and the `trireg` net starts to hold charge. It ends whenever either one of the following two conditions is satisfied: the charge-decay time elapses and the net makes a transition to x or the drivers turn on and propagate a 1, 0 or x into the net.

The following is an example illustrating the effect of charge decay.

EXAMPLE 2.20 *An Example Illustrating the Effect of Charge Decay*

This example declares a `trireg` net, `cap1`, with `medium` storage strength. The delay specifications for both rise and fall delays are 0 and the charge-decay time is 10 time units. After simulation time 25, the `trireg` net, `cap1` has decayed to an unknown value x.

```
module capacitor_decay;
reg x, a;
// trireg declaration with a charge decay time of 10 time
// units
trireg (medium) #(0, 0, 10) cap1;
```

```
nmos nmos1 (cap1, x, a); // nmos that drives the trireg
initial begin
    $monitor("%0d x=%v a=%v cap1=%v", $time, x, a, cap1);
    x = 1;
    // Toggle the driver of the control input to the nmos
    // switch
    a = 1;
    #05 a = 0;
    #05 a = 1;
    #05 a = 0;
    #25 a = 1;
    #25 $finish;
end
endmodule
```

■

The simulation results are as follows:

```
# 0   x=St1 a=St1 cap1=St1
# 5   x=St1 a=St0 cap1=Me1
# 10  x=St1 a=St1 cap1=St1
# 15  x=St1 a=St0 cap1=Me1
```

Review Questions

Q2.27   What are the two MOS switches?

Q2.28   Describe the meanings of `pullup` and `pulldown` nets.

Q2.29   Describe the meanings of `supply0` and `supply1` nets.

Q2.30   Explain why the delay specification cannot be applied to the bidirectional switches, `tran/rtran`.

Q2.31   Describe the function of the `trireg` net.

SUMMARY

In structural style, a module is described as a set of interconnected components. The components can be modules, UDPs, gate primitives and switch primitives.

Gate-level modeling describes a design which only uses gate primitives. In Verilog HDL, there are 12 predefined gate primitives. These primitives can be cast into two groups: and/or gates and buf/not gates. The group of and/or gates includes `and`, `nand`, `or`, `nor`, `xor` and `xnor`. The group of buf/not gates includes `buf`, `not`, `bufif0`, `bufif1`, `notif0` and `notif1`.

Due to the existence of resistance (R) and capacitance (C) in logic circuits, all of the actual logic gates have finite propagation delays. In addition, in every interconnection between any two circuits there always exists an RC delay. This RC delay is called a *transport delay*. To model these two types of delays, inertial and transport delay models are usually adopted by HDLs to model the propagation delay of actual logic circuits and the interconnection wire delay, respectively. With the inertial delay model, any signal event does not persist long enough (longer than the gate propagation delay), will be filtered out and not propagated to the output of gates. With the transport delay model, any signal event appearing at the input will be propagated to the output.

A hazard is an unwanted short-width output pulse, sometimes called a *glitch*, when inputs to a combinational circuit change. These unwanted signals are produced when different paths from an input to the output have different propagation delays. Hazards can be divided into *static hazard* and *dynamic hazard*. A static hazard represents the situation where a circuit output may momentarily go to 0 (or 1) when it should remain at a constant 1 (or 0) during the transient period. Static hazards can be further divided into static-0 hazard and static-1 hazard. A dynamic hazard is a situation where the output of a combinational circuit changes from 0 to 1 and then to 0 (or 1 to 0 and then to 1); namely, the output changes three or more times. Because three or more signal changes are required to have a dynamic hazard, there must exist at least three paths from a signal input to the output of the underlying combinational circuit.

When a logic circuit is modeled at the switch level, all MOS transistors of it are regarded as *ideal* or *nonideal* switches. For ideal switches, there are no signal degradations when they are turned on, whereas for nonideal (resistive) switches, they have a certain finite amount of signal degradation. In Verilog HDL, there are 16 built-in switch primitives. These switch primitives can be grouped into MOS switches, CMOS switches, bidirectional switches, `pulllup` and `pulldown`, and `suppluO` and `supply1`.

The signal propagation delay from any gate input to the gate output can be specified by gate delays. Propagation delays of gate primitives can be specified in no-value delay, one-value delay, two-value delay and three-value delay specifications. The one-value delay specification specifies propagation delay only, the two-value delay specification specifies both rise and fall times and the three-value delay specification specifies rise, fall and turn-off times.

Like gate primitives, all switch primitives, excepting `tran` and `rtran`, may specify delays.

REFERENCES

1. J. Bhasker, *A Verilog HDL Primer,* 3rd Edn, Star Galaxy Publishing, 2005.
2. P. P. Chu, *RTL Hardware Design Using VHDL: Coding for Efficiency, Portability and Scalability,* John Wiley & Sons, Inc., Hoboken, NJ, USA, 2006.
3. M.-B. Lin, *Digital System Design: Principles, Practices and ASIC Realization,* 3rd Edn, Chuan Hwa Book Company, Taipei, Taiwan, 2002.
4. S. Edn Palnitkar, *Verilog HDL: A Guide to Digital Design and Synthesis,* 2nd Edn, SunSoft Press, 2003.
5. IEEE 1364-2001 Standard, *IEEE Standard Verilog Hardware Description Language,* 2001.

PROBLEMS

2.1 Model the following switching expression at gate level in structural style:

$$f(x, y, z) = [(x + y)' + x'z]'$$

2.2 Model the following switching expression at gate level in structural style:

$$f(w, x, y, z) = [(wx + y'z)'xy]'$$

2.3 Model the logic circuit shown in Figure 2.30 at gate level in structural style.

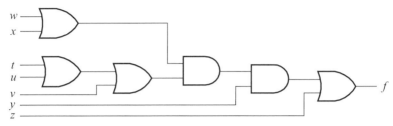

FIGURE 2.30 A logic diagram for problem 2.3

2.4 Model the logic circuit shown in Figure 2.31 at gate level in structural style.

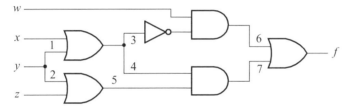

FIGURE 2.31 A logic diagram for problem 2.4

2.5 An excess-3 code to a BCD (binary-coded decimal) code converter is a device which converts an excess-3 code input, represented by w, x, y and z, into its equivalent BCD code, represented by a, b, c and d. Derive each output, a, b, c and d, as a function of the inputs, w, x, y and z. Model the converter circuit at gate level in structural style.

2.6 A complex CMOS logic gate implements the following switching expression:

$$f(w, x, y, z) = [(x + y)z + w]'$$

  (a) Draw the logic circuit.

  (b) Model the logic circuit at switch level.

2.7 A complex CMOS logic gate implements the following switching expression:

$$f(w, x, y, z) = [xy + y(z + w)]'$$

  (a) Draw the logic circuit.

  (b) Model the logic circuit at switch level.

2.8 A complex CMOS logic gate implements the following switching expression:

$$f(w, x, y, z) = (wx + yz)'$$

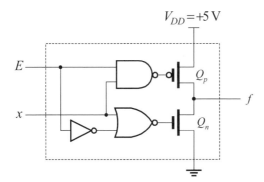

FIGURE 2.32 A logic diagram for problem 2.9

    (a) Draw the logic circuit.

    (b) Model the logic circuit at switch level.

  2.9 Model the logic circuit shown in Figure 2.32 at gate level in structural style.

2.10 A pseudo-nMOS logic circuit implements the following switching expression:

$$f(x, y, z) = (x + yz)'$$

    (a) Draw the logic circuit.

    (b) Model the logic circuit at switch level.

2.11 A pseudo-nMOS logic circuit implements the following switching expression:

$$f(w, x, y, z) = [w(x + y) + z]'$$

    (a) Draw the logic circuit.

    (b) Model the logic circuit at switch level.

2.12 Use ideal MOS switches to model a switch with both rise and fall times of 0 and a turn-off time of 2 time units.

## CHAPTER 3

DATAFLOW MODELING

THE RATIONALE behind dataflow modeling is on the observation that any digital system can be constructed by interconnecting registers and a combinational logic put between them for performing the required functions. A design modeled on the dataflow style has the following features:

1. Dataflow provides a powerful way to implement a design.

2. Automated tools are used to create a gate-level circuit from the description of a dataflow design. This process is often called *logic synthesis*.

3. RTL (register-transfer level) is a combination of dataflow and behavioral modeling.

Continuous assignment is the most basic statement of dataflow modeling. It is used to continuously drive a value onto a net. Continuous assignment begins with the keyword `assign` and is always active. It is usually used to model a combinational logic.

3.1 DATAFLOW MODELING

The essential element of dataflow modeling is the continuous assignment. Hence, in this section, we introduce the basic syntax of continuous assignment and give some insight into it.

3.1.1 Continuous Assignment

A continuous assignment is the most basic statement of dataflow modeling. It continuously drives a value onto a net. The assignment begins with the keyword `assign` and has the following syntax:

```
assign net_lvalue = expression;
assign net1 = expr1, net2 = expr2, ..., netn = exprn;
```

where `net_lvalue` is a scalar or vector net, or their concatenation. The second statement is a compact form of *n* continuous assignments. The operands used in the

Digital System Designs and Practices   Ming-Bo Lin
© 2008 John Wiley & Sons (Asia) Pte Ltd

expression can be variables or nets or function calls. Variables or nets can be scalar or vectors. Any logic function can be realized with continuous assignments. Continuous assignments can only update values of net data types such as `wire`, `triand` and so on.

Continuous assignments are always active. That is, whenever an operand in the right-hand-side expression changes value, the expression is evaluated and the new value is assigned to `net_lvalue`. Thus, it is quite appropriate to model the behavior of combinational logics. For example:

```
assign {c_out, sum[3:0]} = a[3:0] + b[3:0] + c_in;
```

The above continuous assignment carries out a 4-bit addition with carry input. The braces ({ }) in the left-hand side is a concatenation operator, which will be described in detail later.

3.1.1.1 Net Declaration Assignment

Usually, we declare a net and then use a continuous assignment to assign a value onto it. In Verilog HDL, there is a shortcut for this. A continuous assignment can be placed on a net when it is declared – we call it *net declaration assignment* for convenience, such as the following example:

```
wire out;                    // normal continuous assignment
assign out = in1 & in2;
```

which is equivalent to

```
wire out = in1 & in2;    // net declaration assignment
```

Like regular continuous assignment, net declaration assignment is always active. Moreover, there can only be one net declaration assignment per net since a net can only be declared once.

3.1.1.2 Implicit Net Declaration

Implicit net declaration is a feature of Verilog HDL. An implicit net declaration will be inferred for a signal name when it is used on the left-hand side of a continuous assignment. For example:

```
wire in1, in2;
assign out = in1 & in2;
```

Note that `out` is not declared as a `wire`, but an implicit `wire` declaration for `out` is done by the simulator. It is a good practice to declare the net type explicitly and use regular continuous assignment to avoid any unintentional errors. However, implicit net declarations are often used with input and output ports of modules (see Section 6.1.1).

3.1.2 Expressions

The essence of dataflow modeling is using expressions that consist of sub-expressions, operators and operands, instead of primitive gates, or instantiations of modules or UDPs used in structural style. Expressions are constructs that combine operators and operands to produce results. In general, an expression has the following format:

```
expression = operators + operands
```

where operands can be any of the allowed data types and operators act on the operands to product the desired results.

The allowed operands in an expression can be one of the following: constants, net, `reg`, `integer`, `time`, `real`, `realtime`, bit-select, part-select, array element and function calls. We will describe these operands in the next section in greater detail.

Like C programming language, Verilog HDL has a rich set of operators, which include arithmetic, shift, bitwise, case equality, reduction, logic, relational, equality and miscellaneous operators. We will discuss these operators in a dedicated section of this chapter. Table 3.1 summarizes the operators, which are grouped together as similar operators, in Verilog HDL.

The precedence of operators in Verilog HDL is listed in Table 3.2. Operators on the same row have the same precedence. Rows are arranged in order of decreasing precedence for the operators. For example, *, / and % operators have the same precedence and are higher than the binary + and − operators. The left-to-right associativity applies to all operators except the conditional operator, which associates from right to left. Associativity refers to the order in which the operators having the same precedence are evaluated. Thus, in the following example y is added to x and then z is subtracted from the result of $x + y$:

$$x + y - z$$

When operators differ in precedence, the operators with the higher precedence associate first. In the following example, y is divided by z (division has higher precedence than

TABLE 3.1 The summary of operators in Verilog HDL

Arithmetic	Bitwise	Reduction	Relational
+: add	~: NOT	&: AND	>: greater than
−: subtract	&: AND	|: OR	<: less than
*: multiply	|: OR	~ &: NAND	>=: greater than or equal
/: divide	^: XOR	~ |: NOR	<=: less than or equal
%: modulus	~^, ~: XNOR	^: XOR	
**: exponent		~^, ^ ~ : XNOR	Miscellaneous
	Case equality		{ , }: concatenation
Shift		Logical	{c{ }}: replication
	===: equality		? : conditional
<<: left shift	! ==: inequality	&&: AND	
>>: right shift		||: OR	
<<<: arithmetic left shift	Equality	!: NOT	
>>>: arithmetic right shift			
	==: equality		
	! =: inequality		

TABLE 3.2 The precedence of operators in Verilog HDL

Operators	Symbols	Precedence
Unary Exponent Multiply, divide, modulus	$+$! $\sim$ $**$ $* / \%$	Highest
Add, subtract Shift	$+-$ $<<\ >>\ <<<\ >>>$	
Relational Equality	$<\ <=\ >\ >=$ $==\ !=\ ===\ !==$	
Reduction	$\&\ \sim\&$ $\wedge\ \wedge\sim$ $\|\ \sim\|$	
Logical	$\&\&$ $\|\|$	
Conditional	?:	Lowest

addition) and then the result is added to *x*:

$$x + y/z$$

Parentheses can be used to change the operator precedence. For instance, $(x + y)/z$ is not the same as $x + y/z$. It is good practice to use parentheses whenever there may be ambiguities.

3.1.3 Delays

The time between when any operand in the right-hand side changes and when the new value of the right-hand side is assigned to the left-hand side is controlled by the delay value. The delay value is specified after the keyword `assign`. The inertial delay model is used as the default model. The delay value of a continuous assignment can be specified by using the following syntax:

```
assign #delay net_lvalue = expression;
assign #delay1 net1 = expr1,
       #delay2 net2 = expr2,
       ...,
       #delayn netn = exprn;
```

where `#delay`, `#delay1`, ... `#delayn` are specified exactly in the same way as that of gate primitives. That is, there may be zero-value, one-value, two-value or three-value delay specifications using a delay specifier in the form of `min:typ:max`.

For example:

```
wire in1, in2, out;
assign #10 out = in1 & in2; // delay is 10 time units
```

A net declaration assignment is used to specify both the delay and assignment on the net. In this case, the delay is inertial delay rather than transport delay because it is equivalent to first declaring a net and then doing a continuous assignment with regular delay.

```
// net declaration continuous assignment with delay
wire #10 out = in1 & in2;

// regular continuous assignment with delay
wire out;
assign #10 out = in1 & in2;
```

A net can be declared associated with a delay value. The transport delay model is used by default in this case. Net declaration delays can also be used in gate-level modeling.

```
// net delays
wire #10 out;
assign out = in1 & in2;

// regular assignment delay
wire out;
assign #10 out = in1 & in2;

// regular assignment delay
wire #5 out;
assign #10 out = in1 & in2;
```

Review Questions

Q3.1 Explain the meaning of dataflow.

Q3.2 What is net declaration assignment?

Q3.3 What is the basic statement in dataflow style?

Q3.4 Describe how to assign delays to continuous assignments.

Q3.5 Explain the meaning of implicit net declaration.

3.2 OPERANDS

The operands in an expression can be any of constants, parameters, nets, variables (`reg`, `integer`, `time`, `real`, `realtime`), bit-select, part-select, array element and function calls. The various net data types have been discussed in the previous chapters and so we do not repeat them here again. A parameter is like a constant and is declared using a `parameter` or `localparam` declaration (referring to Section 6.1.2 for details). A function call may also be used as an operand in an expression. It can be either a user-defined function or a system function (see Sections 5.2 and 5.3 for details). In this section, we address constants, variable data types, bit-selects and part-selects, array and memory elements.

3.2.1 Constants

There are three types of constant in Verilog HDL. These are *integer*, *real* and *string*.

3.2.1.1 Integer Constants Integer constants can be specified in decimal, hexadecimal, octal or binary format. There are two forms to express integer constants: *simple decimal form* and *base format notation*.

Simple Decimal Form In the simple decimal form, a number is specified as a sequence of digits 0 through 9, optionally beginning with a plus (+) or minus (−) unary operator. For example:

```
-123   // is decimal -123
12345  // is decimal 12345
```

An integer value in this form represents a signed number. A negative number is represented in two's complement form.

Base Format Notation When the base format notation is used, a number is composed of up to three parts: an optional size constant, a single quote followed by a base format character and the digits representing the value of the number.

```
[size]'[s/S][base_format]base_value
```

where " ' " is a single quote. The `size` specifies the size of the constant in number of bits. It is specified as a non-zero unsigned decimal number. For example, the size specification for two hexadecimal digits is 8, because one hexadecimal digit requires 4 bits.

The `base_format` consists of a case-insensitive character specifying the base for the number, optionally preceded by the single character s (or S) to indicate a signed (in two's complement form) quantity, preceded by the single quote character ('). The allowed bases are decimal (d or D), hexadecimal (h or H), octal (o or O) and binary (b or B).

The base_value consists of digits (0 to 9 and a to f) that are legal for the specified base format. The base_value should immediately follow the base format, optionally preceded by white space. The hexadecimal digits a(A) to f(F) are case-insensitive. To represent a signed (two's complement) number, a single character "s or S" must be preceded with the base_format.

```
    4'b1001     // a 4-bit binary number
  16'habcd      // a 16-bit hexadecimal number
      2006      // unsized number -- a 32-bit decimal
                // number by default
     'habc      // unsized number -- a 32-bit hexadecimal
                // number
  4'sb1001      // a 4-bit signed number, it represents
                // -7.
 -4'sb1001      // a 4-bit signed number, it represents
                // -(-7) = 7.
```

In order to improve readability, Verilog HDL allows us to use the underscore character (_) anywhere in a number except as the first character. The underscore character is ignored.

```
16'b1101_1001_1010_0000 // a 16-bit number in binary form
8'b1001_0001            // an 8-bit number in binary form
```

An x represents the unknown value and a z represents the high-impedance value. An x (z) represents 4 x (z) bits in the hexadecimal base, 3 x (z) bits in the octal base, and a 1 x (z) bit in the binary base. In Verilog HDL, the question mark (?) can be used to improve readability in the case where the high-impedance value is a 'don't-care' condition. Hence, when used in a number, the question-mark character is an alternative for the z character.

```
16'hxxbc        // the number is 16'bxxxx_xxxx_1011_1100
16'hzzbc        // the number is 16'bzzzz_zzzz_1011_1100
16'b01??_1001_11?0_??00 //a 16-bit number in binary form
8'b01??_11??            //equivalent to an 8'b01zz_11zz
```

If the size specified for the constant is larger than the size of the base_value, the base_value will be padded to the left with zeros or a sign bit, depending on whether the base_value is an unsigned or a signed quantity. If the left-most bit in the base_value is an x or a z, then the left bits of the base_value will be padded with xs or zs.

```
  8'bx001       // the number is 8'bxxxx_x001
  8'bzz011      // the number is 8'bzzzz_z011
 16'hx8         // the number is 16'bxxxx_xxxx_xxxx_1000
```

```
16'b1101_1001    // the number is 16'b0000_0000_1101_1001
16'sb1001_0001   // the number is 16'b1111_1111_1101_1001
```

In the last one, the left bits are padded with sign bit ("1") since the base_value is a two's complement number, declared by s.

If the size specified for the constant is smaller than the size of the base_value, the left-most bits of base_value are truncated in order to fit the specified size.

```
8'b1110_1101_1001    // the number is 8'b1101_1001
10'sb1001_1001_0001  // the number is 10'b01_1001_0001
```

For the unsized constant, the size of the base_value is by default at least 32 bits. Hence, the base_value will be padded to the left with something like that for when the size is specified.

```
'b1110_1101_1001    // 32'b0000_0000_0000_0000_0000_
                    // 1110_1101_1001
'sb1001_1001_0001   // 32'b1111_1111_1111_1111_1111_
                    // 1001_1001_0001
```

3.2.1.2 Real Constants
Real numbers can be specified in either decimal notation or in scientific notation. Real numbers expressed with a decimal point must have at least one digit on each side of the decimal point.

```
1.5            //
 .3            // illegal ---
1294.872       //
1.44E9         // the exponent symbol can be e or E
1.50e-7
0.1e-0
15E12
32E-6
26.176_45_e-12  // underscores are ignored)
```

The conversion of real numbers to integers is implicitly defined by the language. Real numbers are converted to integers by rounding the real number to the nearest integer. Implicit conversion takes place when a real number is assigned to an integer. For examples:

```
    1.5     // yields 2 when converted into an integer
    0.3     // yields 0 when converted into an integer
  23.445    // yields 23 when converted into an integer
 -245.56    // yields -246 when converted into an
            // integer
```

3.2.1.3 *String Constants*
A string is a sequence of characters enclosed by double quotes (" ") and may not be split into multiple lines. One character is represented as an 8-bit ASCII code. Hence, a string is considered as an unsigned integer constant represented by a sequence of 8-bit ASCII codes.

String variables are variables of `reg` type with widths equal to eight times the number of characters in the string, such as in the following example.

EXAMPLE 3.1 *An Example of String Manipulation*

In this example, assume that a string `str1` is employed to store the string "`Welcome to Digital World!`". Hence, it requires a `reg` 25*8 or a width of 200 bits. Another string `str2` is used to store the string "`Hello!`". It needs a `reg` 6*8 or a width of 48 bits. After the execution of the program, the output is:

```
Welcome to Digital World!
Hello! is stored as: 48656c6c6f21

module string_test;
// internal signal declarations:
reg [25*8:1] str1;
reg [6*8:1] str2;
initial begin
    str1 = ''Welcome to Digital World!'';
    $display(''%s\n'', str1);
    str2 = ''Hello!'';
    $display(''%s is stored as: %h\n'', str2, str2);
end
endmodule
```

The backslash (\) character can be used to escape certain special characters.

```
\n      \\ new line character
\t      \\ tab character
\\      \\ \ character
\''     \\ '' character
\ddd    \\ a character specified in 3 octal digits.
```

Review Questions

Q3.6 What is the meaning of a parameter?

Q3.7 Can a function call be used as an operand in the expression of a continuous assignment?

Q3.8 What are the three constants provided by Verilog HDL?

Q3.9 Explain the meaning of an `s` qualifier in base format notation of integer constants.

Q3.10 At least how many bits are defaulted when a constant is unsized?

3.2.2 Data Types

Verilog HDL has two classes of data types: *nets* and *variables*. Nets mean any hardware connection points and variables represent any data storage elements. Note that net is not a reserved word but represents a class of data types, as we have described in the previous chapter. A net variable can be referenced anywhere in a module and must be driven by a primitive, continuous assignment `assign`, `force/release` statement or module ports. The details of net data types have been described in Chapter 2; hence, we omit them here.

3.2.2.1 *Variable Data Types*   A variable represents a data storage element. It stores a value from one assignment to the next. The variable data types include five different kinds: `reg`, `integer`, `time`, `real` and `realtime`. All `reg`, `time` and `integer` variable data types are initialized with an unknown value, `x`, and `real` as well as `realtime` variables data types are initialized with 0.0.

The `reg` Variable   A `reg` variable holds a value between assignments. It may be used to model hardware registers such as edge-sensitive (i.e. flip-flops) and level-sensitive (i.e. latches) storage elements. However, a `reg` variable need not actually represent a hardware storage element because it can also be used to represent combinatorial logic. The declaration of a `reg` variable may use the following form:

```
reg [signed] [[msb:lsb]] id1[= const_expr1],...,
    idn[= const_exprn];
```

where the initialization part `const_expr` is optional. A `reg` variable may be declared to store a signed value by using the keyword `signed` following `reg`. The default value of a `reg` variable is unsigned; `[msb:lsb]` is an optional part and is used to specify the range of the `reg` variable. The use and meaning of `msb` and `lsb` are the same as in the case of net so we do not repeat here. If no range is specified, it defaults to a 1-bit reg.

```
reg a, b, c_in;    // reg a, b, and c_in are 1-bit reg
                   // variables
reg [7:0] data_a;  // data_a is an 8-bit reg, the msb
                   // is bit 7
reg [0:7] data_b;  // data_b is an 8-bit reg, the msb
                   // is bit 0
reg signed [7:0] d;// d is an 8-bit signed reg
```

Vector Versus Scalar   It is more convenient to describe a bundle of signals as a basic unit when describing a hardware module. This bundle of signals is usually called a vector (multiple bit width) in Verilog HDL. All net data types and `reg` variables can be declared as vectors. The default type of net and `reg` variable data types is a 1-bit vector or is called scalar.

The `integer` Variable   An `integer` variable contains integer values and can be used as a general-purpose variable used for modeling high-level behavior. The syntax for declaring `integer` is as follows:

```
integer id1[= const_expr1],..., idn[= const_exprn];
```

where the initialization part `const_expr` is optional. The `integer` variables use the same assignment rules as `reg` variables. An `integer` variable is treated as a signed `reg` variable with the lsb being bit 0 and has at least 32 bits, regardless of implementations. Arithmetic operations performed on integer variables produce two's complement results.

```
integer  i,j;        // declare two integer variables
integer  data[7:0]; // array of integer
```

The `time` Variable.   A `time` variable is used for storing and manipulating simulation time quantities. It is typically used in conjunction with the `$time` system task. The syntax for declaring `time` is as follows:

```
time id1[= const_expr1],..., idn[= const_exprn];
```

where the initialization part `const_expr` is optional. A `time` variable holds only unsigned value and is at least 64 bits, with the least significant bit being bit 0.

```
time events;        // hold one time value
time current_time;  // hold one time value
```

The `real` and `realtime` Variables   The Verilog HDL supports `real` and `realtime` variable data types in addition to `integer` and `time`. The `real` and `realtime` are identical and can be used interchangeably. The syntax for declaring `real` or `realtime` is as follows:

```
real|realtime identifier1, ... ,identifiern;
```

Both `real` and `realtime` variables cannot use range declaration and their initial values are defaulted to zero (0.0).

```
real     events;        // declare a real variable
realtime current_time;  // hold current time as real
```

Review Questions

Q3.11   What can drive a net?

Q3.12   What variable data type can be used to model a hardware storage element?

Q3.13   What are the differences between `reg` and `integer` variables?

Q3.14 What are the features of `time` variable?

Q3.15 Describe the features of `real` and `realtime` variables.

3.2.3 Bit-Select and Part-Select

A bit-select extracts a particular bit from a vector. As mentioned previously, all net data types and `reg` variable data type can be declared as a vector. Although `integer` and `time` are not allowed to be declared as a vector, they can also be accessed by bit-select or part-select. However, bit-select or part-select of `real` and `realtime` is not allowed. The bit-select has the form:

```
vector_name [bit_select_expr]
```

where `vector_name` can be any vector of net, `reg`, `integer` and `time` data types. The `bit_select_expr` can be an expression. If the `bit_select_expr` evaluates to an `x` or a `z`, or if it is out of bounds, then the bit-select value is an `x`.

In a part-select, a contiguous sequence of bits of a vector is selected. There are two forms of part-select: a *constant* part-select and an *indexed* part-select. A constant part-select has the following form:

```
vector_name [msb_const_expr:lsb_const_expr]
```

where `vector_name` can be any vector of net, `reg`, `integer` and `time` data types. The `msb_const_expr` and `lsb_const_expr` must be constant expressions. The following are some examples:

```
reg   [15:0] data_bus; // declarations
wire [7:0] a;
integer response_time;

data_bus[3:0]       // reg variable part-select
a [4:3]             // net part-select
response_time[5:0] // integer variable part-select
```

An indexed part-select has the following form:

```
vector_name [<starting_bit>+:const_width]
vector_name [<starting_bit>-:const_width]
```

where `vector_name` can be any vector of net, `reg`, `integer` and `time` data types. The `starting_bit` may be a variable but the `width` has to be constant. The range of bits selected is the index specified by `starting_bit` plus or minus the numbers specified by `const_width`; "+:" indicates that part-select increases from the `starting_bit`; "−:" indicates that part-select decreases from the `starting_bit`.

For example:

```
data_bus[8+:8]      // select data_bus[15:8]
data_bus[15-:4]     // select data_bus[15:12]
```

Like the case of bit-select, if either the range index is out of bounds or evaluates to an x or a z, the part-select value is an x.

The following example shows a simple application of the indexed part-select.

EXAMPLE 3.2   *Conversion of a Big Endian to a Little Endian*

By properly setting the starting bits and ranges, the following module converts a big endian into its corresponding little endian and vice versa.

```
module swap_bytes (in, out);
input  [31:0] in;
output [31:0] out;
// using indexed part-select
assign out [31 -: 8] = in [0  +: 8],
       out [23 -: 8] = in [8  +: 8],
       out [15 -: 8] = in [16 +: 8],
       out [7  -: 8] = in [24 +: 8];
endmodule
```

■

Note that a net or variable may be declared with the keyword signed as a signed data type. The default is unsigned. However, bit-select and part-select results are unsigned regardless of the operands.

Vectored and Scalared   Usually, a vector net can be accessed by bit-select and part-select. However, when a vector net does not allow it to be accessed by bit-select or part-select, the keyword vectored may be specified in the net declaration. The syntax is as follows:

```
net_name [vectored|scalared] [signed] [range] [delay]
    identifiers;
```

When no keyword vectored is specified, the default is scalered, which is the same as when the keyword scalared is specified, and bit-selects and part-selects are permitted to access the vector net. For example:

```
wire scalared [63:0] bus64; // a bus that will be
                            // expanded
tri  vectored [31:0] data;  // data is dealt with as
                            // a unit
```

Review Questions

Q3.16   Explain the meanings of bit-select and part-select.

Q3.17   Can we disable a vector to be accessed by bit-select or part-select?

Q3.18 What is a constant part-select?

Q3.19 What is an indexed part-select?

Q3.20 Can bit-select and part-select be applied to `integer` and `time` variables?

Q3.21 Can bit-select and part-select be applied to `real` and `realtime` variables?

3.2.4 Array and Memory Elements

Arrays can be used to group elements into multi-dimensional objects. Although only nets and the `reg` variable can be declared as vectors, all net and variable data types are allowed to be declared as multi-dimensional arrays. A multi-dimensional array is declared by specifying the address ranges after the declared identifier, called *dimension*, one for each. In addition, an array element can be a scalar or a vector if the element is a net or `reg` data type. The vector size (defined by `range`) specifies the number of bits in each element and the dimensions (i.e. address ranges) specify the number of elements in each dimension of the array. The syntax is as follows:

```
net_type [signed][range]id[msb:lsb]{[msb:lsb]};
reg       [signed][range]id[msb:lsb]{[msb:lsb]};
integer  id[msb:lsb]{[msb:lsb]};
time     id[msb:lsb]{[msb:lsb]};
```

where `msb` and `lsb` are constant-valued expressions that indicate the range of indices of a dimension. If no dimensions are specified, each net or variable only stores a value. For example:

```
wire    a[3:0]; // a scalar wire array of 4 elements
reg     d[7:0]; // a scalar reg array of 8 elements
wire    [7:0]  x[3:0];   // an 8-bit wire array of
                         // 4 elements
reg     [31:0] y[15:0];  // a 32-bit reg array of
                         // 16 elements
integer states [3:0];    // an integer array of
                         // 4 elements
time    current[5:0];    // a time array of 6 elements
```

To access an element from an array, we use the following format:

```
array_name[addr_expr]{[addr_expr]}        .
```

where `addr_expr` can be any expression. As with bit-select or part-select, if `addr_expr` is out of bounds, or if any bit in the `addr_expr` is an x or a z, then the reference value is an x. Only an element of an array can be assigned a value in a single assignment; an entire or partial array dimensions cannot be assigned to another using a single assignment. However, a bit-select or part-select of an element of an array may be accessed and assigned. To assign a value to an element of an array, it needs to specify an index for every dimension. The index can be an expression.

```
states[3] = 33559; // assign decimal number to integer
                   // in array
current[t_index] = $time; // assign current simulation
                          // time to element indexed
                          // by integer index
```

3.2.4.1 *Memory* Memory is a basic module in any digital system; in Verilog HDL, it is simply declared as a one-dimensional `reg` array. A memory can be used to model a read-only memory (ROM), a random access memory (RAM) and a register file. Reference to a memory may be made to a whole word or a portion of a word of memory.

To access a memory word, we use the following format:

```
mem_name[addr_expr]
```

where `addr_expr` can be any expression. For example:

```
reg  [7:0] mema [7:0];       // one-dimensional array
                             // of 8-bit vector
reg  [7:0] memb [3:0][3:0]; // two-dimensional array
                             // of 8-bit vector
wire sum [7:0][3:0];         // two-dimensional array
                             // of scalar wire

mema = 0;      // illegal -- attempt to write to entire
               // array
memb[1] = 0; // illegal -- it needs two indices
memb[1][3:1] = 0; // illegal -- attempt to write to
                  // partial array

mema[1] = 0;       // assigns 0 to the second element
                   // of mema
memb[1][0] = 3;    // assigns 3 to the element[1][0]
```

Bit-select and part-select of a memory element or an array element of an array are allowed. To do this, the desired element is first selected by using normal array access. Then bit-select and part-select are applied to the selected element in the same manner as in the cases of vectors.

```
mema[4][3]       // the 3rd bit of the 4th element
mema[5][7:4]     // the higher four bits of the 5th
                 // element
```

```
memb[3][1][1:0]   // the lower two bits of the [3][1]th
                  // element
sum[5][0]         // the [5][0]th element
```

Note that a memory of *n* 1-bit reg variables is different from an *n*-bit vector reg.

```
reg [1:n] rega; // an n-bit registers is not the same
reg mema [1:n]; // as a memory of n 1-bit registers
```

Memory indirection is allowed and can be specified in a single expression, for example:

```
mem_name[mem_name[23]] // use memory indirections
```

In this example, mem_name[23] addresses the 23rd word of the memory mem_name. The value at the word is then used as the address to access mem_name.

Review Questions

Q3.22 What kinds of data types can be declared as an array?

Q3.23 How would you declare a multi-dimensional array in Verilog HDL?

Q3.24 How would you distinguish between vector size and address range in an array declaration?

Q3.25 What is a memory from the viewpoint of Verilog HDL?

Q3.26 Can an array be assigned to another in a single statement?

Q3.27 How would you access a bit in a word of a memory?

3.3 OPERATORS

As shown in Table 3.1, Verilog HDL has a rich set of operators, including arithmetic operators, bit-wise operators, reduction operators, concatenation and replication operators, logical operators, relational operators, equality operators, shift operators and conditional operators. In this section, we describe these operators in detail and give some examples showing how to use them to model actual logic circuits.

3.3.1 Bit-wise Operators

Bit-wise operators perform a bit-by-bit operation on two operands and produce a vector result. In bit-wise operations, a z is treated as an unknown x. When two operands are not of equal length, the shorter operand is zero-extended to match the length of the longer operand.

TABLE 3.3 The bitwise operators

Symbol	Operation
~	Bitwise negation
&	Bitwise and
|	Bitwise or
^	Bitwise exclusive or
~^,^~	Bitwise exclusive nor

The bit-wise operators include five operators: & (and), | (or), ^ (xor), ^~(xnor) and ~(negation), as shown in Table 3.3. The functions of these operators are as follows:

- & (and): if any bit is 0, the result is 0, or else if both bits are 1, then the result is 1; otherwise the result is an x.
- | (or): if any bit is 1, the result is 1, or else if both bits are 0, the result is 0; otherwise the result is an x.
- ^ (xor): if one bit is 1 and the other is 0, the result is 1, or else if both bits are 0 or 1, the result is 0; otherwise the result is an x.
- ^~(xnor): if one bit is 1 and the other is 0, the result is 0, or else if both bits are 0 or 1, the result is 1; otherwise the result is an x.
- ~(negation): if the input bit is 1 the result is 0, or else if the input bit is 0, the result is 1; otherwise the result is an x.

The following module uses bit-wise operators to model a 4-to-1 multiplexer.

EXAMPLE 3.3 *A Dataflow Model for a 4-to-1 Multiplexer*

In this example, we use continuous assignment to directly realize the switching expression derived from Figure 2.5. The resulting expression uses & (and), | (or) and ~(negation) bit-wise operators.

```
module mux41_dataflow(i0, i1, i2, i3, s1, s0, out);
// port declarations
input i0, i1, i2, i3;
input s1, s0;
output out;
// using basic and, or, not logic operators.
   assign out = (~s1 & ~s0 & i0) |
                (~s1 &  s0 & i1) |
                ( s1 & ~s0 & i2) |
                ( s1 &  s0 & i3) ;
endmodule
```

The synthesized result of the above module mux41_dataflow is shown in Figure 3.1.

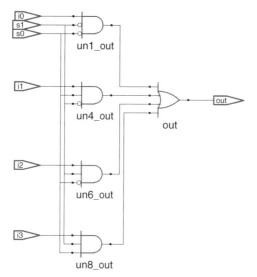

FIGURE 3.1 A dataflow model for a 4-to-1 MUX

Review Questions

Q3.28 Describe the basic operations of the set of bit-wise operators.

Q3.29 What operations are taken when two operands are of unequal length?

Q3.30 Describe the operation of & (and) using the four-value set.

Q3.31 Describe the operation of ^ (xor) using the four-value set.

Q3.32 Describe the operation of ˜(negation) using the four-value set.

3.3.2 Arithmetic Operators

The set of arithmetic operators contains six members $\{+, -, *, /, \%, **\}$, as shown in Table 3.4. The operators + and − perform addition and subtraction of two numbers, respectively. In Verilog HDL, negative numbers are represented in two's complement form. The operators + and − can also be used as unary operators to represent signed numbers. The operators * and / compute the multiplication and division of two numbers, respectively. The integer division (/) truncates any fractional part toward zero, while the modulus operator (%) produces the remainder from the division of two numbers.

TABLE 3.4 The arithmetic operators

Symbol	Operation
+	Addition
−	Subtraction
*	Multiplication
/	Division
**	Exponent (power)
%	Modulus

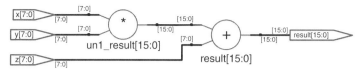

FIGURE 3.2 A multiply and accumulate unit

The exponent operator computes the power of a number. The result of the exponent operator ** is `real` if either operand is real, integer or signed.

The following example is a simple application of arithmetic operators.

EXAMPLE 3.4 *A Multiply and Accumulate Unit*

In this example, we directly use the arithmetic operators, + and *, to construct a multiply and accumulate unit. Depending on the synthesis tool and the target technology library, the multiply operator * may be synthesized by using a hardware macro, basic gates or cells organized by an internal algorithm inherent to the tool.

```
// an example to illustrate arithmetic operators
module multiplier_accumulator(x, y, z, result);
input   [7:0] x, y, z;
output [15:0] result;
   assign result = x * y + z ;
endmodule
```

The synthesized result of the preceding example is shown in Figure 3.2.

The following example is a simple application of divide arithmetic operator.

EXAMPLE 3.5 *An Unsigned Divider*

In this example, we directly use the divide arithmetic operator to construct an unsigned divider. Depending on the synthesis tool and the target technology library, the divide operator / may be synthesized by using a hardware macro, basic gates or cells organized by an internal algorithm inherent to the tool.

```
// an example to illustrate the divide operator
module divide_operator(x, y, result);
input   [7:0] x, y;
output [7:0] result;
   assign result = x / y;
endmodule
```

The following module segment shows the use of the exponent operator (**):

```
parameter ADDR_SIZE = 4;
localparam ROM_SIZE = 2 ** ADDR_SIZE - 1;
```

The exponent (power) operator is usually not supported by synthesis tools except that it can be evaluated to an integer power of 2. The details of `parameter` and `localparam` will be dealt with in Section 6.1.2.

A `reg` variable is treated as an unsigned value unless explicitly declared to be signed. An `integer` variable is treated as signed. In addition, for an arithmetic operation, if any bit of an operand is an `x` or a `z`, then the entire result value would be an `x`.

The size of the result of an arithmetic expression is determined by the size of the largest operand and the left-hand-side target as well if it is an assignment. This rule is also applied equally well to all intermediate results of an expression.

EXAMPLE 3.6 *The Result Size of an Expression*

Both sizes of result and intermediate results of the first continuous assignment are determined by inputs a and b, and net variable c, which is 4 bits. The result size of the second continuous assignment is 8 bits because the largest operand is 8 bits.

```
// an example to illustrate arithmetic operators
module arithmetic_operator(a, b, e, c, d);
input   [3:0] a, b;
input   [6:0] e;
output  [3:0] c;
output  [7:0] d;
    assign c = a + b;
    assign d = a + b + e;
endmodule
```

The synthesized result of the preceding example is shown in Figure 3.3.

In general, an unsigned value is stored in a net, a `reg` variable, and an integer in base format without a signed qualifier (the `s`). A signed value is stored in a signed net, a signed `reg` variable, an `integer` variable, an integer in decimal form and an integer in base format with a signed qualifier (the `s`).

In an expression with mixed signed and unsigned operands, all operands are converted to unsigned before any operation take places. The conversion of a number between signed and unsigned format can also be accomplished by system functions: `$signed` and `$unsigned`, which are dealt with in Section 5.3.4.

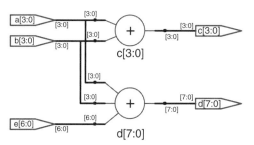

FIGURE 3.3 The synthesized result of the `arithmetic_operator` module

EXAMPLE 3.7 *Mixed Signed and Unsigned Operands*

The first expression is carried out in two's complement but the second expression is performed in unsigned numbers because input e is an unsigned number.

```
// an example to illustrate arithmetic operators
module arithmetic_signed(a, b, e, c, d);
input  signed [3:0] a, b;
input  [6:0] e;
output signed [3:0] c;
output [7:0] d;

   assign c = a + b;
   assign d = a + b + e;
endmodule
```

Review Questions

Q3.33 Describe the operation of the % (modulus) arithmetic operator.

Q3.34 Describe the operation of the ** (exponent) arithmetic operator.

Q3.35 Which data types are unsigned by default?

Q3.36 What will be the result type (signed or unsigned) when mixed signed and unsigned operands are used in an expression?

Q3.37 How would you determine the result size of an arithmetic expression?

3.3.3 Concatenation and Replication Operators

A concatenation operator is expressed by { and }, with commas separating the expressions within it, as shown in Table 3.5. The operands of a concatenation operator must be sized. Operands can be scalar nets or reg variables, vector nets or reg variables, bit-select, part-select or sized constants. For example:

```
y = {a, b[0], c[1]};
```

TABLE 3.5 The concatenation and replication operators

Symbol	Operation
{,}	Concatenation
{const_expr { } }	Replication

concatenates a, b[0] and c[1] into a group in that order and assigns to y. Another example is as follows:

```
y = {a, b[0], 4'b1};
```

which concatenates a, b[0] and 4'b1 into a group and assigns to y. Here, 4'b1 corresponds to 1'b1, 1'b1, 1'b1 and 1'b1 (see Section 3.2.1).

Another form of concatenation is the replication operation, which is expressed as {const_expr{}}, as shown in Table 3.5. The first expression, const_expr, must be a non-zero, non-x and non-z constant expression, while the second expression follows the rules for concatenation. For instance, {4{a}} replicates "a" 4 times. A replication operator specifies how many times to replicate the number inside the braces. For example:

```
y = {a, {4{b[0]}}, c[1]};
```

The right-hand side has six bits in total.

The following example shows how the concatenation operator is used to concatenate a scalar and a vector net operand.

EXAMPLE 3.8 *A 4-bit Adder Illustrates the Use of the Concatenation Operator*

In this example, a 4-bit adder is described in dataflow style. The left-hand side of the continuous assignment is a concatenation of a scaler c_out and a vector sum and forms a 5-bit result.

```
module four_bit_adder(x, y, c_in, sum, c_out);
// I/O port declarations
input  [3:0] x, y;  // declare as a 4-bit array
input  c_in;
output [3:0] sum;   // declare as a 4-bit array
output c_out;

// specify the function of a 4-bit adder.
    assign {c_out, sum} = x + y + c_in;
endmodule
```

The following example deals with the use of a replication operator.

EXAMPLE 3.9 *A 4-bit Two's Complement Adder*

This example uses a replication operator to make four copies of input carry c_in, then combines with an input operand y through an xor operator to form a true/one's complement generation circuit. By using this circuit, a 4-bit two's complement adder is constructed.

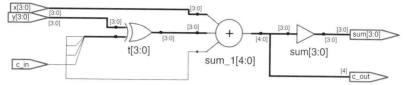

FIGURE 3.4 A two's complement 4-bit adder

```
module twos_adder(x, y, c_in, sum, c_out);
// I/O port declarations
input  [3:0] x, y;  // declare as a 4-bit array
input  c_in;
output [3:0] sum;   // declare as a 4-bit array
output c_out;
wire   [3:0] t;     // outputs of xor gates

// specify the function of a two's complement adder
   assign t = y ^ {4{c_in}};
   assign {c_out, sum} = x + t + c_in;
endmodule
```

The synthesized result of the above `twos_adder` module is shown in Figure 3.4.

3.3.4 Reduction Operators

The set of unary reduction operators carries out a bit-wise operation on a single vector operand and yields a 1-bit result. Reduction operators only perform on one vector operand and work in a bit-by-bit way from right to left. The reduction operators are shown in Table 3.6.

A simple application of reduction operator for modeling the 9-bit parity generator described in the previous chapter is as in the following example.

EXAMPLE 3.10 *A 9-bit Parity Generator Using a Reduction Operator*

As mentioned above, the operand of a reduction operator must be a vector. In this example, the vector x is reduced to a 1-bit result by the reduction operator ^ (xor). The synthesized result is shown in Figure 3.5.

TABLE 3.6 The reduction operators

Symbol	Operation
&	Reduction and
~ &	Reduction nand
		Reduction or
~		Reduction nor
^	Reduction exclusive or
~^, ^~	Reduction exclusive nor

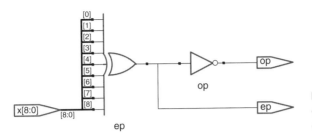

FIGURE 3.5 A 9-bit parity generator using a reduction operator

```
module parity_gen_9b_reduction(x, ep,op);
// I/O port declarations
input   [8:0] x;
output ep, op;
// dataflow modeling using reduction operator
   assign ep = ^x;    // even parity generator
   assign op = ~ep;   // odd parity generator
endmodule
```

From the preceding example, we can see the power of reduction operators. By using these operations, we may write a very compact and readable module. Another simple but useful application is to detect whether all bits in a byte are zeros or ones. In the following example, we assume that both results are needed.

EXAMPLE 3.11 *An All-bit-zero/One Detector*

The outputs `zero` and `one` are assigned to 1 if all bits of the input vector x are zeros and ones, respectively. These two detectors are easily implemented by the reduction operators | (or) & (and). The synthesized result is shown in Figure 3.6.

```
module all_bit_01_detector_reduction(x, zero,one);
// I/O port declarations
input   [7:0] x;
output  zero, one;
// dataflow modeling
   assign zero = ~(|x);  // all-bit zero detector
   assign one = &x;      // all-bit one detector
endmodule
```

Review Questions

Q3.38 Describe the operation of the concatenation operator {}.

Q3.39 Describe the operation of the replication operator {const_expr{}}.

Q3.40 What are the differences between bit-wise operators and reduction operators?

Q3.41 Describe the operation of the reduction operator & (and).

Q3.42 Describe the operation of the reduction operator ^ (xor).

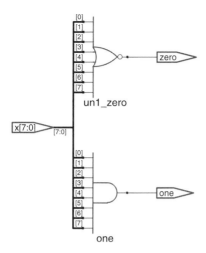

FIGURE 3.6 An all-bit-zero/one detector

3.3.5 Logical Operators

There are three logical operators, (&&) (and), (||) (or) and ! (negation), as shown in Table 3.7. They operate on the logical values 0 or 1 and produce a 1-bit value 0, 1 or an x. If any operand bit is x or z, it is treated to x and is treated as a false condition by simulators. For vector operands, a non-zero vector is treated as a 1.

For example, in the following example, if reg c holds the integer value 123 and d holds the value 0, then a is 1 and b is 0.

```
reg a, b;
reg [7:0] c, d;

a = c || d;   // a is set to 1
b = c && d;   // b is set to 0
```

A comment about ! (negation) is that it often uses a construct like if (!reset) to represent an equivalent one: if (reset == 0).

3.3.6 Relational Operators

Relational operators include four operators, > (greater than), < (less than), >= (greater than or equal to) and <= (less than or equal to), as shown in Table 3.8. Relational operators return the logical value 1 if the expression is true and 0 if the expression is

TABLE 3.7 The logical operators

Symbol	Operation
!	Logical negation
&&	Logical and
||	Logical or

TABLE 3.8 The relational operators

Symbol	Operation
>	Greater than
<	Less than
>=	Greater than or equal
<=	Less than or equal

false. The expression results in a value x if there are any unknown (x) or z bits in the operands.

When two operands are not of equal length and any operand is unsigned, the smaller operand is zero-extended to match the size of the larger operand and the operation is performed between two unsigned values. If both operands are signed, the smaller operand is sign-extended and the operation is performed between signed values. When either operand is a 'real', then the other operand is converted to an equivalent real value and the operation is performed between two real values.

All relational operators have the same precedence and their precedence is lower than that of arithmetic operators.

3.3.7 Equality Operators

The set of equality operators include four operators, == (logical equality), != (logical inequality), === (case equality) and !== (case inequality), as shown in Table 3.9. Equality operators return the logical value 1 if the expression is true and 0 if the expression is false. These operators compare the two operands bit by bit, zero-extended if the operands are not of equal length. The logical equality operators (==, !=) yield an unknown x if either operand has x or z in its bits. The case equality operators (===, !==) yield a 1 if the two operands match exactly and 0 if the two operands do not match exactly.

A magnitude comparator is often used in digital systems to compare two unsigned numbers. In the following example, we employ relational and equality operators to describe a 4-bit magnitude comparator in dataflow style.

EXAMPLE 3.12 *A 4-bit Magnitude Comparator*

Because the operation of a magnitude comparator is to compare and indicate the relative magnitude of two input numbers, logical equality (==) and relational operators, greater than (>) and less than (<) are used to construct the desired circuit. In order to be

TABLE 3.9 The equality operators

Symbol	Operation
==	Logical equality
!=	Logical inequality
===	Case equality
!==	Case inequality

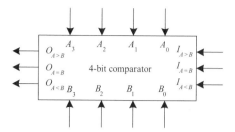

FIGURE 3.7 A 4-bit comparator

cascadable, three additional inputs are also incorporated into the module, as shown in Figure 3.7. Consequently, the overall result is equality if both results from the current and the preceding modules are equal. The overall result is greater (less) than if the current result is greater (less) than or the current result is equal to but the result from the preceding modules is greater (less) than.

```
module four_bit_comparator(Iagtb, Iaeqb, Ialtb, a, b,
 Oagtb, Oaeqb, Oaltb);
// I/O port declarations
input   [3:0] a, b;
input  Iagtb, Iaeqb, Ialtb;
output Oagtb, Oaeqb, Oaltb;
// dataflow modeling using relation operators
    assign Oaeqb = (a == b) && (Iaeqb == 1);  // equality
    assign Oagtb = (a > b) || ((a == b) && (Iagtb == 1));
                                     // greater than
    assign Oaltb = (a < b) || ((a == b) && (Ialtb == 1));
                                     // less than
endmodule
```

The synthesized result of the above `four_bit_comparator` module is shown in Figure 3.8.

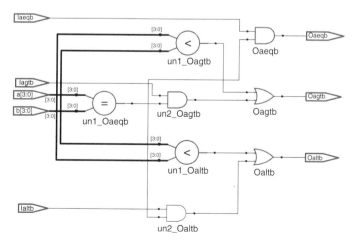

FIGURE 3.8 The synthesized result of a cascadable 4-bit comparator

TABLE 3.10 The shift operators

Symbol	Operation
>>	Logical right shift
<<	Logical left shift
>>>	Arithmetic right shift
<<<	Arithmetic left shift

Although we may depend on the precedence of operators and write a correct expression, such as in the following example:

```
a < s - 1 && b != c && d != f;
```

it is good practice to use parentheses to indicate very clearly the precedence intended for improving readability, as in the following rewriting example:

```
(a < s - 1) && (b != c) && (d != f);
```

3.3.8 Shift Operators

There are two types of shift operators, the logical shift operators, $<<$ and $>>$, and the arithmetic shift operators, $<<<$ and $>>>$, as shown in Table 3.10. Operators $<<$ and $>>$ are logical left and right shifts, respectively, while operators $<<<$ and $>>>$ are arithmetic left and right shifts, respectively. The shift operation shifts the left operand by the number specified by the right operand. The vacant bit positions are filled with zeros for logical and arithmetic left-shift operations and filled with the MSBs (sign bits) for arithmetic right-shift operation.

The following example shows a simple application of a logical left-shift operator.

EXAMPLE 3.13 *A Simple Application of a Logical Left-shift Operator*

In this example, both `start` and `result` are declared as `reg` variables and a logical left-shift operator is applied to `start`. The `result` variable is assigned the binary value `0100`, which is `0001` shifted to the left two positions and zero-filled.

```
module logical_shift;
reg [3:0] start, result;
initial begin
    start = 4'b0001;
    result = (start << 2);
end
endmodule
```

The following example shows a simple application of an arithmetic right-shift operator.

EXAMPLE 3.14 *A Simple Application of an Arithmetic Right-shift Operator*

In this example, both `result` and `start` are declared as `signed reg` variables and the arithmetic right-shift operator is applied to `start`. The `result` variable is assigned the binary value `1110`, which is `1000` shifted to the right two positions and sign-extended.

```
module arithmetic_shift;
reg signed [3:0] start, result;
initial begin
    start = 4'b1000;
    result = (start >>> 2);
end
endmodule
```

■

The following example compares the differences between logical and arithmetic right shifts. You are encouraged to write a test bench, simulate the program and see what happens.

EXAMPLE 3.15   *An Example Which Illustrates Logical and Arithmetic Shifts*

This example explains the differences between logical and arithmetic right shifts. The logical right shift fills the vacant bits with 0 and the arithmetic right shift fills the vacant bits with the sign bit, known as a sign-bit extension.

```
// an example illustrates logic and arithmetic shifts
module arithmetic_shift(x, y, z);
input  signed [3:0] x;
output [3:0] y;
output signed [3:0] z;
    assign y = x >> 1;   // logical right shift
    assign z = x >>> 1;  // arithmetic right shift
endmodule
```

■

Note that the net variables x and z must be declared with the keyword `signed`. It is advised to replace the net variable with an unsigned net (i.e. remove the keyword `signed`) and see what happens.

3.3.9 Conditional Operator

The conditional operator selects an expression based on the value of the condition expression. It has the form:

```
condition_expr ? true_statement: false_statement
```

where `condition_expr` is evaluated first. If the result is true then the `true_statement` is executed; otherwise, the `false_statement` is evaluated. This operator is equivalent to the following:

```
if  (condition_expr) true_statement
else false_statement
```

A simple example for describing a 2-to-1 multiplexer is as follows:

```
assign out = selection ? input1: input0;
```

The input `input1` is passed to the output end `out` if the selection is 1; otherwise, the input `input0` is passed to the output end `out`.

A more complicated example of using a conditional operator to model a 4-to-1 multiplexer is shown in the following example.

EXAMPLE 3.16 *A 4-to-1 Multiplexer Constructed by Using the Conditional Operator*

In this example, we use nested conditional operators; two are inside the outmost conditional operator. You are asked to check the correctness of this module by using a synthesizer and observing the synthesized result. Although a construct like this is quite concise, it is not easy to understand for a naive reader.

```
module mux4_to_1_cond (i0, i1, i2, i3, s1, s0, out);
// port declarations from the I/O diagram
input   i0, i1, i2, i3;
input   s1, s0;
output out;
// using conditional operator (?:)
   assign out = s1 ? ( s0 ? i3 : i2) : (s0 ? i1 : i0) ;
endmodule                                                       ∎
```

The following example rewrites the nested conditional operator used in the preceding example in another more readable form. You may compare both constructs and see the difference between them. Actually, both constructs produce exactly the same gate-level circuits. Check them on your system.

EXAMPLE 3.17 *A 4-to-1 Multiplexer Constructed by Using the Case Equality and Conditional Operators*

In this example, we explicitly use the case equality operator ($===$) to determine which input will be routed to the ouput. The selection signal `sel` is compared with the constant i, where i runs from 0 to 3, in order. When the comparison is exactly equal to each other, the input is assigned to `out`. Indeed, this is exactly the same as in the use of the `case` statement, which will be introduced in Section 4.4.2. If we change the case equality operator ($===$) into a logical equality ($==$), what will happen?

```
module mux41_equality(i0, i1, i2, i3, sel, out);
// port declarations
input   i0, i1, i2, i3;
input   [1:0] sel;
output out;
// using case and conditional operators.
assign out = (sel === 0) ? i0 :
             (sel === 1) ? i1 :
             (sel === 2) ? i2 :
             (sel === 3) ? i3 : 4'bz;
endmodule                                                       ∎
```

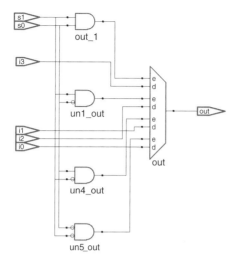

FIGURE 3.9 A 4-to-1 MUX constructed by using the conditional operator

The synthesized results of both modules described previously are the same, as depicted in Figure 3.9.

Review Questions

Q3.43 Describe the operation of logical operators.

Q3.44 Describe the operation of relational operators.

Q3.45 Describe the operation of the set of equality operators.

Q3.46 What is the basic difference between logical equality (==) and case equality (===)?

Q3.47 What is the basic difference between an arithmetic right shift ($>>>$) and a logical right shift ($>>$)?

Q3.48 Describe the operations of the conditional operator (? :).

SUMMARY

The rationale behind dataflow modeling is on the observation that any digital system can be constructed by interconnecting registers and put between them a combinational logic for performing the desired functions. Continuous assignment is the most basic statement of dataflow modeling. It is used to continuously drive a value onto a net. Continuous assignments are always active.

A continuous assignment begins with the keyword `assign` followed by an expression. An expression is a construct that combines operators and operands to produce a result. An expression consists of sub-expressions, operators and operands. The operands in an expression can be any of the following: constants, parameters, nets, variables (`reg`, `integer`, `time`, `real`, `realtime`), bit-select, part-select, array element and function calls. The set of operators includes arithmetic, shift, bit-wise, case equality, reduction, logic, relational, equality and miscellaneous operators.

Most of the operators can be accepted by synthesis tools. Only a few of them cannot, or only a limited form can be accepted. The exponent and modulus might be two such examples – check on your system. Through using the dataflow style, a combinational logic circuit can be modeled by using continuous assignments rather than structural connections between instances of modules.

REFERENCES

1. J. Bhasker, *A Verilog HDL Primer*, 3rd Edn, Star Galaxy Publishing, 2005.
2. S. Palnitkar, *Verilog HDL: A Guide to Digital Design and Synthesis*, 2nd Edn, SunSoft Press, 2003.
3. IEEE 1364-2001 Standard, *IEEE Standard Verilog Hardware Description Language*, 2001.

PROBLEMS

3.1 Simplify the following switching expression and use bit-wise operators to model it:

$$f(w, x, y, z) = \Sigma(5, 6, 7, 9, 10, 11, 13, 14, 15)$$

3.2 Simplify the following switching expression and use bit-wise operators to model it:

$$f(w, x, y, z) = \Sigma(0, 4, 5, 7, 8, 9, 13, 15)$$

3.3 Using bit-wise operators, model the logic circuit shown in Figure 3.10.

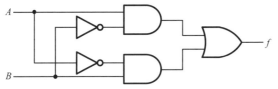

FIGURE 3.10 A logic diagram for problem 3.3

3.4 Using bit-wise operators, model the logic circuit shown in Figure 3.11.

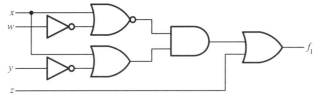

FIGURE 3.11 A logic diagram for problem 3.4

3.5 Using the multiply ($*$) operator, write a module to compute $x * y$, where x and y are two signed numbers. Synthesize the module on your system and check the synthesized result. Write a test bench to verify whether the module behaves correctly.

3.6 Using the divide ($/$) operator, write a module to compute x/y, where x and y are two signed numbers. Synthesize the module on your system and check the synthesized result. Write a test bench to verify whether the module behaves correctly.

3.7 Using the modulus ($\%$) operator, write a module to compute $x\%y$, where x and y are two unsigned numbers. Synthesize the module on your system and check the synthesized result. Write a test bench to verify whether the module behaves correctly.

3.8 By assuming that both numbers are signed, re-do problem 3.7.

3.9 Suppose that an arithmetic and logic unit (ALU) has the function shown in Table 3.11. The output is a function of six mode-selection lines, $m5$ to $m0$.

TABLE 3.11 A function table for problem 3.9

m5	m4	m3	m2	m1	m0	ALU output
0	0	0	1	0	0	A .and B
0	0	1	0	0	0	A .and .not B
0	0	1	0	0	1	A .xor B
0	1	1	0	0	1	A plus .not B plus carry
0	1	0	1	1	0	A plus B plus carry
1	1	0	1	1	0	.not A plus B plus carry
0	0	0	0	0	0	A
0	0	0	0	0	1	A .or B
0	0	0	1	0	1	B
0	0	1	0	1	0	.not B
0	0	1	1	0	0	zero

(a) Derive the 1-bit slice switching expression of the output as the function of the mode selection lines $m5$ to $m0$.

(b) Write a module at the dataflow level to perform the output function. Write a test bench to verify whether the module behaves correctly.

(c) By instantiating the above module, construct a 4-bit ALU. Write a test bench to verify whether the module behaves correctly.

3.10 Using the conditional operator, write a module to shift the input `data` right logically by the number of positions specified by another input `shift`, ranging from 0 to 3.

3.11 Using the conditional operator, write a module to shift the input `data` right arithmetically by the number of positions specified by another input `shift`, ranging from 0 to 3.

3.12 Using the conditional operator, write a two's complement adder. Write a test bench to verify whether the module behaves correctly.

3.13 Use the conditional and case equality operators, respectively, as the selector to model the functions shown in Table 3.12. Write a test bench to verify whether the module behaves correctly.

TABLE 3.12 Table for problem 3.13

Function	Operation
0	z=x+y
1	z=x-y
2	z=x*y

3.14 A 4-to-2 priority encoder is a device with four inputs, in[0] to in[3], and a two-bit output code, out[1] and out[0]. In addition, a signal valid is used to signal that an input signal is valid. The two-bit output code {out[1], out[0]} is i if the input in[i] is 1 and no other inputs in[j] are 1, where $j < i$. The signal valid is 0 if all inputs are 0; it is 1, otherwise. Describe the module in dataflow style. Write a test bench to verify whether the module behaves correctly.

## CHAPTER 4

BEHAVIORAL MODELING

WE HAVE discussed structural style at both gate and switch levels, and dataflow style using continuous assignments in the previous two chapters. In this chapter, we address the behavioral style. To make use of the full power of Verilog HDL, a design usually contains a mix of these three modeling styles.

Like any other programming languages, such as C, Verilog HDL also provides the three basic structures: assignments, selection statements and iterative (loop) statements. Assignment statements include continuous assignment (`assign` expression), procedural assignments, including blocking assignment (=) and nonblocking assignment (<=), and procedural continuous assignments, including `assign-deassign` and `force-release`.

Timing controls provide a way to specify the simulation time at which procedural statements will execute. In Verilog HDL, if there are no timing control statements, the simulation time will not advance. Hence, it is necessary to include some kinds of timing control statements within Verilog HDL modules. There are two timing control methods provided in Verilog HDL: delay timing control and event timing control.

Selection statements are used to make a selection according to a given condition. They contain two statements: `if-else` and `case`, including `case`, `casex` and `casez`. Iterative (loop) statements are used to repeatedly execute a set of statements. There are four iterative (loop) statements: `repeat`, `for`, `while` and `forever`. All these four loop statements are discussed in detail.

4.1 PROCEDURAL CONSTRUCTS

An HDL must have the capability of capturing and describing the hardware features: *continuity* and *parallelism*. In Verilog HDL, this capability is facilitated by `initial` and `always` statements. An `initial` or an `always` statement is usually called an `initial` or an `always` block. `initial` statements are used to initialize variables and set values into variables or nets and `always` statements are used to model the continuous operations required in the hardware modules. All other behavioral statements must be within an `initial` or an `always` block.

In a module, an arbitrary number of `initial` and `always` statements may be used. All `initial` and `always` statements execute concurrently with respect to each other; namely, their relative order in a module is not important. Each of the `initial` and `always` statements represents a separate activity flow. Each activity starts at simulation time 0. Notice that each `always` statement corresponds to a piece of logic circuit. In addition, `initial` and `always` blocks cannot be nested. Each `initial` or `always` statement must form its own block.

4.1.1 `initial` **Block**

An `initial` block is composed of all statements inside an `initial` statement. An `initial` block starts at simulation time 0 and executes exactly once during simulation. It is used to initialize signals, or monitor waveforms, and so on. The `initial` statement has the following syntax:

```
initial [timing_control] procedural_statement
```

The `timing_control` can be either a delay control or an event control, which is discussed in more detail in Section 4.3. The `procedural_statement` can be any expression that evaluates to a value. As a consequence, it can be any of the following:

```
procedural_assignment (blocking and nonblocking)
selection_statement
case_statement
loop_statement
wait_statement
event_trigger_statement
task_enable (user or system)
procedural_continuous_assignment
disable_statement
sequential_block
parallel_block
```

For example, in the following two `initial` statements are declared. The first `initial` statement is a single statement containing only one procedural assignment and the second statement is a complex statement containing six procedural assignments enclosed by a `begin-end` block.

```
reg  x, y, z;
  initial x = 1'b0; // single statement
  initial begin     // complex statement
      x = 1'b0; y = 1'b1; z = 1'b0;
  #10 x = 1'b1; y = 1'b1; z = 1'b1;
end
```

If a delay specification #delay appears before a procedural assignment, it indicates that the expression will be executed #delay time units after the current simulation time. Hence, the x in the preceding example will be set to 1 after (#10) time units.

Each initial block starts to execute concurrently at simulation time 0 and finishes their execution independently when multiple initial blocks exist. For example, both of the following initial blocks start to execute at simulation time 0. The first one ends after 10 time units but the second one ends after 30 time units.

```
reg  x, y, z;
   initial #10 x = 1'b0; // finish at simulation time 10
   initial begin
      #10 y = 1'b1;        // finish at simulation time 10
      #20 z = 1'b1;        // finish at simulation time 30
   end
```

4.1.1.1 *Combined Variable Declaration and Initialization*   Although the initial values of variables can be set by using an initial statement, it is common to combine variable declaration and initialization into one statement. For example:

```
reg  clk;     // regular declaration
   initial clk = 0;
// which can be declared as a single statement instead
reg clk = 0; // can be used only at module level
```

The same approach can be applied to the port list of a module, which is called the port list style. For example:

```
module adder(x, y, c , sum, c_out);
input  [3:0] x, y;
input  c_in;
output reg [3:0] sum = 0; // initialize sum
output reg c_out = 0;      // initialize c_out
```

Of course, this can be applied to the port list declaration style (also called ANSI C style) as well:

```
module adder(input [3:0] x, y,
             input c_in,
             output reg [3:0] sum = 0,
             output reg c_out = 0
);   // ANSI C style
```

Note that a variable declared with initialization is usually not accepted by synthesis tools. Try it on your system and see what happens. Moreover, initial blocks are also not accepted by synthesis tools.

4.1.2 `always` Block

An `always` block consists of all behavioral statements within an `always` statement. The `always` statement starts at simulation time 0 and repeatedly executes the statements within it in a loop fashion during simulation. The `always` statement has the following syntax:

```
always [timing_control] procedural_statement
```

The `timing_control` and `procedural_statement` are the same as that of the `initial` statement described in the previous section.

An `always` block is used to model a block of activities that are continuously repeated in a digital circuit. For example, the following program segment:

```
reg  clock;                 // a clock generator
initial clock = 1'b0;       // initial clock = 0

always #5 clock = ~clock; // period = 10
```

models a clock generator, which produces a symmetric clock signal with a period of 10 time units. Note that we need to set the initial value of clock to either 0 or 1; otherwise, the clock signal produced will be always an unknown value, x.

The following example is a 4-to-1 multiplexer modeled in behavioral style.

EXAMPLE 4.1   *An Example Illustrating the Use of the* `always` *Block*

In this example, the entire body of the module is only an `always` statement. A procedural assignment within the `always` statement is used to implement the operations of the 4-to-1 multiplexer, described in Section 2.1.1. The procedural assignment is the same as the continuous assignment used in Chapter 3, except that the keyword `assign` is absent.

```
module mux41_behavioral(i0, i1, i2, i3, s1, s0, out);
// port declarations
input i0, i1, i2, i3;
input s1, s0;
output reg out;
// using basic and, or, not logic operators.
always @(i0 or i1 or i2 or i3 or s1 or s0)
    out = (~s1 & ~s0 & i0)|
          (~s1 &  s0 & i1)|
          ( s1 & ~s0 & i2)|
          ( s1 &  s0 & i3);
endmodule
```

What follows is another example of `always` block. Here, it is employed to realize a 4-bit adder.

EXAMPLE 4.2 *An Example Illustrating the Use of the* `always` *Block*

In this example, a 4-bit adder is modeled in behavioral style. Due to its intuitive simplicity, we will not explain it furthermore here.

```
module four_bit_adder(x, y, c_in, sum, c_out);
// I/O port declarations
input  [3:0] x, y;      // declare as a 4-bit array
input  c_in;
output reg [3:0] sum;   // declare as a 4-bit array
output reg c_out;
// specify the function of a 4-bit adder.
always @(x or y or c_in)// event timing control
   {c_out, sum} = x + y + c_in;
endmodule
```

Review Questions

Q4.1 Explain the operations of an `initial` block.

Q4.2 Explain the operations of an `always` block.

Q4.3 Distinguish between the port list style and port list declaration style.

Q4.4 Explain the operations of the following `always` statement:

```
always begin   // including inside an initial
               // statement
   initial clock = 1'b0;
   #5 clock =  ~clock;
end
```

Q4.5 Explain the operations of the following `always` statement:

```
always begin   // without an initial value of clock
   #5 clock =  ~clock;
end
```

4.2 PROCEDURAL ASSIGNMENTS

As described before, a continuous assignment is used to continuously assign values onto a net in a manner similar to the way that a logic gate drives a net. In contrast, a procedural assignment puts values in a variable, which holds the value until the next procedural assignment updates that variable. Procedural assignments occur within procedures such as `initial`, `always`, `task` and `function`.

4.2.1 Procedural Assignments

Procedural assignments are placed inside `initial` or `always` statements. They update the values of variable data types, `reg`, `integer`, `time`, `real`, `realtime`, or array (memory) elements. The general syntax of procedural assignments is as follows:

```
variable_lvalue = [timing_control] expression
[timing_control] variable_lvalue =  expression
```

where `variable_lvalue` can be a `reg`, `integer`, `time`, `real`, `realtime`, or an array (memory) element, a bit select, a part select or a concatenation of any of the above. The `timing_control` can be either a delay control or an event control.

As in continuous assignments, the bit widths of both left-hand and right-hand sides need not be the same. The right-hand side is truncated by keeping the least significant bits to match the width of the left-hand-side variable when the right-hand side has more bits. The right-hand side is zero-extended in the most significant bits before assigning to the left-hand-side variable when the right-hand side has fewer bits.

Procedural Versus Continuous Assignments   There is an essential difference between procedural assignments and continuous assignments. Continuous assignments drive nets, which are evaluated and updated whenever any input operand changes its value. In contrast, procedural assignments update the values of variables under the control of the procedural flow constructs that surround them. As a consequence, the left-hand side of continuous assignments are nets but the left-hand side of procedural assignments are variables.

In summary, the procedural assignments have several distinct features in comparison with continuous assignments:

1. They do not use the keyword `assign`.
2. They can only update variable data types, `reg`, `integer`, `time`, `real` and `realtime`, or array (memory) elements.
3. They can only be used within `always` and `initial` blocks.

Two types of procedural assignments in Verilog HDL are *blocking* assignment, using operator =, and *nonblocking* assignment, using operator <=. We discuss each of these in a separate sub-section in more detail.

4.2.2 Blocking Assignments

Blocking assignment statements use the operator "=" and are executed in the order that they are specified. In other words, a blocking assignment is executed before the execution of the statements that follow it in a *sequential* block (statements grouped with `begin` and `end` keywords). However, a blocking assignment will not block the execution of statements following in a *parallel* block (statements grouped with `fork` and `join` keywords). For now, we only consider the case of sequential blocks. The parallel block will be discussed in Section 7.1.2.

The following example illustrates the use and operations of blocking assignments.

EXAMPLE 4.3   *An Example Illustrating the Use of Blocking Assignments*

Due to the sequential features of blocking assignments, the x, y and z are assigned new values at simulation time 5, 8 and 14, respectively.

```
// an example illustrates blocking assignments
module blocking;
reg  x, y, z;
// blocking assignments
initial begin
   x = #5 1'b0;   // x will be assigned 0 at time 5
   y = #3 1'b1;   // y will be assigned 1 at time 8
   z = #6 1'b0;   // z will be assigned 0 at time 14
end
endmodule
```

The following example describes how to construct a 4-bit two's complement adder by using blocking assignments.

EXAMPLE 4.4   *A 4-bit Two's Complement Adder in Behavioral Style*

In this example, a two's complement adder is constructed by using the principle that x minus y is equal to x plus the two's complement of y. The two's complement of y can be further obtained by complementing y and then add 1 to it. As a consequence, the two-step adder is immediately followed. In the first step, we compute the complement of y if the mode control input c_in is 1 and the true value of y if c_in is 0. In the second step, we add the complement of y and 1 or the true value of y and 0 with x. The complete operations are as the following module.

```
module twos_adder_behavioral(x, y, c_in, sum, c_out);
// I/O port declarations
input  [3:0] x, y;      // declare as a 4-bit array
input  c_in;
output reg [3:0] sum;   // declare as a 4-bit array
output reg c_out;
reg    [3:0] t;         // outputs of xor gates
// specify the function of a two's complement adder
always @(x, y, c_in) begin // define two's adder function
   t = y ^ {4{c_in}};      // what is wrong with: t = y ^
                           // c_in ?
   {c_out, sum} = x + t + c_in;
end
endmodule
```

Note that a `reg` variable does not necessarily correspond to a register element when the circuit is synthesized. Actually, whether it is synthesized into a register or not totally depends on the contents of the `always` block containing it.

The following is an example about how to model a 4-bit comparator in behavioral style.

EXAMPLE 4.5 *A 4-bit Comparator in Behavioral Style*

A comparator is a device used to determine the relationship between two numbers. Thus, it is easily realized by using relational operators (> and <) and the logical equality operator (==). The resulting module is shown in the following.

```
module four_bit_comparator(Iagtb, Iaeqb, Ialtb, a, b,
   Oagtb, Oaeqb, Oaltb);

// I/O port declarations
input    [3:0] a, b;
input    Iagtb, Iaeqb, Ialtb;
output reg Oagtb, Oaeqb, Oaltb;


// behavioral modeling using relational operators
always @(*) begin
   Oaeqb = (a == b) && (Iaeqb == 1);  // equality
   Oagtb = (a > b) || ((a == b)&& (Iagtb == 1));
                                      // greater than
   Oaltb = (a < b) || ((a == b)&& (Ialtb == 1));
                                      // less than
end
endmodule
```

Note that the "=" operator used by blocking assignments is also used by procedural continuous assignments (see Section 7.2 for details) and continuous assignments.

4.2.3 Nonblocking Assignments

Nonblocking assignments use the <= operator and are executed without blocking the other statements in a sequential block. Nonblocking assignments provide a method to model several concurrent data transfers that take place after a common event. In other words, a nonblocking assignment statement is used whenever several variable assignments within the same time step need be made, regardless of the order or the dependence on each other.

The following example illustrates the use and operations of nonblocking assignments. The reader should compare the operations and results of this example with those of the example using blocking assignments given in the preceding sub-section.

EXAMPLE 4.6   *An Example Illustrating the Use of Nonblocking Assignments*

Due to the parallel features of nonblocking assignments, the x, y and z are assigned new values at simulation time 5, 3 and 6, respectively.

```
// an example illustrates nonblocking assignments
module nonblocking;
reg  x, y, z;
// nonblocking assignments
initial begin
   x <= #5 1'b0;  // x will be assigned 0 at time 5
   y <= #3 1'b1;  // y will be assigned 1 at time 3
   z <= #6 1'b0;  // z will be assigned 0 at time 6
end
endmodule
```

Applications of nonblocking assignments include the modeling of pipeline and several mutually exclusive data transfers. It is recommended to use nonblocking assignments rather than blocking assignments whenever concurrent data transfers take place after a common event. Nonblocking assignments are executed last in the time step during simulation in which they are scheduled when in mixed use with blocking assignments.

A mixed use of blocking and nonblocking assignments is as in the following example.

EXAMPLE 4.7   *An Example of the Mixed Use of Blocking and Nonblocking Assignments*

When the first procedural assignment is encountered, the variable a is scheduled to assign value of 0 at time 5, and then the simulator executes the blocking assignment to assign b at time 3. The third statement is encountered at time 3 and the c is scheduled to assign a value 0 at 6 time units later, i.e. at time 9. At the same time step, the last blocking assignment is executed so as to assign a value 1 to d at time 10.

```
// an example of mixed use of blocking and nonblocking
// assignments
module blocking_nonblocking_mixed;
reg  a, b, c, d;
// blocking assignments
initial begin
   a <= #5 1'b0;  // a will be assigned 0 at time 5
   b  = #3 1'b1;  // b will be assigned 1 at time 3
   c <= #6 1'b0;  // c will be assigned 0 at time 9
   d  = #7 1'b1;  // d will be assigned 1 at time 10
end
endmodule
```

What would happen when we interchange the operators of nonblocking and blocking in the preceding example? Try this on your system and explain the results.

The following is an example of a 4-bit shift register constructed by using a nonblocking assignment.

EXAMPLE 4.8 *An Example of the Right-Shift Register Without Reset*

In this example, we use a nonblocking assignment coupled with a concatenation operator to perform the right-shift operation.

```
// an example of right-shift register without reset.
module shift_reg_4b(clk, din, qout);
input   clk;
input   din;
output reg [3:0] qout;
// the body of a 4-bit shift register
always @(posedge clk)
   qout <= {din, qout[3:1]};  // right shift
endmodule
```

What would happen when the nonblocking operator is replaced with a blocking operator? Nothing will happen. Explain this. However, we prefer to use a nonblocking operator in this case. The reason will be explained later in this sub-section.

In general, the simulators perform the following three steps when executing nonblocking statements during simulation:

1. *Read*: read the values of all right-hand-side variables.
2. *Evaluation*: evaluate the right-hand-side expressions and store in temporary variables that are scheduled to assign to the left-hand-side variables later.
3. *Assignment*: assign the values stored in the temporary variables to the left-hand-side variables.

4.2.3.1 Race Problems
If we want to swap the contents of two registers, we may possibly use the following two `always` statements with a blocking assignment within each at the first attempt:

```
// using blocking assignment statements
always @(posedge clock) // has race condition
   x = y;
always @(posedge clock)
   y = x;
```

However, this will result in a *race* problem, which means that the final result will depend on the order that both `always` statements are executed. If the first `always` statement is executed first and then the second, the final result will be the case that both registers x and y have the same content as the original y. If the reverse order is executed, the result is that both registers x and y have the same content as the original x. To avoid this race problem, we may replace the blocking assignments with nonblocking assignments within always statements, as shown in the following.

```
// using nonblocking assignment statements
always @(posedge clock) // has no race condition
    x <= y;
always @(posedge clock)
    y <= x;
```

Here, at the positive edge of each clock, the contents of registers x and y are read, writen into temporary variables, and finally, the contents of the temporary variables are assigned to registers x and y, respectively. Consequently, we properly swap the contents of two registers. As a matter of fact, the above nonblocking assignments can be emulated by using blocking assignments as the following program segment:

```
always @(posedge clock)begin
    temp_x = x; // read operation
    temp_y = y;

    y = temp_x; // write operation
    x = temp_y;
end
```

4.2.4 Blocking versus Nonblocking Assignments

In order to make the differences between blocking and nonblocking assignments clearer, we first use a shift-register as an example to illustrate this. The key point is that we need always keep in mind that nonblocking assignments are always executed according to the three steps described previously.

Suppose we want to model the 4-bit shift register depicted in Figure 4.1. An incorrect module using blocking assignments is shown in the following example.

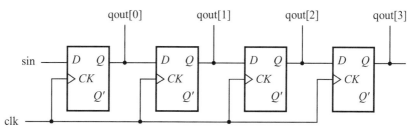

FIGURE 4.1 The basic structure of a 4-bit shift register

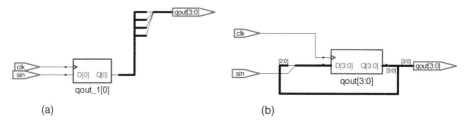

(a)   (b)

FIGURE 4.2 A shift register generated by using (a) blocking and (b) nonblocking assignments

EXAMPLE 4.9   *An Incorrect Version of the Shift Register Module*

According to the logic diagram shown in Figure 4.1, it is quite straightforward to directly employ four blocking assignments. However, the resulting module cannot work properly. When we first examine the synthesized result shown in Figure 4.2(a), we may wonder about the result obtained. However, after we examine Figure 4.2(a) and the module written more carefully, we can realize that the operation of the synthesizer is correct although it produced an unexpected result. The reason is that blocking assignment statements are executed in the order specified, and after they are optimized, the final result is exactly the same as the one shown in Figure 4.2(a).

```
// shift register module example --- an incorrect
// implementation
module shift_reg_blocking(clk, sin, qout);
input   clk;
input   sin;   // serial data input
output reg [3:0] qout;
// the body of a 4-bit shift register
always @(posedge clk)
    begin      // using blocking assignments
        qout[0] = sin;
        qout[1] = qout[0];
        qout[2] = qout[1];
        qout[3] = qout[2];
    end
endmodule
```

Note that when we replace the four blocking assignments within the begin and end block by a single assignment, qout = {qout[2:0], sin}, in the above module, the result is correct. Why is this and how would you explain it?

To solve the above problem, we return to Figure 4.1 and examine the operation of the shift register in greater detail. From this figure, we know that the essential operation of the shift register is that at each positive edge of the clock, each flip-flop is assigned to a new value, which is the output of its previous stage. Thus, the value to be assigned to a flip-flop at each positive edge of the clock is like the one taken from a temporary variable, which stores the output of its previous stage in the previous clock cycle.

Consequently, it should use a nonblocking assignment statement, as in the following module.

EXAMPLE 4.10   *A Correct Version of the Shift Register Module*

In this example, we replace the four blocking assignments with nonblocking assignments. The synthesized result is shown in Figure 4.2(b). Of course, an even better approach is to used a single nonblocking assignment: qout <= {qout[2:0], sin}.

```
// shift register module example ---a correct
// implementation
module shift_reg_nonblocking(clk, sin, qout);
input  clk;
input  sin;  // serial data input
output reg [3:0] qout;
// the body of a 4-bit shift register
always @(posedge clk)
   begin                    // using nonblocking
                            // assignments
      qout[0] <= sin;       // it is even better to use
      qout[1] <= qout[0];   // qout <= {qout[2:0], sin};
      qout[2] <= qout[1];
      qout[3] <= qout[2];
   end
endmodule
```

One further insight into the difference between blocking and nonblocking assignments is considered as follows. First, consider the following always block. Assume that the value of count is 1 and finish is 0 before entering the always block.

```
always @(posedge clk) begin: block_a
   count = count - 1;
   if (count == 0) finish = 1;
end
```

The result is finish = 1 when the count is equal to 0, which is different from that of the gate level. Why? The reason is that according to the inherent features of blocking assignments the two statements within the always block are executed one after the other as the order specified, as shown in Figure 4.3. However, the operations of actual gates are executed concurrently.

However, if we replace the blocking assignments with nonblocking assignments as follows:

```
always @(posedge clk) begin: block_b
   count <= count - 1;
   if (count == 0) finish <= 1;
end
```

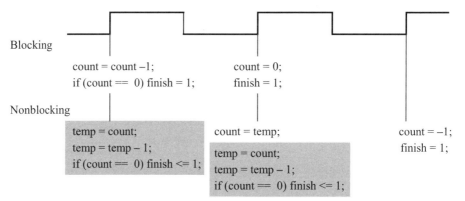

FIGURE 4.3 An example illustrating the differences between blocking and nonblocking assignments

then the result is `finish = 0` when the `count` equals 0, which is the same as that of the gate level. Actually, in this case, the `finish = 1` when the `count` equals −1. The reasons are as follows. Based on the features of nonblocking assignments, the two statements within the `always` block are executed as follows. When `block_b` is entered, the `count` is 1 and at the next positive edge of `clk`, the `count` is assigned to 0. At that time, the condition is true, and the `finish` will be assigned to 1 and the `count` to −1 at the next positive edge of `clk`. So, the above result is obtained. Figure 4.3 shows the detailed operations of `block_b`.

Coding Styles

1. It should not mix blocking and nonblocking assignments in the same `always` block.

2. In the `always` block, it is good practice to use nonblocking operators (<=) when it is a piece of sequential logic. Otherwise, the result of the RTL behavioral may be inconsistent with that of the gate level.

3. In the `always` block, it is a good practice to use blocking operators (=) when it is a piece of combinational logic.

Review Questions

Q4.6   What are the differences between procedural assignments and continuous assignments?

Q4.7   What are the differences between blocking assignments and nonblocking assignments?

Q4.8   Which of the procedural assignments is more appropriate when modeling a pipelined data transfer?

Q4.9   Describe the three steps that simulators will perform when they encounter a nonblocking statement.

Q4.10   Write a correct program segment to swap the contents of two `reg` variables.

4.3 TIMING CONTROL

Timing controls provide a way to specify the simulation time at which procedural statements will execute. In Verilog HDL, if there are no timing control statements, the simulation time will not advance. Hence, it is necessary to include some kinds of timing control statements within the Verilog HDL modules. There are two timing control methods provided in Verilog HDL:

1. delay timing control;

2. event timing control.

In the rest of this section, we discuss these two methods in greater detail.

4.3.1 Delay Timing Control

Delay timing control in an expression specifies the time duration between when the statement is encountered and when the statement is executed. It is specified by # and has either of the following two forms:

```
#delay_value
#min:typ:max_expr
```

Delay timing control can be divided into two types according to the position of the delay specifier in a statement: *regular delay control* and *intra-assignment delay control*.

Regular Delay Control   The regular delay control defers the execution of the entire statement by a specified number of time units. It is specified by putting a non-zero delay to the left of a procedural statement. The regular delay control has the following form:

```
#delay procedural_statement
```

where `#delay` can be a constant or an expression. If no procedural statement is specified, it only causes a wait for the specified delay before the next statement is executed. Thus:

```
#25;       // wait 25 time units
x = a + 6; // execute immediately
```

is equivalent to

```
#25 x = a + 6; // wait 25 time units and execute
```

The "<=" operators in the following statements can be replaced with the "=" operator without affecting the results because the regular delay control defers the execution of the entire statement so that there is no difference whichever "<=" or "=" is used.

```
reg  x, y, z;
integer d, e;
integer count;
   // The "<=" operators in the following statements
   // can be
   // replaced with "=" without affecting the results.
   #25  y <= ~x;                // execute at time 25
   #15  count <= count + 1;   // execute at time 40
   #((d+e)/2) x =  y;          // execute at time 40 +
                               // (d + e)/2
   #z  z = z + 1;              // execute at time 40 +
                               // (d + e)/2 + z
```

Intra-assignment Delay Control   Intra-assignment delay control defers the assignment to the left-hand-side variable by a specified number of time units but the right-hand-side expression is evaluated at the current simulation time. Intra-assignment delay control has the following forms:

```
variable_lvalue = #delay expression
variable_lvalue <= #delay expression
```

where #delay can be a constant or an expression. Depending on which kind of assignment, blocking or nonblocking, is used, the effect is different. The following example illustrates the results when using blocking assignments.

```
// both statements are evaluated at time 0
    y = #25 ~x;         // assign to y at time 25
count = #15 count + 1; // assign to count at time 40
```

The results will be different when using nonblocking assignments, such as in the following example.

```
// both statements are evaluated at time 0
    y <= #25 ~x;          // assign to y at time 25
count <= #15 count + 1; // assign to count at time 15
```

The three-step rules may be used to explain the above results successfully.

Note that if the delay expression evaluates to an unknown, x, or a high-impedance value, z, it is the same as zero delay. If the delay expression evaluates to a negative value, the two's complement unsigned integer is used as the delay.

4.3.2 Event Timing Control

A procedural statement can be executed in synchronism with an event. An event is the value change on a net or variable or the occurrence of a declared event, which is called a *named event* and will be addressed later in this section. In other words, an event control with a procedural statement defers the execution of the statement until the occurrence of the specified event. There are two kinds of event control:

1. *edge-triggered* event control;
2. *level-sensitive* event control.

Edge-triggered Event Control   In the edge-triggered event control, a procedural statement defers its execution until a specified transition of a given signal occurs. The symbol @ is used to specify an edge-triggered event control and has the following form:

```
@event procedural_statement;
```

where `procedural_statement` is executed once the specified `event` occurs. For example:

```
@(enable) a = b; // controlled by enable
```

where the procedural assignment `a = b` is controlled by the event `enable`. It is executed once the value of `enable` changes, such as from 1 to 0 or from 0 to 1.

In some applications, we need to specify the values to be changed (i.e. the event) at positive edge or negative edge. The keyword `posedge` is used for a positive (rising) transition and the keyword `negedge` for a negative (falling) transition. The behavior of `posedge` and `negedge` is described as follows:

1. A `negedge` means the transition from 1 to x, z or 0, and from x or z to 0.
2. A `posedge` means the transition from 0 to x, z or 1, and from x or z to 1.

For example, the following `always` block:

```
always @(posedge clock) begin // intra-assignment delay
                              // control
   reg1 <= #25 in_1;
   reg2 <= @(negedge clock) in_2 ^ in_3;
end
```

means the value of `in_1` is assigned to `reg1` after 25 time units whenever the positive edge of the `clock` signal occurs; the evaluated result of `in_2^in_3` is assigned to `reg2` at the oncoming negative edge of `clock` after the positive edge of the `clock` signal occurs.

Named Event Control   In Verilog HDL, a new data type called *event* can be declared in addition to nets and variables. This provides users a capability to declare an event and then to trigger and recognize it.

An identifier declared with an event is called a *named event*. The event does not hold any data and has no time duration. A named event is triggered explicitly by using the symbol -> and the triggering of the event is recognized by using the symbol @.

A named event has the same usage as event control. A named event should be declared explicitly before it is used. To declare a named event, the following syntax can be used:

```
event identifier1, identifier2, ..., identifiern;
```

where `event` is a keyword and `identifiier1` to `identifiiern` are identifiers to be used as named events. The event trigger uses the following form:

```
->name_event;
```

which means the `name_event` is triggered. The following scenario demonstrates the three steps of using a named event: declaration, triggering and recognition.

```
// step 1: declare an event received_data
event  received_data;
// step 2: trigger the event received_data
always @(posedge clock)
   if (last_byte)-> received_data;

// step 3: recognize the event
always @(received_data) begin
   // put the required operations here
end
```

The following example demonstrates the use of a named event.

EXAMPLE 4.11   *A Simple Example Illustrating the Use of a Named Event*

In this example, a counter `count` counts up its value by 1 every 2 time units. It triggers an event `ready` whenever its value is a multiple of 5. Once the event `ready` is triggered, we display the `count` value. Note that what would happen if the `$display` system task is replaced by the `$monitor` system task?

```
'timescale 1 ns/100 ps
// the use of named event
module named_event;
reg [7:0] count;
event ready;
initial begin
  count = 0;
  #400 $finish;
end
// trigger the event on condition
always begin
    #2 count <= count + 1;
    if (count % 5 == 0) ->ready;
end
// receive the event and do something
```

```
always @(ready) // wait for ready event
   $display($realtime,,"The count is %d", count);
endmodule
```

The combined use of named event and event control provides a powerful and efficient means to synchronize two or more concurrently processes, such as handshaking control, within the same module, or between two modules with different clock domains.

The following example simply demonstrates the combined use of named events and event control within the same module. The case that crosses different modules is left as an exercise for the reader.

EXAMPLE 4.12 *A Simple Handshaking Example*

In this example, a sender repeatedly sends data generated randomly to the receiver. After data are produced, it triggers event `ready` to tell the receiver that the data are ready. The receiver having once received the data, triggers event `ack` to notify the sender that it has accepted the data and is ready for the next data. To initiate the handshaking operations, we trigger the event `ack` at the startup during simulation.

```
`timescale 1 ns/100 ps
// an example illustrates the use of named event
module handshaking_event;
reg [7:0] data, data_output;
event ready, ack;
integer seed = 1;
initial begin
   #5 ->ack;  // initiate the operation
   #50 $finish;
end
// source device part starts from here
always @(ack) begin: sender
     // once event ack is triggered
     data =  $random(seed) % 13;
     $display ($realtime,, "The source data is : %d",
     // data);
     seed = seed + 7;
   #5 ->ready;
end
// destination device part starts from here
always @(ready) begin: receiver
     // once event ready is triggered
     data_output = data;
     $display ($realtime,, "The output data is : %d",
     // data_output);
   #5 ->ack;
end
endmodule
```

What would happen if all blocking operators within the `always` blocks are replaced with nonblocking ones? (Answer: the data actually sent by the sender will start from the second event.)

Event or Control   In many applications, any one of multiple signals or events can trigger the execution of a procedural statement or a block of procedural statements. The signals or events are represented as logical or are also called a *sensitivity list* or *event list*. The keyword `or` or a comma (`,`) is used to specify multiple triggers in a sensitivity list. For example, the following `always` statement:

```
always @(a or b or c_in) // use the keyword or
   {c_out, sum} = a + b + c_in;
```

is equivalent to:

```
always @(a, b, c_in)      // use comma (,)
   {c_out, sum} = a + b + c_in;
```

Another example of an event or control which describes a *D*-type flip-flop with an asynchronous reset is as follows:

```
always @(posedge clock or negative reset_n) // event or
                                            // control
begin
   if (!reset_n) q <= 1'b0; // asynchronous reset.
   else          q <= d;
end
```

The output of the *D*-type flip-flop is 0 when the `reset_n` is set to 0; otherwise, the output is a sample of its input.

Implicit Event List   When the number of input variables for a combinational logic is very large, users often forget to add some of the nets or variables in the sensitivity list. The wildcard `@*` or `@(*)` is a shorthand that eliminates this problem by adding all nets and variables which are read by any statement within the associated block.

The following example uses the wildcard character "*" to mean a change on any signal (a, b or c_in) in the right-hand side of the procedural assignment within the `always` block. The use of `@*` and `@(*)` means the same thing.

```
always @(*)      // use *
   {c_out, sum} = a + b + c_in;
```

Another example is shown below. Here, the wildcard `@*` means `@(a, b, c, d, x, z)`.

```
always @* begin // equivalent to @(a or b or c or d or
                // x or z),
```

```
        x = a & b;     // @(a, b, c, d, x, z), or @(*)
        z = c | d;
        y = x ^ z;
    end
```

Level-sensitive Event Control In the level-sensitive event control, the execution of a procedural statement is delayed until a condition is true. The level-sensitive event control uses the keyword `wait` and has the following form:

```
    wait (condition) procedural_statement
```

The `procedural_statement` is executed when `condition` is true; otherwise, it is deferred until `condition` becomes true. The `procedural_statement` can also be a null statement. Note that if `condition` is already true when it is encountered, then the `procedural_statement` is executed without any additional delay.

Consider the following `always` block:

```
    always
        wait (count_enable) count = count - 1;
```

If `count_enable` is 0, the `wait` statement delays the execution of statement `count = count - 1` until `count_enable` becomes 1. Of course, the `procedural_statement` controlled by a `wait` statement can also contain a delay specification, such as:

```
    always
        wait (!count_enable) #10 count = count - 1;
```

If `count_enable` is 1, the `wait` statement delays the execution of statement `count = count - 1` until `count_enable` becomes 0, then defers 10 time units and executes the statement `count = count - 1`. If `count_enable` is already 0 when the `always` block is entered, then the assignment `count = count - 1` is executed after a delay of 10 time units.

The level-sensitive event control can be used to model the handshaking operations between two asynchronous devices abstractly. Consider Figure 4.4(a), where the source device wants to send data to the destination device asynchronously. The source device first places its data on the data bus and then activates the ready signal to notify the destination device of this situation. Once it has received the `ready` signal, the destination device activates the signal `ack` to indicate that it has acknowledged the case. Then, the source device deactivates its `ready` signal and the destiantion device uses this event to latch the data on the data bus into its internal register. Finally, the destination device deactivates the `ack` signal to signal the completion of one transaction. The detailed timing is shown in Figure 4.4(b).

The following example describes how to model the handshaking operations between two devices in accordance with the timing diagram shown in Figure 4.4(b) by using `wait` statements.

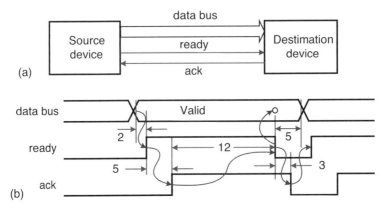

FIGURE 4.4 An example illustrates the handshaking operations between two asynchronous devices: (a) block diagram; (b) timing diagram

EXAMPLE 4.13 *A Handshaking Module Illustrates the Use of the* `wait` *Statement*

Two `always` blocks corresponding to source and destination, respectively, are used in the module. It is straightforward to see that both source and destination blocks exactly model the handshaking operations between two devices in accordance with the timing diagram shown in Figure 4.4(b).

```
`timescale 1ns/100 ps
// an example illustrates the use of wait
module handshaking_wait;
reg [7:0] data, data_output;
reg ready = 0, ack = 0;
integer seed = 5;
initial begin
  #400 $finish;
end
// source device part starts from here
always begin: source
    #5 data <= $random(seed) % 13;
    $display ("The source data is : %d", data);
    #2 ready <= 1;
    wait (ack);     // wait until ack is active
    #12 ready <= 0;
    seed = seed + 3;
end
// destination device part starts from here
always begin: destination
    wait (ready); // wait until ready is active
    #5 ack <= 1;
```

```
@(negedge ready) data_output <= data;
    $display ("The output data is : %d", data_output);
    #3 ack <= 0;
end
endmodule
```

■

Review Questions

Q4.11   Explain why regular delay control has the same effects on blocking and non-blocking assignments.

Q4.12   Explain why intra-assignment delay control causes different results on blocking and nonblocking assignments.

Q4.13   Define sensitivity list and event list.

Q4.14   What are the three ways that can be used to represent an event or control?

Q4.15   What are the meanings of the keywords `posedge` and `negedge`?

Q4.16   What does named event mean?

Q4.17   Describe the three-step scenario when using a named event.

4.4 SELECTION STATEMENTS

Selection statements are used to make a selection according to a given condition. Two types of selection statements provided by Verilog HDL are `if-else` and `case` statements. The `if-else` statement can also be nested so as to provide both two-way and multiway selection. The `case` statement supports the multiway selection and has two variants: `casex` and `casez`. These `case` statements are usually used to model multiplexers or the like. In this section, we will deal with these statements in detail.

4.4.1 `if-else` Statement

The conditional statement (`if-else` statement) is used to make a decision according to a given condition. It has the following syntax:

```
if (<condition>) true_statement

if (<condition>) true_statement else false_statement

if (<condition1>) true_statement1
{else if (<condition2>) true_statement2}
[else false_statement]
```

where `condition` can be a constant or an expression and must always be in parentheses. If the `condition` is true (has a nonzero known value), the `true_statement` is executed. If it is false (has a 0, an x or a z), the `true_statement` is not executed.

If there is an `else` statement and `condition` is false, the `false_statement` associated with `else` is executed. Note that the `else` part is always associated to the closest previous `if` that lacks an `else` except when using a `begin-end` block to force the proper association.

Basically, an `if-else` statement is used to perform a two-way selection according to a given condition. However, it also allows the nested statement so that it can carry out a multiway selection as well.

The following module uses an `if-else` statement to construct a 4-to-1 multiplexer. The reader is encouraged to explain the operations of this module. In addition, it is instructive to synthesize it and compare the result with that obtained from using dataflow style.

EXAMPLE 4.14 *A 4-to-1 Multiplexer Using a Selection Statement*

A 4-to-1 multiplexer is constructed by using an `if-else` statement. Due to its straightforward simplicity, we do not explain its operations furthermore.

```
module mux4_to_1_ifelse(i0, i1, i2, i3, s1, s0, out);
// port declarations
input   i0, i1, i2, i3;
input   s1, s0;
output reg out;
// using conditional operator if-else statement
always @(*)  // triggered for all signals used in the
             // if-else statement
   if (s1) begin
      if (s0) out = i3; else out = i2; end
   else begin
      if (s0) out = i1; else out = i0; end
endmodule
```

The following example demonstrates a 4-bit synchronous binary counter constructed with an `if-else` statement.

EXAMPLE 4.15 *A 4-bit Synchronous Binary Counter with a Synchronous* `Clear`

In this example, the counter is cleared whenever a postive-edge `clear` signal is activated; otherwise, the counter counts up at every negative edge of the `clock` signal. The counter is a modulo-16 counter because `qout` is declared to be 4 bits. In addition, the clear operation is done asynchronously since the `clear` signal is put inside the sensitivity list.

```
module counter(clock, clear, qout);
input   clock, clear;
output reg [3:0] qout;
// the body of the 4-bit binary counter.
```

```
always @(negedge clock or posedge clear) begin
   if (clear)
      qout <= 4'd0;
   else
      qout <= (qout + 1); // qout = (qout + 1) % 16;
end
endmodule
```

■

How can we achieve this if the clear operation needs to be done synchronously? (see Section 9.1.1).

4.4.2 case **Statement**

A case statement is used to perform a multiway selection according to a given input condition and equivalent to a nested if-else statement. The case statement has the following general form:

```
case (case_expression)
   case_item1 {,case_item1}: procedural_statement1
   case_item2 {,case_item2}: procedural_statement2
      ...
   case_itemn {,case_itemn}: procedural_statementn
   [default: procedural_statement]
endcase
```

where the default statement is optional. Only one default statement can be placed inside one case statement. In addition, if there is no default statement and all comparisons fail, then none of the case_item associated statements are executed. Furthermore, multiple case_item expressions can be specified in one branch.

The case_expression is evaluated first. Then, the case_item expressions are evaluated and compared in the order given. The procedural_statement associated with the case_item that first matches the case_expression is executed. When none of the case_item expressions matches the case_expression, the default statement is executed if it exists.

The case statement acts like a multiplexer. Consequently, it is often used to construct a multiplexer, such as the following 4-to-1 multiplexer. You may compare this with the one that is modeled in dataflow style described in the previous chapter.

EXAMPLE 4.16 *A 4-to-1 Multiplexer Using the* case *Construct*

In this example, a case statement is used to construct a 4-to-1 multiplexer. You can see that it is quite straightforward when using a case statement to construct an *n*-to-1 multiplexer.

```
// a 4-to-1 multiplexer using case statement
module mux_4x1_case (i0, i1, i2, i3, s, y);
```

```
input i0, i1, i2, i3;
input [1:0] s;      // declare s as a two-bit selection
                    // signal.
output reg  y;
always @(i0 or i1 or i2 or i3 or s) // it can use
                                     // always @(*).
   case (s)
      2'b00: y = i0;
      2'b01: y = i1;
      2'b10: y = i2;
      2'b11: y = i3;
   endcase
endmodule
```
■

It is good practice to include a `default` statement as the last statement of a `case` statement to cover all other possible cases even when a `case` statement is already completely specified. The following module demonstrates this.

EXAMPLE 4.17 *A 4-to-1 Multiplexer Using the* `case` *and* `default` *Constructs*

This example is the same as the preceding one except that we add a `default` case item as the last item of the `case` statement. Therefore, when none of the case items matches the `case_expression`, the `default` case item is executed. Can you give some examples when this situation will occur? (*Hint*: remember that each net or variable data type can have four values.)

```
// a 4-to-1 multiplexer using case and default statements.
module mux4_to_1_case_default (i0, i1, i2, i3, s1, s0,
                                 out);
input   i0, i1, i2, i3, s1, s0;
output reg out;        // output declared as register
always @(s1 or s0 or i0 or i1 or i2 or i3)
   case ({s1, s0})    // as a two-bit selection signals
      2'b00: out = i0;
      2'b01: out = i1;
      2'b10: out = i2;
      2'b11: out = i3;
      // using default to include all other possible cases.
      default: out = 1'bx;
   endcase
endmodule
```
■

The following module demonstrates the situation that a `default` statement is required in the `case` statement to completely specify the statement; otherwise, a latch will be inferred when it is synthesized.

EXAMPLE 4.18 *A 3-to-1 Multiplexer Using the* `case` *Construct—an Incorrect Version*

An *n*-to-1 multiplexer, where $n \neq 2^k$, and *k* is a positive integer, is quite commonly used in practice. In this case, the common error is just to write the required case items and ignore the rest ones. The resulting circuit will be a latch-inferred one because for those case items not specified explicitly must leave the output unchanged. Fortunately, the powerful `default` statement may be used to compensate this flaw and removes the unwanted latch.

```
// a 3-to-1 multiplexer using case statement
// this is an incompletely specified version.
module mux_3x1_case2 (i0, i1, i2, s, y);
input i0, i1, i2;
input [1:0] s;     // declare s as a two-bit selection
                   // signal.
output reg  y;
always @(i0 or i1 or i2 or s) // it can use always @(*).
   case (s)
      2'b00: y = i0;
      2'b01: y = i1;
      2'b10: y = i2;
   endcase
endmodule
```

You are encouraged to synthesize this module and explain the result obtained from the synthesizer.

The following module demonstrates how the `default` statement is used to completely specify the `case` statement when the number of desired `case_item` expressions cannot be combined into the whole space spanned by `case_expression`. As a consequence, it avoids the unwanted latch inferred by the synthesis tools.

EXAMPLE 4.19 *A 3-to-1 Multiplexer Using the* `case` *Construct—A Correct Version*

We add the `default` statement as the last case item of the `case` statement so that it makes up for the flaw of the preceding example. The procedural assignment used to assign a value to the output y when the `default` statement is executed must take the actual requirement into consideration such as to leave the output an unknown x, a z or a definite known value 0 or 1.

```
// a 3-to-1 multiplexer using case statement
// using default to completely specify the case statement
module mux_3x1_case3 (i0, i1, i2, s, y);
input i0, i1, i2;
input [1:0] s;    // declare s as a two-bit selection
                  // signal.
```

```
output reg  y;
always @(i0 or i1 or i2 or s) // It can use always @(*).
    case (s)
        2'b00: y = i0;
        2'b01: y = i1;
        2'b10: y = i2;
        default: y = 1'b0;
    endcase
endmodule
```

The following module demonstrates the situation where the multiple case_item expressions are used in one branch.

EXAMPLE 4.20 *A 3-to-1 Multiplexer Using Multiple* `case_items` *in One* `case` *Branch*

In this example, three branches, but including all four combinations of case_expression, are specified; hence, no latch is inferred by the synthesizer even though we do not include the default statement.

```
// a 3-to-1 multiplexer using multiple case items
module mux_3x1_case (i0, i1, i2, s, y);
input i0, i1, i2;
input [1:0] s; // declare s as a two-bit selection signal.
output reg  y;
always @(i0 or i1 or i2 or s) // It can use always @(*).
    case (s)
        2'b00: y = i0;
        2'b01: y = i1;
        2'b10, 2'b11: y = i2;
    endcase
endmodule
```

A block of statements must be grouped by `begin` and `end` keywords. The `case` statement compares 0, 1, x and z values in the case_expression and the case_item bit for bit. The zero-extended approach is used to match the unequal bit widths between the case_expression and the case_item. The following example explains this.

EXAMPLE 4.21 *A 5-to-1 Multiplexer Using the* `case` *Construct*

In this example, a 5-to-1 multiplexer is to be constructed. There are five inputs named i0 to i4. The input i4 is routed to the output y whenever the LSB of selection s is 1. The other four inputs are routed to the output y according to the next two higher bits of selection s. If we code this idea as the module shown below, then according to the zero-extended approach used in Verilog HDL, the 5th case_item that includes only 1 bit is extended to as 001. The result is that input i4 is selected when selection input s is 3'b001 rather than what we have expected. You may check this from Figure 4.5.

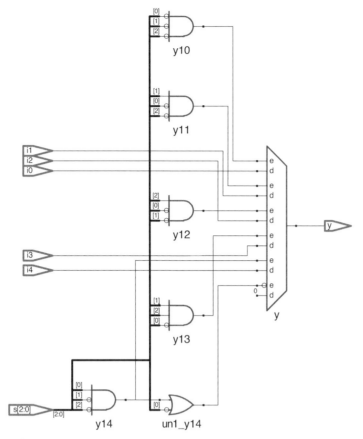

FIGURE 4.5 The synthesized result of the module mux_5x1_case

```
// a 5-to-1 multiplexer using case statement
module mux_5x1_case(i0, i1, i2, i3, i4, s, y);
input i0, i1, i2, i3, i4;
input [2:0] s;     // declare s as a three-bit selection
                   // signal.
output reg  y;
always @(i0 or i1 or i2 or i3 or i4 or s) // it can use
                                          // always @(*).
   case (s)
      3'b000: y = i0;
      3'b010: y = i1;
      3'b100: y = i2;
      3'b110: y = i3;
      1'b1  : y = i4;
      default: y = 1'b0; //include all other possible
      cases.
   endcase
endmodule
```

Note that in the above module, the input i4 is only selected when selection input s is 3'b001. If what we really want is that whenever the LSB is 1, the input i4 is routed to the output y, the most straightforward way is to use multiple case_item expressions as we have done in the module mux_3x1_case. Due to its intuitive simplicity, this is left as an exercise for the reader.

4.4.3 casex and casez Statements

Both casex and casez statements are used to perform a multiway selection like that of the case statement except that the casez statement treats all z values as 'don't cares' while the casex treats all x and z values as 'don't cares'. These two statements only compare non-x or z positions in the case_expression and the case_item expressions.

The following module illustrates how the features of casex statement are used to model a circuit that counts trailing zeros in a nibble (i.e. a half byte).

EXAMPLE 4.22 *Counting the Trailing Zeros in a Nibble*

This example describes how the features of the casex statement are used to count the trailing zeros in a nibble. Because the basic feature of the casex statement is that it only compares non-x or z positions in the case_expression and the case_item expressions, so by arranging each case_item and its associated expression in the order as shown in the module, the job is done.

```
// an example illustrates how to count the trailing zeros
// in a nibble.
module trailing_zero_4b(data, out);
input  [3:0] data;
output reg [2:0] out;  // output declared as register
always @(data)
   casex (data)      // treat both x and z as don't care
                     // conditions.
      4'bxxx1: out = 0;
      4'bxx10: out = 1;
      4'bx100: out = 2;
      4'b1000: out = 3;
      4'b0000: out = 4;
      default: out = 3'b111; //include all other possible
                             //cases.
   endcase
endmodule
```

In Chapter 8, we give more examples to address the useful features of the casex statement to model priority encoders and many others.

Review Questions

Q4.18 Explain the basic syntax of the `if-else` statement.

Q4.19 Can a block of statements be used as a case-item expression?

Q4.20 Explain the role of the `default` case-item in a `case` statement.

Q4.21 What are the differences between `case` and `casex` statements?

Q4.22 What are the differences between `casex` and `casez` statements?

4.5 ITERATIVE (LOOP) STATEMENTS

Like any general-purpose high-level programming languages, Verilog HDL also provides a set of powerful and efficient loop statements. These include four types of iterative statements: `while`, `for`, `repeat` and `forever`. These loop statements provide a means of controlling the execution of a statement or a block of statements zero, one or more times. The `while` loop executes a statement until a given conditional expression becomes false. The `for` loop repeatedly executes a statement a fixed number of times or until a given condition becomes false. The `repeat` loop executes a statement a fixed number of times. The `forever` loop continuously executes a statement infinitely. The general features of loop constructs include that they may appear only inside an `initial` or `always` block and may contain delay expressions.

4.5.1 `while` Loop Statement

The `while` loop is used to perform a procedural statement. It executes the procedural statement until the given condition becomes false. The `while` loop statement uses the keyword `while` and has the form:

```
while (condition_expr) procedural_statement
```

If the `condition_expr` starts out false, the `procedural_statement` would not be executed at all.

For example, a simple condition is used in the following `while` loop:

```
while (count < 12) count <= count + 1;
```

The procedural assignment `count <= count + 1` is carried out continuously until the condition `count < 12` is no longer valid.

Another example with a more complicated condition is shown in the following `while` loop:

```
while (count <= 50 && flag) // complex condition
begin
    // put statements wanted to be carried out here.
end
```

Here, a complex condition with logical and (`&&`) of `count <= 50` and `flag == 1` is used to control the loop. The statements inside the block enclosed with the keywords `begin` and `end` are executed only if both conditions are true; otherwise, they are not executed at all.

The following module computes the zeros in a byte by using a `while` loop statement.

EXAMPLE 4.23   *An Example of Counting the Zeros in a Byte*

In this example, we want to count the zeros in a byte. A `while` loop is used to control the number of iterations and an `if` selection statement is used to check and sum up the number of nonzero bits. The loop is terminated when the maximum number of iterations is reached. The bit value is checked in sequence from bit 0 to bit 7, which is controlled by an `integer` variable i. Note that the `if` statement may be replaced with `out = out + ~data[i]`. Why? Give the reasons for this.

```
// an example illustrates how to count the zeros in a
// byte.
module zero_count_while (data, out);
input   [7:0] data;
output reg [3:0] out;      // output declared as register
integer i;
always @(data) begin
   out = 0; i = 0;
   while (i <= 7) begin    // simple condition
      // the following statement may be replaced with
      // out = out + ~data[i].
      if (data[i] == 0) out = out + 1;
      i = i + 1; end
end
endmodule
```

Note that if we replace the blocking assignments by nonblocking assignments, the resulting module would not work properly. Why? Try this on your system and explain it.

The following example shows how to count the trailing zeros in a byte by using a `while` loop statement.

EXAMPLE 4.24   *An Example of Counting the Trailing Zeros in a Byte*

In this example, we want to count the trailing zeros in a byte. A `while` loop is used to schedule the related operations: checking the bit value in sequence from bit 0 to bit 7 and counting the number of zero bits. The loop is terminated whenever a nonzero bit is found or the maximum number of iterations is reached. Please compare this example with that using the `casex` statement described in the previous section.

```
// an example illustrates how to count the trailing zeros
// in a byte.
module trailing_zero_while (data, out);
input  [7:0] data;
output reg [3:0] out;  // output declared as register
integer i;             // loop counter
always @(data) begin
   out = 0; i = 0;
   while (data[i] == 0 && i <= 7) begin // complex
                                        // condition
      out = out + 1;
        i = i + 1;
   end
end
endmodule
```

Note that if we replace the blocking assignments by nonblocking assignments, the resulting module would not work properly. Why? Try this on your system and explain it. Note the differences between this example and the previous one.

4.5.2 `for` Loop Statement

The `for` loop is usually used to perform a counting loop although it is much powerful than this. A counting loop is a loop repeated with a fixed number of times, or a known number of times. The behavior of the `for` loop is much like the *for* statement in C programming language. A `for` loop statement contains three parts: an initial condition, termination condition checking and the control variable updating. The `for` loop uses the keyword `for` and has the following syntax:

```
for (init_expr; condition_expr; update_expr)
    procedural_statement
```

where `init_expr` sets the initial condition, `condition_expr` checks the termination condition and `update_expr` updates the control variable.

The `for` loop is equivalent to the `while` loop with the following construct:

```
init_expr
while (condition_expr) begin
   procedural_statement
   update_expr
end
```

What follows are two very simple application examples of the `for` loop statement. Due to its intuitive simplicity, we do not need to explain their operations any further.

```
for (i = 0; i < 32; i = i + 1)
   state[i] = 0; // initialize to zeros
for (i = 1; i < 32; i = i + 2) begin
   state[i] = 0;
   count = count +1;
end
```

The following example is the `for` loop version of the module that computes the zeros in a byte using the `while` loop described previously. It is instructive to compare both modules and explore the insight into both loop statements.

EXAMPLE 4.25 *An Example of Counting the Zeros in a Byte*

It is quite straightforward to convert the `while` loop of the module in the previous subsection into an equivalent `for` loop. The resulting module is shown in the following. Due to the intuitive simplicity of the module, we do not further explain its operations here.

```
// an example illustrates how to count the zeros in
// a byte.
module zero_count_for (data, out);
input   [7:0] data;
output reg [3:0] out; // output declared as register
integer i;
always @(data) begin
   out = 0;
   for (i = 0; i <= 7; i = i + 1) // simple condition
       // the following statement may be replaced with
       // out = out + ~data[i].
       if (data[i] == 0) out = out + 1;

end
endmodule
```

Note that if we replace the blocking assignments by nonblocking assignments, the resulting module would not work properly. Why? Try it and explain why this is the case.

The following example is the `for` loop version of the module that computes the trailing zeros in a byte using the `while` loop described previously. It is instructive to compare both modules and explore the insight into both loop statements.

EXAMPLE 4.26 *An Example of Counting the Trailing Zeros in a Byte*

It is quite straightforward to convert the `while` loop of the module in the previous subsection into an equivalent `for` loop. The resulting module is shown in the following. Due to the intuitive simplicity of the module, we do not further explain its operations here.

```
// an example illustrates how to count the trailing zeros
// in a byte.
module trailing_zero_for (data, out);
input  [7:0] data;
output reg [3:0] out;  // output declared as register
integer i;             // loop counter
always @(data) begin
   out = 0;
   for (i = 0; data[i] == 0 && i <= 7; i = i + 1) // com-
                                        // plex condition
       out = out + 1;
end
endmodule
```
■

Note the differences between this example and the previous one.

4.5.3 `repeat` Loop Statement

The `repeat` loop is used to perform a procedural statement a specified number of times. The `repeat` loop uses the keyword `repeat` and has the following syntax:

```
repeat (counter_expr) procedural_statement
```

where `counter_expr` can be a constant, a variable or a signal value. The `counter_expr` is evaluated only once before starting the execution of the statement. In addition, if the `counter_expr` is an x or a z, then it is treated as a 0.

In the following program segment, the first `repeat` loop repeatedly executes the `begin` and `end` block 32 times and the second `repeat` loop repeatedly executes the `begin` and `end` block `cycles` times, where `cycles` must be evaluated to a number before entering the loop.

```
i = 0;
repeat (32) begin
   state[i] = 0;       // initialize to zeros
   i = i + 1;          // next item
end
repeat (cycles) begin // cycles must be evaluated to a
                      // number before entering
   @(posedge clock) buffer[i] <= data; // the loop.
    i <= i + 1;       // next item
end
```

The `repeat` statement shown below means to wait for the positive edge on the clock and then increases the `count` for the `loop_count` times.

```
repeat (loop_count)
   @(posedge clock) count <= count + 1;
```

Repeat Event Control   In addition to being used as a loop control, the `repeat` statement can also be used as an intra-assignment event control, called *repeat event control*. The `repeat` event control has the following syntax:

```
repeat (expression) @ (event_expression)
```

This form of control is usually used to specify a delay that is based on the number of occurrences of one or more events. The following `repeat` statement means to compute `count + 1` first, then wait for the positive edge on the clock for `loop_count` times, and finally assigns to `count`.

```
count = repeat (loop_count) @(posedge clock) count + 1;
```

The equivalent form of this `repeat` event control is as follows:

```
temp = count + 1;
@(posedge clock); // repeat loop_count times
@(posedge clock);
   ...
@(posedge clock);
count = temp;
```

4.5.4 `forever` Loop Statement

The `forever` loop continuously performs a procedural statement until the `$finish` system task is encountered. The `forever` loop uses the keyword `forever` and has the following syntax:

```
forever procedural_statement
```

It is equivalent to a `while` loop with an always true expression such as `while (1)`. A `forever` loop can be exited by the use of a `disable` statement. In addition, some form of timing control must be used inside the `procedural_statement` to prevent the `forever` loop from looping in zero delay infinitely.

A simple `forever` loop example is used to generate a clock signal with a period of 10 time units and a duty cycle of 50 %:

```
initial begin
   clock <= 0;
   forever begin
      #5 clock <= ~clock;
   end
end
```

The `forever` loop is usually used with timing control statements, such as the following program segment:

```
reg clock, x, y;

initial
    forever @(posedge clock) x <= y;
```

which means that at every positive edge of the clock the value of the variable `y` is assigned to the `x` variable.

Review Questions

Q4.23 What are the four types of loop statements provided in Verilog HDL?

Q4.24 Describe the operations of the `while` loop statement.

Q4.25 Describe the operations of the `for` loop statement.

Q4.26 Describe how to replace the operations of a `for` loop with a `while` loop.

Q4.27 Describe the operations of the `repeat` loop statement.

Q4.28 What is repeat event control?

Q4.29 Describe the two basic ways that are usually used to exit a `forever` loop.

SUMMARY

In this chapter, we have discussed procedural constructs: `initial` and `always` statements. An `initial` or an `always` statement is usually called an `initial` or an `always` block: `initial` statements are used to initialize variables and set values into variables or nets, while `always` statements are used to model the continuous operations required in the hardware modules. All other behavioral statements must be within an `initial` or an `always` block.

In a module, an arbitrary number of `initial` and `always` statements may be used. All `initial` and `always` statements execute concurrently with respect to each other; in other words, their relative order in a module is not important. Each of the `initial` and `always` statements represents a separate activity flow. Each activity starts at simulation time 0. Notice that each `always` statement corresponds to a piece of logic circuit. In addition, `initial` and `always` blocks cannot be nested. Each `initial` or `always` statement must form its own block.

Procedural assignments occur within procedures such as `initial`, `always`, `task` and `function`. Procedural assignments put values in variables under the control of the procedural flow constructs that surround them. A variable holds the value assigned by a procedural assignment until the next procedural assignment to that variable again. In contrast, continuous assignments drive nets and are evaluated and updated whenever any input operand changes value. In other words, continuous assignments assign values onto nets like the way that logic gates drive nets. As a consequence, the left-hand side of continuous assignments are nets but procedural assignments are variables.

There are two procedural assignments: blocking assignment and nonblocking assignment. Blocking assignment statements use the operator "=" and are executed in the order they are specified. In other words, a blocking assignment is executed before the execution of the statements that follow it in a sequential block. Nonblocking assignments use the <= operator and are executed without blocking the other statements in a sequential block. Nonblocking assignments provide a method to model several concurrent data transfers that take place after a common event.

Timing controls provide a way to specify the simulation time at which procedural statements will execute. In Verilog HDL, if there are no timing control statements, the simulation time will not advance. Hence, it is necessary to include some kinds of timing control statements within Verilog HDL modules. Delay timing control and event timing control are the two timing control methods provided in Verilog HDL.

Delay timing control in an expression specifies the time duration between when the statement is encountered and when the statement is executed. It is specified by the symbol #. According to the position of the delay specifier, delay timing control can be classified into two types: regular delay control and intra-assignment delay control. The regular delay control defers the execution of the entire statement by a number of specified time units. The intra-assignment delay control defers the assignment to the left-hand-side variable by a specified number of time units but the right-hand-side expression is evaluated at the current simulation time.

An event control with a procedural statement defers the execution of the statement until the occurrence of the specified event. There are two kinds of event control: edge-triggered event control and level-sensitive event control. In the edge-triggered event control, a procedural statement defers its execution until a specified transition of a given signal occurs. It uses the symbol @. In the level-sensitive event control, the execution of a procedural statement is delayed until a condition is true. The level-sensitive event control uses the keyword `wait`.

Like any other programming languages, such as C, Verilog HDL also provides selection statements and iterative (loop) statements. Selection statements are used to make a selection according to a given condition. They contain two constructs: `if-else` and `case`, including `case`, `casex` and `casez`. These statements are often used to model multiplexers or the like.

Iterative (loop) statements include the following four types: `repeat`, `for`, `while` and `forever`. These loop statements provide a means of controlling the execution of a statement or a block of statements zero, one, or more times. The `while` loop executes a statement until a given conditional expression becomes false. The `for` loop repeatedly executes a statement a fixed number of times or until a given condition becomes false. The `repeat` loop executes a statement a fixed number of times. The `forever` loop continuously executes a statement infinitely.

REFERENCES

1. J. Bhasker, *A Verilog HDL Primer,* 3rd Edn, Star Galaxy Publishing, 2005.
2. S. Palnitkar, *Verilog HDL: A Guide to Digital Design and Synthesis,* 2nd Edn, SunSoft Press, 2003.
3. IEEE 1364-2001 Standard, *IEEE Standard Verilog Hardware Description Language,* 2001.

PROBLEMS

4.1 Consider the six basic logic gates: AND, NAND, OR, NOR, XOR, and XNOR.

(a) Model these basic logic gates in behavioral style.

(b) Write a Verilog HDL module for each of the gates in behavioral style.

4.2 The following is an example of mixed use of blocking and nonblocking assignments:

```
// an example of mixed use of blocking and
// nonblocking assignments
module blocking_nonblocking_mixed;
reg  a, b, c, d, e;
// blocking assignments
initial begin
    a  = #3 1'b0;
    b <= #4 1'b1;
    c <= #6 1'b0;
    d  = #7 1'b1;
    e  = #8 1'b0;
end
endmodule
```

(a) At what simulation time is each statement executed?

(b) Interchange the blocking and nonblocking operators and re-do part (a).

(c) Replace all nonblocking assignment with blocking assignments and re-do part (a).

(d) Replace all blocking assignment with nonblocking assignments and re-do part (a).

4.3 Without using intra-assignment delay control, rewrite the following two intra-assignment delay control constructs:

```
a = #25 b;
c = #12 d;
```

4.4 Without using intra-assignment delay control, rewrite the following two intra-assignment delay control constructs:

```
a = @(positive clk) b;
c = @(negative clk) d;
```

4.5 Without using intra-assignment delay control, rewrite the following two intra-assignment delay control constructs:

```
a = repeat(3) @(positive clk) b;
c = repeat(4) @(negative clk) d;
```

4.6 Consider the following simple Verilog HDL module:

```
module xyz(din, dout);
input[1:0] din;
output reg [3:0] dout;
always @(din)
    dout = {{3{1'b0}}, 1'b1} << din;
endmodule
```

(a) What does the above module function?

(b) Rewrite the above module with a `case` statement.

4.7 A majority circuit is a device that outputs 1 whenever more than half of its inputs are 1.

(a) Describe the behavior of a majority circuit.

(b) Model the majority circuit in behavioral style. Assume that there are 8 inputs.

(c) If the inputs are checked only when the `ready` signal is 1, re-do part(b).

4.8 A minority circuit is a device that outputs 1 whenever less than half of its inputs are 1.

(a) Describe the behavior of a minority circuit.

(b) Model the minority circuit in behavioral style. Assume that there are 8 inputs.

(c) If the inputs are checked only when the `ready` signal is 1, re-do part(b).

4.9 Assume that a squared-root circuit has an 8-bit input, x, and a 4-bit output, y. The output value is the squared root of its input value, i.e. $y = \sqrt{x}$. Describe the behavior of this circuit and model it in behavioral style on condition that your module must be synthesizable.

4.10 Assume that a square circuit has a 4-bit input, x, and an 8-bit output, y. The output value is the square of its input value, i.e. $y = x^2$. Describe the behavior of this circuit and model it in behavioral style on condition that your module must be synthesizable.

4.11 Assume that a binary-to-BCD (binary coded decimal) code converter is to be designed. Describe the behavior of this circuit and model it in behavioral style. Write a test bench to verify it.

4.12 Assume that a BCD-to-binary code converter is to be designed. Describe the behavior of this circuit and model it in behavioral style. Write a test bench to verify it.

4.13 Assume that an excess-3 code checker is to be designed. The output of the circuit is 1 if its input is a valid excess-3 code. Describe the behavior of this circuit and model it in behavioral style. Write a test bench to verify it.

TABLE 4.1 The relationship between the BCD code and the 2-out-of-5 code

BCD	2-out-of-5 code	BCD	2-out-of-5 code
0	11000	5	01010
1	00011	6	01100
2	00101	7	10001
3	00110	8	10010
4	01001	9	10100

4.14 The relationship between BCD code and 2-out-of-5 code is shown in Table 4.1. Describe a BCD code to 2-out-of-5 code converter and model it in behavioral style. Write a test bench to verify it.

4.15 Describe a 2-out-of-5 code to BCD code converter and model it in behavioral style. Write a test bench to verify it.

4.16 Assume that a clock generator is required. The clock output `clk` is initialized to 0 and toggles its value every 25 time units. Describe this clock generator without using the `always` statement.

4.17 Assume that a clock generator with burst output is required. When the `enable` control signal is raised to high, the clock generator produces 50 pulses with a period of 20 time units and a duty cycle of 50 %. The output pulses should start from 0. Model this clock generator.

4.18 Assume that a clock generator is required. The clock output `clk` is initialized to 0 and toggles its value every 10 time units. Describe this clock generator using a `while` loop.

4.19 Assume that a clock generator is required. The clock output `clk` is initialized to 0 and has a period of 20 time units and a duty cycle of 25 %. Describe this clock generator using a `for` loop.

4.20 Assume that a clock generator is required. The clock output `clk` is initialized to 0 and has a period of 100 time units and a duty cycle of 40 %. Describe this clock generator using a `forever` loop.

4.21 Assume that a reset signal is required. The reset signal has to last for 3 clock cycles and starts from the negative edge of clock signal. Design and model the reset signal generator.

4.22 Assume that we want to design a module that counts the number of ones in a byte. Describe the module in behavioral style and write a test bench to verify its functionality.

4.23 Assume that we want to design a module that counts the number of trailing ones in a byte. Describe the module in behavioral style and write a test bench to verify its functionality.

4.24 Rewrite the handshaking examples in Section 4.3.2 using two modules, namely, place each `always` block into a separate module.

4.25 In general, the data transfer operations between two asynchronous devices using the handshaking control can be initiated by the source or the destination device. Figure 4.4 shows the source-initiated handshaking control. Give a destination-initiated handshaking control and draw its timing diagram.

## CHAPTER 5

TASKS, FUNCTIONS AND UDPs

THERE ARE three behavioral ways that are widely used to model designs. These include tasks, functions and user-defined primitives (UDPs). Tasks and functions provide the ability to re-use the same piece of code from many places in a design. In addition, they provide a means of breaking up a large design into smaller ones so as to make it easier to read and debug the source descriptions.

In addition to user-defined tasks and functions, there is a rich set of predefined tasks and functions called system tasks and system functions, whose identifiers are prefixed with a character "$". This kind of tasks and functions, although cannot be accepted by synthesis tools, are useful when modeling abstractly a design in behavioral style or writing test benches for designs.

UDPs provide a means to model a design with a truth table. The truth table is one of the three most widely used approaches to designing a digital logic circuit. These three approaches are schematic representation, Boolean expression and a truth table, and correspond to the structural style, dataflow styles and UDPs in Verilog HDL, respectively.

5.1 TASKS

A procedure in high-level programming languages provides the ability to re-use a common piece of code from several different places in a design. This common piece of code is embodied as a task when using the task definition in Verilog HDL. In general, a task is used when the procedure has to contain timing controls, has at least one output arguments or has no input arguments. The features of the tasks are listed in Table 5.1.

5.1.1 Task Definition and Call

A task can contain timing controls, and it can call other tasks and functions as well. Tasks are declared with the keywords `task` and `endtask` and can be declared by one

Digital System Designs and Practices   Ming-Bo Lin
© 2008 John Wiley & Sons (Asia) Pte Ltd

TABLE 5.1 Feature comparisons of tasks and functions

Item	Tasks	Functions
Arguments	May have zero or more *input*, *output* and *inout*	At least one *input*, and cannot have *output* and *inout*
Return value	May have multiple values via *output* and *inout*	Only a single value via function name
Timing control statements	Yes	No
Execution	In non-zero simulation time	In 0 simulation time
Invoke functions/tasks	Functions and tasks	Functions only

of the following two forms:

```
// port list style
task [automatic] task_identifier;
   [declarations] // include arguments
   procedural_statement
endtask

// port list declaration style
task [automatic] task_identifier ([argument_
      declarations]);
   [other_declarations]  // exclude arguments
   procedural_statement
endtask
```

where the keyword `automatic` declares an automatic task with all declarations allocated dynamically for each concurrent task entry. The argument declarations specify the following: `input`, `output` or `inout` arguments. A task definition must be written within a module definition and it can have zero, one or more arguments. Values are passed to and from a task through arguments.

5.1.1.1 *Task Call* A task is called (or enabled) by a task enabling statement, which is a procedural statement that can only appear inside an `initial` or `always` statement. The task enable statement has the following syntax:

```
task_identifier [(argument1, ..., argumentn)];
```

The task enable statement specifies the arguments that pass values to the called task and the arguments that receive the results. The argument list is an ordered list that must match the order of the argument list in the task definition of the called task. Note that all arguments are passed by value rather than by reference.

The following module demonstrates the task declaration in port list style. It counts the number of zeros in a byte.

EXAMPLE 5.1 *A Task Example Used to Count the Zeros in a Byte*

In this example, we rewrite the `for` loop of the example described in Section 4.5.2 by a task definition. From this example, we may learn how to define and call (enable) a task.

```
// an example illustrates how to count the zeros in a
// byte.
module zero_count_task (data, out);
input [7:0] data;
output reg [3:0] out;    // output declared as register
always @(data)           // task calling
   count_0s_in_byte(data, out);

// task declaration starts from here
task count_0s_in_byte;
input [7:0] data;
output reg [3:0] count;
integer i;
begin   // task body
   count = 0;
   for (i = 0; i <= 7; i = i + 1)
      if (data[i] == 0) count = count + 1;
end
endtask
endmodule
```

In the above module, the following statement:

```
if (data[i] == 0) count = count + 1;
```

can be replaced by:

```
count = count + ~data[i];
```

Why? The reason is as follows. Because the register `count` increases its value by 1 when the current testing bit `data[i]` is zero, it is equivalent to add the complement of the testing bit to `count`. This idea is further demonstrated by the following module, which also illustrates the use of port list declaration style associated with a task.

EXAMPLE 5.2 *A Task Example Used to Count the Zeros in a Byte*

In this example, we rewrite the preceding example by changing the task declaration from port list style to port list declaration style. In addition, in the task definition, the `if` statement in the preceding example is replaced with the statement described above.

```
// an example illustrates how to count the zeros in a
// byte.
module zero_count_task2 (data, out);
input  [7:0] data;
output reg [3:0] out;  // output declared as register
always @(data)          // task calling
    count_0s_in_byte(data, out);

// task declaration starts from here
task count_0s_in_byte(input [7:0] data, output reg [3:0]
  count);
integer i;
begin    // task body
   count = 0;
   for (i = 0; i <= 7; i = i + 1)
      count = count + ~data[i];
end
endtask
endmodule
```
∎

A task may have timing controls associated with its procedural statements. However, an assignment to an output argument is not passed to the calling argument until the task exits. The actual simulation time required for a task depends on the timing controls used within it.

EXAMPLE 5.3   *A Task Example of Using Timing Controls*

The delay timing controls are used within the task. Because there are 10 time units that are required for the task to complete its operations, the first $display system task is executed at simulation time 3 but the second $display system task at simulation time 19.

```
// the use of a static task
module static_task;
reg [3:0] result;
initial begin
   #3 check_counter(1'b1, result);
   $display ($realtime,,"The value of count is
     %d", result);
   #6 check_counter(1'b0, result);
   $display ($realtime,,"The value of count is
     %d", result);
end
// task definition starts from here
task check_counter;
```

```
input reset;
output reg [3:0] count;
// the body of the task
begin
   if (reset) count = 0;
   else begin
      # 2 count = count + 1;
      # 3 count = count + 1;
      # 5 count = count + 1;
   end
end
endtask
endmodule
```

■

The simulation results of the above module are as follows:

```
# 3 The value of count is 0
# 19 The value of count is 3
```

5.1.2 Types of Tasks

Tasks can be cast into the following two types: a (static) task, declared without the key-word `automatic`, and an automatic (also called re-entrant, dynamic) task, declared with the keyword `automatic`.

Static tasks have the feature that all declared items are statically allocated. Static tasks items can be shared across all uses of the task that execute concurrently.

Automatic tasks have the feature that all declared items are dynamically allocated for each recursive call. In addition, automatic task items cannot be accessed by hier-archical references but automatic tasks can be called through use of their hierarchical names.

A task can be called more than once concurrently. All variables of an automatic task are duplicated on each concurrent call to store state specific to that call. All variables of a static task are static in that there is a single variable corresponding to each declared local variable in a module instance, regardless of the number of concurrent calls of the task. However, static tasks in different instances of a module have separate storage from each other. Variables declared within static tasks retain their values between calls. Variables declared within automatic tasks are initialized to the default value x whenever execution enters their scope and are deallocated at the end of the task call.

EXAMPLE 5.4   *An Example Illustrating the Use of a Task*

In this example, the task needs 7 time units to complete its operations at each call. In the test bench, the task is called at simulation time 5, while the `reset` signal is deactivated at 10 time units later after the task returns, which is at simulation time 12. Therefore, the `reset` signal is assigned 0 at time 22. At simulation time 32, the task is called once again. Hence, the simulation results are:

```
# 5 At the beginning of task, count = x
# 5 After reset, count = 0
# 12 At the end of task, count = 2
# 32 At the beginning of task, count = 2
# 39 At the end of task, count = 4

// the use of a task
module non_automatic_task;
reg reset;
initial begin
        reset = 1'b1;   // reset count
    #5   check_counter;
    #10 reset = 1'b0;
    #10 check_counter;
end
// task definition starts from here
task check_counter;
reg [3:0] count;
// the body of the task
begin
    $display ($realtime,,"At the beginning of task,
      count = %d", count);
    if (reset) begin
      count = 0;
        $display ($realtime,,"After reset, count = %d",
          count);
    end
    # 2 count = count + 1;
    # 5 count = count + 1;
    $display ($realtime,,"At the end of task, count = %d",
      count);
end
endtask
endmodule
```

From the simulation results, we can see that the value of variable count is not destroyed after the task returned. This is also a basic feature of static tasks: variables declared within static tasks retain their values between calls. Hence, the second call to task check_counter uses the value from the previous call to task check_counter. However, if we use automatic task, the situation is different, such as in the following example.

EXAMPLE 5.5 *An Example Illustrating the Use of an Automatic Task*

In this example, the task is declared to be automatic by using the keyword automatic. It still needs 7 time units to complete its operations at each call. The rest of the timing information is exactly the same as in the preceding example. However, for each task call, a separate `reg` variable `count` is allocated and is initialized to the default value x. Hence, the simulation results are:

```
# 5 At the beginning of task, count = x
# 5 After reset, count = 0
# 12 At the end of task, count = 2
# 32 At the beginning of task, count = x
# 39 At the end of task, count = x

// the use of an automatic task
module automatic_task;
reg reset;
initial begin
      reset = 1'b1;  // reset count
   #5  check_counter;
   #10 reset = 1'b0;
   #10 check_counter;
end
// task definition starts from here
task automatic check_counter;
reg [3:0] count;
// the body of the task
begin
   $display ($realtime,,"At the beginning of task,
     count = %d", count);
   if (reset) begin
      count = 0;
      $display ($realtime,,"After reset, count = %d",
        count);
   end
   # 2 count = count + 1;
   # 5 count = count + 1;
   $display ($realtime,,"At the end of task, count = %d",
     count);
end
endtask
endmodule
```

Review Questions

Q5.1 When is a task usually used to describe a piece of code?

Q5.2 What are the differences between static and automatic tasks?

Q5.3 Can timing controls be used within tasks?

Q5.4 Can `always` or `initial` statements be used within the definition of tasks?

Q5.5 Can a task contain no argument?

Q5.6 Can a task call another task?

5.2 FUNCTIONS

A function is an alternate way besides tasks to provide a means of re-using a common code. It is used when the procedure has no timing controls (i.e. any statement introduced with #, @ or `wait`), needs to return a single value, has at least one `input` argument, has no `output` or `inout` arguments, or has no nonblocking assignments. A function is often used to model a combinational logic circuit. Note that a task can call tasks or functions but a function can only call functions. The features of functions are listed in Table 5.1.

5.2.1 Function Definition and Call

The function definition implicitly declares an internal variable and defaults to a 1-bit `reg`, with the same name as the function. This variable is used to return the result from the function by assigning the function result to it within the function.

Functions are declared with the keywords `function` and `endfunction` and can be one of the following forms:

```
// port list style
function [automatic] [signed] [range_or_type]
  function_identifier;
    input_declaration
    other_declarations
    procedural_statement
endfunction

 // port list declaration style
 function [automatic] [signed] [range_or_type]
    function_identifier (input_declarations);
    other_declarations
    procedural_statement
 endfunction
```

where `range_or_type` is optional and can be a range `[m:n]` or a return type of any one of `integer`, `time`, `real` and `realtime`. A function returns a 1-bit `reg` value by default if no range or type is specified. A function must have at least one input argument but cannot have any output or inout arguments.

5.2.1.1 Function Call A function call is an operand within an expression and has the following syntax:

```
function_identifier (argument1, ..., argumentn);
```

where `argument1, ..., argumentn` are input arguments and the `function_identifier` is the variable with which a result is returned. As a consequence, there must have a procedural assignment that assigns the result to the `function_identifier` within the function.

The following example demonstrates the function declaration in port list style. It counts the number of zeros in a byte. In this example, we replace the task definition by a corresponding function definition.

EXAMPLE 5.6   *A Function Example Used to Count the Zeros in a Byte*

In this example, we replace the task definition of the example described in the previous section by a function definition. From this example, we may learn how to define and call a function.

```
// an example illustrates how to count the zeros in a
// byte.
module zero_count_function (data, out);
input  [7:0] data;
output reg [3:0] out;  // output declared as register
always @(data)
    out = count_0s_in_byte(data);
// function declaration starts from here
function [3:0] count_0s_in_byte;
input [7:0] data;
integer i;
begin
   count_0s_in_byte = 0;
   for (i = 0; i <= 7; i = i + 1)
      if (data[i] == 0) count_0s_in_byte
         = count_0s_in_byte + 1;
end
endfunction
endmodule
```

With the same reason as that of using the task definition, the above module can be rewritten as the following example. This module also illustrates the use of the port list declaration style of a function.

EXAMPLE 5.7   *A Function Example Used to Count the Zeros in a Byte*

In this example, we rewrite the preceding example by changing the function declaration from using the port list style to using the port list declaration style. In addition, in the function definition, the `if` statement in the preceding example is replaced with the statement `count_0s_in_byte = count_0s_in_byte + ~data[i];`.

```
// an example illustrates how to count the zeros in a
// byte.
module zero_count_function (data, out);
input  [7:0] data;
output reg [3:0] out;  // output declared as register
always @(data)
    out = count_0s_in_byte(data);
// function declaration starts from here
function [3:0] count_0s_in_byte(input [7:0] data);
integer i;
begin
   count_0s_in_byte = 0;
   for (i = 0; i <= 7; i = i + 1)
      count_0s_in_byte = count_0s_in_byte + ~data[i];
end
endfunction
endmodule
```
■

5.2.2 Types of Functions

Functions can be classified into the following two types: (static) function, which is declared without the keyword `automatic`, and automatic (also called recursive, dynamic) function, which is declared with the keyword `automatic`.

Static functions have the feature that all declared items are statically allocated. Automatic functions have the feature that all function items are allocated dynamically for each recursive call. In addition, automatic function items cannot be accessed by hierarchical references but automatic functions can be called through use of their hierarchical names.

The following example describes how an automatic function can be declared and called.

EXAMPLE 5.8 *An Example Illustrating the Use of Automatic Function*

In this example, an automatic function `fact` is declared and called recursively to compute the factorial of an integer n.

```
// to illustrate the use of automatic function
module factorial(input [15:0] n, output reg [15:0]
  result);
// instantiate the fact function
always @(n)
   result = fact(n);

 // define fact function
```

```
function automatic [15:0] fact;
input [15:0] n;
// the body of function
   if (n == 0) fact = 1;
   else fact = n * fact(n - 1);
endfunction
endmodule
```

5.2.3 Constant Functions

Constant function calls are evaluated at elaboration time. They are used to support the building of complex calculations of values at elaboration time. The execution of constant function calls has no effect on the initial values of the variable used either at simulation time or among multiple calls of a function at elaboration time.

A constant function is local to the calling module where the arguments to the function are constant expressions. Constant functions contain no hierarchical references and can only call constant functions local to the current module. They cannot call system functions. All system tasks, except $display, within a constant function are ignored. However, the $display system task is ignored at elaboration time.

The following module describes how a constant function is used to compute $\log_2(n)$, where n is usually a number to the power of 2. This kind of constant function may be useful when writing a parameterized module. The interested reader is encouraged to synthesize this module and check out the results from your synthesizer.

EXAMPLE 5.9 *An Example Illustrates the Use of Constant Function*

This example describes the use of a constant function to help compute the local parameter M. Maybe one major purpose of using a constant function is to reduce the number of module parameters. For example, with the help of the constant function count_log_b2(), module zero_count_constant only needs a single module parameter SIZE; otherwise, it requires two. From the synthesized result, you can see that the constant function is only used to compute the local parameter as it is expected. It does not matter with the synthesized results of the module.

```
// an example illustrates the use of a constant function.
module zero_count_constant (out, data);
parameter SIZE = 8;
localparam M = count_log_b2(SIZE);
output reg [M:0] out;  // output declared as register
input  [SIZE-1:0] data;
integer i;
always @(data) begin
   out = 0;
   for (i = 0; i <= 7; i = i + 1)
```

```
        if (data[i] == 0)
            out = out + 1;
end
// define a constant function for computing log_2 n.
function integer count_log_b2(input integer depth);
// function body
begin
    count_log_b2 = -1;
    while (depth) begin
        count_log_b2 = count_log_b2 + 1;
        depth = depth >> 1;
    end
end
endfunction
endmodule
```

5.2.4 Sharing Tasks and Functions

Although tasks and functions are defined within a module declaration, they can be shared among different modules by one of the following ways:

1. Define tasks and functions to be shared publicly in a separate module, say `package`, then use the hierarchical name to reference the specific task or function.
2. Define tasks and functions to be shared publicly in a separate text file, say `package.v`, then use `include` compiler directive to include it inside the module which wants to share them.

We demonstrate how these two approaches are used in practice by way of examples. The first example uses a hierarchical name to refer the required function or task.

EXAMPLE 5.10 *Sharing Tasks and Functions in Different Modules Through Use of a Hierarchical Name*

In this example, the function `fact` is declared in module `factorial` and it is called in another module `tasks_function_sharing`. To be properly called, the hierarchical name `factorial.fact` needs to be used.

```
`timescale 1 ns/100 ps
// an example illustrates the sharing of tasks and
// functions using compiler directive include
module tasks_function_sharing;
reg [15:0] result;

initial begin
```

```
     result = factorial.fact(5);
     $display ("The result is %d", result);
 end
 endmodule
 // file factorial.v
 module factorial
  // define fact function
 function automatic [15:0] fact;
 input [15:0] n;
 // the body of function
    if (n == 0) fact = 1;
    else fact = n * fact(n - 1);
 endfunction
 endmodule
```

■

The following module shows how the `include compiler directive is used to include the required function fact into the current module. Note that the `include statement must be placed after the keyword module since the function definitions in the file package.v are not bounded by a module declaration and they should be declared within a module. The contents of file package.v are as follows:

```
 // file package.v
 // define fact function
 function automatic [15:0] fact;
 input [15:0] n;
 // the body of function
    if (n == 0) fact = 1;
    else fact = n * fact(n - 1);
 endfunction
```

EXAMPLE 5.11 *Sharing Tasks and Functions through Use of the* `include *Compiler Directive*

As described, the function fact is included in the module when compiling. Hence, both the function call and the called function are in the same module. So they proceed their operations as usual.

```
`timescale 1 ns/100 ps
// an example illustrates the sharing of tasks and
// functions using compiler directive include
module tasks_function_sharing2;
`include "package.v" // in working directory
reg [15:0] result;

initial begin
```

```
        result = fact(5);
        $display ("The result is %d", result);
    end
endmodule
```

Review Questions

Q5.7 When is a function usually used to describe a piece of code?

Q5.8 What are the differences between static and automatic functions?

Q5.9 Can timing controls be used within functions?

Q5.10 Can `always` or `initial` statements be used within functions?

Q5.11 Can a function contain no arguments?

Q5.12 Can a function call another function?

Q5.13 Can a function call tasks?

Q5.14 What are the features of constant functions?

5.3 SYSTEM TASKS AND FUNCTIONS

In Verilog HDL, there are many built-in system tasks and functions. These tasks and functions are predefined in the language and can be used for various applications in writing test benches, describing designs abstractly in behavioral style and so on. The system tasks and functions provided by Verilog HDL can be divided into the following types:

1. Display system tasks.
2. Timescale system tasks.
3. Simulation time system functions.
4. Simulation control system tasks.
5. File I/O system tasks.
6. String formatting system tasks.
7. Conversion system functions.
8. Probabilistic distribution system functions.
9. Stochastic analysis system tasks.
10. Command line input.
11. PLA modeling system tasks.

The first four types are often collected as simulation-related system tasks and functions. Except that the final type is discussed in Chapter 10, all of the rest are discussed in this section.

5.3.1 Simulation-Related System Tasks

There are four types of system tasks and functions that are widely used during simulation. These are display system tasks, timescale system tasks, simulation time system functions and simulation control system tasks.

5.3.1.1 Display System Tasks

The display system tasks are used to display information about a design for debugging or other useful purposes. Verilog HDL provides a group of system tasks for displaying information on standard output. This group of system tasks is known as $display system tasks. It includes the following system tasks:

```
$display   $displayb   $displayh   $displayo
$write     $writeb     $writeh     $writeo
$monitor   $monitorb   $monitorh   $monitoro
$strobe    $strobeb    $strobeh    $strobeo
```

For convenience, we further divide them into the following three types:

1. Display and write system tasks.
2. Continuous monitoring system tasks.
3. Strobed monitoring system tasks.

Display and Write System Tasks   The display and write system tasks include $display and $write. The $display and $write system tasks display values of variables, strings or expressions in the same order as they appear in the argument list. The difference between $display and $write system tasks is that after they print out the specified information on the standard output, the $display system task also outputs an end-of-line character but the $write system task does not. The $display and $write system tasks have the form:

```
task_name[(arguments)];
```

where the task_name is one of $display, $displayb, $displayo, $displayh, $write, $writeb, $writeo and $writeh, and the arguments are quoted strings or variables expressions.

   To print information with a specified format, an escape sequence is used. Each of the following escape sequences, when included in a string argument, specifies the display format for a subsequent expression.

```
%h or %H // hexadecimal
%d or %D // decimal
%o or %O // octal
%b or %B // binary
%c or %C // ASCII character
%l or %L // library binding information
```

```
%v or %V // net signal strength
%m or %M // hierarchical name
%s or %S // string
%t or %T // current time format
```

Any expression with no format specification is displayed using the following default formats:

```
$display  and $write   // decimal
$displayb and $writeb  // binary
$displayo and $writeo  // octal
$displayh and $writeh  // hexadecimal
```

Escape sequences for printing special characters are:

```
\n    // the newline character
\t    // the tab character
\\    // the \ character
\"    // the " character
\ddd  // a character specified by 1 to 3 octal digits
%%    // the % character
```

Continuous Monitoring System Tasks   The continuous monitoring system task $monitor continuously monitors and displays the values of any variables or expressions. It has the form:

```
task_name[(arguments)];
```

where task_name is one of $monitor, $monitorb, $monitoro and $monitorh. The arguments for this system task group are specified in exactly the same way as for the $display system task.

Only one $monitor system task display list can be active at any time; however, a new $monitor system task with a new display list can be issued any number of times during simulation. In addition to the $monitor system task group, there are two related system tasks, $monitoron and $monitoroff, which are widely used to control the monitoring operations. The $monitoron and $monitoroff system tasks enable and disable the monitoring operations, respectively.

Strobed Monitoring System Tasks   The strobed monitoring system task $strobe displays the values of any variables or expressions at a specified time but at the end of the time step. It has the following form:

```
task_name[(arguments)];
```

where task_name is one of the following system tasks: $strobe, $strobeb, $strobeo and $strobeh. The arguments for this system task group are specified in exactly the same way as for the $display system task.

5.3.1.2 *Timescale System Tasks*

There are two timescale-related system tasks, $printtimescale and $timeformat, that can be used to display and set the timescale information. The $printtimescale system task displays the time unit and precision for a particular module. It has the syntax:

```
$printtimescale [(hierarchical_identifier)];
```

where the hierarchical_identifier specifies the module whose time unit and precision are to be displayed. When no argument is specified, the $printtimescale system task displays the time unit and precision of current module. The timescale information is displayed in the following format:

```
Time scale of (module_name) is unit / precision
```

For example, in the following example, module test_a calls the $print timescale system task to display timescale information about another module test_b.

```
'timescale 1 ms/1 us
module test_a;
initial
    $printtimescale(test_b);
endmodule
'timescale 10 ns/1 ns
module test_b;
   .....  ;
endmodule
```

The information about test_b is displayed in the following format:

```
Time scale of (test_b) is 10 ns/1 ns
```

The $timeformat system task specifies how the %t format specification reports time information. It has the syntax:

```
$timeformat[(units_number, precision, suffix,
    numeric_field_width)];
```

where units_number is the power of 10^a second, i.e., a, where a is an integer and $-15 \leq a \leq 0$. For instance, to represent 1 μs, the units_number is set to -6 since 1 μs $= 10^{-6}$. The default value of units_number is the smallest time precision argument of all 'timescale compiler directives. The precision defaults to 0. The default value of suffix is a null and of numeric_field_width is 20.

The following system task call:

```
$timeformat(-9, 3, " ns.", 5);
$display("The current time is %t", $realtime);
```

will display the %t specifier value in the display task as:

```
The current time is 5.102 ns.
```

5.3.1.3 Simulation Time System Functions There are three system functions that provide access to current simulation time: $time, $stime and $realtime. The $time system function returns a 64-bit integer of time, the $stime system function returns an unsigned 32-bit integer of time and the $realtime system function returns a real number of time. All of the returned values from these three system functions are scaled to the time unit of the module that invokes them. For example:

```
'timescale 10 ns/ 1 ns
module time_usage;
reg a;
initial begin
    $monitor($time,, "a = ", a);
    #2.55 a = 0;
    #2.55 a = 1;
end
endmodule
```

The result is:

```
#0  a = x
#3  a = 0
#5  a = 1
```

The simulation times are 26 ns and 52 ns. They are scaled to 3 ns and 5 ns, respectively, because the time unit of the module is 10 ns. The usage of the $stime system function is the same as that of the $time system function except that it returns a 32-bit unsigned integer rather than a 64-bit integer. It will have the same result for the above example as the $time system function.

The $realtime system function returns a real number. Hence, if we replace the first argument, $time, with $realtime of the $monitor system task in the above module, the simulation results would be as follows:

```
#0   a = x
#2.6 a = 0
#5.2 a = 1
```

5.3.1.4 Simulation Control System Tasks There are two simulation control system tasks: $stop and $finish. The $stop system task suspends the simulation and the $finish system task terminates the simulation.

5.3.2 File I/O System Tasks

The file input/output system tasks for file-based operations are classified into the following categories: *opening and closing file system function/task, file output system tasks* and *file input system tasks*. We describe their operations in detail in this sub-section.

5.3.2.1 *Opening and Closing File System Function/Task*

Before a file can be read, it must be opened by the $fopen system function. The file also need be closed after it has no longer been used. The $fclose system task is used for this purpose. In Verilog HDL, there are two types of output files: *file descriptor* (fd) file and *multiple channel descriptor* (mcd) file. The fd file is like the files used in most programming languages such as C/C++ and Java. The mcd file is unique in Verilog HDL; it allows us to write to multiple files the same data simultaneously. We first describe the fd file.

The $fopen system function and the $fclose system task have the following syntax:

```
integer fd = $fopen(file_name, mode);
$fclose(fd);   // close a file opened previously
```

The $fopen system function returns an integer value, called a file descriptor (fd), after opening the file with the specified mode. The returned file descriptor is 0 if it fails to open the specified file. The file descriptor is a 32-bit value and its MSB (bit 32) is always set to 1, to distinguish it from a multiple channel descriptor (mcd) file, which will be described later. Like C programming language, three files are pre-opened: STDIN (32'h8000_0000), STDOUT (32'h8000_0001) and STDERR (32'h8000_0002). STDIN is pre-opened for reading and STDOUT and STDERR are pre-opened for appending.

The allowed modes for specifying the type of a file to be opened are as follows:

- "r" or "rb": open file for reading at the beginning of file. Return 0 if it fails.
- "w" or "wb": open file for writing at the beginning of file or create file if it does not exist.
- "a" or "ab": open file for appending at the end of file or create file if it does not exist.
- "r+", "r+b" or "rb+": open file for update (reading and writing) at the beginning of file. Issue error if it does not exist.
- "w+", "w+b" or "wb+": open file for update (reading and writing) at the beginning of file or create file if it does not exist.
- "a+", "a+b" or "ab+": open file for update (reading and writing) at the end of file or create file if it does not exist.

where the "b" refers to opening binary files.

5.3.2.2 *File Output System Tasks*

The file output system tasks, $fdisplay, $fwrite, $fmonitor and $fstrobe, are used to write information to a file. They have the same type of arguments with their counterparts, $display, $write, $monitor and $strobe, except that the first parameter should be a file descriptor.

$fdisplay	$fdisplayb	$fdisplayh	$fdisplayo
$fwrite	$fwriteb	$fwriteh	$fwriteo
$fmonitor	$fmonitorb	$fmonitorh	$fmonitoro
$fstrobe	$fstrobeb	$fstrobeh	$fstrobeo

$fflush

The file output system tasks have the form:

```
file_output_task(fd, argument1, argument2, ...,
  argumentn);
```

where the file_output_task can be any of the following system tasks: $fdisplay, $fwrite, $fmonitor and $fstrobe.

The $fflush system task flushes the output buffer to the specified output file. It has the following syntax:

```
$fflush (fd);
$fflush ();
```

When no argument is specified, the $fflush system task flushes any buffered output to all open files.

The following example illustrates the use of file output system tasks.

EXAMPLE 5.12 *The Use of File Output System Tasks*

In this example, three file output system function and tasks, $fopen, $fdisplay and $fclose, are used to completely store the results from the output of the function fact.

```
module fact_file;
// test the fact function
integer n, result, fd;
initial begin
   fd = $fopen("fact_result.dat", "w");
   if (fd == 0)
      $display ("File open error !");
   for (n = 0; n <= 12; n = n + 1) begin
      result = fact(n);
      $fdisplay (fd, "%0d factorial = %0d", n, result);
   end
   $fclose (fd);
end
```

```
 // define fact function
function automatic integer fact;
input [31:0] n;
// the body of function
   if (n == 0) fact = 1;
   else fact = n * fact(n - 1);
endfunction
endmodule
```

■

Multiple Channel Descriptor (mcd)   The mcd is a 32-bit reg in which each set bit corresponds an opened output file. The MSB (bit 32) is set to 0 in order to distinguish it from fd whose MSB is 1. In addition, the LSB (bit 0) of mcd is always reserved for standard output. As a consequence, there are at most 30 files that can be opened for output through mcd. The use of mcd allows us to write multiple files simultaneously by bit-wise ORing together their mcds and writing to the resultant value.

To open an mcd file, the following syntax is used:

```
integer mcd = $fopen("file_name");
```

where mcd is returned from the $fopen system function. It has a value of 0 if the $fopen system function fails to open the specified file. To close file(s) specified by the mcd, the following syntax is used:

```
$fclose(mcd);
```

The file output system tasks when using mcd files have the form:

```
file_output_tasks(mcd, argument1, argument2, ...,
   argumentn);
```

where file_output_tasks can be any of the following system tasks: $fdisplay, $fwrite, $fmonitor and $fstrobe.

The $fflush system task flushes the output buffer to the specified mcd output file(s). It has the following syntax:

```
$fflush (mcd);
$fflush ();
```

When no argument is specified, the $fflush system task flushes any buffered output to all open files.

The following example illustrates the use of file output system tasks in the case of mcd files.

EXAMPLE 5.13 *The Use of* mcd *Files*

This example illustrates how to set up mcds. First, two files are opened. Then their mcds are combined together in a bit-wise OR operation and assigned to the integer variable cluster. In order to output to standard output as well, a constant 1 is also logical OR into the cluster. The first $fdisplay system task writes current simulation time into all three files, including the standard output, the second $fdisplay system task writes to the two files, mcd1 and mcd2, while the third $fdisplay system task writes to the mcd2 file only.

```
integer mcd1, mcd2, cluster, all;
initial begin
   if ((mcd1 = $fopen("cpu.dat")) == 0)
      $display ("File open error !");
   if ((mcd2 = $fopen("dma.dat")) == 0)
      $display ("File open error !");
   cluster = mcd1 | mcd2;
   // includes standard output
   all = cluster | 1'b1;
   // possible uses of opened files
   $fdisplay(all, "System restart at time %d",$time);
   $fdisplay(cluster, "Error occurs at time %d,
            address = %h", $time, addr);
   forever @(posedge clock)
      $fdisplay(mcd2, "dma_mode = %h address = %h",
            dma_mode, address);
end
```

5.3.2.3 File Input System Tasks File input system tasks and functions provide the capabilities of reading a character or a line at a time, reading formatted data and reading binary data from a specified file. They also provide a mechanism to control the position to be read of or return I/O status from a specified file. The file input system tasks include the following system tasks:

$fgetc	$ungetc	$fgets	$fscanf
$fread	$readmemb	$readmemh	$ftell
$fseek	$rewind	$ferror	

These system tasks can be divided into the following types: reading a character/line at a time, reading formatted data, reading binary file, file positioning and I/O error status.

Reading a Character at a Time The $fgetc system function reads a byte from the file specified by fd. It has the form:

```
c = $fgetc(fd);
```

If an error occurs when reading a file, it returns −1 (EOF).

To undo the effect of the $fgetc system function, i.e. to insert the character back to the file buffer specified by fd, the $ungetc system function may be used. It has the form:

```
code = $ungetc(char, fd);
```

If an error occurs, it returns −1 (EOF).

Reading a Line at a Time   The $fgets system function reads characters from the file specified by fd. It has the form:

```
code = $fgets(str, fd);
```

The characters read are stored into str and the operation is terminated when either the buffer is full, a newline character is read or an end-of-file is encountered. If an error occurs when reading a file, it returns 0, indicating that no character is read.

Reading Formatted Data   The $fscanf system task reads formatted data from the file specified by fd. It has the form:

```
code = $fscanf(fd, format, arguments);
```

For example, if a file test.dat contains the following data:

```
test_data_1 34 25
```

then, after the following statements are executed:

```
integer code, fd, x, y;
reg [87:0] name;

fd = $fopen("test.dat", "r");
code = $fscanf(fd, "%s %d %d", name, x, y);
```

The reg variable name has test_data_1, integer variable x is 34 and integer variable y is 25.

Reading Binary Files   There are three system tasks that can be used to read data from a file: $fread, $readmemb and $readmemh. The $fread system task reads binary data from a file specified by fd into memory or a reg variable. It has the form:

```
code = $fread(mem, fd, [start], [count]);
code = $fread(reg, fd);
```

where the arguments start and count are optional. If the start presents, it specifies the address of the first element in the memory to be loaded; otherwise, the lowest address in the memory will be used. If the count presents, it specifies the maximum number of locations in mem that will be loaded; otherwise, the mem is filled with what is data available.

The $readmemb and $readmemh system tasks read binary and hexadecimal data from a specified file into memory, respectively. They have the same form shown as follows:

```
$readmemb("<file_name>",<memory_name>, [start], [end]);
$readmemh("<file_name>",<memory_name>, [start], [end]);
```

where the arguments start and end are used to optionally specify the start and end addresses of memory, respectively. If both start and end are not specified and no address specifications appear within the file, the default start address is the lowest address of the memory. If only the start argument is specified, the file is read into the memory starting at the specified address and toward the higher address. If both start and end arguments are specified, the file is read into the memory beginning at the start address and through the end address, regardless of how the range of the memory is declared.

The following example demonstrates the use of the $readmemb system task. Here, the cache_mem is declared as an array of 256 vectors with 8 bits each.

```
reg  [7:0] cache_mem[255:0];

initial $readmemb("cache_mem.data", cache_mem);
initial $readmemb("cache_mem.data", cache_mem, 24);
initial $readmemb("cache_mem.data", cache_mem, 156, 1);
```

The first $readmemb system task reads the file cache_mem.data into cache_mem, starting at location 0. The second $readmemb system task loads cache_mem, starting at address 24 and continues on towards address 256. The third $readmemb system task loads cache_mem, starting at address 156 and continues through address 1.

The text file can contain white space (spaces, new lines, tabs and form-feeds), comments (both types of comment are allowed) and binary (for $readmemb) or hexadecimal (for $readmemh) numbers. Numbers can contain x, z or underscore (_) and are separated by white spaces and/or comments. Moreover, addresses can be specified in the file as hexadecimal numbers using @<hexa_address>. Uninitialized locations default to x. For example:

```
@002   // from location 2
0000 1011
0001 11zz
@008   // from location 8
1111 zz1z
```

The locations 6 and 7 will be initialized to the unknown x.

File Positioning The file positioning system functions include: $ftell, $fseek and $rewind. The $ftell system function returns the offset from the beginning of the file specified by fd. It has the form:

```
position = $ftell(fd);
```

The $fseek system function sets the position of the next input or output position on the file specified by fd. It has the form:

```
position = $fseek(fd, offset, operation);
```

The offset can be counted from the beginning, from the current position, or from the end of the file, according to an operation value of 0, 1 or 2.

The $rewind system function is equivalent to the $fseek(fd, 0, 0) system function. It has the form:

```
code = $rewind(fd);
```

Note that the operations of the $fseek and $rewind system functions will cancel any $ungetc system function operations.

I/O Error Status The $ferror system function returns more information about the causes of the file I/O operations specified by fd. It has the form:

```
code = $ferror(fd, str);
```

where str must be at least 640 bits. The $ferror system function returns 0 and the str is cleared if the most recent operation did not result in an error.

5.3.3 String Formatting System Tasks

The string system tasks provide ways to convert the formats between strings. This set of system tasks include the following:

```
$swrite      $swriteb     $swriteh      $swriteo
$sformat     $sscanf

$sdf_annotate
```

The $swrite system task provides ways to convert the formats between strings. Their functions are similar to the $fwrite system task except that instead of writing to a file, the $swrite system task writes to a string of reg variable data types, output_reg. The $swrite system task has the following syntax:

```
str_tasks_name(output_reg, arguments);
```

where str_tasks_name can be one of the following system tasks: $swrite, $swriteb, $swriteh and $swriteo. If no format specification exists for an argument, the $swrite system task defaults to decimal. The $swriteb, $swriteh and $swriteo system tasks default to binary, hexadecimal and octal, respectively.

The $sformat system task is similar to the $swrite system task, except that the format must be specified only as the second argument. It has the form:

```
$sformat(output_reg, format, arguments);
```

where the second argument is the format, and the rest are variables to be formatted. The format used in the $sformat system task is exactly the same as the $display system task.

The $sscanf system function reads a line from a string, input_reg and stores the values read in the specified arguments. It has the form:

```
integer code = $sscanf(input_reg, format, arguments);
```

where the second argument is the format, and the rest are variables to be formatted. It returns 0 if it fails to read the string, input_reg.

The following example illustrates the use of file input system tasks and the $sscanf system function.

EXAMPLE 5.14 *The Use of File Input System Tasks*

In this example, two file input system tasks, $fgets and $sscanf, are used to read a line of characters from standard input and then convert it into an integer. Finally, it calls the function fact to compute the factorial of the integer and displays the result on standard output.

```
module fact_file_input;
// an example illustrates the use of file input system
// tasks
`define STDIN 32'h8000_0000 // standard input
integer i, n, result, code;
reg [15:0] number;
initial begin
   for (i = 0; i < 10; i = i + 1) begin
      $display ("Please input a number: ");
      code = $fgets(number, `STDIN);
      code= $sscanf (number, "%d", n);
      $display ("The input number is =%d", n);
      result = fact(n);
      $display ("The factorial of %d is %d", n, result);
   end
end
 // define fact function
```

```
function automatic integer fact;
input [31:0] n;
   // the body of function
   if (n == 0) fact = 1;
   else fact = n * fact(n - 1);
endfunction
endmodule
```

The $sdf_annotate system task reads timing data from a standard delay format (SDF) file and put its into a specified region of the design. One most widely used form of it is as follows:

```
$sdf_annotate("sdf_file", module_instance);
```

where sdf_file is the SDF file path and module_instance specifies the scope to which to annotate the information in the SDF file. We will discuss SDF files and their applications in greater detail in the chapter concerning 'verification'.

5.3.4 Conversion System Functions

The following utility system functions convert numbers between different types:

$signed	$unsigned	$rtoi	$itor
$realtobits	$bitstoreal		

1. The $signed system function interprets the argument as a signed quantity:

   ```
   integer $signed(value);
   ```

2. The $unsigned system function interprets the argument as an unsigned quantity:

   ```
   integer $unsigned(value);
   ```

3. The $rtoi system function converts a real to an integer by truncating the real value. For example, 324.35 becomes 324:

   ```
   integer $rtoi(real_value);
   ```

4. The $itor system function converts an integer to a real. For example, 1 becomes 1.0:

   ```
   real $itor(int_value);
   ```

5. The $realtobits system function passes bit patterns across module ports and converts from a real to the 64-bit vector of that real number:

   ```
   [63:0] $realtobits(real_value);
   ```

6. The `$bitstoreal` system function is the reverse of the `$realtobits` system function; it converts from the bit pattern to a real, rounding the result to the nearest real:

```
real $bitstoreal(bit_value);
```

The following example shows how the `$realtobits` and `$bitstoreal` system functions are used to pass a real value across modules.

EXAMPLE 5.15 *The Use of* `$realtobits` *and* `$bitstoreal` *System Functions*

This example describes how to pass a real number between two modules. Because a real number cannot be assigned to an output port, it needs to be converted into a bit pattern by using the `$realtobits` system function before it is sent to the port `net_real`. At the receiver, it is converted back to the real from the bit pattern received by another `$bitstoreal` system function.

```
// module 1: the driver
module driver (net_real);
output net_real;
real r;
wire [64:1] net_real = $realtobits(r);
endmodule
// module 2: the receiver
module receiver (net_real);
input net_real;
wire [64:1] net_real;
real r;
initial assign r = $bitstoreal(net_real);
endmodule
```
∎

5.3.5 Probability Distribution System Functions

The probability distribution system functions are random number generators that return integer values distributed in accordance with standard probabilistic functions. These functions include the following:

`$random`	`$dist_uniform`	`$dist_normal`
`$dist_exponential`	`$dist_poisson`	`$dist_chi_square`
`$dist_t`	`$dist_erlang`	

They can be divided into two groups: random number generator and distribution functions.

5.3.5.1 *The* `$random` *System Function* The `$random` system function is the most widely used random number generator of the set of the probability distribution system functions. It returns a 32-bit random number of signed integer each time it is called. It has the following form:

```
$random [(seed)];
```

The `seed` controls the random numbers returned by the `$random` system function. The `seed` can be either a `reg`, `integer` or `time` variable. The `$random` system function always returns the same value with the same seed.

In general, the `$random` system function returns a 32-bit signed integer. However, we may use the concatenation ({}) operator to interpret the signed integer returned by the `$random` system function as an unsigned number. For example:

```
reg [4:0] random_number;
random_number = {$random} % 23;
```

5.3.5.2 *Distribution(`$dist_`) System Functions* Each of these system functions returns a pseudo-random number with characteristics described by the function name. For example, the `$dist_uniform` system function returns random numbers uniformly distributed in the interval specified by its parameters: `seed`, `start` and `end`. The probabilistic distribution system functions have the forms:

```
$dist_uniform (seed, start, end);
$dist_normal (seed, mean, standard_deviation);
$dist_exponential (seed, mean);
$dist_poisson (seed, mean);
$dist_chi_square (seed, degree_of_freedom);
$dist_t (seed, degree_of_freedom);
$dist_erlang (seed, k_stage, mean);
```

where all parameters to the system functions are integer values. The system functions always return the same value with the same seed. The `seed` is an in–out parameter. It should be an integer variable that is initialized by the user and only updated by the system functions.

Details of these probabilistic distribution functions can be found in the *IEEE Language Reference Manual* [3].

5.3.6 Stochastic Analysis System Tasks

Stochastic analysis system tasks and functions are used to manage queues and generate random numbers with specific distributions. The set of stochastic analysis system tasks and functions includes:

```
$q_initialize  $q_add         $q_remove
$q_full        $q_exam
```

These system tasks and functions facilitate a mechanism to implement stochastic queuing models. In what follows, we describe their functions of these system tasks in greater detail.

The $q_initialize system task. The $q_initialize system task creates a new queue and has the form:

```
$q_initialize (q_id, q_type, max_length, status);
```

where all parameters are integers. The q_id is an integer queue identifier. The q_type specifies which type of the queue to be used: 1 for first-in, first-out (FIFO) queue and 2 for last-in, first out (LIFO) queue, i.e. stack. The max_length is an integer number specifying the maximum allowed number of entries on the queue.

The $q_add system task. The $q_add system task adds an entry onto a specified queue. It has the following form:

```
$q_add (q_id, job_id, inform_id, status);
```

where job_id is an integer job identifier. The inform_id is an integer input associated with the queue entry. Its meaning is defined by the users.

The $q_remove system task. The $q_remove system task removes an entry from a queue. It has the following form:

```
$q_remove (q_id, job_id, inform_id, status);
```

The $q_full system function. The $q_full system function checks whether the specified queue is full or not. It returns 1 when the queue is full; otherwise, it returns 0. The $q_full system function has the following form:

```
$q_full (q_id, status);
```

The $q_exam system task. The $q_exam system task provides statistical information about activity at the specified queue. It has the following form:

```
$q_exam (q_id, q_stat_code, q_stat_value, status);
```

where the q_stat_value is the returned value requested by the input request code: q_stat_code. The relationship between q_stat_code and q_stat_value is presented in Table 5.2.

The status codes. All of the above queue management tasks and functions return a status to give some information about the associated operation requested. The status codes and their meanings are given in Table 5.3.

5.3.7 Command Line Arguments

In Verilog HDL, there are two system functions, $test$plusargs and $value$plusargs, which allow access to arguments and their values from the

TABLE 5.2 The relationship between
`q_stat_code` **and** `q_stat_value`

q_stat_code	q_stat_value
1	Current queue length
2	Mean interarrival time
3	Maximum queue length
4	Shortest wait time ever
5	Longest wait time for jobs still in the queue
6	Average wait time in the queue

command line. These arguments begin with the plus (+) character and hence are referred to as *plusargs*.

The `$test$plusargs` system function tests whether the plusargs exists or not. It has the syntax:

```
$test$plusargs (string);
```

where `string` is the one to be tested. It returns 0 when no plusarg from the command line matches the string provided; otherwise, it returns 1. For example, if the command line contains:

```
+FINISH=10000
```

then the following two `$test$plusargs` statements will return 1:

```
$test$plusargs ("FINISH");
$test$plusargs ("FIN");
```

because the provided strings `FINISH` and `FIN` match the prefixes of the plusarg in the command line.

The `$value$plusargs` system function returns the value of a specified plusarg, using a format string to access the value in plusargs. The format is much

TABLE 5.3 The status codes and their meanings

Status	Meanings
0	Successful
1	Queue full
2	Undefined q_id
3	Queue empty
4	Unsupported queue type
5	Specified length ≤ 0
6	Duplicate q_id
7	System memory full

like the `$display` statement. The `$value$plusargs` system function has the syntax:

```
$value$plusargs (string, variable);
```

where the format `string` is the same as the `$display` system task. It returns 0 if no string is found matching; otherwise, it returns a nonzero value, such as:

```
$value$plusargs("FINISH=%d", stop_clock);
$value$plusargs("START=%d", start_clock);
```

The variable `stop_clock` receives the value of 10 000 because the string `FINISH` exactly matches with that on the command line. However, the variable `start_clock` remains unchanged because no +START is specified on the command line.

Review Questions

Q5.15 What are the differences between `fd` and `mcd`?

Q5.16 Describe the operations of the `$fflush` system task.

Q5.17 Which system task or function can be used to read a character from standard input?

Q5.18 Describe the operations of the `$sscanf` system function.

Q5.19 Describe how to pass a real number between two modules.

Q5.20 Describe the features of the `$random` system function.

5.4 USER-DEFINED PRIMITIVES

In Verilog HDL, in addition to built-in gate and switch primitives, two types of UDPs are provided for users to build their own primitives. These two UDPs are combinational UDPs and sequential UDPs. Combinational UDPs are defined when the output is uniquely determined by the combination of the inputs. Sequential UDPs are defined when the output is determined by the combination of the current output and the inputs. UDPs are instantiated in exactly the same ways as gate or switch primitives. In this section, we describe these kinds of primitives in greater detail.

5.4.1 UDP Basics

UDPs are defined as an entity independent of modules, that is, they are defined at the same level as module definitions in the syntactic hierarchy. A UDP is defined by using the keywords `primitive` and `endprimitive` and can use either of the following two forms:

```
// port list style
primitive udp_name(output_port, input_ports);
output output_port;
input  input_ports;
reg    output_port;                      // only for
                                         // sequential UDP
initial output-port = expression;  // only for
                                         // sequential UDP
table     // define the behavior of UDP
   <table rows>
endtable
endprimitive


// port list declaration style
primitive udp_name(output output_port,
                input input_ports);
reg  output_port;                        // only for
                                         // sequential UDP
initial output-port = expression;  // only for
                                         // sequential UDP
table     // define the behavior of UDP
   <table rows>
endtable
endprimitive
```

where the first argument must be the output port and the other arguments are input ports.

A UDP can have exactly one output port but can have multiple input ports. The maximum number of inputs is limited by implementations, but must have at least 9 inputs for sequential UDPs and 10 inputs for combinational UDPs. All ports of a UDP are scalar; they cannot be bidirectional `inout` ports or vector ports. The output can only have a single-bit value with one of three states: 0, 1 or x. The high-impedance z is not supported. The z values passed to UDP inputs are treated as x values.

The behavior of a UDP is defined by a *state table*. The state table begins with the keyword `table` and is terminated with the keyword `endtable`. Each row of the table defines the output for a particular combination of the input values and is terminated by a semicolon. The permitted values for inputs and output are 0, 1 and x; high-impedance is not allowed. The order of the input ports of each row must be exactly the same as they are in the port list.

The instantiation of UDPs is exactly the same as that of gate primitives. It has the following form:

```
udp_id [delay] udp_instance[range](output, input1,
   ..., inputn);
```

where `delay` may only use up to two delays because the output of a UDP can never have a value of z. As a consequence, there is no turn-off delay. Because no delay specifications can be specified within the definition of UDPs, the only way to use delay specifications is through instantiations of UDPs.

5.4.2 Combinational UDP

A combinational UDP is used when its output is determined by only the current inputs of it. Whenever an input changes, the UDP is evaluated and the output is set to the output field of the table entry that matches all input values. All combinations of the inputs that are not explicitly specified will drive the output to an x.

As described previously, a UDP can be defined as one of the two forms: port list style and port list declaration style. In what follows, we use the port list style as an example to further illustrate how a combinational UDP is defined. The general form in port list style of a combinational UDP is as follows:

```
// port list style
primitive udp_name(output_port, input_ports);
output output_port;
input   input_ports;

// UDP state table
table  // the state table starts from here
   <table rows>
endtable
endprimitive
```

Each row of the state table has the following form:

```
<input1><input2> ... <inputn>:<output>;
```

The `<input#>` values appearing in each row of the state table must be in the same order as in the input port list. Inputs and output are separated by a ":". Each row of the state table must end with a ";". All possible combinations of inputs must be specified to avoid unknown output values.

What follows is a simple example of combinational UDP.

EXAMPLE 5.16 *A Simple Example of Combinational UDPs*

In this example, we define a two-input OR gate using a combinational UDP construct. Here, we only assume that 0 and 1 are the two valid values of inputs. Therefore, only four combinations need to be considered.

```
// a simple example of combinational UDPs.
primitive udp_or(f, a, b);
output out;  // declaration for output
```

```
input   a, b; // declarations for inputs.
table // state table starts from here
//     a   b   :   f;
       0   0   :   0;
       0   1   :   1;
       1   0   :   1;
       1   1   :   1;
endtable      // end state table
endprimitive // end of udp_or definition
```

As a matter of fact, using a combinational UDP to define a combinational logic circuit corresponds to directly describing the behavior of the logic circuit in table form. In order to provide the capability of describing a large state table in a concise manner, a set of shorthand symbols is defined. The details of these symbols can be referred to in Table 5.4 below.

The following example shows how the shorthand symbols can be used to shorten the table. Here, the character "**?**" is used to denote 0, 1 or x.

EXAMPLE 5.17   *The Use of Shorthand Notation*

From the behavior of a two-input AND gate, the output is 0 whenever one of its input is 0, no matter what the value of the other input is. Hence, the state table can be shorten to three rows. In this example, we assume that the inputs can have three values: 0, 1 and x.

```
// an example of combinational UDPs.
primitive udp_and(f, a, b);
output f;
input   a, b;
table
  //   a   b   :   f;
       1   1   :   1;
       0   ?   :   0;
       ?   0   :   0; // ? expanded to 0, 1, x
endtable
endprimitive
```

As mentioned previously, a combinational UDP is virtually a truth table implementation of a logic circuit. The following example gives some further insight into this viewpoint by modeling a switching expression:

$$f(x, y, z,) = \tilde{\ }(\tilde{\ }(x \mid y) \mid \tilde{\ }x \mathbin{\&} z)$$

To model this function using a combinational UDP, we need to compute its truth table on which the state table of UDP is constructed.

EXAMPLE 5.18 *A Simple Example of a Combinational UDP*

Based on the expression described above, the truth table is obtained as shown in the state table below.

```
primitive my_function(f, x, y, z);
output f;
input x, y, z;
table  // truth table for f(x, y, z,) = ~(~(x | y) | ~x
       // & z);
//   x y z : f
     0 0 0 : 0;
     0 0 1 : 0;
     0 1 0 : 1;
     0 1 1 : 0;
     1 0 0 : 1;
     1 0 1 : 1;
     1 1 0 : 1;
     1 1 1 : 1;
endtable
endprimitive
```
∎

UDPs can be instantiated exactly like gate or switch primitives, such as in the following example.

EXAMPLE 5.19 *Instantiations of Combinational UDPs*

This example uses the UDPs defined previously, except udp_xor, to construct a full adder. Of course, this is a structural modeling. It is interesting to write a test bench to verify it. Both the udp_xor UDP and the test bench are left to the reader as exercises.

```
// an example illustrates the instantiations of UDPs.
module UDP_full_adder(sum, cout, x, y, cin);
output sum, cout;
input  x, y, cin;
wire   s1, c1, c2;
// instantiate udp primitives
   udp_xor (s1, x, y);
   udp_and (c1, x, y);
   udp_xor (sum, s1, cin);
   udp_and (c2, s1, cin);
   udp_or  (cout, c1, c2);
endmodule
```
∎

5.4.3 Sequential UDP

Sequential UDPs are used to model sequential circuits such as flip-flops and latches, including both level-sensitive and edge-sensitive behavior. As described previously, a sequential UDP can be defined by one of the following two forms: port list style and port list declaration style. In what follows, we use the port list style as examples to further illustrate how a sequential UDP is defined. The general form in port list style of a sequential UDP is as follows:

```
// port list style
primitive udp_name(output_port, input_ports);
output  output_port;
input   input_ports;
reg     output_port;            // unique for
                                // sequential UDP
initial output-port = expression; // optional for
                                // sequential UDP

// UDP state table
table  // keyword to start the state table
   <table rows>
endtable
endprimitive
```

The output port also needs to be declared as a reg data type and an initial statement can be used to initialize the output port. The initial statement begins with the keyword initial and specifies the value of the output port when simulation begins. Unlike in modules, the declaration of the output port in sequential UDPs cannot be combined with the reg data type; each of them must be declared in a separate statement. The assignment statement must assign a single-bit value to the output port.

Each row of the state table has the following form:

```
<input1><input2>...<inputn>:<current_state>:
   <next_state>;
```

The <input#> values appearing in each row of the state table must be in the same order as in the input list. Inputs, current state and next state are separated by a colon ":". Each row of the state table must end with a ";". The input specifications can be either level or edge transitions. As in combinational UDPs, all possible combinations of inputs must be specified to avoid unknown output values.

The following simple example models a level-sensitive D-type latch.

EXAMPLE 5.20   *A Level-Sensitive D-type Latch*

As described before, the output q has to be declared as reg type and there is an additional field in each row of the state table. This new field represents the current state of the sequential UDP while the output field in a sequential UDP represents the next state.

```
// define a level-sensitive latch using UDP.
primitive d_latch(q, d, gate, clear);
output q;
input   d, gate, clear;
reg     q;
initial  q = 0; // initialize output q
//state table
table
// d  gate  clear : q :  q+;
   ?    ?     1    : ? :  0 ;  // clear
   1    1     0    : ? :  1 ;  // latch q = 1
   0    1     0    : ? :  0 ;  // latch q = 0
   ?    0     0    : ? :  - ;  // no change
endtable
endprimitive
```

In a level-sensitive latch, the output value is determined by the values of the inputs and the current state; in an edge-sensitive flip-flop, changes in the output are triggered by specific transitions of the inputs. In order to model the transitions of inputs, a pair of values in parenthesis such as (10) or a symbol such as f is used. The shorthand symbols used in UDPs are shown in Table 5.4. Each table entry can have a transition specification on at most one input. In general, all transitions that do not affect the output should be explicitly specified. Otherwise, they cause the output to change to x. In addition, all unspecified transitions default to the output value x.

The following simple example models an edge-sensitive D-type flip-flop.

EXAMPLE 5.21 *A Positive-edge Triggered D-type Flip-flop*

This example describes a positive-edge triggered D-type flip-flop using a sequential UDP. In the state table, the symbols r and f are used to represent 01 and 10, respectively.

```
// define a positive-edge triggered flip-flop using UDP.
primitive D_FF(q, clk, d);
```

TABLE 5.4 The shorthand symbols used in the state table of UDPs

Symbols	Meaning	Explanation
?	0, 1, x	Cannot be specified in an output field
b	0, 1	Cannot be specified in an output field
−	No change in state value	Can use only in a sequential UDP output field
r	(01)	Rising edge of a signal
f	(10)	Falling edge of a signal
p	(01), (0x), or (x1)	Potential rising edge of a signal
n	(10), (1x), or (x0)	Potential falling edge of a signal
*	(??)	Any value change in signal

```
output q;
input   clk, d;
reg     q;
initial q = 1'b1; // initialize output q
table
// clk d   q    q+
    r   0 : ? : 0 ; // latch q = 0
    r   1 : ? : 1 ; // latch q = 1
    f   ? : ? : - ; // no change
    ?   * : ? : - ; // no change
endtable
endprimitive
```

Mixing Level-sensitive and Edge-sensitive UDPs In order to model the actual hardware flip-flops with asynchronous clear or reset, UDP definitions allow a mixing of the level-sensitive and the edge-sensitive constructs in the same state table. When any input changes, the edge-sensitive cases are processed before the level-sensitive ones. Hence, the level-sensitive entries override the edge-sensitive entries when they specify different output values.

The following example addresses a positive-edge triggered T-type flip-flop with asynchronous clear input.

EXAMPLE 5.22   *A Positive-edge Triggered T-type Flip-flop*

Due to the inherent features of sequential UDPs, the level-sensitive entries override the edge-sensitive entries. The first row of the state table overrides the rest rows of the state table. As a consequence, the output is 0 once the input clear is raised to 1, regardless of the status of clock input clk or the previous output value.

```
// define a positive-edge triggered T-type flip-flop
// using UDP.
primitive T_FF(q, clk, clear);
output q;
input   clk, clear;
reg     q;
// define the behavior of edge-triggered T_FF
table
  //  clk    clear :  q  : q+;
        ?      1   :  ?  : 0 ;  // asynchronous clear
        ?     (10) :  ?  : - ;  // ignore negative edge of
                                // clear
      (01)     0   :  1  : 0 ;  // toggle at positive edge
      (01)     0   :  0  : 1 ;  // of clk
      (1?)     0   :  ?  : - ;  // ignore negative edge of
                                // clk
endtable
endprimitive
```

The following example instantiates four T-type flip-flops defined above and constructs a 4-bit ripple counter.

EXAMPLE 5.23   *An Example of Sequential UDP Instantiations*

In this example, four T-type flip-flops are instantiated to construct a 4-bit ripple counter. Because of being triggered in the positive edge, the counter is a down counter. Why? Try to explain this by drawing a timing diagram.

```
// an example of sequential UDP instantiations.
module ripple_counter(clock, clear, q);
input  clock, clear;
output [3:0] q;

// instantiate the T_FFs.
T_FF tff0(q[0], clock, clear);
T_FF tff1(q[1],  q[0], clear);
T_FF tff2(q[2],  q[1], clear);
T_FF tff3(q[3],  q[2], clear);
endmodule
```

In summary, both combinational and sequential UDPs have the following general features:

1. UDPs model functionality only; they do not model timing or process technology.
2. They have exactly one output terminal and are implemented as a lookup table in memory.
3. UDPs are not the appropriate method to design a block because they are usually not accepted by synthesis tools.
4. The UDP state table should be specified as completely as possible.
5. It should use shorthand symbols to combine table entries wherever possible.

Although both types of UDPs have almost the same syntax of declaration, sequential UDPs have the following unique features. First, they must contain a `reg` declaration for the output port. Secondly, the output port always has the same value as the internal state (next state). Thirdly, the initial value of the output port can be specified in an `initial` statement.

Review Questions

Q5.21   What are two declaration styles of UDPs, regardless of combinational or sequential UDPs?

Q5.22   Describe the features of combinational UDPs.

Q5.23   Describe the features of sequential UDPs.

Q5.24   What is the difference of the state table between combinational and sequential UDPs?

Q5.25   How can a UDP be instantiated?

SUMMARY

There are three behavioral ways that are widely used to model designs. These include tasks, functions and UDPs. Tasks and functions provide the ability to re-use the same piece of code from many places in a design. In addition, they provide a means of breaking up a large design into smaller ones so as to make it easier to read and debug the source descriptions.

Tasks are declared with the keywords `task` and `endtask`. A task can contain timing controls, has at least one output arguments, or has no input arguments. Tasks can call other tasks and functions as well. A task can be either static or automatic (also called reentrant, dynamic), depending on whether the keyword `automatic` appears in the declaration or not. Static tasks have the feature that all declared items are statically allocated. Automatic tasks have the feature that all declared items are dynamically allocated for each call.

Functions are declared with the keywords `function` and `endfunction`. A function is used when the procedure has no timing controls, needs to returns a single value, has at least one input argument, has no output or inout arguments or has no nonblocking assignments. Functions can only call functions. A function can be either a (static) function, declared without the keyword `automatic`, or an automatic function, declared with the keyword `automatic`. Static functions have the feature that all declared items are statically allocated. Automatic functions have the feature that all function items are allocated dynamically for each recursive call.

In addition to user-defined tasks and functions, there is a rich set of predefined tasks and functions called system tasks and functions, whose identifiers are prefixed with a character "$." System tasks and functions can be used for various applications in writing test benches, describing designs abstractly in behavioral style, and so on. The system tasks and functions can be divided into the following types: display system tasks, timescale system tasks, simulation time system functions, simulation control system tasks, file I/O system tasks, string formatting system tasks, conversion system functions, probabilistic distribution system functions, stochastic analysis system tasks, command line input and PLA modeling system tasks.

UDPs provide a means to model a design in a truth table. Two types of UDPs are provided for users to build their own primitives. These two UDPs are combinational UDPs and sequential UDPs. Combinational UDPs are defined when the output is uniquely determined by combination of the inputs. Sequential UDPs are defined when the output is determined by combination of the current output and the inputs. A UDP is instantiated in exactly the same way as a gate primitive.

REFERENCES

1. J. Bhasker, *A Verilog HDL Primer,* 3rd Edn, Star Galaxy Publishing, 2005.
2. S. Palnitkar, *Verilog HDL: A Guide to Digital Design and Synthesis,* 2nd Edn, SunSoft Press, 2003.
3. IEEE 1364-2001 Standard, *IEEE Standard Verilog Hardware Description Language*, 2001.

PROBLEMS

5.1 Write a task that counts the number of ones in a byte. In addition, write a test bench to apply stimuli to the task and check the results.

5.2 Write a task that counts the number of trailing ones in a byte. In addition, write a test bench to apply stimuli to the task and check the results.

5.3 Write a task that counts the number of leading zeros in a byte. In addition, write a test bench to apply stimuli to the task and check the results.

5.4 Write a task that counts the number of leading ones in a byte. In addition, write a test bench to apply stimuli to the task and check the results.

5.5 Write a task to compute the even parity of a 16-bit number. In addition, write a test bench to apply stimuli to the task and check the results.

5.6 Write a task to generate a reset signal that is raised to 1 for three clocks when it is called each time. The reset signal starts at the first negative edge of the clock signal after it is called. Write a test bench to apply stimuli to the task and check the results.

5.7 Write a task to generate a reset signal that is cleared to 0 for n clocks when it is called each time, where n is passed by the calling statement. The reset signal starts at the first negative edge of the clock signal after it is called. Write a test bench to apply stimuli to the task and check the results.

5.8 Write a task to dump the contents of a memory starting from a specified region. In addition, write a test bench to apply stimuli to the task and check the results.

5.9 Write a function that counts the number of ones in a byte. In addition, write a test bench to apply stimuli to the function and check the results.

5.10 Write a function that counts the number of trailing ones in a byte. In addition, write a test bench to apply stimuli to the function and check the results.

5.11 Write a function that counts the number of leading zeros in a byte. In addition, write a test bench to apply stimuli to the function and check the results.

5.12 Write a function that counts the number of leading ones in a byte. In addition, write a test bench to apply stimuli to the function and check the results.

5.13 The greatest common divisor (GCD) of two non-negative integer numbers m and n is defined recursively as follows:

$$GCD(m, n) = \begin{cases} m & \text{if } n = 0 \\ GCD(n, m \bmod n) & \text{if } m > n \end{cases}$$

   (a) Write an automatic function that computes the $GCD(m, n)$.

   (b) Use the `$random` system function to generate two random numbers m and n as the inputs and then use the `$display` system task to display the result on the standard output.

   (c) Use the `$fgets` and `$sscanf` system tasks to read two numbers m and n from standard input and use the `$display` system task to display the result on the standard output.

5.14 Write a function to compute the odd parity of a 16-bit number. In addition, write a test bench to apply stimuli to the function and check the results.

5.15 Write a function that carries out an arithmetic left shift of a 16-bit vector.

5.16 Write a function that carries out an arithmetic right shift of a 16-bit vector.

5.17 Write a combinational UDP to model the logic circuit shown in Figure 5.1.

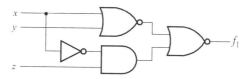

FIGURE 5.1 The figure for Problem 5.17

5.18 Write a combinational UDP to model the logic circuit shown in Figure 5.2.

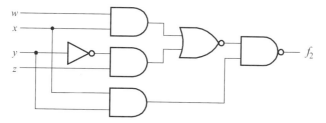

FIGURE 5.2 The figure for Problem 5.18

5.19 Use a combinational UDP to model the following switching expressions:

   (a) $f(x, y, z) = y'z' + x'y + xz$

   (b) $f(x, y, z) = \Sigma(0, 2, 4, 5, 6)$

   (c) $f(w, x, y, z) = \Sigma(1, 3, 6, 7, 11, 12, 13, 15)$

   (d) $f(w, x, y, z) = \Sigma(0, 1, 6, 9, 10, 11, 14, 15)$

5.20 The operations of an SR flip-flop are as follows. At the positive edge of the clock, the output sets to 1 if the input S is 1; the output clears to 0 if the input R is 1.

The output remains unchanged if both R and S are 0; the output is undetermined if both R and S are 1. Consider each of the following modifications and write a sequential UDP to model the resulting flip-flop:

(a) Suppose that the SR flip-flop is modified as a clear-dominated type, that is, the output is cleared to 0 if both R and S are 1.

(b) Suppose that the SR flip-flop is modified as a set-dominated type, that is, the output is set to 1 if both R and S are 1.

HIERARCHICAL STRUCTURAL MODELING

IN THE previous chapters, we have discussed a variety of modeling styles, including structural style at gate level as well as switch level, dataflow style and behavioral style. In this chapter, we consider the three closely related issues of hierarchical structural modeling. These include instantiations, generate statements and configurations.

The first issue is the mechanism of instantiation through which the hierarchical structure is formed by embedding modules into other modules. These modules can be gate and switch primitives, UDP primitives and/or modules. Higher-level modules create instances of lower-level modules and communicate with them through input, output and bidirectional ports, which can be scalar or vector.

The second issues is the `generate` statement, which can conditionally generate declarations and instantiations into a design. Almost what can be put inside a module can also be placed within `generate` statements. The `generate` statement can be employed to build a large regular design, such as n-bit ripple carry adders, $m \times n$ array multipliers and $m \times n$ array dividers, among others.

The final issue is the configuration. A configuration is a set of rules that maps instances to cells, which include UDPs, configurations and modules. By using configurations, we may specify a new set of target libraries so as to change the mapping of a design without having to change the source description.

6.1 MODULE

The basic units of Verilog HDL are modules. Like a hardware module or an IC (integrated circuit), each module has two major parts: *interface* and *body*. The interface provides a way through which the module can communicate with the outside world and the body defines the functionality of the module. In the previous chapters, we have discussed how to describe the body of a module in various modeling styles. In this section, we discuss how to define the interface of a module and how to use the interface to interconnect with other modules.

Digital System Designs and Practices   Ming-Bo Lin
© 2008 John Wiley & Sons (Asia) Pte Ltd

6.1.1 Module Definition

A module is defined by using the keywords module and endmodule and can be one of the following two forms:

```
// port list style
module module_name [#(parameter_declarations)]
  [port_list];
    parameter_declarations; // if parameter ports are
                            // used
    port_declarations;
    other_declaration;
    statements;
endmodule

// port list declaration style
module module_name [#(parameter_declarations)]
  [port_declarations];
    parameter_declarations; // if parameter ports are
                            // used
    other_declarations;
    statements;
endmodule
```

The parameter is optional and, if presents, is used to specify an ordered list of the parameters for the module. It can be declared within the module or at the place between the module_name and port_list or port_declarations. The port_list gives the list of ports through which the module can communicate with other modules. A port is declared as input, output or inout. Ports declared in the port list cannot be redeclared within the module. The statements define the functionality of a module. The number of ports that can be specified in a module is limited by the implementation but is at least 256.

6.1.1.1 Port Declaration   The interface signals (including supply and ground) of any Verilog HDL module can be cast into three types: input, output and inout. In what follows, we describe these three port declarations in detail.

The input Port Declaration   The keyword input is used to declare a group of signals as input ports. It has the following form:

```
input [net_type] [signed] [range] port_names;
```

where net_type can be any of net data types. The net type is often omitted if the net type is wire. The keyword signed is used to declare that the net is signed. As shown in Figure 6.1, an input port must be a net data type but it can be driven by a net or a variable port. The variable data types include reg, integer and time.

FIGURE 6.1 Port declaration

The output Declaration   The keyword output is used to declare a group of signals as output ports. It has the form:

```
output [net_type] [signed] [range] port_names;
output [reg] [signed] [range] port_names;

output [port_var_type] port_names;
```

where port_var_type can be integer or time variable data types. As shown in Figure 6.1, an output port can be a net or variable data type but it only drives a net port, which of course is an input port of another module. The variable data types include reg, integer and time.

The inout Declaration   The keyword inout is used to declare a group of signals as bidirectional ports. It has the form:

```
inout [net_type] [signed] [range] port_names;
```

As shown in Figure 6.1, an inout port can only be a net data type as well as it can drive or be driven by a net port.

A complete port declaration of a port must include two parts: a port type and its associated data type. The port types include input, output or inout and the data types include nets or variables. A complete port declaration cannot be redeclared again within the module. The example shown in the following declares three input ports as nets, x, y and c_in, as well as two output ports as reg variables: sum and c_out.

```
module adder(x, y, c_in, sum, c_out);
input [3:0] x, y;    // no explicit declaration-- net is
                     // unsigned
input c_in;          // no explicit declaration-- net is
                     // unsigned
output reg [3:0] sum;
output reg c_out;
```

Implicit nets are considered unsigned. Nets connected to ports without an explicit net declaration are considered unsigned, unless the port is declared as signed.

```
module test(x, y, z, w, v);
input  [3:0] x, y;  // no explicit declaration-- net
                    // is unsigned
```

```
input   signed z;      // explicit declaration-- net is
                       // signed
output signed w;       // explicit declaration-- w is
                       // signed
output reg v;          // no explicit declaration-- v is
                       // unsigned
reg     w;             // reg w inherits signed attribute
                       // from port
```

An implicit continuous assignment is assumed to exist between a port and a port expression. When a port and the port expression are of different widths, the same rules used for a continuous assignment are applied to match the port.

6.1.2 Parameters

Parameters are constants, which can be used throughout the module defining them, and are often used to specify delays and widths of variables. Two types of parameters are *module parameters* and *specify parameters*. The module parameters can be defined by using the keyword `parameter` or `localparam` within a module. The major difference between these two parameters is that a parameter defined by `parameter` can be modified by using the `defparam` parameter or through module instantiations but those defined by `localparam` cannot. These two module parameters combined with the `` `define `` compiler directive comprise the three ways of defining constants in Verilog HDL. In other words, a constant can be defined in any of the following: `` `define `` compiler directive, `parameter` declaration and `localparam` declaration. The details of the `` `define `` compiler directive will be discussed in Section 7.4.1 and the specify parameters are addressed in Section 7.3.2.

6.1.2.1 `parameter` ***Declaration***   The `parameter` is the most common approach used to define module parameters that can be overridden by `defparam` or module instance parameter value assignment. It has the form:

```
parameter [signed] [range] param_assignments
parameter var_types param_assignments
```

where `var_types` can be one of the following: `integer`, `time`, `real` and `realtime`. Note that no range specification or the keyword `signed` is allowed when the second form is used.

In the following, we give some examples of the usage of `parameter`.

```
parameter SIZE = 7;
parameter WIDTH_BUSA = 24, WIDTH_BUSB = 8;
parameter signed [3:0] mux_selector = 4'b0;
```

6.1.2.2 `localparam` ***Declaration***   The `localparam` is used to define parameters local to a module. It is identical to `parameter` except that it cannot be overridden by the `defparam` statement or by module instance parameter value assignment. The declaration of a local parameter is exactly like that of the `parameter` except that the keyword `localparam` is used instead.

The following are some examples of the usage of `localparam`.

```
localparam SIZE = 7;
localparam WIDTH_BUSA = 24, WIDTH_BUSB = 8;
localparam signed [3:0] mux_selector = 4'b0;
```

In summary, the `localparam` is used to define parameters when their values are not changed whereas the `parameter` is used to define parameters when their values may be changed at module instantiation or by using the `defparam` statement.

Parameter Dependence   In some applications, it is useful to define a parameter expression containing another parameter so that an update of a parameter, either by the `defparam` statement or in an module instantiation statement for the module that defined these parameters, will automatically update another parameter. For example, in the following parameter declaration, an update of `word_size` will automatically update `memory_size`.

```
parameter  word_size = 16,
localparam memory_size = word_size * 512;
```

Since `memory_size` depends on the value of `word_size`, a modification of `word_size` changes the value of `memory_size` accordingly.

6.1.2.3 *Parameter Ports*
As described previously, module parameters declared with `parameter` can be placed after port list/port list declarations or between module name and port list/port list declarations. However, when the module is declared in port list declaration style, the module parameters often need be declared before the declarations of port list because they may be referenced in the port list declarations.

When a `parameter` is placed between module name and port list/port list declarations, it is called a *parameter port*. Like the declarations of regular module parameters, a parameter port can have both the type and range specifications and has the following form:

```
module module_name
#(parameter [signed] [range] param_assignments
  parameter var_types param_assignments
  )
(port list or port list declarations)
   ...
endmodule
```

What follows are some examples of the usage of parameter ports.

```
module module_name
#(parameter SIZE = 7,
  parameter WIDTH_BUSA = 24, WIDTH_BUSB = 8,
  parameter signed [3:0] mux_selector = 4'b0
  )
(port list or port list declarations)
   ...
endmodule
```

Note that there is no ";" after parameter declarations in the parameter ports.

6.1.3 Module Instantiation

In Verilog HDL, module definitions cannot be nested. However, much like hardware modules, a module can incorporate a copy (called an instance) of another module into itself through instantiation. To instantiate an instance of a module, the following syntax is used:

```
module_name [#(parameters)] instance_name [range]
  ([ports]);
module_name [#(parameters)] instance_name [{,instance_
  name}]([ports]);
```

where `range` specification means that the array of instances can be created through instantiation. In addition, one or more module instances can be created in a single module instantiation statement.

6.1.3.1 *Port Connection Rules* The port connection rules of Verilog HDL modules are consistent with that of actual hardware modules. Connecting ports to external signals can be done by one of the following two methods: *named association* and *positional association*.

Named Association In this method, the ports are connected by listing their names. The port identifiers and their associated port expressions are explicitly specified. The form is as follows:

```
.port_id1(port_expr1),..., .port_idn(port_exprn)
```

An unconnected port is just skipped or places an empty, such as "`.port_id()`".

Positional Association In this method, the ports are connected by the ordered list of ports, each corresponding to a port. The form is as follows:

```
port_expr1, ..., port_exprn
```

An unconnected port is just skipped, such as "x, , y", where a port is skipped between x and y.

Regardless of which method is employed, each port_expr can be any of the following:

1. an identifier (a net or a variable);
2. a bit-select of a vector;
3. a part-select of a vector;
4. a concatenation of the above;
5. an expression for input ports.

Note that the two port connection methods cannot be mixed in the same module. Moreover, Verilog HDL primitives, including built-in and user-defined primitives, can only be connected by positional association.

EXAMPLE 6.1 *Named Association*

The following module demonstrates how to connect ports through named association. Using this approach, the order of the port list is unimportant.

```
`timescale 1 ns / 100 ps
module ripple_counter_generate_tb;
// internal signals declarations:
reg  clk, clear;
wire [0:3] qout;
parameter clk_period = 20;
// Unit Under Test port map
   ripple_counter_generate UUT (
          .clk(clk), .clear(clear), .qout(qout));
initial begin
   repeat (3) @(posedge clk) clear = 1'b1;
   clear = 1'b0;
end
initial begin
   clk = 1'b0;
   forever  #(clk_period/2) clk = ~clk;
end
initial
   #500 $finish;
initial
   $monitor($realtime,"ns %b %b %h", clk, clear, qout);
endmodule
```

It is advisory that named association should be used for port connections at the top-level module because synthesis tools may change the order of ports in the port list of the module they generated.

Unconnected Ports   Unconnected inputs are driven to the "z" state; unconnected outputs are not used. The unconnected ports in an instantiation can be specified by leaving the port_expr blank.

Real Numbers in Port Connections   The real data type is not permitted to pass directly to a port. To do this, both $realtobits and $bitstoreal system functions are used to pass the bit patterns across module ports. An example has been shown in Section 5.3.4.

6.1.4 Module Parameter Values

When a module instantiates other modules, the higher-level module can change the values of parameters defined by the keyword parameter within the lower-level modules. This operation is called *parameter override*. The parameter override provides a mechanism to design a module that can be parameterized. In other words, the features or operations of the module can be changed by modifying the parameters defined within it. The following example illustrates how this idea is done.

EXAMPLE 6.2   *A Parameterized Module Example*

The module parameter N of the following parameterized module adder_nbit can be modified from its default value of 4 to any other value so that the adder_nbit can then be functioned as a new adder with new word size.

```
module adder_nbit(x, y, c_in, sum, c_out);
// I/O port declarations
parameter N = 4;  // set default value
input    [N-1:0] x, y;
input    c_in;
output   [N-1:0] sum;
output   c_out;
// specify the function of an n-bit adder using generate
assign {c_out, sum} = x + y + c_in;
endmodule                                        ■
```

There are two ways to override parameter values:

1. defparam statement;
2. module instance parameter value assignment.

In what follows, we describe these two ways in detail.

Using the `defparam` Statement A `defparam` statement is used to redefine the parameter values defined by the keyword `parameter` in any module instance throughout the design using the hierarchical name of the parameter and has the syntax:

```
defparam hierarchical_path_name1 = value1,
         hierarchical_path_name2 = value2,
         ...
         hierarchical_path_namen = valuen;
```

The `value` can be a constant expression involving only numbers and references to parameters. The referenced parameters must be defined in the same module as the `defparam` statement. This approach is particularly useful for grouping together the parameter value override assignments in the same module.

The following example illustrates how the `defparam` statement is used to redefine the parameter values defined within the lower-level module.

EXAMPLE 6.3 *Parameter Overridden: `defparam` Statement*

The following module demonstrates how the `defparam` statement is used to change the parameter values defined within the lower-level module, `counter_nbits`. The default value of N within the module `counter_nbits` is 4. The result is that two instances of `counter_nbits` with bits of 4 and 8, respectively, are created.

```
// define top level module
module two_counters(clock, clear, qout4b, qout8b);
input clock, clear;
output [3:0] qout4b;
output [7:0] qout8b;
// instantiate two counter modules
defparam cnt_4b.N = 4, cnt_8b.N = 8;
counter_nbits cnt_4b (clock, clear, qout4b);
counter_nbits cnt_8b (clock, clear, qout8b);
endmodule

module counter_nbits (clock, clear, qout);
parameter N = 4;  // define counter size
input clock, clear;
output reg [N-1:0] qout;
always @( posedge clear or negedge clock)
begin              // qout <= (qout + 1) % 2^N;
   if (clear) qout <= {N{1'b0}};
   else       qout <= (qout + 1) ;
end
endmodule
```
■

Module Instance Parameter Value Assignment   In this approach, the parameters defined by using the keyword `parameter` within a module are overridden by the parameters passed through parameter ports whenever the module is instantiated. There are two approaches in which parameter values can be specified:

1. **Positional association.** In this form, the order of the assignments must follow the order of declaration of the parameters within the module. It is not necessary to assign values to all of the parameters declared within a module and not possible to skip over a parameter. As a consequence, for those module parameters that are only local to the module are often declared as local parameters with the keyword `localparam`.

2. **Named association.** This form is similar to connecting module ports by named association. It explicitly links the names specified in the instantiated module and the associated value. The advantage of this apporach is that we only need to specify those parameters that are assigned new values.

The above two parameter assignment approaches cannot be mixed.

In the following, we give three examples to illustrate the method of module instance parameter value assignment.

EXAMPLE 6.4 *Module Instance Parameter Value Assignment: Positional Association*

In this example, two instances of module `counter_nbits` are instantiated through module instance parameter value assignment. One of them is a4-bit counter and the other is an 8-bit counter.

```
// define top level module
module two_counters(clock, clear, qout4b, qout8b);
input clock, clear;
output [3:0] qout4b;
output [7:0] qout8b;
// instantiate two counter modules
counter_nbits #(4) cnt_4b (clock, clear, qout4b);
counter_nbits #(8) cnt_8b (clock, clear, qout8b);
endmodule

module counter_nbits (clock, clear, qout);
parameter N = 4;  // define counter size
input clock, clear;
output reg [N-1:0] qout;
always @(posedge clear or negedge clock)
begin               // qout <= (qout + 1) % 2^N;
   if (clear) qout <= {N{1'b0}};
   else       qout <= (qout + 1) ;
end
endmodule
```
■

The following example describes how positional association is used to pass module instance parameter to the instantiated module.

EXAMPLE 6.5   *Module Instance Parameter Value Assignment: Positional Association*

In this example, an instance of module `hazard_static` is instantiated through module instance parameter value assignment. Two module parameters are passed to the instantiated module `hazard_static` through the use of the positional association approach.

```
// define top level module
module parameter_overriding_example(x, y, z, f);
input  x, y, z;
output f;
hazard_static #(4, 8) example (x, y, z, f);
endmodule

module hazard_static (x, y, z, f);
parameter DELAY1 = 2, DELAY2 = 5;
input x, y, z;
output f;
// internal declaration
wire  a, b, c;      // internal net
// logic circuit body
   and #DELAY2 a1 (b, x, y);
   not #DELAY1 n1 (a, x);
   and #DELAY2 a2 (c, a, z);
   or  #DELAY2 o2 (f, b, c);
endmodule
```

The following is an example of using named association to pass a module instance parameter to the instantiated module.

EXAMPLE 6.6   *Module Instance Parameter Value Assignment: Named Association*

In this example, an instance of module `hazard_static` is instantiated through module instance parameter value assignment. Two module parameters are passed to the instantiated module `hazard_static` through the use of the named association approach.

```
// define top level module
module parameter_overriding_example(x, y, z, f);
input  x, y, z;
output f;
hazard_static #(.DELAY2(4), .DELAY1(6))
  example (x, y, z, f);
```

```
endmodule

module hazard_static (x, y, z, f);
parameter DELAY1 = 2, DELAY2 = 5;
input x, y, z;
output f;
// internal declaration
wire  a, b, c;      // internal net
// logic circuit body
    and #DELAY2 a1 (b, x, y);
    not #DELAY1 n1 (a, x);
    and #DELAY2 a2 (c, a, z);
    or  #DELAY2 o2 (f, b, c);
endmodule
```

The advantage of parameter value assignment by named association is that it can minimize the chance of error.

6.1.5 Hierarchical Names

In Verilog HDL, an identifier can be defined within one of the following four elements:

1. Modules.
2. Tasks.
3. Functions.
4. Named blocks (See Section 7.1.3).

Within each element, an identifier must uniquely declare one item. Any identifier can be accessed directly within the elements defining it. To refer an identifier defined within the other elements, a hierarchical path name must be used. The hierarchical path name is formed by using identifiers separated by a period character for each level of hierarchy and has the form:

```
hierarchical_branch[{.simple_hierarchical_branch}]
```

Based on this notation, each identifier in a Verilog HDL description has a unique hierarchical path name. The complete hierarchical path name of any identifier starts from the top-level module and can be used in any level in a description. The top-level module is a module that is not instantiated by any other modules.

The following example shows some hierarchical path names of Figure 1.5. Each hierarchical path name is formed by appending the lower-level name on the right side of the higher-level name separated by a period. For instance, the 4-bit adder is constructed by four full adders: fa_1, ..., fa_4. Thus, the complete hierarchical path name of fa_1 is 4bit_adder.fa_1. To refer the net S at the output of the xor gate within fa_1, the complete hierarchical path name 4bit_adder.fa_1.ha_1.xor1.S is used.

```
4bit_adder                        // top level --- 4bit_
                                  // adder
4bit_adder.fa_1                   // fa_1 within 4bit_adder
4bit_adder.fa_1.ha_1              // ha_1 within fa_1
4bit_adder.fa_1.ha_1.xor1         // xor1 within ha_1
4bit_adder.fa_1.ha_1.xor1.S  // net s within xor1
```

Review Questions

Q6.1   Describe the port connection rules.

Q6.2   What is the difference between a gate instantiation and a module instantiation?

Q6.3   What are the values of ports when they are left unconnected?

Q6.4   Describe the two methods that can be used to override module parameter values defined by `parameter`.

Q6.5   What does a parameterized module mean?

Q6.6   What are the differences between `parameter` and `localparam`?

Q6.7   What are the differences between `defparam` and `parameter`?

Q6.8   What is the meaning of parameter ports?

Q6.9   Explain what a top-level module is.

Q6.10   Explain the meaning of a hierarchical name.

6.2   GENERATE STATEMENT

A `generate` statement allows selection and replication of some statements during elaboration time. The elaboration time is the time after a design has been parsed but before simulation begins. The `generate` statement uses the keywords `generate` and `endgenerate` and has the following general form:

```
generate
   generate-declarations
   generate-loop statements
   generate-conditional statements
   generate-case statements
   generate-block
   nested generate statements
endgenerate
```

The statements can be used within the `generate` statement, named *generate-statements*, and are one or more of the following: declarations, modules, parameter override, user-defined primitives, gate primitives, continuous assignments, `initial` blocks and `always` blocks. The declarations include data types, events, genvars, functions and tasks. Both net and variable data types are allowed within `generate` statements. These include net, `reg`, `integer`, `real`, `time` and `realtime`. In addition, the `event` data type is also permitted within `generate` statements.

The power of `generate` statements is that they can conditionally generate declarations and instantiations into a design. Almost what can be put inside a module can also be placed within `generate` statements except the following: parameters, local parameters and port declarations: `input`, `output` and `inout`. There are three kinds of statements that can be used within a `generate` statement: *generate-loop*, *generate-conditional*, and *generate-case*.

6.2.1 Generate-Loop Statement

A generate-loop is formed by using a `for` statement within a `generate` statement. The generate-loop allows statements to be duplicated at elaboration time. The generate-loop is of the form:

```
for (init_expr; condition_expr; update_expr)
   begin: block_name
       generate_statements
   end
```

To use the `for` statement within a `generate` statement, it is necessary to declare an index variable by using the keyword `genvar`. This index variable is an integer and often referred to as a *genvar*. Since a genvar is used only in the evaluation of a generate-loop, it cannot have a negative value. A genvar can be declared either inside or outside of the generate statement using it. The `genvar` declaration is of the form:

```
genvar genvar_id1, ..., genvar_idn;
```

Genvars are only defined during the evaluation of the generate blocks. They do not exist during simulation of the design. In general, generate-loops can be nested. However, two generate-loops using the same genvar as an index cannot be nested.

The `block_name` of a block enclosed by `begin-end` is required and is used to refer any instance names that are local to the generate-loop.

The following example demonstrates how a continuous assignment is used within a generate-loop statement.

EXAMPLE 6.7 *The Generate-loop and* `assign`

This example uses a continuous assignment within a generate-loop for converting Gray code into binary code. To convert a Gray code into its equivalent binary code, we may count the number of '1s' from MSB to the current position, i. If it is odd, the binary bit is 1; otherwise, the binary bit is 0. Therefore, the reduction operator ˆ is applied to an indexed sub-range, counting from $SIZE - 1$ to i, of the `gray` vector and reduces it into a single binary bit. By repeating this process from bit 0 to bit $SIZE - 1$, the conversion is completed.

```
//an example illustrates the usage of generate statement.
//an example of converting Gray code into binary code.
module gray2bin1 (gray, bin);
```

```
parameter SIZE = 8;   // this module is parameterized
input   [SIZE-1:0] gray;
output [SIZE-1:0] bin;
genvar i;              // generate statement variable.
generate for (i = 0; i < SIZE; i = i + 1) begin: bit
   assign bin[i] = ^gray[SIZE-1:i];
end endgenerate
endmodule
```

During elaboration time, the generate-loop is expanded. The body of the `for` statement is replicated once for each value of the iteration. For instance, the `assign` statement in the preceding example will be expanded into the following statements:

```
assign bin[0] = ^gray[SIZE-1:0];
assign bin[1] = ^gray[SIZE-1:1];
assign bin[2] = ^gray[SIZE-1:2];
assign bin[3] = ^gray[SIZE-1:3];
assign bin[4] = ^gray[SIZE-1:4];
assign bin[5] = ^gray[SIZE-1:5];
assign bin[6] = ^gray[SIZE-1:6];
assign bin[7] = ^gray[SIZE-1:7];
```

The `always` block can also be used within a generate statement, such as in the following example.

EXAMPLE 6.8 *The Generate-loop and* `always`

This example simply replaces the continuous assignment statement in the preceding example with an `always` block. In this example, it is better to use the wildcard character (*) in the sensitivity list of the `always` statement to avoid warning from synthesis tools.

```
// an example of converting Gray code into binary code.
module gray2bin2 (gray, bin);
parameter SIZE = 8;   // this module is parameterized.
input   [SIZE-1:0] gray;
output [SIZE-1:0] bin;
reg     [SIZE-1:0] bin;
genvar i;              // generate variable
//generate loop
generate for (i = 0; i < SIZE; i = i + 1) begin:bit
   always @(*) // using always @(*) to avoid synthesizer
               // warning.
      bin[i] = ^gray[SIZE - 1: i];
end endgenerate
endmodule
```

6.2.2 Generate-Conditional Statement

A generate-conditional statement allows modules, gate primitives, UDPs, continuous assignments, `initial` blocks, `always` blocks, tasks and functions, to be instantiated into another module based on an `if-else` conditional expression. The generate-conditional statement can be used alone or inside a generate-loop. It is of the form:

```
if (condition) generate_statements
[else generate_statements]

if (condition1) generate_statements
[else if(condition2) generate_statements
    [else generate_statements]]
```

The `condition` must be a static condition; namely, it must be computable at elaboration time. This implies that the `condition` must only be a function of constants and parameters. The `generate_statements` can be any set of statements that are allowed within `generate` statements. Based on the condition, the appropriate set of statements are selected for being expanded during elaboration time.

In the following, we give three examples for illustrating the usage of `generate` statements coupled with modules, continuous assignments and `always` statements.

EXAMPLE 6.9 *A Parameterized n-bit Ripple-carry Adder*

In this example, a conditional expression `if-else` is employed to set up the boundary cells, the LSB and the MSB, as well as the rest of the bits. The LSB cell needs to accept the external carry input `c_in` and the MSB cell is required to send the carry `c_out` out of the module. Each of the rest cells accepts the carry from its preceding cell and sends the carry to its succeeding cell.

```
// using module instantiations inside generate block
module adder_nbit(x, y, c_in, sum, c_out);
// I/O port declarations
parameter N = 4;  // define N as a parameter
input   [N-1:0] x, y;
input   c_in;
output [N-1:0] sum;
output c_out;
// specify the function of an N-bit adder using generate
genvar i;
wire    [N-2:0] c;    // internal carries declared as nets
generate
for (i = 0; i < N; i = i + 1) begin: adder
   if (i == 0)         // specify LSB
      full_adder fa (x[i], y[i], c_in, sum[i], c[i]);
   else if (i == N-1) // specify MSB
```

```
          full_adder fa (x[i], y[i], c[i-1], sum[i], c_out);
   else                 // specify other bits
          full_adder fa (x[i], y[i], c[i-1], sum[i], c[i]);
end endgenerate
endmodule
// define a full adder at data flow level.
module full_adder(x, y, c_in, sum, c_out);
// I/O port declarations
output  sum, c_out;
input   x, y, c_in;
// specify the function of a full adder
   assign {c_out, sum} = x + y + c_in;
endmodule
```
∎

The following example demonstrates the combined use of if-else conditional expression and continuous assignments.

EXAMPLE 6.10 *A Parameterized n-bit Ripple-carry Adder*

This example simply replaces module instantiations in the preceding example with continuous assignments. The rest operations remain the same and so that we do not explain this any further.

```
// using continuous assignments inside generate block
module adder_nbit(x, y, c_in, sum, c_out);
// I/O port declarations
parameter N = 4;  // define N as a parameter
input  [N-1:0] x, y;
input  c_in;
output [N-1:0]  sum;
output c_out;
// specify the function of an N-bit adder using generate
genvar i;
wire   [N-2:0] c;    // internal carries declared as nets
generate for (i = 0; i < N; i = i + 1) begin: adder
   if (i == 0)          // specify LSB
      assign {c[i], sum[i]} =  x[i] + y[i] + c_in;
   else if (i == N-1) // specify MSB
      assign {c_out, sum[i]} =  x[i] + y[i] + c[i-1];
   else                 // specify other bits
      assign {c[i], sum[i]} =  x[i] + y[i] + c[i-1];
end endgenerate
endmodule
```
∎

The following example demonstrates the combined use of if-else conditional expression and always blocks.

EXAMPLE 6.11 *A Parameterized n-bit Ripple-carry Adder*

This example simply replaces continuous assignments in the preceding example with always blocks. The rest operations remain the same so that we do not explain it furthermore.

```
// using always blocks inside generate block
module adder_nbit(x, y, c_in, sum, c_out);
// I/O port declarations
parameter N = 4;  // define N as a parameter
input  [N-1:0] x, y;
input  c_in;
output reg [N-1:0]  sum;
output reg c_out;
// specify the function of an N-bit adder using generate
genvar i;
reg    [N-2:0] c;    // internal carries declared as nets
generate for (i = 0; i < N; i = i + 1) begin: adder
   if (i == 0)          // specify LSB
      always @(*) {c[i], sum[i]} =  x[i] + y[i] + c_in;
   else if (i == N-1) // specify MSB
      always @(*) {c_out, sum[i]} =  x[i] + y[i] + c[i-1];
   else                 // specify other bits
      always @(*) {c[i], sum[i]} =  x[i] + y[i] + c[i-1];
end endgenerate
endmodule
```
■

In a synthesizable module, an always block corresponds to a piece of logic. However, how can we specify a combinational logic or a sequential logic? We will answer this and return to this problem in later chapters.

The following example demonstrates how an n-bit two's complement adder can be easily constructed by using the powerful generate statement.

EXAMPLE 6.12 *An n-bit Two's Complement Adder*

In this example of an n-bit two's complement adder, two generate blocks are used to construct an n-bit one's complement generator circuit and an n-bit ripple-carry adder, respectively. The two's complement adder is then performed by using the fact that the two's complement of a number is equivalent to its one's complement plus one. Based on this, the two's complement adder can then be constructed by routing the output of the one's complement generator to one of the inputs of the adder and connecting the control signal of the one's complement generator together with the carry input, named mode, of the adder. The result is a subtracter when mode is 1 and an adder when mode is 0.

```
module twos_adder_nbit(x, y, mode, sum, c_out);
// I/O port declarations
```

```
parameter N = 4;   // define N as a parameter
input   [N-1:0] x, y;
input   mode;       // mode = 1: subtraction; =0: addition
output [N-1:0] sum;
output c_out;
// specify the function of an n-bit adder using generate
genvar i;
wire [N-2:0] c;    // internal carries declared as nets.
wire [N-1:0] t;    // true/one's complement outputs
generate for (i = 0; i < N; i = i + 1)
   begin: ones_complement_generator
     xor xor_ones_complement (t[i], y[i], mode);
end endgenerate
generate for (i = 0; i < N; i = i + 1) begin: adder
   if (i == 0)        // specify LSB
     full_adder fa (x[i], t[i], mode, sum[i], c[i]);
   else if (i == N-1) // specify MSB
      full_adder fa (x[i], t[i], c[i-1], sum[i], c_out);
   else               // specify other bits
      full_adder fa (x[i], t[i], c[i-1], sum[i], c[i]);
end endgenerate
endmodule
//
// define a full adder at data flow level.
module full_adder(x, y, c_in, sum, c_out);
// I/O port declarations
output sum, c_out;
input   x, y, c_in;
// specify the function of a full adder
   assign {c_out, sum} = x + y + c_in;
endmodule
```

Figure 6.2 shows the result of the n-bit two's complement adder generated from a synthesis tool. From Figure 6.2(a), you can see that there are four XOR gates and four full-adder modules. Figure 6.2(b) shows the result after dissolving the second and third full-adder modules.

In the rest of this sub-section, we use some examples to illustrate how tasks, functions and UDPs are used within generate statements. The first example is to demonstrate the combined use of a task and a generate ststement.

EXAMPLE 6.13   *Combined use of a Task and a* generate *Statement*

This example explains how to combine the use of a task and a generate statement. Here, we define a task to carry out the operations of the full-adder. The task full_adder is then invoked within the always blocks inside the generate statement to complete an n-bit adder.

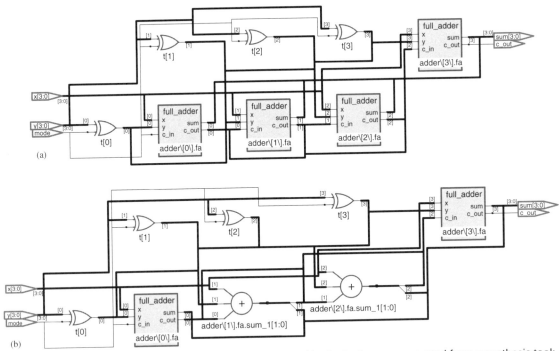

FIGURE 6.2 An *n*-bit two's complement adder logic diagram generated from a synthesis tool: (a) the RTL schematic from 'Symplify'; (b) after dissolving the second and the third full-adder modules

```
// using tasks inside generate block
module adder_nbit(x, y, c_in, sum, c_out);
// I/O port declarations
parameter N = 4;   // define N as a parameter
input   [N-1:0] x, y;
input   c_in;
output reg [N-1:0]   sum;
output reg c_out;
// specify the function of an N-bit adder using generate
genvar i;
reg     [N-2:0] c;     // internal carries declared as nets
generate for (i = 0; i < N; i = i + 1) begin: adder
    if (i == 0)            // specify LSB
        always @(*) full_adder (x[i], y[i], c_in, {c[i],
            sum[i]});
    else if (i == N-1) // specify MSB
        always @(*) full_adder (x[i], y[i], c[i-1], {c_out,
            sum[i]});
```

```
      else                 // specify other bits
        always @(*) full_adder (x[i],  y[i], c[i-1], {c[i],
          sum[i]});
end endgenerate
// define full adder task here
task full_adder;
input x, y, c_in;
output reg [1:0] sum;
    sum = x + y + c_in;
endtask
endmodule
```

■

The following example demonstrates the combined use of a function and a generate statement.

EXAMPLE 6.14 *Combined use of a Function and a* generate *Statement*

This example explains how to combine the use of a function and a generate statement. Here, we define a function to carry out the operations of the full-adder. The function full_adder is then called within the always blocks inside the generate statement to complete an *n*-bit adder.

```
// using functions inside generate block
module adder_nbit(x, y, c_in, sum, c_out);
// I/O port declarations
parameter N = 4;  // define N as a parameter
input  [N-1:0] x, y;
input  c_in;
output reg [N-1:0]  sum;
output reg c_out;
// specify the function of an N-bit adder using generate
genvar i;
reg    [N-2:0] c;    // internal carries declared as nets
generate for (i = 0; i < N; i = i + 1) begin: adder
   if (i == 0)         // specify LSB
     always @(*) {c[i], sum[i]} = full_adder (x[i], y[i],
       c_in);
   else if (i == N-1) // specify MSB
     always @(*) {c_out, sum[i]} = full_adder(x[i], y[i],
       c[i-1]);
   else                   // specify other bits
     always @(*) {c[i], sum[i]} = full_adder(x[i],  y[i],
       c[i-1]);
end endgenerate
```

```
// define full adder function here
function [1:0] full_adder;
input x, y, c_in;
   full_adder =  x + y + c_in;
endfunction
endmodule                                                    ■
```

The following example demonstrates the combined use of a UDP and a generate ststement.

EXAMPLE 6.15   *Combined use of a UDP and a* generate *Statement*

This example explains how to combine the use of a UDP and a generate statement. Here, we define a UDP to carry out the function of a T-type flip-flop. The UDP T_FF is then instantiated within the if-else conditional expression inside the generate statement to complete an n-bit ripple counter.

```
// an example of sequential UDP instantiations
module ripple_counter_generate(clk, clear, qout);
parameter N = 4;
input  clk, clear;
output [N-1:0] qout;
// specify the function of an N-bit ripple counter using
// generate
genvar i;
generate for (i = 0; i < N; i = i + 1)
begin: ripple_counter
   if (i == 0) // specify LSB
      T_FF tff(qout[i], clk, clear);
   else        // specify the rest bits
      T_FF tff(qout[i], qout[i-1], clear);
end endgenerate
endmodule                                                    ■
```

6.2.3 Generate-Case Statement

Like a generate-conditional statement, a generate-case statement allows modules, user-defined primitives, gate primitives, continuous assignments, initial blocks, always blocks, tasks and functions, to be conditionally instantiated into another module. The generate-case statements can be used alone or inside a generate-loop. The general form of it is:

```
case(case_expr)
case_item1: generate_statements
   ...
case_itemn: generate_statements
   [default: generate_statements]
```

The `case_expr` must be computable at elaboration time. This implies that the `case_expr` must only be a function of constants and parameters. Based on this value, the appropriate set of statements on a selected branch are expanded during elaboration time. The `generate_statements` can be any set of statements that are allowed within a `generate` statement. The `default` statement is optional.

What follows is a simple use of the generate-case statement.

EXAMPLE 6.16 *An n-bit Adder*

In this example, we use a `case` statement to construct an *n*-bit adder. The case items 0 and $n - 1$ correspond to the boundary cases of LSB and MSB, respectively. The rest bits are matched by the default case.

```
// using continuous assignments inside generate block
module adder_nbit(x, y, c_in, sum, c_out);
// I/O port declarations
parameter N = 4;  // define N as a parameter
input   [N-1:0] x, y;
input   c_in;
output [N-1:0]  sum;
output c_out;
// specify the function of an N-bit adder using generate
genvar i;
wire    [N-2:0] c;  // internal carries declared as nets
generate for (i = 0; i < N; i = i + 1) begin: adder
  case (i)
        0: assign {c[i], sum[i]}  = x[i] + y[i] + c_in;
      N-1: assign {c_out, sum[i]} = x[i] + y[i] + c[i-1];
  default: assign {c[i], sum[i]}  = x[i] + y[i] + c[i-1];
  endcase
end endgenerate
endmodule
```

The following example explains how the parameter can control the expansion of a generate-case statement.

EXAMPLE 6.17 *The Generate-case Statement*

This example explains how the parameter port can be used to control the actual implementation of the *n*-bit adder. The *n*-bit adder is implemented by instantiating the `adder_nbit` when the `WIDTH` has 4 and 8, and by instantiating the `adder_cla` when the `WIDTH` has other values.

```
module adder_nbit(x, y, c_in, c_out, sum);
// I/O port declarations
parameter WIDTH = 4;  // define n as a parameter
```

```
input  [WIDTH-1:0] x, y;
input  c_in;
output [WIDTH-1:0]  sum;
output c_out;
// specify the function of an n-bit adder using generate
   generate
// instantiate an appropriate instance based on WIDTH
   case (WIDTH)
            4: adder_nbit #(4)  adder4(x, y, c_in, sum,
               c_out);
            8: adder_nbit #(8)  adder8(x, y, c_in, sum,
               c_out);
      default: adder_cla  #(WIDTH) cla(x, y, c_in, sum,
               c_out);
   endcase
endgenerate
endmodule                                              ■
```

Review Questions

Q6.11 What statements can be used within a `generate` statement?

Q6.12 Describe the features of generate-loops.

Q6.13 Describe the features of generate-conditional statements.

Q6.14 Describe the features of generate-case statements.

Q6.15 Can generate-loops be nested?

Q6.16 What elements cannot be used within `generate` statements?

6.3 CONFIGURATIONS

A configuration is a set of rules that maps instances to cells. A cell can be a UDP, a configuration or a module. The collection of compiled cells are stored in a place called a *library*. The operation of mapping an instance to a cell is referred to as *binding*. Hence, a configuration is a set of instance binding rules. The advantage of using configurations is that it may specify a new set of instance binding rules without having to change the source description. In this section, we discuss how to specify a target library for source descriptions and how to use configurations to bind a design to cells associated with different libraries.

6.3.1 Library

A library is a logical collection of compiled source descriptions. It can be a logical or a symbolic library. A list of such library declarations is specified in a library map file. Multiple library map files are allowed. The library map file is read by the Verilog HDL parser before parsing any source code. A tool that is compatible to Verilog HDL must provide a mechanism such as a command line to specify one or more library map files. If multiple map files are specified, then they are read in the order specified.

For convenience, suppose that a file called lib.map already exists in the current working directory, which is readily read by the parser before parsing any source files. A library specification is used to sepecify a target library into which cells in a source file can be stored after they are compiled. A library specification uses the keyword library and has the following syntax:

```
library lib_name file_path;
```

where the file_path can use an absolute or relative path of files. Single character (?) and multiple character (*) wildcards are allowed. In addition, ". . ." denotes matching to any number of hierarchical directories, ". ." specifies the parent directory and "." specifies the current directory, namely, the directory containing the lib.map. A source file which does not specify a target library is by default compiled into the work library.

The include command is used to insert a library map file into another file during parsing. It has the following form:

```
include <lib_map_file_path>;
```

The following is an example of a library map file.

```
// file: my_lib.map
library my_lib    "*.v";
library gate_lib "./*.vg";
include "../my_lib_definitions";
library homework_lib "./homework*.v";
```

The library map file my_lib.map contains three library declarations and one include statement. The three libraries declared are my_lib, gate_lib and homework_lib. All files in the current directory with a suffix ".v" are compiled into my_lib, all files in the current directory with a suffix ".vg" are compiled into gate_lib and all files in the current directory with a form of "homework*.v" are compiled into homework_lib.

6.3.2 Basic Configuration Elements

A configuration is simply a set of rules that are used to search for library cells to which instances bind. Similar to a module or a UDP, a configuration is a design element. It is at the top level. A configuration comprises four basic elements: design, default, instance and cell statements. The design statement specifies the top-level module in the design and the source description to be used. The default statement coupled with the liblist statement specifies, by default, which library or libraries are used to bind instances. The instance and cell statements override the default rule and specify from which the source description of a particular instance and cell are taken.

A configuration is defined outside of a module by using the keywords `config` and `endconfig`. It has the form:

```
config config_identifier;
    design    {[lib_name.]cell_id};
    default  liblist lib_list;
    instance instance_name use lib.cell [:config];
    instance instance_name liblist lib_list;
    cell cell_name use lib.cell [:config];
    cell cell_name liblist lib_list;
endconfig
```

The optional `:config` extension is used explicitly to refer to a `config` that has the same name as a `module`.

In a configuration, the `design` statement must be unique, but it can include multiple top-level modules. The `design` statement specifies the top-level module or modules and should appear before any statements in the `config`. If the `lib_name` is omitted, then the library which contains the `config` is used to search for the cell.

The `default` statement selects all instances which do not match a more specific selection statement. For simple design configurations, it might be sufficient to specify a `default liblist`. The `liblist` clause defines an ordered set of libraries to be searched to find the current instance.

The `instance` statement provides an exception to the default binding rules. It is used to map the specific instance to a cell in `lib.cell` or in a library searching from the `lib_list`. The `instance_name` associated with the instance statement is the name of the cell in the design statement.

The `cell` statement specifies the cell to which it applies. Like in the `instance` statement, it can be specified in one of the two forms, `use` clause and `liblist` clause. The `use` clause specifies a specific binding for the selected cell. It can only be used in conjunction with an `instance` or `cell` statement.

The following is an example used for illustrating the various configurations that follow.

EXAMPLE 6.18   *A Configuration Example*

This program example includes three files: `eight_bit_adder.v`, `four_bit_adder_gate.v` and `four_bit_adder.v`. The first file is the top module and instantiates two instances of `four_bit_adder`. The following two files are the gate and RTL representations of `four_bit_adder`, respectively. The `four_bit_adder` module in the gate representation file instantiates `full_adder` and `half_adder` in a descendant manner. The `four_bit_adder` module in the RTL representation file instantiates `full_adder`, which is modeled in dataflow style.

```
// top file: eight_bit_adder.v
// RTL modeling: using module instantiations
module eight_bit_adder(x, y, c_in, c_out, sum);
```

```verilog
// I/O port declarations
input   [7:0] x, y;
input   c_in;
output c_out;
output [7:0] sum;
wire    c_out_1;  // internal net
// instantiate two four_bit_adder
   four_bit_adder adder_a (x[3:0], y[3:0], c_in, c_out_1,
     sum[3:0]);
   four_bit_adder adder_b (x[7:4], y[7:4], c_out_1, c_out,
     sum[7:4]);
endmodule

// file: four_bit_adder_gate.v
// gate-level hierarchical description of 4-bit adder
module four_bit_adder (x, y, c_in, c_out, sum);
input   [3:0] x, y;
input   c_in;
output [3:0] sum;
output c_out;
wire    c1, c2, c3; // intermediate carries
// -- four_bit adder body-- //
// instantiate the full adder
   full_adder fa_1 (x[0], y[0], c_in, c1, sum[0]);
   full_adder fa_2 (x[1], y[1], c1, c2, sum[1]);
   full_adder fa_3 (x[2], y[2], c2, c3, sum[2]);
   full_adder fa_4 (x[3], y[3], c3, c_out, sum[3]);
endmodule
// gate-level description of full adder
module full_adder (x, y, c_in, c_out, sum);
input   x, y, c_in;
output sum, c_out;
wire    s1, c1, c2;  // outputs of both half adders
// -- full adder body-- //
// instantiate the half adder
   half_adder ha_1 (x, y, c1, s1);
   half_adder ha_2 (c_in, s1, c2, sum);
   or (c_out, c1, c2);
endmodule
// gate-level description of half adder
module half_adder (x, y, c, s);
input   x, y;
output c, s;
// -- half adder body-- //
// instantiate primitive gates
   xor (s,x,y);
```

```
    and (c,x,y);
endmodule

// file: four_bit_adder.v
// RTL modeling: using module instantiations inside
// generate block
module four_bit_adder(x, y, c_in, c_out, sum);
// I/O port declarations
input  [3:0] x, y;
input  c_in;
output c_out;
output [3:0] sum;
// specify the function of an n-bit adder using generate
genvar i;
wire   [2:0] c;  // internal carries declared as nets
generate
for (i = 0; i < 4; i = i + 1) begin: adder
   if (i == 0)     // specify LSB
      full_adder fa (x[i], y[i], c_in, c[i], sum[i]);
   else if (i == 3) // specify MSB
      full_adder fa (x[i], y[i], c[i-1], c_out, sum[i]);
   else            // specify other bits
      full_adder fa (x[i], y[i], c[i-1], c[i], sum[i]);
end endgenerate
endmodule
// define a full adder at data flow level.
module full_adder(x, y, c_in, c_out, sum);
// I/O port declarations
output c_out, sum;
input  x, y, c_in;
// specify the function of a full adder
   assign {c_out, sum} = x + y + c_in;
endmodule
```

The library map file `mylib.map` is shown in the following:

```
// file mylib.map
// In ModelSim, the following steps are used to build
// a library:
// step 1: create a new library, e.g., my_lib_top
// step 2: vlog -work my_lib_top  ../chapter06/src/
// eight_bit_adder.v
library my_lib_top  ../chapter06/src/eight_bit_adder.v
library my_lib_gate ../chapter06/src/four_bit_adder_
    gate.v
library my_lib_rtl  ../chapter06/src/four_bit_adder.v
```

my_lib_top	my_lib_gate	my_lib_rtl
eight_bit_adder.v	four_bit_adder_gate.v	four_bit_adder.v
eight_bit_adder()	four_bit_adder()	four_bit_adder()
four_bit_adder --- adder_a	full_adder --- fa_1	full_adder()
four_bit_adder --- adder_b	full_adder --- fa_2	
	full_adder --- fa_3	
	full_adder --- fa_4	
	full_adder()	
	half_adder()	

FIGURE 6.3 The relationship between files, modules and libraries of the underlying example

The relationship of files, modules and libraries of the underlying example is depicted in Figure 6.3 for reference in the rest of this section.

In the rest of this section, we discuss how to specify a configuration to meet the specific requirement of a particular application. A configuration can be in the same file as the library map file or as a separate file.

Default Configuration The libraries are searched in accordance with the library declaration order in the library map file when no configuration is specified. This means all instances of module `eight_bit_adder` use the cells from library `my_lib_gate` (since `my_lib_gate` is placed before `my_lib_rtl`).

Using the `Default` Statement As mentioned previously, the `default liblist` statement can be used to change the library search order. An example of using the `default` statement is shown in the following.

```
// configuration 1: using the default clause
config cfg_1;
   design  my_lib_top.eight_bit_adder
   default liblist my_lib_rtl my_lib_gate;
endconfig
```

The `default liblist` statement overrides the library search order in the `mylib.map` file and so `my_lib_rtl` is always searched before `my_lib_gate`. Based on this configuration, the cells used in the top module `eight_bit_adder` are as follows:

```
# Loading work.cfg_1
# Loading my_lib_top.eight_bit_adder
# Loading my_lib_rtl.four_bit_adder
# Loading my_lib_rtl.full_adder
```

All descendant cells of the top module are taken from the library `my_lib_rtl`.

Using the `cell` Statement The `cell` statement selects a specified cell instance and explicitly binds it to a particular library cell or the first found cell from the specified library list. For instance, in the following configuration `cfg_2`, the `cell` statement selects all cells named `full_adder` and explicitly binds them to the gate representation in `my_lib_gate`.

```
// configuration 2: using the cell clause
config cfg_2;
   design  my_lib_top.eight_bit_adder
   default liblist my_lib_rtl my_lib_gate;
   cell full_adder use my_lib_gate.full_adder;
endconfig
```

Using this configuration, the cells used in the top module `eight_bit_adder` are as follows:

```
# Loading work.cfg_2
# Loading my_lib_top.eight_bit_adder
# Loading my_lib_rtl.four_bit_adder
# Loading my_lib_gate.full_adder
# Loading my_lib_gate.half_adder
```

Since the `liblist` is inherited, all of the descendants of `full_adder` inherit its `liblist` from the `cell` selection statement.

Using the `instance` Statement The `instance` statement selects a specified instance and explicitly binds it to a particular library cell or the first found cell from the specified library list. For instance, in the following configuration `cfg_3`, the `instance` statement selects the instance named `adder_b` and explicitly binds it to the gate representation in `my_lib_gate`. All descendants of this instance are taken from the same library. The other instance and its descendant are bound to the library `my_lib_rtl` specified by the `default liblist` statement.

```
// configuration 3: using the instance clause
config cfg_3;
   design  my_lib_top.eight_bit_adder
   default liblist my_lib_rtl my_lib_gate;
   instance eight_bit_adder.adder_b liblist my_lib_gate;
endconfig
```

Using this configuration, the cells used in the top module `eight_bit_adder` are as follows:

```
# Loading work.cfg_3
# Loading my_lib_top.eight_bit_adder
# Loading my_lib_rtl.four_bit_adder
# Loading my_lib_gate.four_bit_adder
```

```
# Loading my_lib_gate.full_adder
# Loading my_lib_gate.half_adder
# Loading my_lib_rtl.full_adder
```

Using the `instance` and `cell` Statements   When both `instance` and `cell` statements are used, the specified instance and cell instance are explicitly bound to a particular library cell or the first found cell from the specified library list. For instance, in the following configuration `cfg_4`, the instance named `adder_b` and the cells named `full_adder` in both instances are explicitly bound to the gate representation in `my_lib_gate`. However, the instance `adder_a` is bound to the library `my_lib_rtl` except the cell named `full_adder` instantiated within it.

```
// configuration 4: using the instance and cell
// statements
config cfg_4;
   design  my_lib_top.eight_bit_adder
   default liblist my_lib_rtl my_lib_gate;
   instance eight_bit_adder.adder_b liblist my_lib_gate;
   cell full_adder use my_lib_gate.full_adder;
endconfig
```

Based on this configuration, the cells used in the top module `eight_bit_adder` are as follows:

```
# Loading work.cfg_4
# Loading my_lib_top.eight_bit_adder
# Loading my_lib_rtl.four_bit_adder
# Loading my_lib_gate.four_bit_adder
# Loading my_lib_gate.full_adder
# Loading my_lib_gate.half_adder
```

Hierarchical Configurations   Often, it is desirable to specify a set of configuration rules for a sub-section of a design. Suppose that all of this work has only been on the gate-level module `four_bit_adder_gate` by itself and we want to use the `my_lib_rtl.full_adder` cell for `fa_3`, and the `my_lib_gate` for the other cells. Then a possible configuration, called `cfg_5`, is as follows:

```
// configuration 5: hierarchical configuration--part1
config cfg_5;
   design  my_lib_gate.four_bit_adder
   default  liblist my_lib_gate my_lib_rtl;
   instance four_bit_adder.fa_3 liblist my_lib_rtl;
endconfig
```

Based on this configuration, the cells used in the top module `four_bit_adder` are as follows:

```
# Loading work.cfg_5
# Loading my_lib_gate.four_bit_adder
# Loading my_lib_gate.full_adder
# Loading my_lib_gate.half_adder
# Loading my_lib_rtl.full_adder
```

To use this configuration cfg_5 for the adder_b instance of eight_bit_adder and take the full default my_lib_rtl for the adder_a instance, the following configuration cfg_6 can be used:

```
// configuration 6: hierarchical configuration--part 2
config cfg_6;
   design  my_lib_top.eight_bit_adder_tb  // testbench
           my_lib_top.eight_bit_adder;    // top module
   default liblist my_lib_rtl my_lib_gate;
   instance eight_bit_adder.adder_b use work.cfg_5;
endconfig
```

The instance statement specifies the work.cfg_5 configuration and is used to resolve the bindings of instance adder_b and its descendants. The design statement in configuration cfg_5 defines the exact binding for the adder_b instance itself. The rest of cfg_5 defines the rules to bind the descendants of adder_b. Notice that the instance statement in cfg_5 is relative to its own top-level module, four_bit_adder.

```
# Loading work.cfg_6
# Loading my_lib_top.eight_bit_adder_tb
# Loading my_lib_top.eight_bit_adder
# Loading my_lib_rtl.four_bit_adder
# Loading my_lib_top.cfg_5
# Loading my_lib_rtl.full_adder
```

Review Questions

Q6.17 What is a configuration?

Q6.18 Describe what a library is.

Q6.19 What are the basic elements of a configuration?

Q6.20 What is the meaning of binding?

Q6.21 Describe the function of the design statement in a configuration.

Q6.22 Describe the function of the default statement in a configuration.

Q6.23 Describe the function of the instance statement in a configuration.

Q6.24 Describe the function of the cell statement in a configuration.

SUMMARY

In order to build a large design, a mixed style is usually used. Through the use of instantiations, a hierarchical structure is formed. The instantiation is the process that gate and switch primitives, UDP primitives and modules are embedded into other modules. Higher-level modules create instances of lower-level modules and communicate with them through input, output and bidirectional ports, which can be scalar or vector.

When a module instantiates other modules, the higher-level module can change the values of the parameters defined by the keyword `parameter` within the lower-level modules. This operation is called *parameter override*. Through the parameter override mechanism, the design of modules can be parameterized. That is, the features or operations of the module can be changed by modifying the parameters defined within it. There are two ways that can be used to override parameter values. These are the `defparam` statement and module instance parameter value assignment. In the `defparam` statement method, the parameter values defined by `parameter` in any module are overridden by using the hierarchical name of the parameter whenever the module is instantiated. In the method of module instance parameter value assignment, the parameters defined by `parameter` within a module are overridden by the parameters passed through parameter ports whenever the module is instantiated.

In order to build a large regular design, it is more constructive to use iterative logic modules. For this purpose, a useful as well as powerful statement, `generate`, is supported by Verilog HDL. By using `generate` statements, one or multiple dimensional iterative logic structures, such as n-bit ripple carry adders, $m \times n$ array multipliers and $m \times n$ array dividers, are easily constructed. A `generate` statement allows selection and replication of some statements during elaboration time. The elaboration time is the time after a design has been parsed but before simulation begins.

The power of `generate` statements is that it can conditionally generate declarations and instantiations into a design. Almost what can be put inside a module can also be placed within `generate` statements except the following, parameters, local parameters, and port declarations: `input`, `output` and `inout`. There are three kinds of statements that can be used within a `generate` statement: generate-loop, generate-conditional and generate-case.

A configuration is a set of rules that map instances to cells. A cell can be a UDP, a configuration or a module. The collection of compiled cells are stored in a library. The operation of mapping an instance to a cell is referred to as *binding*. Hence, a configuration is simply a set of rules that are used to search for library cells to which instances bind. The advantage of using configurations is that it may specify a new set of instance binding rules without having to change the source description.

Similar to a module, a configuration is a design element. It is at the top level. A configuration comprises four basic elements: `design`, `default`, `instance` and `cell` statements. The `design` statement specifies the top-level module in the design and the source description to be used. The `default` statement coupled with the `liblist` statement specifies, by default, which library or libraries, are used to bind instances. The `instance` and `cell` statements override the default rule and specify from which the source description of a particular instance and cell are taken.

REFERENCES

1. J. Bhasker, *A Verilog HDL Primer,* 3rd Edn, Star Galaxy Publishing, 2005.
2. M.D. Ciletti, *Modeling, Synthesis and Rapid Prototyping with the Verilog HDL,* Prentice-Hall, Upper Saddle River, NJ, USA, 1999.
3. IEEE 1364-2001 Standard, *IEEE Standard Verilog Hardware Description Language*, 2001.
4. S. Palnitkar, *Verilog HDL: A Guide to Digital Design and Synthesis,* 2nd Edn, SunSoft Press, 2003.

PROBLEMS

6.1 Write a parameterized module that counts the number of ones in a byte. Write a test bench to instantiate it with a parameter port value of 16 and verify the results.

6.2 Write a parameterized module that counts the number of trailing ones in a byte. Write a test bench to instantiate it with a parameter port value of 16 and verify the results.

6.3 Write a parameterized module that counts the number of leading zeros in a byte. Write a test bench to instantiate it with a parameter port value of 16 and verify the results.

6.4 Write a parameterized module that counts the number of leading ones in a byte. Write a test bench to instantiate it with a parameter port value of 16 and verify the results.

6.5 Write a parameterized module that computes the even parity of an 8-bit number. Write a test bench to instantiate it with a parameter port value of 16 and verify the results.

6.6 Write a parameterized module that computes the odd parity of an 8-bit number by using port list declaration style. Write a test bench to instantiate it with a parameter port value of 16 and verify the results.

6.7 Use a generate-loop statement to instantiate the `xor` gate primitive and construct a parameterized even-parity checker module. Assume that a linear structure is used. Write a test bench to instantiate it with a parameter port value of 16 and verify the results.

6.8 Use a generate-loop statement to instantiate the `xor` gate primitive and construct a parameterized odd-parity checker module. Assume that a linear structure is used. Write a test bench to instantiate it with a parameter port value of 16 and verify the results.

6.9 Use a generate-loop statement to instantiate the `xor` gate primitive and construct a parameterized even-parity checker module. Assume that a binary-tree structure is used. Write a test bench to instantiate it with a parameter port value of 16 and verify the results.

6.10 Use a generate-loop statement to instantiate the `xor` gate primitive and construct a parameterized odd-parity checker module. Assume that a binary-tree structure is used. Write a test bench to instantiate it with a parameter port value of 16 and verify the results.

6.11 Assume that an incremented-by-1 circuit is required. That is, the circuit always adds a constant 1 to an operand.

    (a) Derive a one-bit cell of the circuit.

    (b) Use a generate-conditional statement to instantiate the one-bit cell and construct a parameterized incremented-by-1 module. Write a test bench to instantiate it with a parameter port value of 8 and verify the results.

    (c) Re-do (b) by using continuous assignment statements instead of instantiations.

    (d) Re-do (b) by using `always` statements instead of instantiations.

    (e) Define the one-bit cell as a task and re-do (b) by using `always` statements to invoke the task.

    (f) Define the one-bit cell as a function and re-do (b) by using `always` statements to call the function.

6.12 Assume that a decremented-by-1 circuit is required. That is, the circuit always subtracts a constant 1 from an operand.

    (a) Derive a one-bit cell of the circuit.

    (b) Use a generate-conditional statement to instantiate the one-bit cell and construct a parameterized decremented-by-1 module. Write a test bench to instantiate it with a parameter port value of 8 and verify the results.

    (c) Re-do (b) by using continuous assignment statements instead of instantiations.

    (d) Re-do (b) by using `always` statements instead of instantiations.

    (e) Define the one-bit cell as a task and re-do (b) by using `always` statements to invoke the task.

    (f) Define the one-bit cell as a function and re-do (b) by using `always` statements to call the function.

ADVANCED MODELING TECHNIQUES

WE HAVE described the basic features of Verilog HDL in the previous chapters. In this chapter, we discuss additional features of Verilog HDL. These features include block constructs, procedural continuous assignments, specify blocks, timing checks and compiler directives.

The block constructs, including sequential blocks and parallel blocks, are used to group two or more statements together so that they can act as a single statement. The procedural statements within sequential blocks are executed sequentially in the given order but within parallel blocks are executed concurrently regardless of their order.

The procedural continuous assignments allow expressions to be driven continuously onto variables or nets. They can only be used within `initial` and `always` statements. Two kinds of procedural continuous assignments are `assign/deassign` and `force/release`. The `assign` and `deassign` procedural continuous assignments assign values to variables. The `force` and `release` procedural continuous assignments assign values to nets or variables.

The specify blocks are used to define the module paths coupled with delays of cells in ASIC cell libraries. In addition, they also provide mechanisms for checking timing constraints of the modules containing the specify blocks.

The final section describes compiler directives defined in Verilog HDL. With the aid of compiler directives, a macro text may be defined, parts of a design can be compiled conditionally, default net features can be changed and unconnected input ports may be pull-up or pull-down.

7.1 SEQUENTIAL AND PARALLEL BLOCKS

Block statements are used to group two or more statements together so that they can act as a single statement. There are two types of block statements:

Digital System Designs and Practices   Ming-Bo Lin
© 2008 John Wiley & Sons (Asia) Pte Ltd

1. *Sequential block*. A sequential block uses the keywords `begin` and `end` to group statements. It is also called a *begin-end block*. The procedural statements within sequential blocks are executed sequentially in the given order.

2. *Parallel block*. A parallel block uses the keywords `fork` and `join` to group statements. It is also called a *fork-join block*. The procedural statements within parallel blocks are executed concurrently regardless of their relative order.

7.1.1 Sequential Blocks

A sequential block is enclosed with the keywords `begin` and `end`. The procedural statements in sequential blocks are processed in the given sequence, one after another. The delays for each procedural statement is relative to the simulation time of the execution of its previous statement. Once a sequential block completes its execution, execution continues to the procedural statements following the sequential block. In other words, the control is only passed out of a sequential block after the sequential block completes the execution of its last procedural statement.

Sequential blocks must be within `initial` or `always` blocks. The general syntax of a sequential block is as follows:

```
begin [:block_id {block_declaration}]
   procedural_statement(s)
end
```

where `block_declaration` can be one of variables, events, local parameters and parameters. Note that the `block_declaration` can only be used within a named block; namely, the block has a block name. In other words, you need to name the block when you want to use the `block_declaration`.

In the following example, the two assignments are executed in sequence and so the final result is deterministic. That is, c stores the value of b.

```
always @(b)
   begin
      a = b;
      c = a; // c stores the value of b
   end
```

Delay control can be applied in a sequential block to schedule the executions of procedural statements. The following example shows how the delay control is used in a sequential block. Only blocking assignments are used. In this example, x is assigned to 1 at simulation time 0, and y and z are assigned to 1 and 0 at simulation times 12 and 32 (why?), respectively.

```
initial  begin
       x = 1'b1;    // execute at time 0.
   #12 y = 1'b1;    // execute at time 12.
   #20 z = 1'b0;    // execute at time 32.
end
```

The following example shows that both blocking and nonblocking procedural assignments are mixed in a sequential block. In this case, however, there are no differences between blocking and nonblocking procedural assignments because delay controls are used to schedule the execution order of statements.

```
initial begin
        x   = 1'b0;     // execute at time 0.
    #20 w   = 1'b1;     // execute at time 20.
    #12 y  <= 1'b1;     // execute at time 32.
    #10 z  <= 1'b0;     // execute at time 42.
    #25 x   = 1'b1; w = 1'b0; // execute at time 67.
end
```

Review Questions

Q7.1 Describe the features of a sequential block.

Q7.2 How would you define a sequential block?

Q7.3 What is the major purpose of a block identifier?

Q7.4 What kind of elements can be declared within sequential blocks?

Q7.5 Can the intra-assignment delay control be used in sequential blocks?

7.1.2 Parallel Blocks

A parallel block is enclosed with the keywords `fork` and `join`. All procedural statements in a parallel block are executed in parallel, i.e. independent of their writing order. However, delay control can be used to schedule the execution of procedural assignments. The delay for each procedural statement is relative to the simulation time when the block starts its execution. When the last activity, but not necessary the last statement, in a parallel block completes its execution, the execution continues to the procedural statements following the parallel block. In other words, the control is passed out a parallel block only after all procedural statements in the parallel block complete their executions.

Like sequential blocks, parallel blocks must be within `initial` or `always` blocks. The general syntax of a parallel block is as follows:

```
fork [:block_id {block_declaration}]
    procedural_statement(s)
join
```

where `block_declaration` can be one of variables, events, local parameters and parameters. Note that the `block_declaration` can only be used within a named block. In other words, you need to name the block when you want to use the `block_declaration`.

Delay control can be applied in a parallel block to schedule the executions of procedural statements. The following example shows the combination of the parallel block and delay control. In this example, x is assigned to 1 at simulation time 0, and y and z are assigned to 1 and 0 at simulation times 12 and 20, respectively. It is advisory to compare the result of this example with that of the case in sequential block.

```
initial fork
        x = 1'b1;       // execute at time 0.
    #20 z = 1'b0;       // execute at time 20.
    #12 y = 1'b1;       // execute at time 12.
join
```

In the above example, we use blocking assignments. However, the result remains the same if we replace the blocking assignments with nonblocking assignments such as in the following example.

```
initial fork
        x <= 1'b0;      // execute at time 0.
    #12 y <= 1'b1;      // execute at time 12.
    #20 z <= 1'b1;      // execute at time 20.
join
```

In summary, when regular delay controls are used to schedule the execution order of assignment statements within a parallel block, the results are the same regardless of whether assignments are blocking or nonblocking.

Review Questions

Q7.6 Describe the features of a parallel block.

Q7.7 How would you define a parallel block?

Q7.8 What are the differences between sequential and parallel blocks?

Q7.9 What kind of elements can be declared within parallel blocks?

Q7.10 Can intra-assignment delay control be used in parallel blocks?

7.1.3 Special Features of Blocks

In this sub-section, we consider two special features of blocks: *named blocks* and *nested blocks*.

Named Blocks　As mentioned previously, both sequential and parallel blocks can be named by adding :block_id after the keyword begin or fork. These kinds of blocks are called named blocks. A named block has the following features:

1. It allows local variables, parameters, local parameters and named events to be declared within the block. However, all local variables declared are static; namely, their values remain valid during the entire simulation time.

2. It allows the block to be referenced in procedural statements such as the disable statement.

3. A named block can be disabled by the `disable` statement.

4. Local variables declared in named blocks can be referenced through using a hierarchical path name.

In the following example, `test` is the block name. Variables x, y and z are local to the block. Since this is a sequential block, the variable z is assigned a value of 1 at time 22.

```
initial begin: test  // test is the block name.
reg x, y, z;          // local variables
      x = 1'b0;
  #12 y = 1'b1;       // execute at time 12.
  #10 z = 1'b1;       // execute at time 22.
end
```

To refer a `reg` variable local to the block, such as `x`, from outside of the block `test`, a hierarchical path name `test.x` should be used.

Nested Blocks   Blocks can be nested and both sequential and parallel blocks can be mixed. For example, in the following `initial` block, the outer block is sequential since it is a `begin-end` block and the inner block is a parallel block enclosed with the keywords `fork` and `join`. The inner block starts at simulation time 0 and finishes at time 20. As a consequence, the variable x is assigned with 1 at the simulation time 45. Note that the entire inner block is considered as a single (complex) statement from the viewpoint of the outer block.

```
initial begin
    x = 1'b0;              // execute at time 0.
    fork // parallel block -- enter at time 0 and leave
         // at time 20.
       #12 y <= 1'b1;  // execute at time 12.
       #20 z <= 1'b0;  // execute at time 20.
    join
    #25 x = 1'b1;         // execute at time 45.
end
```

Review Questions

Q7.11   Describe the features of named blocks.

Q7.12   Give an example of a nested block by putting a sequential block inside a parallel block and explain its operations.

Q7.13   Give an example of a nested block by putting a parallel block inside a sequential block and explain its operations.

Q7.14   How would you refer to a variable declared within a named block?

Q7.15   Can intra-assignment delay control be used in nested blocks?

7.1.4 The `disable` Statement

A `disable` statement is a procedural statement; namely, it can only be used within `initial` or `always` blocks. The `disable` statement using the keyword `disable`. It is usually used to terminate a task or a named block, sequential or parallel, before which it completes the execution of all its statements. The `disable` statement can be used to get out of loops, handle error conditions, control execution of pieces of code, model global reset or a hardware interrupt, based on a control signal.

The general form of the `disable` statement is:

```
disable task_id;
disable block_id;
```

where `task_id` and `block_id` are the hierarchical path names of the task and the named block to be disabled, respectively. After a `disable` statement is executed, the execution continues with the next statement following the block being disabled or the task call.

7.1.4.1 Disabling Named Blocks   For example, in the following sequential block, the third statement is never executed since it is disabled by the `disable` statement. The statement "`c = b;`" is executed immediately following the execution of the `disable` statement.

```
begin: test
   a = b;
   disable test;
   c = a; // it will never be executed
end
c = b;   // it is executed after the disable statement
```

In the following sequential block, the `disable` statement is executed and disables the `test` block whenever `flag` is true. It has no effect on the block otherwise.

```
initial begin: test             // test is the block name.
   while (i < 10) begin
      if (flag) disable test; // block test is disabled
      i = i + 1;               // if flag is true
   end
end
```

7.1.4.2 Disabling Tasks   Although the `disable` statement can be used to disable a task, it is not recommended to use it for this purpose, especially when the task returns output values, because the values of the output and inout arguments are indeterminate when a task is disabled. For instance, in the following example the `disable` statement is used as an early return from a task. In this case, the return value x is sure to be assigned a value of 164 when the "flag == 0" is true.

```
task test;
input flag;
output [7:0] reg x;
begin: test_block
    x = 164;
    if (flag == 0)
        disable test_block;
end
endtask
```

However, if we replace the disable statement with the following statement:

```
disable test;
```

then the return value x will be indeterminate after the disable statement is executed.

Review Questions

Q7.16 What are the functions of the disable statement?

Q7.17 What is the next statement to be executed after the execution of a disable statement?

Q7.18 Why is it not suggested to disable a task by using a disable statement?

Q7.19 Give an example to show how the disable statement can be used to model a hardware interrupt.

7.2 PROCEDURAL CONTINUOUS ASSIGNMENTS

The procedural continuous assignments allow expressions to be driven continuously onto variables or nets. They are procedural statements; namely, they can only be used within initial and always statements. Two kinds of procedural continuous assignments are:

1. assign and deassign procedural continuous assignments assign values to variables.

2. force and release procedural continuous assignments assign values to nets or variables.

The assign and deassign constructs are now considered as a bad coding style and the force and release constructs should only appear in stimulus or as debug statements. Moreover, because these procedural continuous assignments are not generally acceptable by synthesis tools and only used in test benches, they are often called *test-bench constructs*.

7.2.1 `assign` and `deassign` Statements

The `assign` statement overrides all regular procedural assignments to a variable. Its effect is ended by a `deassign` statement. The variable remains the continuously assigned value after being deassigned until a future procedural assignment changes it. If an `assign` statement is applied to a variable which already has an `assign`, then it deassigns the variable before making a new procedural continuous assignment to the variable.

The `assign` and `deassign` statements have the forms as follows:

```
assign    variable_assignment
deassign variable_lvalue
```

The left-hand side of the `assign` statement must be a variable or a concatenation of variables. It cannot be a memory word (array) or a bit-select or a part-select of a variable. The right-hand side of the `assign` statement can be an expression, which is treated as a continuous assignment.

In summary, the `assign` statement behaves like a continuous assignment except that it is only active a limited period of time set by a pair of `assign` and `deassign` statements.

The `assign` and `deassign` statements allow modeling of asynchronous clear/preset on a *D*-type edge-triggered flip-flop, where the clock is inhibited when the clear or preset is active. The following example demonstrates this type of application.

EXAMPLE 7.1 *The Use of* `assign` *and* `deassign` *Statements*

The following example shows a use of the `assign` and `deassign` statements to describe a *D*-type flip-flop with asynchronous reset in behavioral style. The output q and qbar are assigned to new values by regular procedural assignments at each negative edge of the clock whenever the `reset` input is low. When the `reset` input is high, both q and qbar are assigned to the new values by the `assign` statement. When the `reset` is low, both q and qbar are deassigned so as to allow the values of both q and qbar to be changed at the next negative edge of the clock.

```
// negative-edge triggered D-flip flop with asynchronous
// reset.
module edge_dff(input clk, reset, d, output reg q, qbar);
always @(negedge clk) begin
   q <= d; qbar <= ~d;
end
always @(reset)    // override the regular assignments to
                   // q and qbar.
   if (reset) begin
      assign q = 1'b0;
      assign qbar = 1'b1;
```

```
   end else begin // end the effect of assign
      deassign q;
      deassign qbar;
   end
endmodule
```

Review Questions

Q7.20 What are the two kinds of procedural continuous assignments?

Q7.21 What are the differences between `assign` and `force` statements?

Q7.22 Can `assign` and `deassign` statements assign values to nets?

Q7.23 Can `force` and `release` statements assign values to variables?

Q7.24 Can the right-hand side of an `assign` statement be an expression?

Q7.25 What will happen if a variable is assigned by an `assign` statement but does not end with a `deassign` statement?

Q7.26 Can a concatenation operator be used as the left-hand side of an `assign` statement? What are the constraints in this situation?

7.2.2 `force` and `release` Statements

Another procedural continuous assignment is the `force` and `release` statements. They have a similar effect to the `assign` and `deassign` statements except that they can be applied to nets as well. A `force` statement on a variable overrides any procedural assignments or procedural continuous assignments on the variable until the variable is released by a `release` statement. The variable remains the forced value after being released until a future procedural assignment changes it. Note that `force` and `release` statements override the effect of `assign` and `deassign` statements.

The `force` and `release` statements have the forms as follows:

```
force   variable_assignment
force   net_assignment
release variable_lvalue
release net_lvalue
```

The left-hand side of the `force` statement can be a variable or a net, a constant bit-select of a vector net, a part-select of a vector net or a concatenation of the above nets. It cannot be a memory word (array) or a bit-select or a part-select of a vector variable. The right-hand side of the `force` statement can be an expression, which is treated as a continuous assignment.

A `force` statement on a net overrides all continuous assignments until the net is released by a `release` statement. The net will immediately return to its normal driven value when it is released. In summary, the `force` statement behaves like a continuous assignment except that it is only active a limited period of time set by a pair of `force` and `release` statements.

TABLE 7.1 Comparison of various assignments in Verilog HDL

Data type	Primitive output	Continuous assignment	Procedural assignment	assign deassign	force release
Net	Yes	Yes	No	No	Yes
Variable	Yes (Seq. UDP)	No	Yes	Yes	Yes

Table 7.1 compares various assignments provided by Verilog HDL, including primitive output, continuous assignment, procedural assignment, `assign` and `deassign`, and `force` and `release`.

EXAMPLE 7.2 *The Use of* `force` *and* `release` *Statements*

In this example, an AND gate instance `and_1` is replaced by an OR gate implemented by a `force` statement that forces its outputs to the value of its logical OR of inputs between the time units 10 and 20 during simulation.

```
`timescale 1 ns/100 ps
module force_release_test_tb;
// internal signals declarations:
reg  w, x, y;
wire f;
   and and_1 (f, w, x, y);
initial begin
   $monitor("%d, f = %b", $stime, f);
   w = 1;
   x = 0;
   y = 1;
   #10 force f = (w | x | y);
   #10 release f;
   #10 $finish;
end
endmodule
```

From this example, it is easy to see that the `force` and `release` statements can be used to temporarily replace the assignments on nets or variables with other expressions. As a consequence, it is useful that they can be used to debug a design without changing the actual contents of the design before making sure where the bugs are.

Review Questions

Q7.27 When would you use `force` and `release` statements?

Q7.28 Explain the operations of the `force` statement.

Q7.29   Explain the operations of the `release` statement.

Q7.30   Can the right-hand side of a `force` statement be an expression?

Q7.31   What will happen if a variable is assigned by a `force` statement but does not end with a `release` statement?

Q7.32   Can a concatenation operator be used as the left-hand side of a `force` statement? What are the constraints in this case?

7.3 DELAY MODELS AND TIMING CHECKS

In a realistic hardware module, delays are often associated with the interconnect wires and gates. In order to model these delays inherent in hardware modules, Verilog HDL provides a mechanism called a *specify block*. By using the specify block, the following three functions can be performed: (1) describe various paths across the module, (2) assign delays to these paths and (3) perform necessary timing checks.

In this section, we introduce the delay models provided in Verilog HDL, the definition of a specify block and timing checks in sequence.

7.3.1 Delay Models

There are three delay models that are often used to model various delays in a realistic hardware module: *distributed delay model*, *lumped delay model* and *path delay model*.

7.3.1.1 Distributed Delay Model
In this model, delays are associated with an individual element, gate, cell or interconnect. Hence, distributed delays are specified on a per element basis. The distributed delay model specifies the time when it takes the events to propagate through gates and nets inside the module. An example used to illustrate the distributed delays of a module is shown in Figure 7.1.

The distributed delays can be modeled by assigning delays to an individual gate or by using delays in individual continuous assignment. The following example is the structural modeling of Figure 7.1, coupled with distributed delay model at gate level.

```
module test_gate(input x, y, z, output f);
wire a, b, c;
    and  #5 a1 (a, x, y);
    not  #2 n1 (c, x);
    and  #5 a2 (b, c, z);
```

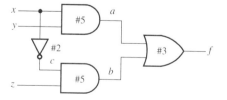

FIGURE 7.1 An example of a distributed delay model

```
      or   #3 o1 (f, a, b);
   endmodule
```

Of course, the above module is quite easily modeled in dataflow style, as shown in the following example.

```
   module test_dataflow(input x, y, z, output f);
   wire a, b, c;
      assign  #5  a = x & y;
      assign  #2  c = ~x;
      assign  #5  b = c & z;
      assign  #3  f = a | b;
   endmodule
```

7.3.1.2 Lumped Delay Model   In this model, delays are associated with the entire module. The cumulative delay of all paths are lumped at the single output. Hence, a lumped delay is specified on a per output basis. It specifies the time when it takes the events to propagate at the last gate along all paths. As a consequence, the lumped delay model uses the worst-case delay obtained from all paths. An example used to illustrate the lumped delay model of a module is shown in Figure 7.2.

The lumped delay can be modeled by assigning a single delay to the output gate or by using a delay only in the output continuous assignment. The following example is the structural modeling of Figure 7.2 along with a lumped delay model at gate level. From this figure, we can see that the delay is only assigned to the output gate. The delay is the worst-case value of all possible paths from any input to the output.

```
   module test_gate(input x, y, z, output f);
   wire a, b, c;
      and a1 (a, x, y);
      not n1 (c, x);
      and a2 (b, c, z);
      or  #10 o1 (f, a, b);
   endmodule
```

Of course, the above module is quite easily modeled in dataflow style, as shown in the following example.

```
   module test_dataflow(input x, y, z, output f);
   wire a, b, c;
      assign     a = x & y;
```

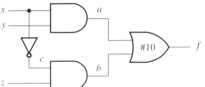

FIGURE 7.2 An example of a lumped delay model

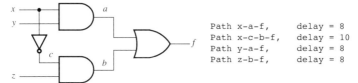

```
Path x-a-f,    delay = 8
Path x-c-b-f,  delay = 10
Path y-a-f,    delay = 8
Path z-b-f,    delay = 8
```

FIGURE 7.3 An example of a path delay model

```
    assign     c = ~x;
    assign     b = c & z;
    assign #10 f = a | b;
endmodule
```

7.3.1.3 Path Delay Model In this model, delays are individually assigned to each module path, i.e. from an input to an output. The input may be an input port or an inout (bidirectional) port and the output may be an output port or an inout port. Hence, a path delay is specified on a pin-to-pin (port-to-port) basis. This kind of path is called a *module path* later in order to distinguish it from other kinds of paths. A module path specifies the time it takes the events to propagate from an input port to an output port along all possible paths. An example used to illustrate the path delays of a module is shown in Figure 7.3. The path delay can be modeled in Verilog HDL by using a specify block, which will be introduced in the next sub-section. Note that each module path may contain many internal paths. In this case, the worst-case internal path delay is taken as the delay of the module path.

In summary, the lumped delay model is the easiest one to use among the three delay models when a module is considered as a unit. However, it only provides the worst-case delay of all possible path combinations of input and output ports. The path delay model is easier to use than the distributed delay model because it need not know details of the internal structure of a module. This feature of without knowing the internal structure of a module also makes the delay calculation much easier than the other two delay models so that the path delay model is the most popular one, in particular, in describing the delays of ASIC cells.

Review Questions

Q7.33 What are the delay models often used to model hardware delays?

Q7.34 What are the features of the distributed delay model? How would you model this in dataflow style?

Q7.35 What are the features of the lumped delay model? How would you model this in dataflow style?

Q7.36 What are the features of the path delay model?

Q7.37 Define module path.

Q7.38 Why does it need to use the worst-case path delay from all paths in the lumped delay model?

7.3.2 Specify Blocks

The specify block is a mechanism in Verilog HDL for providing the following functions:

1. Describe various paths across the module.
2. Assign delays to these paths.
3. Perform the necessary timing checks.

The first two functions of specify blocks are discussed in this sub-section while the timing checks will be discussed in the next sub-section in more detail.

A specify block is defined within a module and enclosed by the keywords specify and endspecify. The general form of a specify block is:

```
specify
    path_declarations
    specparam_declarations
    timing_checks
endspecify
```

By using a specify block, the circuit shown in Figure 7.3 can be described as in the following example.

EXAMPLE 7.3   *An Example of a* specify *Block*

This example explains how a specify block can be used to model module path delays. As shown in Figure 7.3, there are three module paths: x-f, y-f and z-f. The path x-f actually contains two paths, x-a-f and x-c-b-f, with separate delay of 8 and 10 time units. Hence, the worst-case path x-c-b-f is used for the specified module path. All three module paths are specified in the specify block along with their delays.

```
module test_gate_specify(input x, y, z, output f);
wire a, b, c;
// specify block with path delay statements
specify
    (x => f) = 10;
    (y => f) = 8;
    (z => f) = 8;
endspecify
// gate instantiations
    and  a1 (a, x, y);
    not  n1 (c, x);
    and  a2 (b, c, z);
    or   o1 (f, a, b);
endmodule
```

FIGURE 7.4 Examples of simple path connections: (a) parallel connection; (b) full connection

7.3.2.1 Path Declarations

There are three kinds of path declarations that can be declared within a specify block:

1. Simple path.
2. Edge-sensitive path.
3. State-dependent path.

Simple Path   A simple path means that it simply constructs a path from a path source to a path destination. The path source is an input port or an inout port and the path destination is an output port or an inout port. It can be declared in one of the following two forms:

1. Parallel connection (using source => destination).
2. Full connection (using source *> destination).

Both connections are graphically illustrated in Figure 7.4.

1. Parallel connection. The parallel connection uses the symbol "=>" and every path source connects to exactly one path destination. Hence, both source and destination must have the same size. The syntax of parallel connection is as follows:

```
(source [polarity] => destination) [= delay_value;]
```

where the `destination` can be any scalar output or inout port or the bit-select of a vector output or inout port. The `polarity` arbitrarily denotes whether or not a signal is inverted when it propagates from the `source` to the `destination`. However, it does not matter with the actual propagation, depending on the internal logic, of signals through the model. The positive polarity (+) is used when the transition at both source and destination are the same; the negative polarity (−) is used when the transition at both source and destination are different. The default has an unknown polarity (no specification), which means both transitions at the source and destination cannot be predicted. For example:

```
specify
    (a => x) = 10;
    (b => y) = 8;
    (c => z) = 8;
endspecify
```

2. Full connection. The full connection uses "*>" and every path source connects to all path destinations. The source need not have the same size as the destination. It is used to decribe a module path between a vector and a scalar, between vectors of different sizes and with multiple sources or multiple destinations in a single statement. The syntax of full connection is as follows:

```
(source [polarity] *> destination) [= delay_value;]
```

The destination can be a list of one or more of the vector or scalar output and inout ports, bit-selects or part-selects of vector output, and inout ports. For example:

```
specify
    (a, b, c *> x) = 8;
    (a, b, c *> y) = 10;
    (b, c *> z) = 12;
endspecify
```

Multiple module paths can be described in a single statement such as the following example:

```
specify
    (a, b, c *> x, y) = 8;
endspecify
```

is equivalent to:

```
specify
    (a, b, c *> x) = 8;
    (a, b, c *> y) = 8;
endspecify
```

This can be further expanded into the following six individual module path assignments:

```
specify
    (a *> x) = 8;
    (b *> x) = 8;
    (c *> x) = 8;
    (a *> y) = 8;
    (b *> y) = 8;
    (c *> y) = 8;
endspecify
```

Edge-sensitive Path   In an edge-sensitive path, the module path is described based on an edge transition of a specified input.

```
[edge] input_descriptor => destination [polarity]:source
                                  [= delay_value;]
```

```
[edge] input_descriptor *> destination [polarity]:source
                        [= delay_value;]
```

where `edge` can be either `posedge` or `negedge`. The `input_descriptor` may be any input port or inout port. When the `input_descriptor` is a vector port, the edge transition is detected on the LSB. If the `edge` is not specified, it is active on any transition.

The following example demonstrates an edge-sensitive path declaration along with a positive polarity operator.

```
posedge clock => (out +: in) = (8, 6);
```

At the positive edge of `clock`, a module path extends from `clock` to `out` using a rise time of 8 time units and a fall time of 6 time units. The data path is from `in` to `out` and the signal is not inverted.

The next example addresses an edge-sensitive path declaration coupled with a negative polarity operator.

```
negedge clock => (out -: in) = (8, 6);
```

At the negative edge of `clock`, a module path extends from `clock` to `out` using a rise time of 8 time units and a fall time of 6 time units. The data path is from `in` to `out` and the signal is inverted.

Of course, when the path is active on any transition at the input port, it is required to leave the `edge` argument unspecified, for example:

```
clock => (out : in) = (8, 6);
```

At any changes in `clock`, a module path extends from `clock` to `out`. The data path is from `in` to `out` and the signal transition is unknown.

State-dependent Path   In an actual hardware module, the module path delays might be changed when the states of input signals to a circuit change. To reflect this situation, the module path delay can be assigned conditionally, based on the value of the signals in the circuit. The general form of a state-dependent path is as follows:

```
if (cond_expr) simple_path_declaration
if (cond_expr) edge_sensitive_path_declaration
ifnone simple_path_declaration
```

where `cond_expr` controls the module path. The `cond_expr` can contain any logical, bitwise, reduction, concatenation or conditional operators. The operands in the `cond_expr` can be one of the following: scalar or vector input ports or inout ports or their bit-selects or part-selects, locally defined variables or nets or their bit-selects or part-selects, constants and specify parameters. The `ifnone` statement is a default state-dependent path delay when all other conditions for the path are false.

The following example is a state-dependent path whose module path description is a simple path.

```
specify
    if (x)  (x => f) = 10;
    if (~x) (y => f) = 8;
endspecify
```

If input x is true, a module path extends from x to f using a delay of 10 time units; otherwise, a module path extends from y to f using a delay of 8 time units.

The following example demonstrates the use of state-dependent paths for describing an NOR gate.

EXAMPLE 7.4   *The Timing of an NOR Gate*

This example illustrates the use of state-dependent paths. It specifies the delays of an NOR gate. The first two state-dependent paths describe a pair of output rise time and fall time when the NOR gate (nor1) outputs a 0 independent of the other input. The last two state-dependent paths describe another pair of output rise time and fall time when the NOR gate inverts a changing input.

```
module my_nor (a, b, out);
input a, b:
output out;
    nor nor1 (out, a, b);
    specify
        if (a) (b => out) = (1, 2);
        if (b) (a => out) = (1, 2);
        if (~a)(b => out) = (2, 3);
        if (~b)(a => out) = (2, 3);
    endspecify
endmodule
```

The if (cond_expr) can also be combined with an edge-sensitive path. The result is called the *state-dependent edge-sensitive path*, such as the following example:

```
specify
    if (reset_n && clear_n)
        (negative clock => (out +: in) = (8, 6);
endspecify
```

If the negative edge of clock occurs when reset_n and clear_n inputs are high, a module path extends from clock to out using a rise time of 8 time units and a fall time of 6 time units. The data path is from in to out and the signal is not inverted.

The **ifnone** Condition   As mentioned before, the `ifnone` statement is used to specify a default state-dependent path delay when no other conditions are true. The source and destination used in the `ifnone` statement should be the same as the state-dependent module paths. For example, the following are valid state-dependent path descriptions.

```
specify
    (posedge clk => (q +: d)) = (3, 4);
    ifnone (clk => q) = (2, 5);
endspecify
```

However, the following are invalid because the `ifnone` state-dependent path is the same as an unconditional path rather than the state-dependent path above it.

```
specify  // invalid state-dependent path descriptions
    if (~a)(b => out) = (3,3);
    ifnone (a => out) = (2,2); // the same module path as
            (a => out) = (3,3); // the following statement
endspecify
```

Review Questions

Q7.39   What are the functions of specify blocks?

Q7.40   What can be declared inside a specify block?

Q7.41   Define simple path, edge-sensitive path and state-dependent path.

Q7.42   What are the differences between parallel connection and full connection?

Q7.43   When multiple module paths want to be declared as concise as possible, which type of connections, parallel or full, should be used?

Q7.44   What is the meaning of the `ifnone` statement? When would you use it?

7.3.2.2 *Delay Specifications*   As we have used many times in the previous examples, each module path can be assigned a delay. The delay specification may be one, two, three, six or twelve delays according to the actual requirements. The delays must be constant expressions containing literals or specify parameters, and can be a delay expression of the form `min:typ:max`. If the path delay expression results in a negative value, it is treated as zero. The general form of a delay specification is as follows:

```
tpath_expr                          // one delay
trise_expr, tfall_expr              // two delays
trise_expr, tfall_expr, tz_expr // three delays

t01_expr, t10_expr, t0z_expr,   // six delays
tz1_expr, t1z_expr, tz0_expr
```

```
        t01_expr, t10_expr, t0z_expr,   // twelve delays
        tz1_expr, t1z_expr, tz0_expr,
        t0x_expr, tx1_expr, t1x_expr,
        tx0_expr, txz_expr, tzx_expr
```

For one delay, the specified delay is the propagation time of the path. For two delays, the specified delays are rise time and fall time in order. For three delays, an additional delay, called turn-off time, is appended to the two delays. The six and twelve delays include all transistion combinations of the value set, {0, 1, z} and {0, 1, x, z}, respectively.

7.3.2.3 `specparam` *Declarations*

The `specparam` (a specify parameter) statement is used to define specify parameters, which are intended to provide timing and delays. It may be declared within the `specify` block or outside of a specify block but within a module. When a specparam is declared outside a specify block it must be declared before it is used. The `specparam` has the following general form:

```
specparam [range] specparam_assignment;
```

where `specparam_assignment` can be any constant expression which can be composed of other specify parameters or module parameters. The module parameters include `parameter` and `localparam`. A specify parameter can be used as part of a constant expression for a subsequent specify parameter declaration. However, it can only be changed through SDF annotation. The following example demonstrates the use of specify parameters.

```
specify
   // define parameters inside the specify block
   specparam  d_to_q = (10, 12); // two value
                                 // specification
   specparam clk_to_q = (15, 18);
      (  d => q) = d_to_q;
      (clk => q) = clk_to_q;
endspecify
```

The following example demonstrates the combined use of specify parameters and state-dependent paths for describing an NOR gate.

EXAMPLE 7.5   *The Timing of an NOR Gate*

This example illustrates the combined use of the specify parameter `specparam` and state-dependent paths. It specifies the delays of an NOR gate. The first two state-dependent paths describe a pair of output rise time and fall time when the NOR gate (nor1) outputs a 0 independent of the other input. The last two state-dependent paths describe another pair of output rise time and fall time when the NOR gate inverts a changing input.

```
module my_nor (a, b, out);
input a, b:
output out;
   nor nor1 (out, a, b);
   specify
      specparam trise = 1, tfall = 2
      specparam trise_n = 2, tfall_n = 3;
      if (a) (b => out) = (trise, tfall);
      if (b) (a => out) = (trise, tfall);
      if (~a)(b => out) = (trise_n, tfall_n);
      if (~b)(a => out) = (trise_n, tfall_n);
   endspecify
endmodule
```

■

7.3.2.4 *Pulse-Width Limit Control*   Because the inertial delay model is used to model module path delays, any signals with a pulse width (the time between two consecutive transitions) less than the specified module path delays will be filtered out and not present at the output of the module paths. However, there are three ways in Verilog HDL that can be used to change the pulse-width limits from their default values:

1. *Specify block control approach.* The `specparam PATHPULSE$` is provided by Verilog HDL to modify the pulse-width limits from their default values.
2. *Global control approach.* Invocation options (command lines) can specify percentages that apply to all module path delays to form the corresponding pulse-width limits.
3. *SDF annotation appproach.* SDF annotation can individually annotate the pulse-width limits of each module path.

The pulse-width ranges associated with each module path delay are defined by two limit values: *the error limit* and *the rejection limit*. The error limit is always at least as large as the rejection limit. Pulses with a width greater than or equal to the error limit pass unfiltered. Pulses with width less than the error limit, but greater than or equal to the rejection limit, are filtered to an unknown value x. Pulses with a width less than the rejection limit are rejected. By default, both the error limit and the rejection limit are set equal to the delay.

Specify Block Control Approach   Pulse-width limit values may be set with the `specparam PATHPULSE$` within the specify block. The syntax of `specparam PATHPULSE$` is of the form:

```
PATHPULSE$ = (reject_limit[, error_limit]);
PATHPULSE$source$destination = (reject_limit
                                [, error_limit]);
```

If only the reject limit is specified, it applies to both pulse-width limits. The pulse-width limits may be specified for a specific module path by using the second form. When no module path is specified, the pulse-width limits apply to all module paths defined in the module containing the specify block. The path-specific PATHPULSE$ has a higher precedence than the non-path-specific PATHPULSE$.

The following example addresses the use of specparam PATHPULSE$.

EXAMPLE 7.6 *The Use of* Specparam PATHPULSE$.

Two path-specific PATHPULSE$ and one non-path-specific PATHPULSE$ are used to declare the pulse-width limits. The path (clk => q) has a reject limit of 2 and an error limit of 3. The path (reset *> q) has a reject limit of 0 and an error limit of 3. The path (data *> q) receives reject and error limit of 3, as defined by the non-path-specific PATHPULSE$ declaration.

```
specify
// define parameters inside the specify block
specparam    d_to_q = 4:5:6;
specparam  clk_to_q = 4:5:6;
specparam reset_to_q = 3:4:5;
    (clk   => q) = clk_to_q;
    (data  *> q) = d_to_q;
    (reset *> q) = reset_to_q;
specparam
    PATHPULSE$clk$q   = (2, 3),
    PATHPULSE$reset$q = (0, 3),
    PATHPULSE$ = 3;
endspecify
```

Global Control Approach The reject limit and error limit invocation options can specify percentages applying globally to all module path delays. The percentage values are integers between 0 and 100. The default value is 100 %. The PATHPULSE$ values have a higher precedence than the global pulse limit invocation options. The following is a general form of invocation options:

```
>vsim +pulse_int_e/<percent> +pulse_int_r/<percent>
```

For example, consider an interconnect delay of 10 along with a +pulse_int_e/80 option. The error limit is 80 % of 10 and the rejection limit defaults to 80 % of 10. This results in the propagation of pulses greater than or equal to 8, and all other pulses are filtered out.

SDF Annotation Approach SDF annotation can be used to specify the pulse-width limits of module path delays. We will discuss it in Section 13.4.4 in more detail.

Review Questions

Q7.45 What are the differences between specify and module parameters?

Q7.46 List the six transitions from the value set: {0, 1, z}.

Q7.47 List the twelve transitions from the value set: $\{0, 1, x, z\}$.

Q7.48 What is the error limit when only the reject limit is specified to 5 time units?

Q7.49 What are specified in the three delays?

Q7.50 How would you specify a specific module path in PATHPULSE\$?

Q7.51 How would you change the limits of pulse width?

7.3.3 Timing Checks

In Verilog HDL, a set of functions is provided for timing checks. All timing checks must be inside the specify blocks. Although they begin with character "\$", these timing checks are not system tasks. It is instructive to remember that no system task can appear in a specify block and no timing check can appear in procedural code.

The timing checks can be roughly divided into two sets: *window-related timing checks* and *clock and signal timing checks*. The window-related timing checks are used to check the *setup time*, *hold time*, *recovery time* and *removal time* and include \$setup, \$hold, \$setuphold, \$recovery, \$removal and \$recrem. The clock and signal timing checks are used to check the *width*, *period*, *skew* and *change* of signals and include \$width, \$period, \$skew, \$timeskew, \$fullskew and \$nochange.

Every timing check may include an optional notifier which toggles its value whenever a violation is detected. In addition, all timing checks have a reference event and a data event, and boolean conditions can be associated with each.

7.3.3.1 *Conditioned Events*   In the timing checks, every event can be associated with a condition that controls the occurrence of the event conditionally. This kind of event is called a *conditioned event*. The event used in timing checks has the following general form:

```
[timing_event] specify_terminal_descriptor [&&& check_
    condition]
```

where timing_event can be either of posedge, negedge or edge-control specifiers. The edge-control specifiers contain the keyword edge and has the following form:

```
edge [edge_descriptor[, edge_descriptor]]
```

where the edge_descriptor can be one or more transitions from the signal value set $\{0, 1, x\}$, separated by commas:

```
01 Transition from 0 to 1
0x Transition from 0 to x
10 Transition from 1 to 0
1x Transition from 1 to x
```

```
x0 Transition from x to 0
x1 Transition from x to 1
```

where z is treated as x. The posedge is equivalent to edge[01, 0x, x1] and the negedge is equivalent to edge[10, x0, 1x].

The check_condition can be one of the following:

```
expression, or ~expression
expression == scalar_constant
expression === scalar_constant
expression != scalar_constant
expression !== scalar_constant
```

The scalar_constant is only allowed to use 0 (or 1'b0) and 1 (1'b1). For example:

```
$hold(posedge clk &&& (~clr), data, 3);
$hold(posedge clk &&& (clr === 0), data, 3);
```

show two ways to trigger on the positive clk edge only when clr is low.

7.3.3.2 Window-Related Timing Checks For window-related timing checks, two signals named as *reference event* and *data event* are accepted and a window is defined with respect to one signal while checking the transition time of the other with respect to the window. When the data signal transitions within the window, timing violation is reported.

Setup Time Check For a latch- or flip-flop-based circuit, the setup time is defined as the amount of time that data must be stable before they are sampled, as shown in Figure 7.5. The setup time can be checked by timing check, $setup, which has the following form:

```
$setup (data_event, reference_event, limit[, notify_
    reg]);
```

where the data_event is usually a data signal and the reference_event is a clock signal. The limit is a non-negative constant expression and the optional notify_reg is a reg variable. Violation is reported when:

```
t_reference_event - t_data_event < limit.
```

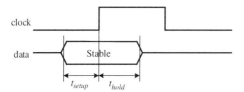

FIGURE 7.5 A timing diagram shows the setup time and hold time

When the limit is zero, the $setup timing check never reports a violation.

The following is a use of the $setup timing check. The violation occurs when the data changes its value within 5 time units before the positive edge of clock.

```
specify
    $setup (data, posedge clock, 5);
endspecify
```

Hold Time Check   For a latch- or flip-flop-based circuit, the hold time is defined as the amount of time that data must continually remain stable after they have been sampled, as shown in Figure 7.5. The hold time can be checked by the timing check $hold, which has the following form:

```
$hold (reference_event, data_event, limit
      [, notify_reg]);
```

where the data_event is usually a data signal and the reference_event is a clock signal. The limit is a non-negative constant expression. Violation is reported when:

```
t_data_event - t_reference_event < limit.
```

When the limit is zero, the $hold timing check never reports a violation.

The following is a use of the $hold timing check. The violation occurs when the data changes its value within 3 time units after the positive edge of clock.

```
specify
    $hold (posedge clock, data, 3);
endspecify
```

The following example addresses how the $setup and $hold timing checks can be used in an actual case to check whether or not the input signals confine to the timing requirements of the instantiated instance.

EXAMPLE 7.7   *Setup and Hold Time Check*

In this example, we use specparam to define the required delays. The propagation delays of clock and reset signals are represented by min:typ:max format. When no specification is given, the simulator defaults to use the typical values. The simulator will report errors when the timing relationship between inputs clk, d and reset_n violates any specification on the timing checks: $setup and $hold. In this example, two module paths are specified and the two-value specification is used to specify the delay of each module path. The reference events in both timing checks are conditioned events. They are triggered only when the reser_n signal is high.

```
module dff_setuphold_check(clk, reset_n, d, q);
output reg q;
input clk, d, reset_n;
specify
// define timing specparam values
specparam tSU = 10, tHD = 2;
// define module path delay rise and fall min:typ:max
// values
specparam tPLHclk = 4:6:9, tPHLclk = 5:8:11;
specparam tPLHreset = 3:5:6, tPHLreset = 4:7:9;
    // specify module path delays
    (clk *> q)     = (tPLHclk, tPHLclk);
    (reset_n *> q) = (tPLHreset, tPHLreset);
    // setup time : data to clock, only when reset_n is 1
    $setup(d, posedge clk &&& reset_n, tSU);
    // hold time: clock to data, only when reset_n is 1
    $hold(posedge clk, d &&& reset_n, tHD);
endspecify
// the body of flip flop
always @(posedge clk or negedge reset_n)
    if (!reset_n) q <= 0; // reset_n signal is active-low
    else          q <= d;
endmodule                                              ■
```

Setup and Hold Time Check   Both setup time and hold time are often checked together and can be performed by using a single timing check $setuphold. The operation of the $setuphold timing check is a combination of the $setup and $hold timing checks when positive limits are used. The $setuphold timing check with positive limits has the following form:

```
$setuphold (reference_event, data_event, setup_limit,
            hold_limit [, notify_reg]);
```

where the data_event is usually a data signal and the reference_event is a clock signal. Both setup_limit and hold_limit are constant expressions; in other words, their values may be less than 0. We return to this case and discuss it in more detail later in this section. Violation is reported when:

```
t_reference_event - t_data_event < setup_limit.
t_data_event - t_reference_event < hold_limit.
```

When both limits are zero, the $setuphold timing check never reports a violation.

The following example illustrates a simple use of the $setuphold timing check:

```
specify
    $setuphold (posedge clock, data, 10, 5);
endspecify
```

which is equivalent to:

```
specify
    $setup(data, posedge clock, 10);
    $hold (posedge clock, data, 5);
endspecify
```

As mentioned previously, conditions can be associated with both the reference and data signals by using the &&& operator. The following example illustrates the use of these conditions.

```
$setup (data, clk &&& condition, tsetup, notifier);
$hold (clk, data &&& condition, thold, notifier);
```

is equivalent to a single $setuphold:

```
$setuphold(clk, data, tsetup, thold, notifier, ,
    condition);
```

where the stamptime_cond is null and the checktime_cond is condition.

Recovery Timing Check   This specifies the minimum time that an asynchronous input must not be asserted prior to the active edge of the clock, as shown in Figure 7.6(a). The $recovery timing check has the following form:

```
$recovery (reference_event, data_event, limit
    [, notify_reg]);
```

where the reference_event is usually an asynchronous control signal such as clear, reset or set, and the data_event is a clock. The limit is a non-negative constant

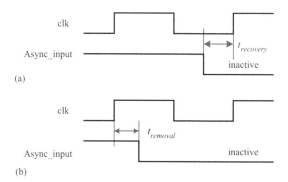

FIGURE 7.6 Timing diagrams showing (a) the recovery and (b) the removal times

expression. Violation is reported when:

```
t_data_event - t_reference_event < limit.
```

When the limit is zero, the $recovery timing check never reports a violation.

The use of the $recovery timing check is shown in the following example.

```
specify
    $recovery (negedge aysyn_input, posedge clk, 5);
endspecify
```

Removal Timing Check   This specifies the minimum time that an asynchronous input such as clear and preset, must remain asserted after the active edge of the clock, as shown in Figure 7.6(b). The $removal timing check has the following form:

```
$removal (reference_event, data_event, limit
  [, notify_reg]);
```

where the reference_event is usually an asynchronous control signal such as clear, reset or set, and the data_event is a clock. The limit is a non-negative constant expression. Violation is reported when:

```
t_reference_event - t_data_event < limit.
```

When the limit is zero, the $removal timing check never reports a violation.

The use of the $removal timing check is shown in the following example.

```
specify
    $removal (negedge aysyn_input, posedge clk, 5);
endspecify
```

Recovery and Removal Timing Checks   Both removal time and recovery time are often checked together. This kind of timing check can be performed by using $recrem, which is simply a combination of the $removal and $recovery timing checks when limits are positive or zero. The $recrem timing check with positive limits has the following form:

```
$recrem (reference_event, data_event, recovery_limit,
    removal_limit [, notify_reg]);
```

where the reference_event is usually an asynchronous input signal and the data_event is a clock signal. Both recovery_limit and removal_limit are constant expressions; namely, their values may be less than 0. Violation is reported when:

```
t_data_event - t_reference_event < recovery_limit
t_reference_event - t_data_event < removal_limit
```

When both limits are zero, the $recrem timing check never reports a violation.

The following example illustrates a simple use of the $recrem timing check:

```
specify
    $recrem (negedge aysyn_input, posedge clk, 5, 5);
endspecify
```

which is equivalent to:

```
specify
    $recovery (negedge aysyn_input, posedge clk, 5);
    $removal  (negedge aysyn_input, posedge clk, 5);
endspecify
```

7.3.3.3 *Negative Timing Checks*   As mentioned before, both $setuphold and $recrem timing checks can accept negative limit values. Their behaviors are identical with respect to negative limit values. Hence, the following descriptions are for the $setuphold timing check but apply equally to the $recrem timing check. The general forms of $setuphold and $recrem timing checks are as follows:

```
$setuphold (reference_event, data_event, setup_limit,
    hold_limit [,[notify_reg][,[stamptime_cond]
    [,[checktime_cond] [,[delayed_ref]
      [,[delayed_data]]]]]]);
$recrem (reference_event, data_event, recovery_limit,
    removal_limit [,[notify_reg][,[stamptime_cond]
    [,[checktime_cond] [,[delayed_ref]
      [,[delayed_data]]]]]]);
```

where stamptime_cond and checktime_cond are conditions for negative timing checks and delayed_ref and delayed_data are delayed reference and data signals for negative timing checks, respectively.

The setup time and hold time define a *violation window* with respect to the reference event during which the data must remain stable. Any change of the data during the specified window causes a timing violation. A positive value for setup time and hold time means the violation window straddles the reference signal while a negative value for hold or setup time implies the violation window is shifted to either left or right of the reference edge, as shown in Figure 7.7.

Due to unequal delays of internal signal paths, a hardware module can have a negative setup time or hold time, but can never have both negative simultaneously. In order to accurately model negative-value timing constraints, delayed copies of the data and reference signals are generated and used internally for timing check evaluation at runtime. The setup time and hold time used internally are adjusted so that the violation window can overlap the reference signal. Delayed data and reference signals can also be explicitly declared within the timing check to ensure accurate simulation. If no delayed

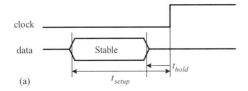

(a)

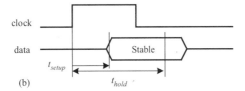

(b)

FIGURE 7.7 Timing diagrams showing both the negative (a) setup time and (b) hold time

signals are declared and a negative setup or hold value is used, then implicit delayed signals are created instead.

For instance, in the following example implicit delayed signals are created for `clk` and `data` because a setup time with negative value of −5 is used. The `data` value will not be properly clocked in if it does not transition after posedge `clk` in 5 time units.

```
$setuphold (posedge clk, data, -5, 10);
```

The following example explicitly declares the delayed signals `dclk` and `ddata`.

```
$setuphold(posedge clk, data, -5, 10,,,, dclk, ddata);
```

The delayed timing check signals are only actually delayed when negative limit values are present.

Review Questions

Q7.52 Explain the basic operations of window-related timing checks.

Q7.53 What is the conditioned event?

Q7.54 Define the following two terms: setup time and hold time.

Q7.55 What is the function of the keyword `edge`?

Q7.56 What is the meaning of violation window?

7.3.3.4 Clock and Signal Timing Checks   The set of clock and signal timing checks accepts one or two signals and verifies transitions on them that are never separated by more than the limit. This set of timing checks is used to check the width, period, skew and change of signals.

Pulse Width Check   The $width timing check tests whether the width of a pulse meets the minimum width requirement, as shown in Figure 7.8. The $width

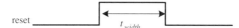

reset t_{width}

FIGURE 7.8 A timing diagram showing the pulse width

timing check has the following general form:

```
$width (reference_event, limit, threshold
    [, notify_reg]);
```

where the data_event is derived from the reference_event. It is the next opposite edge of the reference_event. In other words, the $width checks the time between two consecutive transitions of a signal. Both limit and threshold are non-negative constant expressions. If the comma before the threshold is present, the comma before the notifier must also be present, even though both arguments are optional. Violation is reported when:

```
threshold < t_data_event - t_reference_event < limit.
```

Note that no violation is reported for glitches smaller than the threshold. In order to avoid a timing violation, the pulse width must be greater than or equal to the specified limit. For example, what in the following specifies the pulse width of the reset signal must be at least 10 time units.

```
specify
    $width (posedge reset, 10);
endspecify
```

Period Check   This is used to check whether the period of a signal meets the minimum period requirement, as shown in Figure 7.9.

The $period timing check has the following general form:

```
$period (reference_event, limit[, notify_reg]);
```

where the data_event is derived from the reference_event. It is the next same polarity edge of the reference_event signal. In other words, the $period checks the time between two adjacent transitions of same polarity of a signal. The limit is a nonnegative constant expression. Violation is reported when:

```
t_data_event - t_reference_event < limit,
```

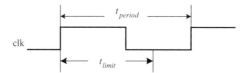

clk

t_{period}

t_{limit}

FIGURE 7.9 A timing diagram showing the period

For example, what in the following specifies the clock signal `clk` must have a period at least of 25 time units.

```
specify
   $period (posedge clk, 25);
endspecify
```

The following is a more complex and complete example of the use of timing checks.

EXAMPLE 7.8   *A More Complex Timing Check Example*

In this example, we add three timing checks, $period, $width and $recovery, to the preceding example. These timing checks are used to check the timing constraints of the period of clock signal, the pulse width of reset signal and the recovery time of reset signal, respectively.

```
module dff_timing_check(clk, reset_n, d, q);
output reg q;
input clk, d, reset_n;
specify
// define timing specparam values
specparam tSU = 10, tHD = 2, tPW = 25,
         tWPC = 10, tREC = 5;
// define module path delay rise and fall min:typ:max
// values
specparam tPLHclk = 4:6:9, tPHLclk = 5:8:11;
specparam tPLHreset = 3:5:6, tPHLreset = 4:7:9;
   // specify module path delays
   (clk *> q)     = (tPLHclk, tPHLclk);
   (reset_n *> q) = (tPLHreset, tPHLreset);
   // setup time : data to clock, only when reset_n is 1
   $setup(d, posedge clk && reset_n, tSU);
   // hold time: clock to data, only when reset_n is 1
   $hold(posedge clk, d && reset_n, tHD);
   // clock period check
   $period(posedge clk, tPW);
    // pulse width : reset_n
   $width(negedge reset_n, tWPC, 0);
    // recovery time: reset_n to clock
   $recovery(posedge reset_n, posedge clk, tREC);
endspecify
// the body of flip flop
always @(posedge clk or negedge reset_n)
   if (!reset_n) q <= 0; // reset_n signal is active-low.
   else          q <= d;
endmodule
```
■

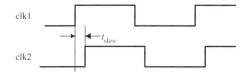

FIGURE 7.10 A timing diagram showing the clock skew

Skew Checks The skew timing checks examine whether the time difference between two signals meets the specified timing constraint. Two timing violation detection mechanisms of skew timing checks are *event-based* and *timer-based*. Event-based skew checking relies on signal transitions whereas timer-based skew checking employs simulation time.

There are three skew timing checks: $skew, $timeskew and $fullskew. The $skew timing check is event-based whereas the $timeskew and $fullskew checks are timer-based by default, and can be altered to event-based. The timing diagram showing the skew between two signals is depicted in Figure 7.10.

$skew Timing Check The $skew timing check examines whether the time skew between two signals meets the maximum requirement based on events. It has the following form:

```
$skew (reference_event, data_event,
      limit[, notify_reg]);
```

where limit is a non-negative constant expression. Violation is reported when:

```
t_data_event - t_reference_event > limit.
```

In the following, a simple example shows that the skew between two clock signals clk1 and clk2 must be within 5 time units; otherwise, timing violation is reported.

```
specify
   $skew (posedge clk1, posedge clk2, 5);
endspecify
```

$timeskew Timing Check The $timeskew timing check examines whether the time skew between two signals meets the maximum requirement based on events or the simulation times. It has the following general form:

```
$timeskew (reference_event, data_event, limit
  [,[ notify_reg] [,[ event_based_flag]
    [,[ remain_active_flag]]]]);
```

where limit is a non-negative constant expression. Both event_based_flag and remain_active_flag are optional and are constant expressions. Violation is reported when:

```
t_data_event - t_reference_event > limit.
```

After reporting the violation, the $timeskew timing check becomes dormant and reports no more violations until the occurrence of next reference event.

A simple example is shown in the following:

```
specify
    $timeskew (posedge clk1, posedge clk2, 5);
endspecify
```

The operation of $timeskew defaults to be timer-based. However, it can be altered to event-based using the event_based_flag. When only the event_based_flag is set, the $timeskew timing check behaves like the $skew timing check except that it becomes dormant after reporting the first violation. To make it behave exactly like the $skew timing check, both the event_based_flag and the remain_active_flag have to be set.

$fullskew Timing Check   The $fullskew timing check is identical to $timeskew except that the reference and data events can transition in either order. It has the following general form:

```
$fullskew (reference_event, data_event, limit1, limit2
    [,[ notify_reg] [,[ event_based_flag]
      [,[ remain_active_flag]]]]);
```

where limit1 and limit2 are non-negative constant expressions. Violation is reported when either of the following two conditions occurs:

```
t_data_event - t_reference_event > limit1
t_reference_event - t_data_event > limit2.
```

A simple example of the use of $fullskew is as follows:

```
specify
    $timeskew (posedge clk1, negedge clk2, 5, 7);
endspecify
```

Timing violation will be reported when the negative edge of clk2 is 5 time units lagged behind the positive edge of clk1 or the positive edge of clk1 is 7 time units lagged behind the negative edge of clk2.

$nochange Timing Check   The $nochange timing check examines whether a signal remains unchanged in a time window defined by the reference event. The time window is defined as the time between the leading edge and the trailing edge of the reference event. In other words, it reports a violation when a data event occurs during the specified level (high or low) of the reference event. It has the following general form:

```
$nochange (reference_event, data_event, start_offset,
           end_offset[,[ notify_reg]]);
```

where `start_offset` and `end_offset` are used to expand or shrink the timing violation region. They are constant expressions and relative to the `reference_event`. The `reference_event` must be an edge-triggered event and specified with the keywords `posedge` or `negedge`.

A violation results if the data event occurs anytime within the time window. Violation is reported when:

```
t_start_window = t_leading_edge_reference_event -
                   start_offeset
 t_end_window   = t_trailing_edge_reference_event +
                   end_offeset
 t_start_window < t_data_event < t_end_window
```

For example, the following `$nochange` timing check will report timing violation if `clear` changes while `reset_n` is high. The time window is the time between the positive edge and negative edge of the `reset_n` signal.

```
specify
   $nochange (posedge reset_n, clear, 0, 0);
endspecify
```

Review Questions

Q7.57 What is the basic principle of checking the width of a pulse? Which timing check can be used for this purpose?

Q7.58 What is the basic principle of checking the period of a periodic signal? Which timing check can be used for this purpose?

Q7.59 What are the differences between event-based and time-based skew detection mechanisms?

Q7.60 What are the differences between `$skew`, `$timeskew` and `$fullskew` timing checks?

Q7.61 Explain the operation of the `$nochange` timing check. Why is it so named?

7.4 COMPILER DIRECTIVES

Like the C programming language, Verilog HDL provides a rich set of compiler directives. All compiler directives begin with the " ` " character, called accent grave, and has the form: `` `<keyword> ``. A compiler directive, when compiled, remains in effect across all files processed until another compiler directive specifies otherwise. The set of compiler directives provided by Verilog HDL is as follows:

1. `` `define `` and `` `undef `` compiler directives.
2. `` `include `` compiler directive.
3. `` `ifdef ``, `` `else ``, `` `elsif ``, `` `endif `` and `` `ifndef `` compiler directives.
4. `` `timescale `` compiler directive.
5. `` `celldefine `` and `` `endcelldefine `` compiler directives.
6. `` `line `` compiler directive.
7. `` `default_nettype `` compiler directive.
8. `` `unconnected_drive `` and `` `nounconnected_drive `` compiler directives.
9. `` `resetall `` compiler directive.

7.4.1 `` `define `` and `` `undef `` Compiler Directives

The `` `define `` compiler directive is used to create a macro for text substitution. It is similar to the `#define` construct in C programming language. Two major features of the `` `define `` compiler directive are as follows: (1) It is usually placed at the head of a file or a separated file and (2) it can be used both inside and outside module definitions. The `` `define `` compiler directive has the general form:

```
`define macro_name [(formal_arguments)] macro_text
```

Once a macro text is defined with the `` `macro_name ``, every occurrence of `` `macro_name `` in the source file will be substituted by the macro text. For example:

```
`define BUS_WIDTH   8
```

Any place where a `` `BUS_WIDTH `` appears will be substituted by 8.

In order to allow the macro to be customized for each use individually, a text macro can also be defined with arguments. A text macro with argument(s) is expanded by substituting each formal argument with the actual argument when it is called. The general form of using a text macro is as follows:

```
`macro_name [(actual_argument1, ..., actual_argumentn)]
```

When a text macro is expanded, each actual argument is substituted for the corresponding formal argument in a way of literal by literal. Therefore, when an expression is used as an actual argument, the entire expression is substituted into the macro text. For example:

```
`define min(a, b)((a) < (b) ? (a) : (b))
y = `min(p+q, r+s);  // macro call
```

will be expanded into:

```
y = ((p+q) < (r+s) ? (p+q) : (r+s));
```

The `undef` compiler directive removes a previously defined text macro. It has the form:

```
`undef macro_name
```

Any attempt to remove an undefined text macro using the `define` compiler directive will result in a warning.

7.4.2 `include` Compiler Directive

The `include` compiler directive allows us to insert an entire source file into the place where the compiler directive of another file appears during compilation. It is usually used to include commonly used definitions, which are often placed in a separate file. The work of the `include` compiler directive is similar to that of `#include` in the C programming language. The `include` compiler directive can be specified anywhere within a source file. The syntax of the `include` compiler directive is as follows:

```
`include "filename"
```

where the `filename` is the name of the file to be included. The filename can be a full (absolute) or relative path name. Only white space or a comment may appear on the same line as the `include` compiler directive. In addition, the `include` compiler directive may be nested. That is, a file included in the source using the `include` compiler directive may contain other `include` compiler directive. The following two examples illustrate the usage of the `include` compiler directive:

```
`include "count.v"
`include "../program/chapter06/count.v"
```

7.4.3 `ifdef`, `else`, `elsif`, `endif` and `ifndef` Compiler Directives

In Verilog HDL, optional parts of a source file can be compiled conditionally based on some macro definitions. These conditionally compiled directives can be classified into two types: check the existence of a macro definition (`ifdef`) and check the lack of a macro definition (`ifndef`). The `ifdef` compiler directive has the following form:

```
`ifdef macro_name
   ifdef_group_of_lines
{`elsif macro_name elsif_group_of_lines}
[`else  else_group_of_lines]
`endif
```

where the `else` compiler directive is optional with the `ifdef` compiler directive. The following example illustrates the usage of the `ifdef` compiler directive.

```
`ifdef DWORD
   parameter WORD_SIZE = 32;
`else
   parameter WORD_SIZE = 16;
`endif
```

During compilation, the WORD_SIZE is set to 32 if the text macro name DWORD is defined; otherwise, the WORD_SIZE is set to 16.

The `ifndef compiler directive has the following form:

```
`ifndef macro_name
    ifndef_group_of_lines
{`elsif macro_name elsif_group_of_lines}
[`else   else_group_of_lines]
`endif
```

where the `else compiler directive is optional with the `ifndef compiler directive. The following example illustrates the usage of the `ifndef compiler directive.

```
module xor_op (a, b, c);
output a;
input b, c;
`ifndef DATAFLOW
    xor a1 (a, b, c);
`else
    wire a = b ^ c;
`endif
endmodule
```

During compilation, if the text macro named DATAFLOW is not defined, an xor gate is instantiated; otherwise, a continuous net assignment is compiled instead.

Review Questions

Q7.62   Explain the meaning and usage of the `define compiler directive.

Q7.63   Explain the meaning and usage of the `include compiler directive.

Q7.64   Explain the meaning and usage of the `ifdef compiler directive.

Q7.65   Explain the meaning and usage of the `ifndef compiler directive.

7.4.4 `timescale Compiler Directive

The `timescale compiler directive is used to specify both the physical unit of measure or the time scale of a numerical time delay and the resolution of the time scale, i.e. the minimum step size of the scale during simulation. The format of the `timescale compiler directive is as follows:

```
`timescale time_unit / time_precision
```

The `time_unit` specifies the unit of measure for times and delays while the `time_precision` specifies how delays are rounded during simulation. Only the integers 1, 10 and 100 may be used to specify `time_unit` and `time_precision`; the valid units are s, ms, μs, ns, ps and fs. For instance:

```
`timescale 10 ns /1 ns

#2.55 a = b + 1; // corresponds to 26 ns
#2.54 b = c + 1; // corresponds to 25 ns
```

The `timescale 10 ns / 1 ns` specifies that the time unit is 10 ns. As a result, the time values in the module are multiples of 10 ns, rounded to the nearest 1 ns. Therefore, the delays 2.55 and 2.54 are scaled rounded to 26 ns and 25 ns, respectively. As another example:

```
`timescale 1 ns /10 ps

#2.55 a = b + 1; // corresponds to 2.55 ns
#2.54 b = c + 1; // corresponds to 2.54 ns
```

Here, all time values are multiples of 1 ns because the `time_unit` is 1 ns. Delays are rounded to real numbers with two decimal places because the `time_precision` is 10 ps.

The `time_precision` must not exceed the `time_unit`. Hence, `timescale 1 ns /10 ns` is illegal.

Review Questions

Q7.66 Explain the meaning of the `timescale` compiler directive.

Q7.67 Explain the meaning of the `time_unit` in the `timescale` compiler directive.

Q7.68 Explain the meaning of the `time_precision` in the `timescale` compiler directive.

7.4.5 Miscellaneous Compiler Directives

In this sub-section, we describe the following compiler directives:

1. `celldefine` and `endcelldefine`.
2. `line`.
3. `unconnected_drive` and `nounconnected_drive`.
4. `default_nettype`.
5. `resetall`.

`celldefine` and `endcelldefine` **Compiler Directives** These compiler directives are used to mark a module as a cell module. That is, their combined use defines the cell boundary. These compiler directives may appear anywhere in the source file, but it is recommended that the compiler directives be specified outside of the module definition. An example of the usage of these two directives is as follows:

```
`timescale 1ns / 10ps
`celldefine
module AN2D1 (a1, a2, f);
input a1, a2;
output f;
    and    (f, a1, a2);
    specify
        (a1 => f)=(0, 0);
        (a2 => f)=(0, 0);
    endspecify
endmodule
`endcelldefine
```

Cell modules are used by some PLI routines, such as delay calculations. They typically come along with a `specify` block, as shown in the above example, which specifies the delays of the cell modules.

`line` **Compiler Directive** The `line` compiler directive is used to reset the line number and file name of the current file to that as specified. It is of the form:

```
`line number "file_name" level
```

where the `number` is the new line number of the next line, the `file_name` is the new name of the file and the value of `level` indicates whether an include file has been entered if it is 1, an include file is exited if it is 2 or neither has been done if it is 0. The compiler directive can be specified anywhere within the Verilog HDL source file. The results of this compiler directive are not affected by the `resetall` compiler directive.

`unconnected_drive` and `nounconnected_drive` **Compiler Directives** Any unconnected input ports of a module that appear between the `unconnected_drive` and `nounconnected_drive` compiler directives are pulled up or pulled down depending on what value is specified: `pull1` or `pull0`. The general form is as follows:

```
// the first example
`unconnected_drive pull0
// all unconnected input ports are pull down
`nounconnected_drive
// the second example
```

```
`unconnected_drive pull1
// all unconnected input ports are pull up
`nounconnected_drive
```

These two compiler directives must be specified in pairs and placed outside of the module declarations.

**`default_nettype Compiler Directive**   The default net data type is `wire` for implicit net declarations. The `` `default_nettype `` compiler directive changes the default net type as specified. It can only be used outside of module definitions. Multiple `` `default_nettype `` compiler directives are allowed. It has the form:

```
`default_nettype net_type
```

where `net_type` is one of the following: `wire`, `tri`, `tri0`, `wand`, `triand`, `wor`, `trior`, `trireg` or `none`. When `none` is specified, all nets must be explicitly declared.

`resetall Compiler Directive   The `` `resetall `` compiler directive resets all compiler directives to their default values. It is of the form:

```
`resetall
```

It is recommended to place the `` `resetall `` compiler directive at the beginning of a source file before any compiler directives.

Review Questions

Q7.69   Explain the meaning and usage of the `` `celldefine `` compiler directive.

Q7.70   Explain the meaning and usage of the `` `unconnected_drive `` compiler directive.

Q7.71   Explain the meaning and usage of the `` `default_nettype `` compiler directive.

Q7.72   Explain the meaning and usage of the `` `resetall `` compiler directive.

SUMMARY

In this chapter, we discuss additional features of Verilog HDL. These features include block constructs, procedural continuous assignments, specify blocks and timing checks.

The block constructs are used to group two or more statements together so that they can act as a single statement, and can be divided into sequential blocks and parallel blocks. The procedural statements within sequential blocks are executed sequentially in the given order but within parallel blocks are executed concurrently regardless of their relative order.

The procedural continuous assignments allow expressions to be driven continuously onto variables or nets. They are procedural statements; in other words, they can only be used within `initial` and `always` statements. Two kinds of procedural continuous assignments are `assign`/`deassign` and `force`/`release`. The `assign` and `deassign` statements assign values to variables. The `force` and `release` statements assign values to nets or variables.

The specify block is a mechanism provided in Verilog HDL for describing various paths across the module, assigning delays to these paths and performing necessary timing checks. The specify blocks are often used to define the cells in ASIC cell libraries.

The timing checks provided in Verilog HDL are used to check various timing constraints of a module. All timing checks must be inside the `specify` blocks. The timing checks can be roughly divided into two sets: window-related timing checks and clock and signal timing checks. For window-related timing checks, two signals, named as reference event and data event, are accepted and a window is defined with respect to one signal while checking the transition time of the other with respect to the window. When the data signal transitions within the window, timing violation is reported. This set of timing checks is used to check the setup time, hold time, recovery time and removal time. The set of clock and signal timing checks accepts one or two signals and verifies transitions on them which are never separated by more than the limit. This set of timing checks is used to check the width, period, skew and change of signals.

Like the C programming language, Verilog HDL provides a rich set of compiler directives. A compiler directive, when compiled, remains in effect across all files processed until another compiler directive specifies otherwise. With the aid of compiler directives, a macro text may be defined, parts of a design can be compiled conditionally, default net features can be changed and unconnected input ports may be pull-up or pull-down.

REFERENCES

1. J. Bhasker, *A Verilog HDL Primer,* 3rd Edn, Star Galaxy Publishing, 2005.
2. M. D. Ciletti, *Modeling, Synthesis, and Rapid Prototyping with the Verilog HDL,* Prentice-Hall, Upper Saddle River, NJ, USA, 1999.
3. IEEE 1364-2001 Standard, *IEEE Standard Verilog Hardware Description Language,* 2001.
4. S. Palnitkar, *Verilog HDL: A Guide to Digital Design and Synthesis,* 2nd Edn, SunSoft Press, 2003.

PROBLEMS

7.1 Explain the operations of the following two parallel blocks:

```
// block a: cannot swap a and b
fork: race
    #5 a = b;
    #5 b = a;
join
```

```
// block b: swap a and b properly
fork: swap
    a = #5 b;
    b = #5 a;
join
```

7.2 Explain the operations of the following two `always` blocks:

```
// block a: cannot swap a and b
always @(*)
    a = b;
always @(*)
    b = a;
// block b: swap a and b properly
always @(*)
    a <= b;
always @(*)
    b <= a;
```

7.3 What are the differences between the following two sequential blocks:

```
// block a
begin: block_a
    a = #5 b;
    b = #5 a;
end
// block b
begin: block_b
    a <= #5 b;
    b <= #5 a;
end
```

7.4 What are the differences between the following two parallel blocks:

```
// block a
fork: block_a
    a = #5 b;
    b = #5 a;
join
// block b
fork: block_b
    a <= #5 b;
    b <= #5 a;
join
```

7.5 Explain the operations of the following nested block and give the time at which each state executes.

```
// block a
fork: block_a
   a = #5 b;
   begin: block_b
      d <= #15 c;
      c <= #15 d;
   end
   b = #5 a;
join
```

7.6 Explain the operations of the following nested block and give the time at which each state executes.

```
// block a
fork: block_a
   a = #5 b;
   begin: block_b
      #15 d <= c;
      #15 c <= d;
   end
   b = #5 a;
join
```

7.7 Explain the operations of the following nested block and give the time at which each state executes.

```
// block a
fork: block_a
   #5 a = b;
   begin: block_b
      d <= #15 c;
      c <= #15 d;
   end
   #5 b = a;
join
```

7.8 Explain the operations of the following nested block and give the time at which each state executes.

```
// block a
fork: block_a
   #5 a = b;
```

```
      begin: block_b
         #15 d <= c;
         #15 c <= d;
      end
      #5 b = a;
  join
```

7.9 Explain the operations of the following nested block and give the time at which each state executes.

```
// block a
begin: block_a
   a <= #5 b;
   fork: block_b
      d = #15 c;
      c = #15 d;
   join
   b <= #5 a;
end
```

7.10 Explain the operations of the following nested block and give the time at which each state executes.

```
// block a
begin: block_a
   a <= #5 b;
   fork: block_b
      #15 d = c;
      #15 c = d;
   join
   b <= #5 a;
end
```

7.11 Use `assign` and `deassign` procedural continuous assignments to define a positive-edge triggered *D*-type flip-flop with asynchronous preset and clear. The output is set to 1 when the preset is active and cleared to 0 when the clear is active.

7.12 Use `force` and `release` procedural continuous assignments to define a negative-edge triggered *D*-type flip-flop with asynchronous preset and clear. The output is set to 1 when the preset is active and cleared to 0 when the clear is active.

7.13 Explain the operations of the following parallel block:

```
fork
   begin: blk1
      @start;
      repeat (3) @trigger;
      #d1 f = x ^ y;
```

```
        end
        @reset disable blk1;
    join
```

7.14 Explain the operations of the following parallel block:

```
    always begin : monostable
        #120 qout = 0;
    end
    always @trigger begin
        disable monostable;
        qout = 1;
    end
```

7.15 Explain the operations of the following specify block:

```
    specify
        if (reset)
            (posedge clk => (q[3:0]-:data)) = (4,5);
    endspecify
```

7.16 Explain why the following two state-dependent path declarations are not valid.

```
    specify
        if (reset)
            (posedge clk => (q[7:0]:data)) = (5,8);
        if (!reset)
            (posedge clk => (q[2]:data)) = (10,6);
    endspecify
```

7.17 Which of the following statements are legal?

```
    $width (posedge clear, limit);
    $width (posedge clear, limit, thresh, notifier);
    $width (negedge clear, limit, notifier);
```

7.18 Which of the following statements are illegal?

```
    $width (negedge clear, limit, 0, notifier);
    $width (posedge clear, limit, , notifier);
    $width (posedge clear, limit, notifier);
```

7.19 Up to now, there are three approaches that can be used to define a constant within a source description of designs. These are the `define compiler directive and the module parameters `parameter` and `localparam`. Describe their meanings and explain the differences when using them to define a constant.

7.20 Explain the meaning of the following timing check:

```
$setup(data, posedge clk, 5);
```

Rewrite the above statement so that the setup timing check is triggered only when the positive edge of the signal `clk` is coming and the signal `clear` is high.

7.21 Consider the figure shown in Figure 7.11.

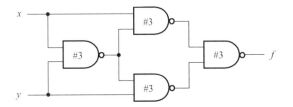

FIGURE 7.11 The figure for problem 7.21

(a) Using the distributed delay model, write a module at gate level to model the circuit in structural style.

(b) Using the distributed delay model, write a module at gate level to model the circuit in dataflow style.

(c) Using the lumped delay model, write a module at gate level to model the circuit in structural style.

(d) Using the lumped delay model, write a module at gate level to model the circuit in dataflow style.

(e) Using the path delay model, write a module at gate level to model the circuit in structural style.

7.22 Consider the positive edge-triggered D-type flip-flop shown in Figure 7.12.

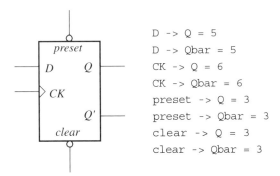

```
D -> Q = 5
D -> Qbar = 5
CK -> Q = 6
CK -> Qbar = 6
preset -> Q = 3
preset -> Qbar = 3
clear -> Q = 3
clear -> Qbar = 3
```

FIGURE 7.12 The figure for problem 7.22

(a) Write a module to describe the D-type flip-flop. Show only the I/O ports and path delay specification. Describe the path delays using parallel connection.

(b) Let all path delays of the D-type flip-flop be 5 time units. Describe the path delays using full connection to `Q` and `Qbar`.

COMBINATIONAL LOGIC MODULES

IN THE previous chapters, we have described most features of Verilog HDL. In this and the next chapters, we will examine some basic combinational and sequential modules which are often used as basic building blocks to construct a complex design. In particular, these modules are the basic building blocks of datapath when using the datapath and controller approach in a complex design, which will be described in more detail in Section 11.2.3.

In this chapter, we are concerned with the most commonly used combinational logic modules. These modules include encoders and decoders, multiplexers and demultiplexers, and magnitude comparators. In addition, a multiplexing-driven seven-segment light-emitting diode (LED) display system which combines the use of decoder as well as a multiplexer is discussed in detail. The other widely used combinational logic modules, including carry-look-ahead (CLA) adders and subtracters, along with other arithmetic logic circuits such as multipliers and dividers, are considered in Chapter 14.

As we have learned, many options can be used to model combinational logic circuits. These options include Verilog HDL primitives, continuous assignment, behavioral statements, functions, tasks without delay or event control, combinational UDPs and interconnected combinational logic modules. We will use the appropriate one whenever it is required.

8.1 DECODERS

A code conversion is a process that transforms a code into another. The circuit used to perform a code conversion is known as a *code converter*. A code conversion is known as an *encoding process* if it transforms an outside world code into the reference code; the reverse process, which transforms the reference code into the outside world code, is then called the *decoding process*. The corresponding circuits are then known as *encoder* and *decoder*, respectively. In this section, we assume that the reference code is binary code and the outside world code is 1-out-of-m code.

Digital System Designs and Practices   Ming-Bo Lin
© 2008 John Wiley & Sons (Asia) Pte Ltd

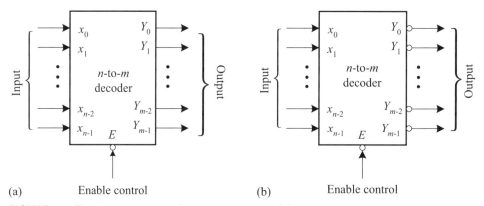

FIGURE 8.1 The block diagrams of a general decoder: (a) noninverted output; (b) inverted output

8.1.1 Decoders

A decoder is a combinational logic circuit that accepts as input an n-bit code and generates as output a 1-out-of-m code such that the activated bit of the output code corresponds to a unique combination of the input code. Usually, $m \leq 2^n$. If $m = 2^n$, the decoder is called *fully decoded*; otherwise; if $m < 2^n$, the decoder is called *partially decoded*.

The general block diagrams of an n-to-m decoder are shown in Figure 8.1. The output may be active-high or active-low, depending on the actual requirement. A decoder usually associates with one or more optional enable controls, which may be either active-high or active-low depending on the actual need. The purpose of enable control is to control the operations of the decoder. The output of the decoder is a 1-out-of-m code when its enable controls are asserted and is all 1s, all 0s or even high-impedance, depending on the actual requirement and design, when its enable controls are deasserted. One advantage of using enable control is that it makes easier the cascading of two or more decoders to form a bigger one.

An example of a 2-to-4 decoder with both active-low output and enable control is shown in Figure 8.2. Figure 8.2(a) is the logic symbol of the circuit and Figure 8.2(b) is its function table. From the function table, it is easy to derive the logic circuit, as depicted in Figure 8.2(c). To model this simple 2-to-4 decoder in Verilog HDL, any style described in Section 1.3 may be used. In the following, we model it in behavioral style with a `case` statement.

EXAMPLE 8.1 *A 2-to-4 Decoder with Active-low Output and Enable Control*

This example uses a `case` statement to code the function table shown in Figure 8.2(b). The output vector y is all 1s if the enable is high; otherwise, it depends on the input vector x. As usual, it is better to add a `default` statement at the end of the `case` statement.

```
// a 2-to-4 decoder with active-low output and enable
// control.
module decoder_2to4_low(x, enable, y);
```

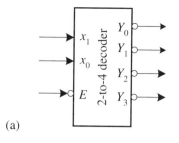

(a)

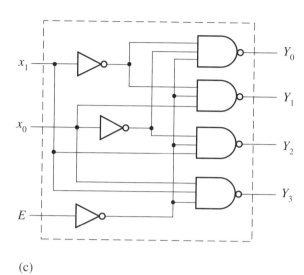

E	x_1	x_0	Y_3	Y_2	Y_1	Y_0
1	ϕ	ϕ	1	1	1	1
0	0	0	1	1	1	0
0	0	1	1	1	0	1
0	1	0	1	0	1	1
0	1	1	0	1	1	1

(b)   (c)

FIGURE 8.2 A 2-to-4 decoder with an active-low enable control: (a) logic symbol; (b) function table; (c) logic circuit

```
input   [1:0] x;
input   enable;
output reg [3:0] y;
// the body of the 2-to-4 decoder
always @(x or enable)
    if (enable) y = 4'b1111; else
        case (x)
            2'b00 : y = 4'b1110;
            2'b01 : y = 4'b1101;
            2'b10 : y = 4'b1011;
            2'b11 : y = 4'b0111;
            default : y = 4'b1111;
        endcase
endmodule
```

The following example is another type of 2-to-4 decoder, whose output and enable controls are both active-high.

EXAMPLE 8.2 *A 2-to-4 Decoder with Active-high Output and Enable Control*

As in the preceding example, a `case` statement is used to carry out the decoding operation of the decoder. In reality, this example has exactly the same structure as the preceding one, except that now the enable control is active-high and the values assigned to the output `y` are complemented.

```verilog
// a 2-to-4 decoder with active-high output.
module decoder_2to4_high(x, enable, y);
input  [1:0] x;
input  enable;
output reg [3:0] y;
// the body of the 2-to-4 decoder
always @(x or enable)
   if (!enable) y = 4'b0000; else
      case (x)
         2'b00 : y = 4'b0001;
         2'b01 : y = 4'b0010;
         2'b10 : y = 4'b0100;
         2'b11 : y = 4'b1000;
         default : y = 4'b0000;
      endcase
endmodule
```

Both of the above examples simply use case statements to implement the desired operations and are not parameterized designs. A design is said to be *parameterized* if its size is specified when it is instantiated. Although using a case statement to design a decoder is easy and straightforward, it is difficult to make the design to be parameterized. In order to design a parameterized decoder, it is necessary to explore the basic operation of decoders in greater insight. As we can see from Figure 8.2(b), the essential operation of a decoder is to place the "0" (active-low output) at the output line specified by the input code. The output line is numbered from 0 to $n - 1$. For example, it puts "0" to output line "1" when the input code x is 2'b01.

The following example illustrates how this idea is used to design a parameterized decoder.

EXAMPLE 8.3   *A Parameterized m-to-n Decoder with Active-high Output*

This example uses the above idea that puts "1" to the output line specified by input code x. To realize this idea, we first put a value "1" in the LSB and a value "0" in all other bits. Next, these values are shifted left logically the number of x positions and then assigned to the output vector y. The resulting program is shown as follows.

```verilog
// a parameterized M-to-N decoder with active-high output.
module decoder_m2n_high(x, enable, y);
parameter M = 3;  // define the number of input lines
parameter N = 8;  // define the number of output lines
input  [M-1:0] x;
input  enable;
output reg [N-1:0] y;
// the body of the M-to-N decoder
always @(x or enable)
   if  (!enable) y = {N{1'b0}};
   else y = {{N-1{1'b0}},1'b1}} << x;
endmodule
```

The interested reader is encouraged to synthesize it and see what happens. In addition, it is advisory to instantiate the above parameterized module with different parameters and observe the synthesized results. For example, you may try to design a 2-to-4 and a 3-to-6 decoder by instantiating the above module.

```
decoder_m2n_high #(2, 4) decoder1(x, enable, y);
decoder_m2n_high #(3, 6) decoder2(x, enable, y);
```

8.1.2 Expansion of Decoders

In many practical applications, it often requires a big decoder such as an 8-to-256 or even bigger. For such a big decoder, at least two ways can be employed to construct it. The straightforward approach is like what we have done for the case of a 2-to-4 decoder, using a `case` statement. The other approach is by instantiating the parameterized decoder. Regardless of which approach is used, the result is a single big decoder module.

Another widely used approach to constructing a big decoder is by cascading small decoder modules hierarchically. Due to this reason, the decoders are usually designed as modules with one or more enable controls so that they can be cascaded to form a bigger one in a hierarchical manner. For instance, we can use two 2-to-4 decoders and one 1-to-2 decoder to construct a 3-to-8 decoder, as shown in Figure 8.3. The following example illustrates how this 3-to-8 decoder works.

EXAMPLE 8.4 *A 3-to-8 Decoder Constructed by Cascading Two 2-to-4 Decoders*

As shown in Figure 8.3, input line x_2 is connected directly to the upper decoder and connected to the lower decoder through an inverter. Therefore, the upper decoder is enabled when x_2 is 0 and the lower decoder is enabled when x_2 is 1. The resulting module is a 3-to-8 decoder. Note that the inverter is acted as a 1-to-2 decoder. ■

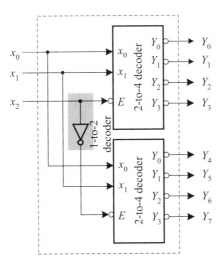

FIGURE 8.3 A 3-to-8 decoder constructed by cascading two 2-to-4 decoders

Review Questions

Q8.1 Define an m-to-n decoder.

Q8.2 Design a 2-to-4 decoder with active-low output and active-high enable control.

Q8.3 Design a 2-to-4 decoder with active-high output and active-low enable control.

Q8.4 Design an m-to-n decoder with active-low output and enable control.

Q8.5 Write a Verilog HDL module to model the logic circuit shown in Figure 8.3.

8.2 ENCODERS

As mentioned previously, the encoding process is the reverse of the decoding process and transforms an outside world code into the reference code. The logic circuit used to perform the encoding process is known as an encoder. In this section, we describe the principles of general encoders and then introduce priority encoders.

8.2.1 Encoders

An encoder has $m = 2^n$ (or fewer) input lines and n output lines. The output lines generate the binary code corresponding to the position of the activated input line. All input lines are arranged and fixed in positions numbered from 0 up to $m - 1$. The general block diagrams of encoders are shown in Figure 8.4, where Figure 8.4(a) is a noninverted output while Figure 8.4(b) is an inverted output. Like a decoder, an encoder usually associates with one or more optional enable controls, which may be either active-high or active-low depending on the actual need. The purpose of enable control is to control the operations of the encoder. The output of the encoder is a binary code when its enable controls are asserted and is all 1s, all 0s or even high-impedance, depending on the actual requirement and design, when its enable controls are deasserted. One advantage of using enable control is that it makes easier the cascading of two or more encoders to form a bigger one.

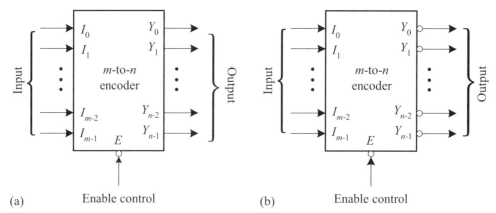

(a)   Enable control
(b)   Enable control

FIGURE 8.4 The block diagrams of a general encoder: (a) noninverted output; (b) inverted output

I_3	I_2	I_1	I_0	Y_1	Y_0
0	0	0	1	0	0
0	0	1	0	0	1
0	1	0	0	1	0
1	0	0	0	1	1

(a)

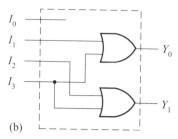

(b)

FIGURE 8.5 The function table (a) and logic diagram (b) of a simple 4-to-2 encoder

A simple 4-to-2 encoder is shown in Figure 8.5. From the function table shown in Figure 8.5(a), it is easy to derive the logic circuit shown in Figure 8.5(b). An encoder is easily constructed by using an `if-else` or a `case` statement. The following gives an example of using a nested `if-else` statement to construct a 4-to-2 encoder.

EXAMPLE 8.5 *A 4-to-2 Encoder Using an `if-else` Structure*

This example uses a nested `if-else` statement to build a 4-to-2 encoder. The conditional expression compares the input code with a constant which determines the output values, as the function table shown in Figure 8.5(a).

```
// a 4-to-2 encoder using if-else structure.
module encoder_4to2_ifelse(in, y);
input   [3:0] in;
output reg [1:0] y;
// the body of the 4-to-2 encoder
always @(in) begin
    if (in == 4'b0001) y = 0; else
    if (in == 4'b0010) y = 1; else
    if (in == 4'b0100) y = 2; else
    if (in == 4'b1000) y = 3; else
                       y = 2'bx;
end
endmodule
```

∎

The `case` statement is also often used to construct an *m*-to-*n* encoder such as in the following example.

EXAMPLE 8.6 *A 4-to-2 Encoder Using the `case` Structure*

The use of the `case` statement to model a 4-to-2 encoder is similar to that of the `if-else` statement. Here, the constants that determine the output values are used as the case items.

```
// a 4-to-2 encoder using case structure.
module encoder_4to2_case(in, y);
```

```
input   [3:0] in;
output reg [1:0] y;
// the body of the 4-to-2 encoder
always @(in)
    case (in)
        4'b0001 : y = 0;
        4'b0010 : y = 1;
        4'b0100 : y = 2;
        4'b1000 : y = 3;
        default : y = 2'bx;
    endcase
endmodule
```

In fact, the above two examples cannot exactly model the 4-to-2 encoder, as depicted in Figure 8.5(b), because we set the output y to an unknown value as the default.

8.2.2 Priority Encoders

The simple encoder discussed above has a disadvantage that the input code must be a 1-out-of-m code; otherwise, the encoded output code might be meaningless. For example, as shown in Figure 8.5(a), when both inputs I_1 and I_2 are 1, the output code Y_1Y_0 is 11, which does not represent the actual input I_1 or I_2, but I_3.

To avoid the above situation when multiple inputs are activated, a *priority* is often associated with the inputs. For example, as shown in Figure 8.6, the priority is associated with the index values of the input lines; namely, the priority of the input lines are $I_3 > I_2 > I_1 > I_0$. As a consequence, when the input I_3 is 1, the output Y_1Y_0 is 11, regardless of the values of the other inputs.

The following is an example of using a nested if-else statement to model a 4-to-2 priority encoder.

EXAMPLE 8.7 *A 4-to-2 Priority Encoder Using the if-else Structure*

Due to the priority inherently associated with the if-else statement, the priority may be set or varied by the order of if-else statements. Because the input I_3 has the highest priority, it is put in the first statement so that it is to be examined firstly. If it fails, then input I_2 is checked. By using in this way, a 4-to-2 priority encoder is constructed.

```
// a 4-to-2 priority encoder using if-else structure.
module priority_encoder_4to2_ifelse(in, valid_in, y);
```

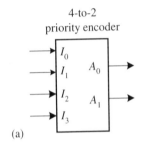

(a)

Input				Output	
I_3	I_2	I_1	I_0	A_1	A_0
0	0	0	1	0	0
0	0	1	ϕ	0	1
0	1	ϕ	ϕ	1	0
1	ϕ	ϕ	ϕ	1	1

(b)

FIGURE 8.6 The block diagram (a) and function table (b) of a 4-to-2 priority encoder

```
input   [3:0] in;
output reg [1:0] y;
output valid_in;
// the body of the 4-to-2 priority encoder
assign valid_in = |in;
always @(in) begin
   if (in[3]) y = 3; else
   if (in[2]) y = 2; else
   if (in[1]) y = 1; else
   if (in[0]) y = 0; else
               y = 2'bx;
end
endmodule
```
■

The priority encoder can also be modeled with a casex statement. Remember that the casex statement only compares non-x positions in the case_expression and case_item. As a consequence, the function table shown in Figure 8.6(b) can be coded into case_item directly.

EXAMPLE 8.8   *A 4-to-2 Priority Encoder Using the* case *Structure*

This example uses a casex statement to model a 4-to-2 priority encoder. A default case item is needed to avoid inferring a latch because the casex statement is incompletely specified. Try it on your own system.

```
// a 4-to-2 priority encoder using case structure.
module priority_encoder_4to2_case(in, valid_in, y);
input   [3:0] in;
output reg [1:0] y;
output valid_in;
// the body of the 4-to-2 priority encoder
assign valid_in = |in;
always @(in) casex (in)
   4'b1xxx: y = 3;
   4'b01xx: y = 2;
   4'b001x: y = 1;
   4'b0001: y = 0;
   default: y = 2'bx;
endcase
endmodule
```
■

Like decoders, a parameterized design of a priority encoder is often preferred in practical applications. A parameterized priority encoder can be easily modeled by exploring the basic operation of priority encoders. It can be easily seen from Figure 8.6(b) that the output code of a priority encoder corresponds to the index of the left-most bit with value one. Hence, it is sufficient to detect this bit and output the index of it as the desired code. The following two examples illuminate this idea.

EXAMPLE 8.9 *An m-to-n Priority Encoder using a `for` Loop*

As mentioned above, the basic operation of a priority encoder is that it outputs as its output code the index of the left-most bit with value one. Consequently, it is sufficient to use a loop running from $M - 1$ to 0 to detect the left-most bit with value one. Once such a bit is detected, the loop breaks and sets the index as the output code. The result is as in the following module.

```
// a parameterized M-to-N priority encoder.
module priencoder_m2n(x, valid_in, y);
parameter M = 8;   // define the number of inputs
parameter N = 3;   // define the number of outputs
input   [M-1:0] x;
output valid_in;   // indicates the data input x is valid.
output reg [N-1:0] y;
integer i;
// the body of the M-to-N priority encoder
assign valid_in = |x;
always @(*) begin: check_for_1
   for (i = M - 1; i >= 0; i = i - 1)
       if  (x[i] == 1) begin y = i;
          disable check_for_1; end
       else y = 0;   // Why need else part?
end
endmodule
```

▪

Can you rewrite the above module without using the `disable` statement? The following example uses a `while` statement to rewrite the preceding example and remove the `disable` statement.

EXAMPLE 8.10 *An m-to-n Priority Encoder Using a `while` Loop*

This example uses the same idea as the preceding one except that a `while` statement is used instead of the `for` statement. The interested reader is encouraged to compare the differences between these two examples.

```
// a parameterized M-to-N priority encoder.
module priencoder_m2n_while(x, valid_in, y);
parameter M = 4;   // define the number of inputs
parameter N = 2;   // define the number of outputs
input   [M-1:0] x;
output valid_in;   // indicates the data input x is valid.
output reg [N-1:0] y;
integer i;
// the body of the M-to-N priority encoder
assign valid_in = |x;
always @(*) begin
```

```
    i = M - 1 ;
    while(x[i] == 0 && i >= 0 ) i = i - 1;
    y = i;
end
endmodule
```

■

Review Questions

Q8.6 Define an m-to-n encoder.

Q8.7 Explain the difficulty of an encoder when multiple inputs are active at the same time.

Q8.8 Define an m-to-n priority encoder.

Q8.9 Design a 4-to-2 priority encoder with active-low output.

8.3 MULTIPLEXERS

Multiplexing and demultiplexing are two widely used techniques in communications. The former selects a channel from a set of channels and routes to its destination and the latter routes the incoming channel to its selected destination. In fact, demultiplexing is the reverse process of multiplexing and vice versa. In this section, we address the device used to perform the multiplexing, known as a *multiplexer*. The device used to carry out demultiplexing is called a *demultiplexer*, which will be discussed in the next section.

8.3.1 Multiplexers

An m-to-1 ($m = 2^n$) multiplexer has m input lines, one output line and n selection lines. The input line I_i, selected by the binary combination of n source selection lines, is routed to the output line, Y. In other words, the input line I_i is routed to the output Y if the value of the combination of the source selection lines is i. The block diagrams of general multiplexers are shown in Figure 8.7. Figure 8.7(a) does not contain an enable control while Figure 8.7(b) includes an active-low enable control. Like decoders and encoders, a multiplexer usually associates with one or more optional enable controls, which may be either active-high or active-low depending on the actual need. The purpose of enable control is to control the operations of the multiplexer. The output of the multiplexer is the value of a selected input when its enable controls are asserted and is 1, 0 or even a high-impedance, depending on the actual requirement and design, when its enable controls are deasserted.

An example of a 4-to-1 multiplexer without an enable control is shown in Figure 8.8. Figure 8.8(a) is the logic symbol and Figure 8.8(b) is the function table. From the function table, it is easy to derive the logic circuit, as shown in Figure 8.8(c). There are many ways that can be used to model a multiplexer. The most straightforward approach is based on the structural style. It instantiates gate primitives and connects them together according to the logic circuit shown in Figure 8.8(c). The details of this approach can be found in Section 2.1.1.

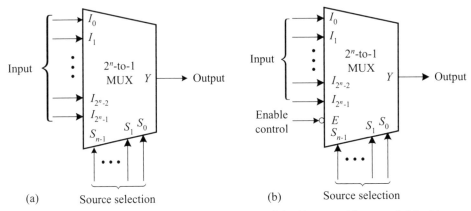

FIGURE 8.7 Block diagrams of a general multiplexer: (a) without enable control; (b) with enable control

The following example models an n-bit 4-to-1 multiplexer without an enable control in dataflow style.

EXAMPLE 8.11 *An n-bit 4-to-1 Multiplexer Using a Conditional Operator*

A single continuous assignment with nested conditional operators is used in this example. If `select[1]` is 1, the inputs `in3` or `in2` are selected. Otherwise, the inputs `in1` or `in0` are selected. The inputs `in3` or `in2` are further selected by the value of `select[0]`. Similarly, the inputs `in1` or `in0` are also further selected by the value of `select[0]`. The resulting behavior is an n-bit 4-to-1 multiplexer.

```
// an N-bit 4-to-1 multiplexer using conditional operator.
module mux_nbit_4to1(select, in3, in2, in1, in0, y);
```

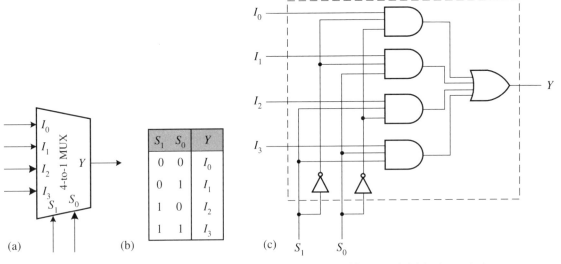

FIGURE 8.8 A simple 4-to-1 multiplexer without enable control: (a) logic symbol; (b) function table; (c) logic circuit

```
parameter N = 4; // define the width of 4-to-1
                 // multiplexer
input  [1:0] select;
input  [N-1:0] in3, in2, in1, in0;
output [N-1:0] y;

// the body of the N-bit 4-to-1 multiplexer
assign y = select[1] ?
           (select[0] ? in3 : in2) :
           (select[0] ? in1 : in0) ;
endmodule
```

A multiplexer usually associates with an enable control. The following is such an example.

EXAMPLE 8.12   *An n-bit 4-to-1 Multiplexer with an Enable Control*

In this example, besides the fact that an enable control is added, the description of the *n*-bit 4-to-1 multiplexer is also changed into behavioral style.

```
// an N-bit 4-to-1 multiplexer with enable control.
module mux_nbit_4to1_en (select, enable, in3, in2, in1,
  in0, y);
parameter N = 4;    // define the default width
input  [1:0] select;
input  enable;
input  [N-1:0] in3, in2, in1, in0;
output reg [N-1:0] y;
// the body of the N-bit 4-to-1 multiplexer
always @(select or enable or in0 or in1 or in2 or in3)
   if (!enable) y = {N{1'b0}};
   else y = select[1] ?
           (select[0] ? in3 : in2) :
           (select[0] ? in1 : in0) ;
endmodule
```

Although the above example can model an *n*-bit 4-to-1 multiplexer properly, it is hard to read, especially for naive readers. A more understandable example is shown in the following, which uses a case statement to list each combination of source selection inputs and its associated output.

EXAMPLE 8.13   *An n-bit 4-to-1 Multiplexer Using the* case *Structure*

Using a case statement to model a multiplexer may be the most straightforward method. In this approach, we only need to use the source selection inputs as the case_expression and list each combination of source selection inputs as the case_item and its associated output as the case_item statement.

```
// an N-bit 4-to-1 multiplexer using case structure.
module mux_nbit_4to1_case(select, in3, in2, in1, in0, y);
parameter N = 8; // define the width of 4-to-1 multiplexer
input   [1:0] select;
input   [N-1:0] in3, in2, in1, in0;
output reg [N-1:0] y;
// the body of the N-bit 4-to-1 multiplexer
always @(*)
   case (select)
      2'b11: y = in3 ;
      2'b10: y = in2 ;
      2'b01: y = in1 ;
      2'b00: y = in0 ;
   endcase
endmodule                                                    ■
```

Although using a case statement to model a multiplexer is easy to understand, it may not be an efficient and useful way to describe a big multiplexer. In addition, when we want to model a multiplexer in a parameterized manner, the method of using a case statement may not be a good choice. Like the case of describing a parameterized decoder, one way to design a parameterized multiplexer is to explore the basic operation of multiplexers: The input I_i is selected by the source selection input of value i and routed to the output Y. The following example illustrates how this idea can be applied to construct a parameterized m-to-1 multiplexer.

EXAMPLE 8.14   *An Example of a Parameterized m-to-1 Multiplexer*

A parameterized multiplexer is easily modeled by exploring the essential operation of multiplexers. That is, the input I_i is selected by the source selection inputs of value i and routed to the output Y. Based on this idea, a loop running from 0 to $M - 1$ is used to detect the values of the current source selection inputs and then assign the input in[i] to the output y. The resulting program is as follows.

```
// an example of parameterized M-to-1 multiplexer.
module mux_m_to_1(select, in, y);
parameter M = 4; // define the size of M-to-1 multiplexer
parameter K = 2; // define the number of selection lines
input   [K-1:0] select;
input   [M-1:0] in;
output reg y;
// the body of the M-to-1 multiplexer
integer i;
always @(*)
   for (i = 0; i < M; i = i + 1)
      if (select == i) y = in[i];
endmodule                                                    ■
```

The interested reader is encouraged to synthesize the above module and try to explain the rationale behind it.

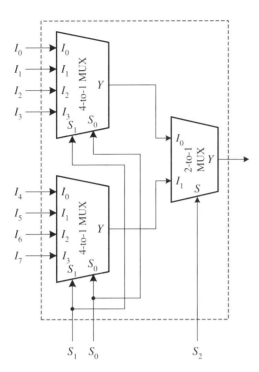

FIGURE 8.9 An 8-to-1 multiplexer constructed by cascading two 4-to-1 and one 2-to-1 multiplexers

8.3.2 Expansion of Multiplexers

Like decoders, in practical applications, it often requires a big multiplexer such as a 256-to-1 or even bigger. At least two ways can be used to construct such a big multiplexer. The straightforward approach is like what we have done for the case of a 4-to-1 multiplexer, using a `case` statement. The other approach is by instantiating the parameterized multiplexer. Regardless of which approach is used, the result is a single big multiplexer module.

Another widely used approach to constructing a big multiplexer is by cascading small multiplexer modules hierarchically. The resulting bigger multiplexer is known as *a multiplexer tree*. For example, we can combine two 4-to-1 multiplexers with one 2-to-1 multiplexer to form an 8-to-1 multiplexer, as shown in Figure 8.9. The following example illustrates how this 8-to-1 multiplexer works.

EXAMPLE 8.15 *An Example of a Multiplexer Tree — an 8-to-1 Multiplexer*

As shown in Figure 8.9, the source selection input lines S_1 and S_0 of both 4-to-1 multiplexers are connected together, as shown in this figure. Both outputs of the two 4-to-1 multiplexers are separately connected to the input lines I_0 and I_1 of the output 2-to-1 multiplexer. The source selection line of the output 2-to-1 multiplexer is denoted as the source selection line S_2. Therefore, the upper 4-to-1 multiplexer is selected when S_2 is 0 and the lower 4-to-1 multiplexer is selected when S_2 is 1. The resulting module is an 8-to-1 multiplexer. ∎

Review Questions

Q8.10   Describe the operation of multiplexers.

Q8.11   How many 4-to-1 multiplexers are needed to construct a 16-to-1 multiplexer?

Q8.12   Describe the meaning of a multiplexer tree.

Q8.13   How many source selection lines are required in an m-to-1 multiplexer?

Q8.14   Design a 4-to-1 multiplexer with active-low output.

8.4 DEMULTIPLEXERS

As mentioned previously, demultiplexing is the reverse process of multiplexing and routes the input to a selected output line. The logic circuit used to perform the demultiplexing is known as a demultiplexer. In this section, we define the general demultiplexers and provide an introduction on how to model demultiplexers in Verilog HDL.

8.4.1 Demultiplexers

A 1-to-m ($m = 2^n$) demultiplexer has one input line, m output lines and n destination selection lines. The input line D is routed to the output line Y_i, selected by the binary combination of n destination selection lines. In other words, the output line Y_i is equal to D if the value of the combination of the destination selection lines is i. Like multiplexers, a demultiplexer usually associates with one or more optional enable controls, which may be either active-high or active-low, depending on the actual need. The purpose of the enable control is to control the operations of a demultiplexer. The output selected by the destination selection lines has the same value as the input when its enable controls are asserted and all outputs are 1s, 0s or even high-impedance, depending on the actual requirement and design, when its enable controls are deasserted. The block diagrams of general demultiplexers are shown in Figure 8.10. Figure 8.10(a) does not contain an enable control while Figure 8.10(b) includes an active-low enable control.

An example of a 1-to-4 demultiplexer is shown in Figure 8.11. Figure 8.11(a) is the logic symbol and Figure 8.11(b) is the function table. From the function table, it is easy to derive the logic circuit, as shown in Figure 8.11(c). There are many ways that can be used to model a demultiplexer. The most straightforward approach is based on the structural style. It instantiates gate primitives and connects them together according to the logic circuit shown in Figure 8.11(c). Due to its intuitive simplicity, we will not describe it furthermore here.

The following example models an n-bit 1-to-4 demultiplexer without an enable control in behavioral style.

EXAMPLE 8.16   *An n-bit 1-to-4 Demultiplexer Using the* `if-else` *Structure*

This example uses `if-else` statements to describe an n-bit 1-to-4 demultiplexer. Due to its intuitive simplicity, we will not further explain it here.

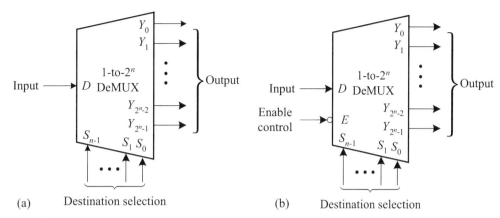

FIGURE 8.10 Block diagrams of a general demultiplexer: (a) without enable control; (b) with enable control

```
// an N-bit 1-to-4 demultiplexer using if-else structure.
module demux_1to4_ifelse(select, in, y3, y2, y1, y0);
parameter N = 4;  // define the width of the demultiplexer
input  [1:0] select;
input  [N-1:0] in;
output reg [N-1:0] y3, y2, y1, y0;

// the body of the N-bit 1-to-4 demultiplexer
always @(select or in) begin
   if (select == 3) y3 = in; else y3 = {N{1'b0}};
```

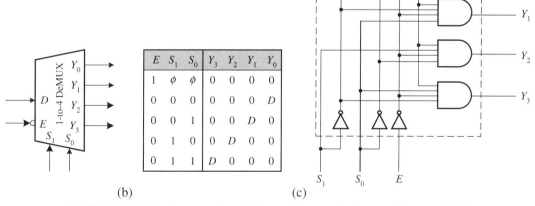

E	S_1	S_0	Y_3	Y_2	Y_1	Y_0
1	ϕ	ϕ	0	0	0	0
0	0	0	0	0	0	D
0	0	1	0	0	D	0
0	1	0	0	D	0	0
0	1	1	D	0	0	0

(a)   (b)   (c)

FIGURE 8.11 A simple 1-to-4 demultiplexer with an active-low enable control: (a) logic symbol; (b) function table; (c) logic circuit

```
      if (select == 2) y2 = in; else y2 = {N{1'b0}};
      if (select == 1) y1 = in; else y1 = {N{1'b0}};
      if (select == 0) y0 = in; else y0 = {N{1'b0}};
   end
endmodule
```

Like multiplexers, a demultiplexer usually associates with an enable control. The following is such an example.

EXAMPLE 8.17   *An n-bit 1-to-4 demultiplexer with an Enable Control*

This example is exactly the same as the preceding one, except that an enable control is added to it.

```
// an N-bit 1-to-4 demultiplexer with enable control.
module demux_1to4_ifelse_en(select, enable, in, y3, y2,
   y1, y0);
parameter N = 4;        // define the width of the
                        // demultiplexer
input   [1:0] select;
input   enable;
input   [N-1:0] in;
output reg [N-1:0] y3, y2, y1, y0;
// the body of the N-bit 1-to-4 demultiplexer
always @(select or in or enable) begin
   if (enable) begin
      if (select == 3) y3 = in; else y3 = {N{1'b0}};
      if (select == 2) y2 = in; else y2 = {N{1'b0}};
      if (select == 1) y1 = in; else y1 = {N{1'b0}};
      if (select == 0) y0 = in; else y0 = {N{1'b0}};
   end
   else begin
      y3 = {N{1'b0}}; y2 = {N{1'b0}};
      y1 = {N{1'b0}}; y0 = {N{1'b0}};
   end
end
endmodule
```

Although the above two examples can model a 1-to-4 demultiplexer properly, they may be hard to read for naive readers. A more understandable example is shown in the following, which uses a `case` statement to list each combination of the destination selection inputs and its associated output.

EXAMPLE 8.18   *An n-bit 1-to-4 Demultiplexer Using the `Case` Structure*

Like the case for multiplexers, using a `case` statement to model a demultiplexer may be the most straightforward method. In this approach, we only need to use the destination selection inputs as the case_expression and list each combination of the destination selection inputs as the case_item and its associated output as the case_item statement.

```
// an N-bit 1-to-4 demultiplexer using case structure
module demux_1to4_case (select, in, y3, y2, y1, y0);
parameter N = 4;  // define the width of the demultiplexer
input     [1:0] select;
input     [N-1:0] in;
output reg [N-1:0] y3, y2, y1, y0;

// the body of the N-bit 1-to-4 demultiplexer
always @(select or in) begin
   y3 = {N{1'b0}}; y2 = {N{1'b0}};
   y1 = {N{1'b0}}; y0 = {N{1'b0}};
   case (select)
      2'b11: y3 = in;
      2'b10: y2 = in;
      2'b01: y1 = in;
      2'b00: y0 = in;
   endcase
end
endmodule
```

Although using a case statement to model a demultiplexer is easy to understand, it may not be an efficient and useful way to describe a big demultiplexer. In addition, when we want to model a demultiplexer in a parameterized manner, the method of using a case statement may not be a good choice. When this is the case, like the cases of describing a parameterized decoder and multiplexer, we may resort to the basic operation of demultiplexers: The output line Y_i is equal to D if the value of the combination of the destination selection inputs is i. The following example illustrates how this idea can be applied to construct a parameterized 1-to-m demultiplexer.

EXAMPLE 8.19 *An Example of a Parameterized 1-to-m Demultiplexer*

A parameterized demultiplexer is easily modeled by exploring the basic operation of demultiplexers. That is, the input line D is routed to the output line Y_i when the value of the combination of the destination selection inputs is i. Based on this idea, a loop running from 0 to $M - 1$ is used to detect the value of the current destination selection inputs and then assign the output line y[i] as the input in. The resulting program is as follows.

```
// an example of parameterized 1-to-M demultiplexer
module demux_1_to_m(select, in, y);
parameter M = 4;  // define the default size of 1-to-m
                  // demultiplexer
parameter K = 2;  // define the number of selection lines
```

```
input   [K-1:0] select;
input   in;
output reg [M-1:0] y;
integer i;
// the body of the 1-to-M demultiplexer
always @(*)
    for (i = 0; i < M; i = i + 1) begin
        if (select == i) y[i] = in; else y[i] = 1'b0; end
endmodule                                              ■
```

8.4.1.1 Demultiplexer versus Decoder As we can see from Figure 8.2(b), the decoder may function as a 1-to-4 demultiplexer if the enable control is used as the data input. Similarly, a 1-to-4 demultiplexer may function as a 2-to-4 decoder when its data input is set to 1 if the demultiplexer has an enable control or when its data input is used as the enable control if the demultiplexer does not have an enable control. For example, as shown in Figure 8.11(b), the demultiplexer functions as an enable-controlled 2-to-4 decoder if its data input is set to 1.

In summary, an enable-controlled n-to-2^n decoder can function as a 1-to-2^n demultiplexer if the enable control is used as the data input. A 1-to-2^n demultiplexer can function as an n-to-2^n decoder when its data input is set to 1 if the demultiplexer has an enable control or when its data input is used as the enable control if the demultiplexer does not have an enable control.

8.4.2 Expansion of Demultiplexers

In practical applications, it often requires a big demultiplexer such as a 1-to-128 or even bigger. There are many ways that can be used to construct such a big demultiplexer. The most straightforward approach is to use a `case` or `if-else` statement as we have done in the preceding examples. The other approach is by instantiating the parameterized demultiplexer. Regardless of which approach is used, the result is a single big demultiplexer module.

Another widely used approach to building a big demultiplexer is by cascading small demultiplexer modules in a hierarchical way. The resulting bigger demultiplexer is known as *a demultiplexer tree*. For example, we can use two 1-to-4 demultiplexers and one 1-to-2 demultiplexer to construct a 1-to-8 demultiplexer, as shown in Figure 8.12. The following example illustrates how this 1-to-8 demultiplexer works.

EXAMPLE 8.20 *An Example of a Demultiplexer Tree — a 1-to-8 Demultiplexer*

As shown in Figure 8.12, the destination selection input lines S_1 and S_0 of both 1-to-4 demultiplexers connected together are as shown in this figure. Their inputs are separately connected to the output lines Y_0 and Y_1 of the input 1-to-2 demultiplexer. The destination selection line of the input 1-to-2 multiplexer is denoted as the source selection line S_2. Therefore, the upper 1-to-4 demultiplexer is selected when S_2 is 0 and the lower 1-to-4 demultiplexer is selected when S_2 is 1. The resulting module is a 1-to-8 demultiplexer.   ■

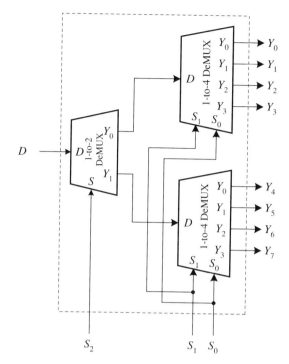

FIGURE 8.12 A 1-to-8 demultiplexer constructed by cascading two 1-to-4 and one 1-to-2 demultiplexers

Review Questions

Q8.15 Describe the operation of demultiplexers.

Q8.16 Explain the differences between a 2-to-4 decoder and a 1-to-4 demultiplexer.

Q8.17 Describe the meaning of a demultiplexer tree.

Q8.18 How many 1-to-4 demultiplexers are needed to construct a 1-to-16 demultiplexer?

Q8.19 How many destination selection lines are required in a 1-to-m demultiplexer?

Q8.20 Design a 1-to-4 demultiplexer with active-low output.

8.5 MAGNITUDE COMPARATORS

It is often needed to compare two numbers for equality in digital systems. A circuit that compares two numbers and indicates whether they are equal is called a *comparator*. When a comparator not only compares two numbers for testing their equality, but also indicates the arithmetic relationship between them, is called a *magnitude comparator*. The numbers input to a magnitude comparator can be signed or unsigned.

The structure of an n-bit comparator is quite simple – it is only an n-bit XOR gate and can be modeled in much the same way as that of a parity generator, as we have introduced in Sections 2.1.1 and 3.3.4. Due to its inherent simplicity, we will not further discuss it further here. As for magnitude comparators, there are two types of circuits that are widely used in practical applications: a magnitude comparator and a cascadable magnitude comparator.

8.5.1 Magnitude Comparators

A magnitude comparator is used to compare two numbers, signed or unsigned, determines and indicates their arithmetic relationship, namely, less than, equal to or greater than. By using relational operators, an n-bit magnitude comparator can be easily modeled, as shown in the following example, which is an unsigned version.

EXAMPLE 8.21   *An Example of an n-bit Magnitude Comparator*

The two input numbers are declared as unsigned quantities, and thus this magnitude comparator is an unsigned comparator.

```
// an example of N-bit comparator module.
// an N-bit comparator module example
module comparator_simple(a, b, cgt, clt, ceq);
parameter N = 4;   // define the size of comparator
// I/O port declarations
input  [N-1:0] a, b;
output cgt, clt, ceq;
// the body of the N-bit comparator
assign cgt = (a > b);
assign clt = (a < b);
assign ceq = (a == b);
endmodule
```

The synthesized result of the above example is shown in Figure 8.13. The interested reader is advised to change the preceding example to a signed version and see what happens after synthesizing it.

8.5.2 Cascadable Magnitude Comparators

Like other combinational logic modules, a magnitude comparator is often designed as a cascadable module. To do this, the magnitude comparator module needs to receive

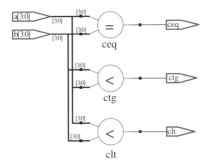

FIGURE 8.13 The synthesized result of a 4-bit magnitude comparator

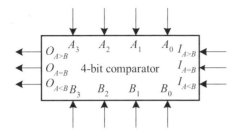

FIGURE 8.14 The block diagram of a cascadable 4-bit comparator

the compared results from its preceding stage and then combines them with the current inputs of the module to determine the output-compared result up to this stage.

The block diagram of a cascadable 4-bit magnitude comparator is shown in Figure 8.14. The inputs $I_{A>B}$, $I_{A=B}$ and $I_{A<B}$ are the compared result from its preceding stage, namely, the lower significant digit. These inputs are called the *input relationship*. The outputs $O_{A>B}$, $O_{A=B}$ and $O_{A<B}$, known as the *output relationship*, indicate the combined result of the current input numbers of this stage and the input relationship, which is the output relationship from its preceding stage.

The following is an example of a cascadable *n*-bit magnitude comparator.

EXAMPLE 8.22 *An Example of an Cascadable n-bit Magnitude Comparator*

This example compares two input numbers and indicates the arithmetic relationship between them. The output relationship is determined not only by the current two input numbers but also by the input relationship of the module. The result is greater than when input number a is larger than b or when both current input numbers are equal and the input relationship is greater. A similar description is applicable to the 'less than' relationship. The equality relationship validates when both current input numbers are equal, as well as the input relationship is also an equality.

```
module comparator_cascadable(Iagtb,Iaeqb,Ialtb,a,b,Oagtb,
   Oaeqb,Oaltb);
parameter N = 4; // define the default size of comparator
// I/O port declarations
input  Iagtb, Iaeqb, Ialtb;
input  [N-1:0] a, b;
output Oagtb, Oaeqb, Oaltb;
// dataflow modeling using relational operators
assign Oaeqb = (a == b) && (Iaeqb == 1); // equality
assign Oagtb = (a > b) || ((a == b) && (Iagtb == 1));
                                      // greater than
assign Oaltb = (a < b) || ((a == b) && (Ialtb == 1));
                                      // less than
endmodule                                              ■
```

The reader interested in this is advisored to synthesize it and see the difference between its result and the preceding example. In addition, it is worth changing the

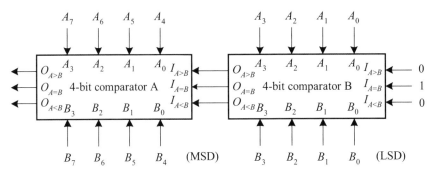

FIGURE 8.15 The block diagram of an 8-bit magnitude comparator constructed by cascading two 4-bit magnitude comparators

unsigned declaration of the inputs numbers to signed and see what happens after synthesizing it.

Figure 8.15 illustrates how to cascade two 4-bit magnitude comparators into an 8-bit magnitude comparator. As described in the preceding example, the most significant digit (MSD) magnitude comparator may solely determine the final arithmetic relationship from its two input numbers unless for the case where both input numbers are equal. In this situation, the final arithmetic relationship is determined by its preceding magnitude comparator, namely, LSD.

Review Questions

Q8.21 What is the function of a comparator?

Q8.22 Design a logic circuit to compare two 1-bit numbers.

Q8.23 Design a logic circuit to compare two 2-bit numbers.

Q8.24 Explain the differences between a comparator and magnitude comparator.

Q8.25 Describe the meaning of a cascadable magnitude comparator.

Q8.26 Explain what will happen when we change the input relationship setting of Figure 8.15 to 000.

8.6 A CASE STUDY: SEVEN-SEGMENT LED DISPLAY

Light-emitting diodes (LEDs) are widely used in most digital or mixed-signal systems as indicators or to present information visually. By appropriately combining LEDs into patterns, such as seven-segment digit or dot-matrix character, digits and characters may be displayed as required. In this section, we are concerned with the issues of basic LED devices, seven-segment LED displays, a BCD-to-seven-segment decoder and a typical multiplexing seven-segment display system.

8.6.1 Seven-Segment LED Display

An LED is a device that can emit visible or invisible light when an appropriate voltage is applied to it. In this book, we only consider visible LEDs. Currently, many types of

TABLE 8.1 Various commercial light-emitting diodes

Color	Material	Forward voltage (V)
Amber	AlInGaP	2.0
Blue	GaN	5.0
Green	GaP	2.2
Orange	GaAsP	2.0
Red	GaASP	1.8
White	GaN	4.0
Yellow	AlInGaP	2.0

visible LEDs with a variety of colors are available. As shown in Table 8.1, the most widely used colors in digital systems are green, orange and red. All LEDs have a forward voltage dropped between their ends, regardless of what color they are. The magnitude of the forward voltage of an LED is determined by the material used to manufacture the device. The forward voltage value of LEDs with green, orange and red is about 2.0 V, as presented in Table 8.1. The forward voltage values of blue and white LEDs are much higher than 2 V. Typically, they are of 5 and 4 V, respectively.

LEDs can be used individually as indicators or combined into a particular pattern, such as the one shown in Figure 8.16(a), which is patterned as a digit and is known as a seven-segment display since it consists of seven LEDs. In practice, there are two types of seven-segment display available for use, as shown in Figures 8.16(b) and (c), respectively. A seven-segment display is known as a *common-anode structure* when all anodes of the seven LEDs are connected together and is known as a *common-cathode structure* when all cathodes of the seven LEDs are connected together.

As mentioned above, to turn on an LED, we need to apply a forward bias on the device. Hence, in Figure 8.17(a) when we activate the input signal a, the LED connected at the output of its buffer is turned on. The resistor with the value of 220 Ω connected in series between the LED and power supply is to limit the current passing through the device in the range of 10 to 20 mA so as to protect it from being damaged due to possible overheating the device.

When we want to display a digit on the seven-segment display, we need to turn on the appropriate sub-set of LEDs according to the digit being displayed. This sub-set comprises the *display code*. In order to specify the codeword corresponds to each digit, we need to define the codeword format, as shown in Figure 8.18(b). Based on this format, we can derive the codeword for each digit to be displayed. For instance, from

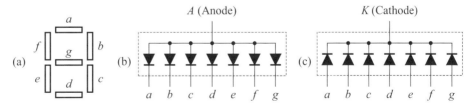

FIGURE 8.16 The structures of seven-segment LEDs: (a) pattern; (b) common anode structure; common cathode structure

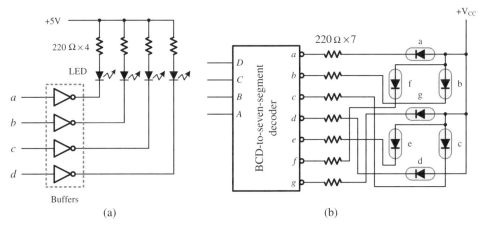

FIGURE 8.17 Examples of LED indicators and a seven-segment LED display circuit: (a) LED indicator; (b) seven-segment LED display

Figure 8.18(a), we know that to display "0", all LEDs, except the one labeled with g, must be turned on. This means that with a common-anode seven-segment LED display all values of a to f must be 0s and g must be 1. As a result, the code for digit "0" is 40 in hexadecimal. The rest digits can be derived in a similar way. The complete code is given in Figure 8.18(c). It is worth noting that this display code is special for use with common-anode seven-segment LED displays. For the devices with a common cathode, the code needs to be complemented. In addition, when the input is only limited to digital

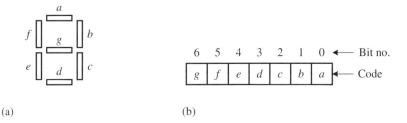

(a)   (b)

Digit	g	f	e	d	c	b	a	Code	Digit	g	f	e	d	c	b	a	Code
0	1	0	0	0	0	0	0	40	8	0	0	0	0	0	0	0	00
1	1	1	1	1	0	0	1	79	9	0	0	1	0	0	0	0	10
2	0	1	0	0	1	0	0	24	A (a)	0	0	0	1	0	0	0	08
3	0	1	1	0	0	0	0	30	B (b)	0	0	0	0	0	1	1	03
4	0	0	1	1	0	0	1	19	C (c)	1	0	0	0	1	1	0	46
5	0	0	1	0	0	1	0	12	D (d)	0	1	0	0	0	0	1	21
6	0	0	0	0	0	1	0	02	E (e)	0	0	0	0	1	1	0	06
7	1	1	1	1	0	0	0	78	F (f)	0	0	0	1	1	1	0	0E

(c)

FIGURE 8.18 The relationship between digit and common-anode seven-segment LED display code (a) seven-segment LED display; (b) seven-segment LED display code format; (c) seven-segment LED display code

"0" to "9", we can only use the first 10 codewords of the display code and ignore the rest of the six codewords.

In the following, we give an example to illustrate how to convert a BCD input to a common-anode seven-segment LED display code.

EXAMPLE 8.23 *An Example of a BCD-to-Seven-Segment Decoder*

In this example, we assume that the input is a BCD code. Hence, when the input is not a valid BCD code, the display will be blank. The circuit used with this decoder is shown in Figure 8.17(b).

```
// converting input data into seven_segment_display code
// with active-low outputs. the input data is assumed in
// the range of 0 to 9.
module BCD-to-seven_segment_decoder(data_in, data_out);
input       [3:0] data_in;
output  reg [6:0] data_out;
// the default case is assumed to be blank.
always @(data_in)
    case (data_in)
        0: data_out = 7'b100_0000;    // h40
        1: data_out = 7'b111_1001;    // h79
        2: data_out = 7'b010_0100;    // h24
        3: data_out = 7'b011_0000;    // h30
        4: data_out = 7'b001_1001;    // h19
        5: data_out = 7'b001_0010;    // h12
        6: data_out = 7'b000_0010;    // h02
        7: data_out = 7'b111_1000;    // h78
        8: data_out = 7'b000_0000;    // h00
        9: data_out = 7'b001_0000;    // h10
        default: data_out = 7'b111_1111;
    endcase
endmodule
```

■

To display multiple digits, the most straightforward approach is to duplicate Figure 8.17(b) up to the desired number of copies. This is known as a *direct-driven approach*. For example, when we want to display 4 digits at the same time, we may duplicate Figure 8.17(b) four times, each for one digit. The big problem of this approach is that, as we have mentioned previously, each turned-on LED has to consume 10 to 20 mA. As a result, the average current required is 392 mA when the turned-on current of the LED is assumed to be 20 mA. This is quite extreme and is often undesirable in most applications.

8.6.2 Multiplexing-Driven Seven-Segment LED Display

It can be shown that the light is quite satisfied when an LED is turned on for 1 ms at a current level of 10 mA and refreshed at least every 10 to 16 ms. This is known as a *multiplexing-driven approach*. Based on this concept, a multiplexing-driven seven-segment LED display can be constructed, as illustrated in Figure 8.19(a).

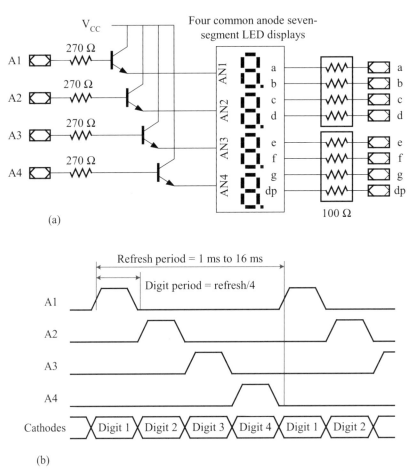

(a)

(b)

FIGURE 8.19 The logic circuit and timing diagram of a four-digit multiplexing-driven seven-segment LED display: (a) a four-seven-segment LED display using the multiplexing technique; (b) timing diagram for the four-seven-segment LED display shown in (a)

In a multiplexing-driven seven-segment LED display, all seven-segment LED displays are connected in such a way that all segments with the same label are connected together and all anodes are left as separate control inputs, each of which is driven by a separate transistor, as shown in Figure 8.19(a). Figure 8.19(b) shows the timing that controls the turned-on period of all four transistors and hence the seven-segment LED displays. From Figure 8.19(b), we can see that each seven-segment LED display is repeatedly enabled to display its own digit for a short time. In other words, all seven-segment LED displays share a common BCD-to-seven-segment decoder and are turned on in a time-division multiplexing manner.

The major feature of a multiplexing-driven seven-segment LED display is that the amount of current consumed is only equivalent to that of one seven-segment LED display when using the direct-driven approach. Therefore, the power consumption is very low when compared to the direct-driven approach.

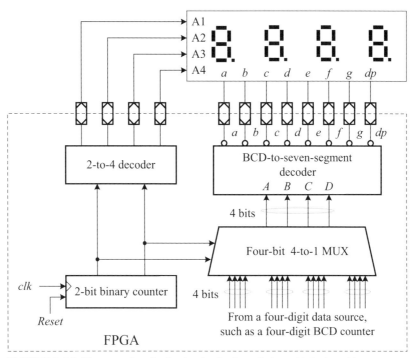

FIGURE 8.20 The complete logic circuit of a four-digit multiplexing-driven seven-segment LED display

A complete four-digit multiplexing-driven seven-segment LED display system is illustrated in Figure 8.20. It consists of four parts: a binary counter and decoder, a four-bit 4-to-1 multiplexer, a BCD-to-seven-segment decoder and the four-digit seven-segment LED display (namely, Figure 8.19(a)). The binary counter and decoder generate the selection signal for the four-bit 4-to-1 multiplexer and the control timing of Figure 8.19(b). The selection signal and the control timing have to be in synchronism with each other. This is accomplished by decoding the selection signals directly into the control timing. The four-bit 4-to-1 multiplexer multiplexes the four-digit data source into the input of the BCD-to-seven-segment decoder through which a BCD input is converted into a seven-segment display code.

The following example generates the selection signals required for the four-bit 4-to-1 multiplexer and the control timing required for the multiplexing-driven LED display.

EXAMPLE 8.24 *An Example of a Multiplexer Timing Generator*

This module generates the selection signals for the four-bit 4-to-1 multiplexer directly through the outputs of a 2-bit binary counter which is described by the first `always` block. Further details of the binary counter will be discussed in the next chapter. The control timing of the multiplexing-driven LED display is then generated through decoding the outputs of the binary counter directly. Thus, both control signals are in synchronism.

```verilog
// binary counter with decoder serve as a timing generator
module mux_timing_generator(clk, reset, mux_sel, addr);
input   clk, reset;
output reg  [3:0] addr; // generates A1 to A4 signals
output wire [1:0] mux_sel;
reg     [1:0] qout;
// the body of binary counter
always @(posedge clk or posedge reset)
   if (reset) qout <= 0;
   else       qout <= qout + 1;
assign mux_sel = qout;
// decode the output of the binary counter
always @(qout)
   case(qout)
      2'b00:   addr = 4'b0001;
      2'b01:   addr = 4'b0010;
      2'b10:   addr = 4'b0100;
      2'b11:   addr = 4'b1000;
      default: addr = 4'b0000;
   endcase
endmodule                                          ■
```

Another module used in the seven-segment LED display system shown in Figure 8.20(a) is the four-bit 4-to-1 multiplexer, which is described as follows.

EXAMPLE 8.25 *A Multiplexer Used to Route Data to a BCD-to-Seven-Segment Display*

The width of data in this multiplexer is 4 bits. Since it is the typical one described with a case statement, we will not discuss it further here.

```verilog
// a four-bit 4-to-1 multiplexer
module seven_segment_mux(mux_sel, data_in3, data_in2,
                         data_in1, data_in0, data_out);
input      [3:0] data_in3, data_in2, data_in1, data_in0;
input      [1:0] mux_sel;
output reg [3:0] data_out;
// the body of the multiplexer
always @(*)
   case (mux_sel)
      2'b00: data_out = data_in3;
      2'b01: data_out = data_in2;
      2'b10: data_out = data_in1;
      2'b11: data_out = data_in0;
   endcase
endmodule                                          ■
```

Review Questions

Q8.27 What is the forward voltage value of a typical blue LED device?

Q8.28 What are the forward voltage values of the most widely used green, orange and red LED devices?

Q8.29 What is the major disadvantage of a direct-driven seven-segment LED display?

Q8.30 What is the rationale behind the multiplexing-driven seven-segment LED display?

Q8.31 Compare the features of direct-driven and multiplexing-driven seven-segment LED displays.

SUMMARY

In this chapter, we have examined the most widely used combinational modules, including encoders and decoders, multiplexers and demultiplexers, and magnitude comparators. In addition, a multiplexing-driven seven-segment LED display system, which combines the use of a decoder, as well as a multiplexer is discussed in detail.

A decoder is a combinational logic circuit that accepts as input an n-bit code and generates as output a 1-out-of-m code such that the activated bit of the output corresponds to a unique combination of the input code. A decoder can be fully decoded or partially decoded, depending on the actual requirement. A decoder usually associates with one or more enable controls. The enable controls may be active-high or active-low. In addition, the outputs of a decoder may be active-high or active-low, depending on the actual need.

The encoders perform the reverse operations of decoders. An encoder has $m = 2^n$ (or fewer) input lines and n output lines. The output lines generate the binary code corresponding to the position of the activated input line. Encoders may be either the simple one in which only one of its input can be active at any time or the one with priority in which each input has a different weight. However, priority encoders are often used in most practical applications. Like decoders, an encoder usually associates one or more enable controls. The enable controls may be active-high or active-low. In addition, the outputs of an encoder may be active-high or active-low, depending on the actual requirement.

Multiplexing is a process that selects a channel from a set of channels and routes to its destination. The device used to perform the multiplexing, is known as a multiplexer. An m-to-1 ($m = 2^n$) multiplexer has m input lines, one output line and n selection lines. The input line, I_i, selected by the binary combination of n source selection lines is routed to the output line, Y. A multiplexer usually associates with one or more enable controls. The enable controls may be active-high or active-low. In addition, the outputs of a multiplexer may be active-high or active-low, depending on the actual requirement.

Demultiplexing is the reverse process of multiplexing and routes the input to a selected output line. The device used to perform the demultiplexing is known as a demultiplexer. A 1-to-m ($m = 2^n$) demultiplexer has one input line, m output lines and n destination selection lines. The input line, D, is routed to the output line Y_i

selected by the binary combination of *n* destination selection lines. Like multiplexers, a demultiplexer usually associates with one or more enable controls. The enable controls may be active-high or active-low. In addition, the outputs of a demultiplexer may be active-high or active-low, depending on the actual requirement.

A circuit that compares two numbers and indicates whether they are equal is called a comparator. When a comparator not only compares two numbers for testing their equality but also indicates the arithmetic relationship between them is called a magnitude comparator. The numbers input to a magnitude comparator can be signed or unsigned.

LEDs are widely used in most digital or mixed-signal systems as indicators or to present information visually. By appropriately combining LEDs into patterns such as a seven-segment digit or dot-matrix character, digits and characters may be displayed as required. Two ways can be used to display multiple digits: direct-driven approach and multiplexing-driven approach. In the direct-driven approach, each seven-segment LED display associates a separate BCD-to-seven-segment decoder and is always turned on. In the multiplexing-driven approach, all seven-segment LED displays share a common BCD-to-seven-segment decoder and are turned on in a time-division multiplexing manner. The direct-driven approach has a simpler control circuit but consumes much more power than the multiplexing-driven approach.

REFERENCES

1. J. Bhasker, *A Verilog HDL Primer*, 3rd Edn, Star Galaxy Publishing, 2005.
2. R.L. Boylestad and L. Nashelsky, *Electronic Devices and Circuit Theory*, 9th Edn, Prentice-Hall, Upper Saddle River, NJ, USA, 2006.
3. M.D. Ciletti, *Modeling, Synthesis and Rapid Prototyping with the Verilog HDL*, Prentice-Hall, Upper Saddle River, NJ, USA, 1999.
4. M.-B. Lin, *Digital System Design: Principles, Practices and ASIC Realization*, 3rd Edn, Chuan Hwa Book Company, Taipei, Taiwan, 2002.
5. M.-B. Lin, *Basic Principles and Applications of Microprocessors: MCS-51 Embedded Microcomputer System, Software and Hardware*, 2nd Edn, Chuan Hwa Book Company, Taipei, Taiwan, 2003.
6. J.F. Wakerly, *Digital Design Principles and Practices*, 3rd Edn, Prentice-Hall, Upper Saddle River, NJ, USA, 2001.

PROBLEMS

8.1 Design a 3-to-8 decoder with the following output polarities:

(a) Noninverted output.

(b) Inverted output.

8.2 Design a 4-to-16 decoder using at most two enable-controlled 3-to-8 decoders.

8.3 Design a 4-to-16 decoder using at most five enable-controlled 2-to-4 decoders.

8.4 Design a 5-to-32 decoder using at most four enable-controlled 3-to-8 decoders and a 2-to-4 decoder.

8.5 Design a 4-to-2 priority encoder. The encoder must have an enable control input to enable the encoder and a valid output to indicate that at least an input is active.

8.6 Design a 4-to-1 multiplexer with enable control. The enable control is active-low. The output is at high-impedance when the enable control is deasserted.

8.7 Use the following specified approaches to design a 32-to-1 multiplexer:

(a) Two 16-to-1 multiplexers and one 2-to-1 multiplexer.

(b) Four 8-to-1 multiplexers and one 4-to-1 multiplexer.

8.8 Use two 2-to-1 multiplexers to design a 3-to-1 multiplexer.

8.9 Design a parameterized n-bit m-to-1 multiplexer. Write a test bench to verify its functionality.

8.10 How many 1-to-4 demultiplexers are required when we use them to design the following specified demultiplexers?

(a) a 1-to-32 demultiplexer;

(b) a 1-to-64 demultiplexer;

(c) a 1-to-128 demultiplexer;

(d) a 1-to-256 demultiplexer.

8.11 Design a parameterized n-bit 1-to-m demultiplexer. Write a test bench to verify its functionality.

8.12 Use 4-bit magnitude comparators to design the following specified magnitude comparators:

(a) an 8-bit magnitude comparator;

(b) a 12-bit magnitude comparator.

8.13 Consider a common-cathode seven-segment LED display.

(a) Define the seven-segment LED display code.

(b) Write a BCD-to-seven-segment decoder module.

8.14 Write a Verilog HDL module to test the function of the multiplexing-driven seven-segment LED display described in Section 8.6.2.

8.15 Design an eight-digit seven-segment LED display system.

(a) Draw the block diagram and control timing of the seven-segment LED display.

(b) Draw the complete seven-segment LED display system.

(c) Write various Verilog HDL modules required for the display system.

(d) Write a test bench to verify the functionality of the display system.

8.16 Suppose that the current passing through a turned-on LED is 10 mA and each decimal digit appears equally likely on the display. Calculate the average current consumed in an eight-digit seven-segment LED display system designed on the following specified approaches:

(a) direct-driven approach;

(b) multiplexing-driven approach.

8.17 When designing a large decoder, either a single-module approach or a hierarchical approach can be used. Using an 8-to-256 decoder as an example, study the performance of both approaches in terms of hardware cost and the propagation delay.

8.18 When designing a large encoder, either a single-module approach or a hierarchical approach can be used. Using a 256-to-8 encoder as an example, study the performance of both approaches in terms of hardware cost and the propagation delay.

8.19 When designing a large multiplexer, either a single-module approach or a hierarchical approach can be used. Using a 256-to-1 multiplexer as an example, study the performance of both approaches in terms of hardware cost and the propagation delay.

8.20 When designing a large demultiplexer, either a single-module approach or a hierarchical approach can be used. Using a 1-to-256 demultiplexer as an example, study the performance of both approaches in terms of hardware cost and the propagation delay.

8.21 When designing a large magnitude comparator, either a single-module approach or a hierarchical approach can be used. Using a 64-bit magnitude comparator as an example, study the performance of both approaches in terms of hardware cost and the propagation delay.

CHAPTER **9**

SEQUENTIAL LOGIC MODULES

IN **THIS** chapter, we examine several basic sequential modules that are widely used in digital systems. These include flip-flops, synchronizers, a switch-debouncing circuit, registers, data registers, register files, shift registers, counters (binary, BCD, Johnson), CRC generators and detectors, clock generators, pulse generators, as well as timing generators.

The basic building blocks of any sequential module are flip-flops and latches. Hence, in this chapter we first consider the flip-flop in detail and give some basic applications of it. These include synchronizers which are used to sample an external asynchronous signal in synchronism with an internal system clock and switch debouncers which remove the bouncing effects of mechanical switches due to the mechanical inertia of switches.

Next, we consider the various combinations of flip-flops, such as registers, shift registers, as well as counters. Registers usually include basic registers and register files. Shift registers are often used to perform arithmetic operations, such as divided and multiplied by 2, as well as data format conversion. Counters we consider include both asynchronous and synchronous types. In addition, shift-register-based counters are also considered in greater detail. Finally, we discuss a variety of clock and timing generators which are widely used in digital systems.

There are many options that can be used to model sequential logic. These options include behavioral statements, tasks with delay or event control, sequential UDPs, instantiated library register cell and interconnected sequential logic modules.

9.1 FLIP-FLOPS

Flip-flops are the basic building blocks of sequential circuits. They can be used to construct registers, data registers, register files, shift registers, counters, clock generators, timing generators, as well as others. In this section, we introduce the fundamentals

Digital System Designs and Practices   Ming-Bo Lin
© 2008 John Wiley & Sons (Asia) Pte Ltd

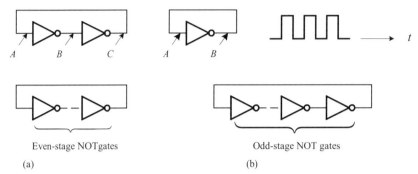

FIGURE 9.1 The basic structures of bistable and unstable devices: (a) bistable circuit; (b) oscillator

of flip-flops, distinguish flip-flops from latches, and describe some basic applications of flip-flops, including synchronizers and a switch-debouncing circuit.

9.1.1 Flip-Flops

A bistable device is a circuit with two stable states, as shown in Figure 9.1(a). A bistable device is usually built from an even number of NOT gates or inverters. For instance, as illustrated in Figure 9.1(a), the bistable circuit is constructed from two NOT gates cascaded together into a loop structure. The signal A propagates to the output of the first NOT gate as an inverted version again, denoted as B, which then propagates to the output of the second NOT gate as an inverted version, denoted as C. Finally, signal C is fed back to the input of the first NOT gate, which is coincident with signal A. Consequently, the circuit is bistable.

Figure 9.1(b) shows an unstable circuit, which consists of an odd number of NOT gates. A conceptual unstable circuit is simply a one-stage loop-connected NOT gate. A signal appearing at its input will be inverted and appeared at its output as an inverted version, which is then fed back to its input. Because the polarities of both the original and feedback signals are different, the feedback signal will be continuously inverted and fed back forever. Hence, the result circuit is unstable, namely, an oscillator. For CMOS technology, it is safe to construct an oscillator by cascading three or more stages of NOT gates.

The fundamental applications of bistable devices are used to construct latches and flip-flops. A latch is a bistable device with the capability of latching or storing the external data. To reach this, a data path must be provided through some mechanism in order to direct the external data into the bistable device. According to the mechanism used to route the external data into the latch, there exist two basic structures that are often used in digital systems to construct various types of latch. The first structure is by using a multiplexer to break the feedback path inherent to the bistable device and route the external data into the latch. This type of latch is known as a *multiplexer-based latch*. An example of a multiplexer-based D-type latch is shown in Figure 9.2(a), which accepts the external data D when the source selection G of the 2-to-1 multiplexer is

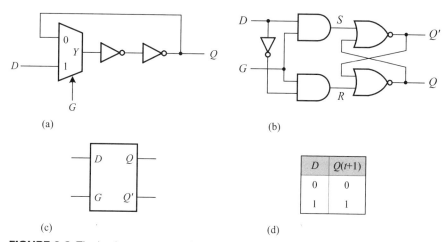

(a)

(b)

(c)

(d)

FIGURE 9.2 The basic structures of (a) a multiplexer-based D-type latch and (b) an SR-latch-based D-type latch, along with (c) the logic symbol and (d) the characteristic table of these D-type latches

set to 1 and latches the sampled data otherwise. The second structure is based on an SR-latch, as shown in Figure 9.2(b), where an SR-latch is combined with two AND gates and one inverter to form the D-type latch. This type of latch is known as a SR-*latch-based D-type latch*. It is not difficult to analyze this circuit by using the concept of a controlled gate, introduced in Section 2.1.1. The detailed analysis is therefore left to the reader as an exercise. The logic symbol of the D-type latch, along with its characteristic table, are also shown in Figures 9.2(c) and (d), respectively.

In CMOS technology, flip-flops are usually built on the basis of latches. This type of flip-flop is called a *master-slave flip-flop*. A *positive-edge triggered* flip-flop constructed from two latches is depicted in Figure 9.3(a). The first latch, also called the *master latch*, is a *negative latch* which accepts the external data when the clock signal (clk) is low and stores the sampled data otherwise. The second latch, also called the *slave latch*, is a *positive latch* which accepts the external data (in this case from the output of the first latch) when its clock signal (clk) is high and stores the sampled data otherwise. As a result, by combining these two latches and appropriately applying the clock signals to both latches, the output of the second latch will present the external data sampled at the positive edge of the clock signal. This kind of circuit is often denoted as a positive-edge triggered flip-flop. The logic symbol of the D-type positive-edge triggered flip-flop, along with its characteristic table and state diagram, are also shown in Figures 9.3(b), (c) and (d), respectively.

Although both latches and flip-flops are often used in digital systems, there exists an essential difference between them. As mentioned above, the output of a latch will follows its external data whenever its enable control G is high. This property is known as *transparent property*. However, the output of a flip-flop is the sampled value of the external data at the specified-edge of the clock signal. In our running example, it is positive edge. As a consequence, a latch is a bistable device with transparent property whereas a flip-flop is a bistable device without transparent property.

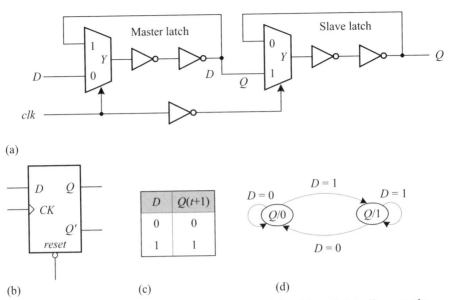

(a)

(b)   (c)   (d)

FIGURE 9.3 The logic circuit, logic symbol, characteristic table and state diagram of a positive-edge triggered *D*-type flip-flop: (a) master-slave *D*-type flip-flop; (b) logic symbol; (c) characteristic table; (d) state diagram

 To further compare the difference between latches and flip-flops, let us consider Figure 9.4. As we can see from this figure, the output of the (positive) latch depicted in Figure 9.2(a) will follow the input data when its enable control G is high and holds the value sampled before the enable control G goes to low. In contrast, the output of the positive-edge triggered flip-flop described in Figure 9.3(a) holds the value sampled at each positive-edge of the clock signal and remains that value during the whole clock cycle. In some applications, the transparent property may cause troubles, such as inducing noise into the system. In addition, in synthesizable digital designs, it is more difficult to handle latches than flip-flops. Therefore, in this book we only consider flip-flops.

 Although there are many types of flip-flops defined and implemented in discrete modules, currently the D-type flip-flop is the most widely used bistable device in sequential logic. A practical D-type flip-flop usually has a *reset* (or *clear*) input, which clears or resets the output of the flip-flop asynchronously or synchronously. The following two examples illustrate how to model such a practical D-type flip-flop.

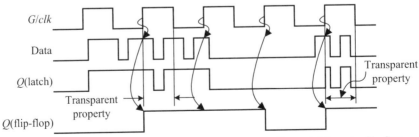

FIGURE 9.4 An illustration of the difference between a *D*-type latch and a flip-flop

EXAMPLE 9.1 *A D-Type Flip-flop with Asynchronous Reset*

The word "asynchronous" means random. In other words, the operation performed by an asynchronous signal occurs in a random and unpredictable way. As a consequence, to model an asynchronous reset signal of a *D*-type flip-flop, it needs to put the signal within the sensitivity list of the `always` block used to describe the flip-flop. The result is as follows. The output q is cleared to 0 whenever the reset signal `reset_n` is active, i.e. 0, regardless of the status of the clock signal.

```
// asynchronous reset D-type flip-flop
module DFF_async_reset(clk, reset_n, d, q);
input  clk, reset_n, d;
output reg q;
// the body of flip flop
always @(posedge clk or negedge reset_n)
   if (!reset_n) q <= 0; // reset_n signal is active-low.
   else          q <= d;
endmodule
```

Of course, the reset operation can also be performed in synchronism with the clock signal, such as in the following example. A flip-flop being operated in this manner is said to be synchronously resettable.

EXAMPLE 9.2 *A D-Type Flip-flop with Synchronous Reset*

The word "synchronous" means under the control of the clock signal. In other words, the operation performed by a synchronous signal occurs in a predictable way. As a consequence, to model a synchronous reset signal of a *D*-type flip-flop, it needs to put the signal outside the sensitivity list of the `always` block used to describe the flip-flop. The result is as follows. At each positive edge of the clock signal `clk`, the reset signal `reset` is checked to see whether it is active or not. If it is active then the output q is cleared to 0; otherwise, the input data d is sampled into the flip-flop.

```
// synchronous reset D-type flip-flop
module DFF_sync_reset(clk, reset, d, q);
input  clk, reset, d;
output reg q;
// the body of flip flop
always @(posedge clk)
   if (reset)  q <= 0; // reset signal is active-high.
   else        q <= d;
endmodule
```

Review Questions

Q9.1 What are the differences between a bistable device and an oscillator?

Q9.2 Explain the operation of the circuit shown in Figure 9.2(b).

Q9.3   What is the meaning of transparent property?

Q9.4   What is the difference between synchronous and asynchronous reset of flip-flops?

Q9.5   What is the basic difference between latches and flip-flops?

Q9.6   Design a negative-edge triggered D-type flip-flop by properly combining two latches.

Q9.7   Describe the basic features of D-type flip-flops.

9.1.2 Metastable State

Any digital system is an interconnected structure of combinational logic and flip-flops. The flip-flops, which we call registers later in this book, are used to store the results computed by the combinational logic. This means that flip-flops used to store the values use a clock signal by which the flip-flops sample the data appearing at their D inputs. In order to be properly sampled by a flip-flop, the input data must be stable at the input for a period of time, called the *setup time*, and must remain stable for another period of time, called the *hold time*, after being sampled.

In summary, when using D-type or other types of flip-flops to design digital systems, the following three basic timing parameters must be considered very carefully:

1. *Setup time* (t_{setup}) is the time before the clock edge (sampling point) that the D input must be stable.

2. *Hold time* (t_{hold}) is the time after the clock edge (sampling point) that the D input must remain stable.

3. *Clock-to-Q delay* (t_q) is the maximum propagation delay from the clock edge (sampling edge) to the new value of the Q output.

The definitions of the above three timing parameters of typical D-flip-flops are illustrated in Figure 9.5.

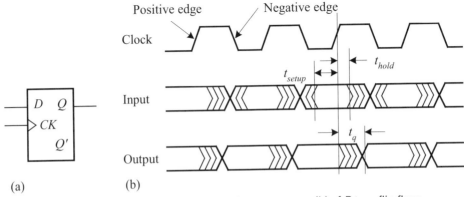

(a)   (b)

FIGURE 9.5 The logic symbol (a) and basic timing parameters (b) of D-type flip-flops

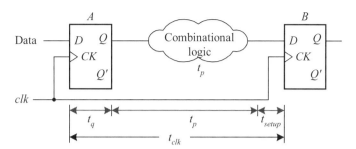

FIGURE 9.6 The basic timing parameters of D-type flip-flops

The above three timing parameters combined with the propagation delay of the combinational logic between the two D-type flip-flops determine the minimum period of the clock signal. More specific, this timing relationship can be represented as the following equation:

$$t_{clk} \geq t_{setup} + t_q + t_p \qquad (9.1)$$

where t_p is the propagation delay of combinational logic between two D-type flip-flops, as shown in Figure 9.6.

For a given clock period, t_{clk}, in order to allow more time for the combinational logic, the setup time and clock-to-Q delay of flip-flops must be as small as possible. Another benefit of a small setup time will manifest itself in the design of synchronizers, which will be discussed in the next sub-section.

As mentioned above, the output of a flip-flop will present the value sampled at the D input after a clock-to-Q delay if both timing constraints, setup time and hold time, are satisfied. Otherwise, it may output a value between "1" and "0" for a period of time. This situation is known as *metastable* state. More precisely, a metastable state is the one that is between two stable states, as shown in Figure 9.7(a). It is temporarily stable at a point between two stable states. Any disturbance will let it go down to either stable states: 1 or 0. In fact, a metastable state can occur in any bistable device and the devices stemmed from it.

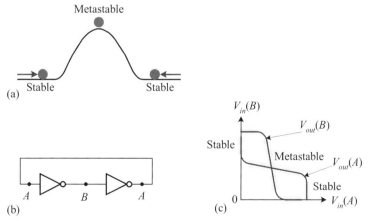

FIGURE 9.7 The logic diagram and voltage-transfer characteristics of a bistable device: (a) metastable state; (b) bistable circuit; (c) VTC

As shown in Figure 9.7(b), when the input signal A does not sustain long enough, both NOT gates may leave their outputs at the midpoints of their voltage-transfer characteristics, as shown in Figure 9.7(c). As a result, the bistable device enters into the metastable state. Fortunately, due to the feature of positive feedback and both NOT gates residing in their high voltage gain regions, the bistable device will come back to either stable states: high or low, after a short time.

Once a bistable device enters into the metastable state, the only way we can proceed is to wait for it to leave the metastable state. The time required for a bistable device to move out of the metastable state is denoted as the *recovery time*, t_{rec}, which is the time elapsed from when the bistable device samples the input data until the bistable device leaves the metastable state and goes back to its normal state.

Review Questions

Q9.8 Define the following two terms: setup time and hold time.

Q9.9 What is the meaning of clock-to-Q delay of a flip-flop?

Q9.10 Describe the basic timing relationship in a flip-flop-based system.

Q9.11 What is the meaning of metastable state of a flip-flop?

Q9.12 Describe the meaning of recovery time.

9.1.3 Synchronizers

In many applications, it is often needed to sample an external signal and then take the desired operations accordingly. As mentioned previously, an external signal can be sampled by using a D-type flip-flop, as shown in Figure 9.8(a). The D-type flip-flop used for this purpose is known as a *synchronizer* since it synchronizes the external signal to the internal system clock.

In some applications this simple synchronizer could actually do its job quite well. Nonetheless, in other applications, this single-stage synchronizer may cause a situation called *synchronization failure* with the synchronizer entering into the metastable state. As illustrated in Figure 9.8(a), when the synchronizer enters the metastable state, both succeeding stages B and C may sample different values of the same output w from the synchronizer. For example, the output of B may be "1" and C may be "0". The result is 'chaos' because of two different recognitions of the same signal. The synchronization failure may cause 'disaster' for a system. This is usually not allowed in any practical system.

As mentioned previously, once a D-type flip-flop enters into the metastable state, it needs a recovery time to go back to its normal state. Hence, the timing relationship between setup time, recovery time and clock period can be represented as the following equation:

$$t_{clk} \geq t_{setup} + t_{rec} + t_p \tag{9.2}$$

where t_{rec} is the recovery time of the D-type flip-flop. For a given system, the clock period is determined by the required performance of the system. Suppose that there is no combinational logic used in the synchronizer. Hence, the value of the propagation

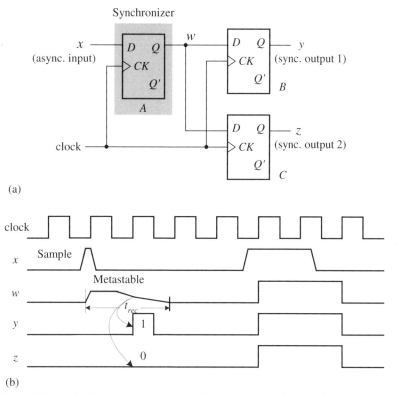

FIGURE 9.8 An illustration of metastable state and synchronization failure: (a) logic circuit; (b) timing diagram

delay, t_p, is 0. The available recovery time, which is $t_{rec} = t_{clk} - t_{setup}$, is then dependent on the setup time of the flip-flop used as the synchronizer. To maximize the available recovery time, we need to use a flip-flop with the setup time as small as possible.

Because whether a D-type flip-flop enters into the metastable state or not is a statistical process, a widely used equation for measuring the probability (mean time between failure, $MTBF$) of a given synchronizer entering into the metastable state is given as follows:

$$MTBF(t_{rec}) = \frac{1}{T_0 fa} \exp\left(\frac{t_{rec}}{\tau}\right) \tag{9.3}$$

where f is the frequency of the flip-flop clock, a is the number of asynchronous input changes per second and T_0 and τ are constants depending on the electrical characteristics of the flip-flop. Some examples are shown in Table 9.1.

The following two examples give some taste of what the $MTBF$ actually means.

EXAMPLE 9.3 *A Numerical Example of MTBF(t_{rec})*

Suppose that $f = 10$ MHz, $a = 100$ kHz and a 74LS74 D-type flip-flop is used. Calculate both $MTBF(20$ ns$)$ and $MTBF(40$ ns$)$.

TABLE 9.1 Parameters T_0 and τ of some example devices [3]

Device	τ(ns)	T_0(s)
74LSxx	1.35	4.8×10^{-3}
74Sxx	2.80	1.3×10^{-9}
74ALSxx	1.00	8.7×10^{-6}
74ASxx	0.25	1.4×10^3
74Fxx	0.11	1.8×10^8
74HCxx	1.82	1.5×10^{-6}

Solution   Because T_0 and τ of the 74LS74 are 4.8×10^{-3} s and 1.35 ns, respectively, the *MTBF* can be computed as follows:

$$MTBF(20 \text{ ns}) = \frac{1}{(4.8 \times 10^{-3})(10 \times 10^6)(100 \times 10^3)} \exp\left(\frac{20 \text{ ns}}{1.35 \text{ ns}}\right)$$

$$= 5.66 \times 10^{-4} \text{ s} \tag{9.4}$$

Consequently, the circuit is almost entirely in the metastable state when it is operating in this environment.

However, when we relax the recovery time to 40 ns, the *MTBF* is increased to the following:

$$MTBF(40 \text{ ns}) = \frac{1}{(4.8 \times 10^{-3})(10 \times 10^6)(100 \times 10^3)} \exp\left(\frac{40 \text{ ns}}{1.35 \text{ ns}}\right)$$

$$= 1537.25 \text{ s} \tag{9.5}$$

which is a much better result than the case of the 20-ns recovery time. However, it still cannot be used in practical systems. ∎

The approaches to increasing the *MTBF* are using flip-flops with smaller setup times and increasing the clock period. The former has to use more advanced technology whereas the latter needs to slow down the system clock and hence the system performance.

EXAMPLE 9.4 *A Numerical Example of MTBF(t_{rec})*

Suppose that $a = 100$ kHz and a 74ALS74 *D*-type flip-flop ($t_{setup} = 10$ ns) is used. Calculate the $MTBF(t_{rec})$ at $f = 20$ MHz and 50 MHz, respectively.

Solution T_0 and τ of 74ALS74 are 8.7×10^{-6} s and 1.0 ns, respectively. At $f = 20$ MHz, $t_{clk} = 50$ ns and $t_{rec} = t_{clk} - t_{setup} = 40$ ns. Hence the *MTBF* is computed as follows:

$$MTBF(40 \text{ ns}) = \frac{1}{(8.7 \times 10^{-6})(20 \times 10^6)(100 \times 10^3)} \exp\left(\frac{40 \text{ ns}}{1.0 \text{ ns}}\right)$$

$$= 1.35 \times 10^{10} \text{ s} \approx 429 \text{ years} \tag{9.6}$$

As a result, the mean time between two metastable states is approximately 429 years.

When the synchronizer operates at $f = 50$ MHz, $t_{clk} = 20$ ns and $t_{rec} = t_{clk} - t_{setup} = 10$ ns.

$$MTBF(10 \text{ ns}) = \frac{1}{(8.7 \times 10^{-6})(20 \times 10^6)(100 \times 10^3)} \exp\left(\frac{10 \text{ ns}}{1.0 \text{ ns}}\right)$$

$$= 1.27 \times 10^{-3} \text{ s} \tag{9.7}$$

Consequently, it is a high probability for this circuit to enter into a metastable state when operating in this environment. ■

Two approaches widely used in practice to attack metastable states are the *cascaded synchronizer* and *frequency-divided synchronizer*, as shown in Figure 9.9. The cascaded synchronizer is just an *n*-stage shift register, as shown in Figure 9.9(a), where the value of *n* is usually 2 or 3. The rationale behind the cascaded synchronizer is that every stage of the synchronizer has the same probability of entering into the metastable state. Hence, the probability of an *n*-stage synchronizer is the product of the individual probabilities of all stages, which is quite small when a *D*-type flip-flop with a good *MTBF* is used as the basic building block.

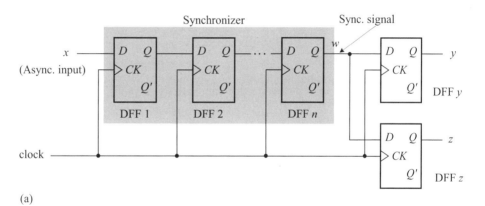

(a)

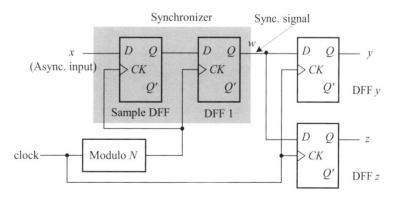

(b)

FIGURE 9.9 Two basic types of synchronizers: (a) cascaded synchronizer; (b) frequency-divided synchronizer

The following example is a cascaded synchronizer with the capability of asynchronous reset.

EXAMPLE 9.5 *A Cascaded Synchronizer with Asynchronous Reset*

As illustrated in Figure 9.9(a), a cascaded synchronizer is basically an *n*-stage shift register. Consequently, a cascaded synchronizer can be easily modeled by using the same style of shift register. The output q_out is cleared to 0 whenever the reset signal reset_n is active, i.e. 0, regardless of the status of the clock signal.

```
// a cascaded synchronizer with asynchronous reset
module cascaded_synchronizer(clk, reset_n, d, q_out);
parameter N = 2;   // default to 2 stages
input  clk, reset_n, d;
output q_out;       // assign output
reg    [N-1:0] q;
// the body of cascaded synchronizer
assign q_out = q[N-1];
always @(posedge clk or negedge reset_n)
   if (!reset_n) q <= 0; // reset_n signal is active-low.
   else          q <= {q[N-2:0], d};
endmodule
```

The frequency-divided synchronizer is shown in Figure 9.9(b). A frequency divider is used to divide the system clock by N and then apply to the synchronizer. As a result, the clock period of the synchronizer is increased N times. The available recovery time is then increased to $t_{rec} = N \times t_{clk} - t_{setup}$. Consequently, the $MTBF$ is improved and a flip-flop with larger setup time can be used to construct the synchronizer.

The following example is a frequency-divided synchronizer with the capability of asynchronous reset.

EXAMPLE 9.6 *A Frequency-Divided Synchronizer with Asynchronous Reset*

The rationale behind the frequency-divided synchronizer is the use of a modulo n frequency divider by which the system clock is divided by n before being sent to the synchronizer. Hence, an always block is used for this purpose. The second always block constructs a two-stage shift register. The output q_out is cleared to 0 whenever the reset signal reset_n is active, i.e. 0, regardless of the status of the clock signal.

```
// a frequency-divided synchronizer with asynchronous
// reset
module frequency_divided_synchronizer(clk, reset_n,
  d, q_out);
parameter N = 1;   // default to modulo 2
input  clk, reset_n, d;
output q_out;       // assign output
reg    [1:0] q;
reg    [N-1:0] clk_q; // clock for synchronizer
```

```
wire clk_sync;
// the body of frequency-divided synchronizer
// a modulo N frequency divider
always @(posedge clk or negedge reset_n)
   if  (!reset_n) clk_q <= 0;
   else clk_q <= clk_q + 1;
assign clk_sync = clk_q[N-1];
// the body of synchronizer
always @(posedge clk_sync or negedge reset_n)
   if (!reset_n) q <= 0; // reset_n signal is active-low.
   else          q <= {q[0], d};
assign q_out = q[1];
endmodule                                              ■
```

Review Questions

Q9.13 Describe the meaning of synchronizer.

Q9.14 What is the meaning of synchronization failure?

Q9.15 Describe the meaning of mean time between failure (*MTBF*).

Q9.16 Describe the rationale behind the cascaded synchronizer.

Q9.17 Describe the rationale behind the frequency-divided synchronizer.

9.1.4 A Switch-Debouncing Circuit

Another basic application of flip-flops is found in constructing a very useful circuit called a *switch-debouncing circuit*, which is used to produce a single pulse whenever a mechanical switch is pressed. Due to the inertia of mechanical switches, an inherent feature of a mechanical switch is that it would bounce forward and back several times during the period of 5 to 20 ms each time it is pressed. As a result, each press of the mechanical switch would be recognized as many pulses instead of only one because the speed of any electronic circuit is much faster than that of the mechanical switch.

Although there are many approaches that can be used to attack the switch-bouncing effects, in this section we introduce a synthesizable scheme, which is based on the idea of a technique often used in software. The idea is that *a switch is recognized as a valid press whenever two samples of a switch status apart 10 ms is the same.* Based on this principle, a simple switch-debouncing circuit constructed with *D*-type flip-flops is illustrated in Figure 9.10. The first two *D*-type flip-flops sample the switch status apart 10 ms. If both samples are the same, then it is recognized as a valid switch press. Otherwise, it is not a valid switch press. Once a valid switch press is detected, the *JK* flip-flop generates a high output pulse until both *D*-type flip-flops sample a low-input, which clears the *JK* flip-flop and terminates the output pulse. Consequently, only one pulse is generated whenever the switch is pressed.

The above switch-debouncing circuit can be modeled as follows.

EXAMPLE 9.7 *A Switch-Debouncing Circuit*

As illustrated in Figure 9.10, two *D*-type flip-flops are cascaded in series to detect the switch status. Next, the *JK* flip-flop is implemented with a *D*-type flip-flop by

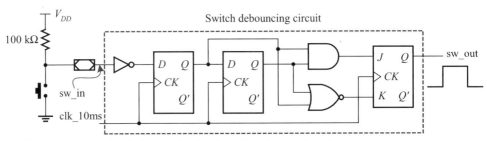

FIGURE 9.10 A simple but synthesizable switch-debouncing circuit

writing its excitation equation: $Q(t+1) = JQ'(t) + K'Q(t)$. Finally, the output of the JK flip-flop is assigned as the output sw_out.

```
// single switch debounce logic
module switch_debounce(clk_10ms, reset, sw_in, sw_out);
input  clk_10ms, reset, sw_in;
output wire sw_out;
wire    d;
reg     [1:0] q_sample;
reg     q_jk;
// remove the bouncing effect associated with mechanical
switches always @(posedge clk_10ms or posedge reset)
   if (reset) q_sample <= 0;
   else begin
      q_sample[1] <= q_sample[0];
      q_sample[0] <= ~sw_in;
   end
// using a JK flip-flop to generate a single pulse when
// the switch is pressed each time
assign d = (((q_sample[0] & q_sample[1]) & ~q_jk) |
           (~(~q_sample[0] & ~q_sample[1])& q_jk));
always @(posedge clk_10ms or posedge reset)
   if (reset)    q_jk <= 1'b0;
   else          q_jk <= d;
assign sw_out = q_jk;
endmodule
```

Review Questions

Q9.18 What is the purpose of the switch-debouncing circuit?

Q9.19 Describe the rationale behind the switch-debouncing circuit.

Q9.20 What will happen if we remove the NOR gate from Figure 9.10?

Q9.21 What is the function of the AND gate in Figure 9.10?

9.2 MEMORY ELEMENTS

In any large digital system, about one-third or more of the hardware are memory elements (or modules), which are used to store the intermediate information of the system. A memory element can be a data register or a random-access memory (RAM). In synthesizable designs, the following three types of memory elements are the most popular: *data register*, *register file* and *synchronous RAM*. In addition, an asynchronous RAM is also often used externally for storing the data required for the system.

9.2.1 Registers

A data register (or register for short) is a set of flip-flops or latches. A single-bit register is just a single flip-flop or a latch. For a synthesizable design, it usually uses flip-flops rather than latches when constructing a register. An *n*-bit register is a set of *n* flip-flops placed in parallel with a common clock and a clear (or reset) signals. Registers are usually used to store small amounts of information in digital systems.

Figure 9.11 shows a 4-bit data register consisting of four *D*-type flip-flops with a common clock signal connected to all flip-flops. The following example illustrates how to model this register.

EXAMPLE 9.8 *An n-Bit Data Register*

It is quite intuitive to model an *n*-bit data register. The result is as follows. At each positive edge of the clock signal `clk`, the input data `din` are sampled and put to the output of the data register `qout`.

```
// an N-bit data register
module register(clk, din, qout);
parameter N = 4;    // number of bits
input  clk;
input  [N-1:0] din;
output reg [N-1:0] qout;
// the body of an N-bit data register
always @(posedge clk) qout <= din;
endmodule
```

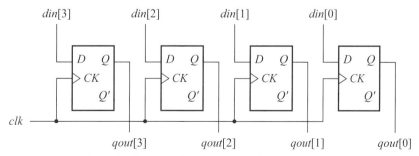

FIGURE 9.11 The logic diagram of a 4-bit data register

In most practical applications, a data register often has the capability of clearing its outputs asynchronously. The following example illustrates such a data register.

EXAMPLE 9.9 *An n-Bit Data Register with Asynchronous Reset*

This example simply expands the single bit *D*-type flip-flop with asynchronous reset to an *n*-bit register by declaring both din and qout as vectors of *n* bits.

```
// an N-bit data register with asynchronous reset.
module register_reset(clk, reset_n, din, qout);
parameter N = 4;    // number of bits
input   clk, reset_n;
input   [N-1:0] din;
output reg [N-1:0] qout;
// the body of an N-bit data register
always @(posedge clk or negedge reset_n)
   if (!reset_n) qout <= {N{1'b0}};
   else          qout <= din;
endmodule
```

A more practical data register often needs to have the capability of loading new data in synchronism with the clock signal and the capability of clearing its outputs asynchronously. The following example illustrates such a data register.

EXAMPLE 9.10 *An n-Bit Data Register with Synchronous Load and Asynchronous Reset*

The output qout is cleared whenever the reset signal reset_n is active. At each positive edge of the clock signal clk, a new input data din is sampled and loaded into the data register qout when load control load is enabled; otherwise, the output remains unchanged.

```
// an N-bit data register with synchronous load and
// asynchronous reset.
module register_load_reset(clk, load, reset_n, din, qout);
parameter N = 4;    // number of bits
input   clk, load, reset_n;
input   [N-1:0] din;
output reg [N-1:0] qout;
// the body of an N-bit data register
always @(posedge clk or negedge reset_n)
   if (!reset_n)    qout <= {N{1'b0}};
   else if (load) qout <= din;
//      else     qout <= qout; // a redundant expression
endmodule
```

Notice that the `else` part is a redundant expression for sequential circuits. Why? Try to explain it.

Coding Styles

1. *We should avoid using latches as much as possible because they are more difficult to test than flip-flops.*
2. *For an active-low signal, it is a good practice to end the signal name with an underscore followed by a lowercase letter* b *or* n, *such as* `reset_n`.

9.2.2 Register Files

Register files often find their applications in data paths (see Section 11.2.3) of most digital systems. A register file is a set of registers with multiple access ports and the capability of random access, which means that the access time of any register in the register file is the same. A register file is usually characterized by the following parameters: the number of registers, the width of a register and the number of ports. Depending on the actual requirements, the most widely used register files have 16 or 32 registers with each containing 16 or 32 bits. As for the number of ports, one read and one write ports, as well as two read and one write ports, are most common.

Register files are usually built from registers, which is composed of flip-flops in turn. However, they can also be built by using synchronous RAMs. Synchronous RAMs will be introduced in the next sub-section. In the following, we give an example to illustrate how to model an n-word register file with one synchronous write port and two asynchronous read ports.

EXAMPLE 9.11 *An n-Word Register File with One-Write and Two-Read Ports*

In most register-file-based data paths, three operands are often required for carrying out an operation, and thus the register file need to have two-read and one-write ports. In the following module, suppose that the access to each read port is unconditional and the access of the write port is controlled by a write enable signal `wr_enable` in synchronism with the clock signal. Each port regardless of read or write has its own access address.

```
// an N-word register file with one-write and two-read
// ports
module register_file(clk, rd_addra, rd_addrb, wr_addr,
                     wr_enable, din, douta, doutb);
parameter M = 4;   // number of address bits
parameter N = 16;  // number of words, N = 2**M
parameter W = 8;   // number of bits in a word
input  clk, wr_enable;
input  [W-1:0] din;
output [W-1:0] douta, doutb;
input  [M-1:0] rd_addra, rd_addrb, wr_addr;
reg    [W-1:0] reg_file[N-1:0];
```

```
// the body of the N-word register file
assign douta = reg_file[rd_addra],
       doutb = reg_file[rd_addrb];
always @(posedge clk)
   if (wr_enable) reg_file[wr_addr] <= din;
endmodule
```

Readers are encouraged to synthesize the above module and examine the results.

9.2.3 Synchronous RAM

Synchronous RAMs are also widely used for storing moderate amounts of information within digital systems. A synchronous RAM is a random-access memory where the data access is in synchronism with a clock signal being applied to the memory module. In general, random-access memory modules can be either asynchronous or synchronous. However, the synchronous RAM modules are particularly popular in synthesizable digital designs due to the ease of use.

A synchronous RAM module may be constructed from flip-flops or compiled RAM cells. However, a flip-flop may take up 10 to 20 times the area of a 6-transistor static RAM cell. Hence, for a synchronous RAM of moderate size, it is better to use the compiled RAM cell in a cell-based design or the block memory (namely, memory macro) built into the device in FPGA-based designs.

A synchronous RAM can be characterized by the following parameters: the word width, the number of words and the number of access ports. In addition, the access time of a synchronous RAM is also an important factor when using a synchronous RAM in a design. The following is an example of a synchronous RAM with one read and one write ports.

EXAMPLE 9.12 *An Example of a Synchronous RAM Module*

This example illuminates how to model a synchronous RAM with one read and one write ports. Since both accesses are in synchronism with the clock, thus they are placed inside the `always` block under the event control of the clock edge.

```
// a synchronous RAM module example
module syn_ram (addr, cs, din, clk, wr, dout);
parameter N = 16;   // number of words
parameter A = 4;    // number of address bits
parameter W = 4;    // number of word size in bits
input  [A-1:0]  addr;
input  [W-1:0]  din;
input  cs, wr, clk; // chip select, read-write control,
                    // and clock signals
output reg [W-1:0] dout;
reg    [W-1:0] ram [N-1:0];   // declare an N * W memory
                              // array
// the body of synchronous RAM
```

```
always @(posedge clk)
   if (cs) if (wr) ram[addr] <= din;
           else    dout <= ram[addr];
endmodule
```
∎

The reader is encouraged to synthesize the above module and examine the result.

Review Questions

Q9.22 What is the relationship between an n-bit register and flip-flops?

Q9.23 What are the features of register files?

Q9.24 Describe the features of synchronous RAMs.

Q9.25 What are the two implementations of synchronous RAMs usually found in synthesizable designs?

Q9.26 What are the possible parameters when describing a register file?

9.2.4 Asynchronous RAM

Asynchronous static random-access memory (SRAM, often called RAM for short) devices are still widely used in low- to middle-end products. Consequently, in this section we introduce the basic features and access timing of typical RAM devices. In addition, we give an example to illustrate how to model these RAM devices in Verilog HDL.

The block diagram of a typical 32-kB RAM device is shown in Figure 9.12, where the memory cells are constituted into a 512×512 memory matrix. As a result, the row address needs nine bits and the column address requires six bits because the data bus is eight bits. Besides row and column decoders, there are a timing generator and a read/write control logic. The timing generator produces all timing required for the memory device to perform read and write operations. The read/write control logic determines the data flow direction and enables read or write operation.

In general, when there is only one access control signal in a RAM device, the control signal is denoted as $R/\overline{W}$ (Read/Write). The RAM is in read mode when $R/\overline{W} = 1$ and in write mode when $R/\overline{W} = 0$, assuming that in both cases the chip is enabled ($\overline{CE} = 0$). When a RAM device has two access control signals, one of them is often named as write enable ($\overline{WE}$) and the other as output enable ($\overline{OE}$). The chip is in write mode when $\overline{WE} = 0$ and in read mode when $\overline{OE} = 0$ and $\overline{WE} = 1$, assuming that the chip is enabled. The output bus is in a high-impedance when both $\overline{WE}$ and $\overline{OE}$ are set to high.

The pin assignments of the two most widely used RAM devices, 6264 (8 kB) and 62256 (32 kB), are shown in Figure 9.13. Both the devices 6264 and 62256 have two access control signals: write enable ($\overline{WE}$) and output enable ($\overline{OE}$). Write enable ($\overline{WE}$) asserts the write operation and output enable ($\overline{OE}$) asserts the read operation by enabling the output buffer. Device 6264 has two chip-select $\overline{CE1}$ and CE2 and the device 62256 has only one chip-select $\overline{CE}$. The chip select controls the entire chip operation. The data bus is in a high-impedance whenever the chip enable is deasserted, regardless of the values of write enable ($\overline{WE}$) and output enable ($\overline{OE}$).

9.2.4.1 Read-Cycle Timing In order to read data from a RAM device, the address and access control signals, $\overline{OE}$ and $\overline{CE}$, must be applied according to the timing

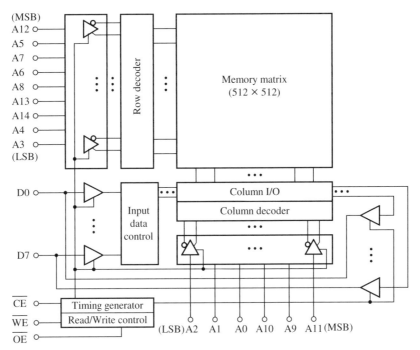

FIGURE 9.12 The block diagram of a typical 32-kB RAM device

62256	6264			6264	62256
A14	NC	1	28	V_{CC}	V_{CC}
A12	A12	2	27	$\overline{WE}$	$\overline{WE}$
A7	A7	3	26	CE2	A13
A6	A6	4	25	A8	A8
A5	A5	5	24	A9	A9
A4	A4	6	23	A11	A11
A3	A3	7	22	$\overline{OE}$	$\overline{OE}$
A2	A2	8	21	A10	A10
A1	A1	9	20	$\overline{CE1}$	$\overline{CE}$
A0	A0	10	19	D7	D7
D0	D0	11	18	D6	D6
D1	D1	12	17	D5	D5
D2	D2	13	16	D4	D4
GND	GND	14	15	D3	D3

$\overline{WE}$	$\overline{CE}$	$\overline{OE}$	Mode	Dn
x	1	x	Unselected	Hi-z
1	0	1	Output inhibited	Hi-z
1	0	0	Read	Data output
0	0	1	Write	Data input
0	0	0	Write	Data input

FIGURE 9.13 The pin assignment of two widely used commercial RAM devices: 6264 and 62256

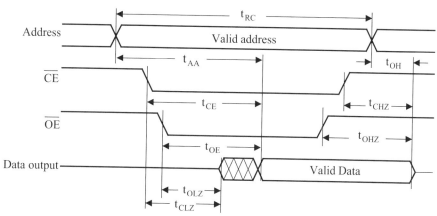

FIGURE 9.14 The read-cycle timing diagram of a typical SRAM device

shown in Figure 9.14. When both $\overline{OE}$ and $\overline{CE}$ control signals are low, the valid data will appear at the output data bus after all three timing constraints – t_{AA}, t_{CE} and t_{OE} are satisfied. In other words, the time required for the valid data to appear at the output data bus is determined by the above three time conditions which is satisfied lastly.

In general, there are four types of access times related to a RAM device:

- *Address access time* (t_{AA}) is the amount of time from when the address is applied to the point that stable data appear at the output data bus.
- $\overline{CE}$ *access time* (t_{CE}) is the amount of time from when the chip enable is asserted to the point that stable data appear at the output data bus.
- $\overline{OE}$ *access time* (t_{OE}) is the amount of time from when the output enable is asserted to the point that stable data appear at the output data bus.
- *Read cycle time* (t_{RC}) is the minimum amount of time needed to separate two consecutive random read operations.

The output data bus begins to be active t_{CLZ} and t_{OLZ} after chip enable ($\overline{CE}$) and output enable ($\overline{OE}$) signals are asserted, respectively. For consecutive read operations, both chip enable ($\overline{CE}$) and output enable ($\overline{OE}$) signals can remain in the asserted status while changing address only. In this case, the data appearing at the output data bus will continually sustain an amount of time known as the hold time, t_{OH}. The output data bus goes to a high-impedance t_{CHZ} or t_{OHZ} after chip enable ($\overline{CE}$) or output enable ($\overline{OE}$) signals are deasserted, respectively.

The parameters related to the read cycle of typical RAM devices are listed in Table 9.2. For most RAM devices, the values of both t_{AA} and t_{CE} are the same. The value of t_{OE} is less than that of t_{AA} or t_{CE}. For RAM devices, read cycle time is usually equal to both address access time and $\overline{CE}$ access time.

9.2.4.2 *Write-Cycle Timing*

Generally speaking, writing data into a RAM device can be controlled by either the $\overline{WE}$ or $\overline{CE}$ signal, as shown in Figure 9.15. Figure 9.15(a) is the $\overline{WE}$-controlled write operation and Figure 9.15(b) is the $\overline{CE}$-controlled write operation. In fact, it can be seen from the figures that the write operation is completed by the control signals $\overline{WE}$ or $\overline{CE}$, which are deasserted earlier.

TABLE 9.2 The read-cycle timing parameters of a typical SRAM device (all times in ns)

Symbol	Parameter	HM6264B-10L		MCM60L256A-C		HM62256B-5	
		Min	Max	Min	Max	Min	Max
t_{RC}	Read cycle time	100	—	100	—	55	—
t_{AA}	Address access time	—	100	—	100	—	55
t_{CE}	Chip select access time	—	100	—	100	—	55
t_{OE}	Output enable to output valid	—	50	—	50	—	35
t_{CLZ}	Chip select to output in low-Z	10	—	10	—	5	—
t_{OLZ}	Output enable to output in low-Z	5	—	5	—	5	—
t_{OH}	Output hold from address change	10	—	10	—	5	—
t_{OHZ}	Output disable to output in high-Z	0	35	0	35	0	20
t_{CHZ}	Chip deselect to output in high-Z	0	35	0	35	0	20

In order to correctly write data into a RAM device, the write timing must be satisfied with the write-cycle specification of the device, especially, data setup time (t_{DW}) and data hold time (t_{DH}). Like the definitions in flip-flops or latches, the data setup time is the amount of time that data must be stable before it is sampled and data hold time is the amount of time that data must remain stable after it is sampled. Here, the sampling point is the positive edge of the $\overline{WE}$ or $\overline{CE}$ signal. Hence, the timing reference point is the positive edge of the $\overline{WE}$ or $\overline{CE}$ signal, depending on whether the underlying write operation is $\overline{WE}$-controlled or $\overline{CE}$-controlled.

In $\overline{WE}$ ($\overline{CE}$)-controlled write operations, the address must be stable for an amount of time known as the *address setup time* (t_{AS}), before the $\overline{WE}$ ($\overline{CE}$) is asserted and must be still stable for an amount of time known as the *address hold time* after the $\overline{WE}$ ($\overline{CE}$) is deasserted. The write recovery time (t_{WR}) is virtually the address hold time. Similarly, the write data must be stable for an amount of time known as the *data setup time* (t_{DW}), before the $\overline{WE}$ ($\overline{CE}$) is deasserted and must still remain stable for an amount of time known as the *data hold time* (t_{DH}) after the $\overline{WE}$ ($\overline{CE}$) is deasserted.

In a $\overline{WE}$-controlled write operation, when the $\overline{OE}$ signal is deasserted before the write cycle begins, the output data bus enters into a high-impedance t_{OHZ} after the $\overline{OE}$ signal is deasserted. When the $\overline{OE}$ signal is asserted before the write cycle begins, the output data bus enters into a high-impedance t_{WHZ} after the $\overline{WE}$ signal is asserted and stays there until t_{OW} after the $\overline{WE}$ signal is deasserted. In a $\overline{CE}$-controlled write operation, the output data bus remains in a high-impedance.

Despite either $\overline{WE}$ or $\overline{CE}$-controlled write operations, $\overline{WE}$ and $\overline{CE}$ signals must separately sustain an amount of time t_{WP} and t_{CW}. In addition, the address must at least sustain an amount of time t_{WC}, known as the *write cycle time*. The parameters related to the write cycle of typical RAM devices are listed in Table 9.3.

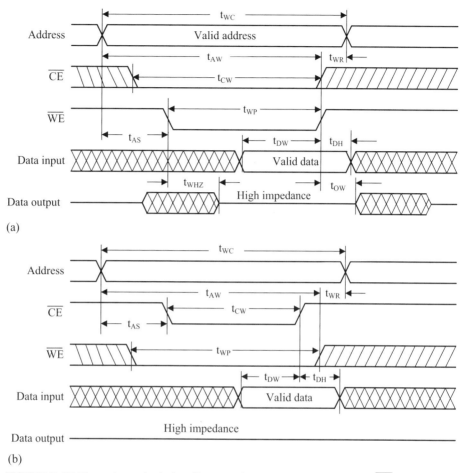

FIGURE 9.15 The write-cycle timing diagram of a typical SRAM device: (a) $\overline{WE}$-controlled write; (b) $\overline{CE}$-controlled write

In the following, we demonstrate how to model an asynchronous static RAM in Verilog HDL by an example.

EXAMPLE 9.13 *An Example of an Asynchronous Static RAM Module*

In this example, we demonstrate how to model a static RAM with timing checks. The details of the timing checks can be referred to in Section 7.3.3.

```
`timescale 1 ns / 100 ps
// an example of modeling a 32k * 8 asynchronous sram
// with timing checks
module sram_timing_check(CE_b, WE_b, OE_b, addr, data);
input CE_b, WE_b, OE_b;
input [14:0] addr;
inout [7:0] data;
```

TABLE 9.3 The write-cycle timing parameters of a typical SRAM device (all times in ns)

Symbol	Parameter	HM6264B-10L Min	HM6264B-10L Max	MCM60L256A-C Min	MCM60L256A-C Max	HM62256B-5 Min	HM62256B-5 Max
t_{WC}	Write cycle time	100	—	100	—	55	—
t_{AS}	Address setup time	0	—	0	—	0	—
t_{AW}	Address valid to end of write	80	—	80	—	40	—
t_{WP}	Write pulse width	60	—	60	—	35	—
t_{DW}	Data setup time	40	—	35	—	25	—
t_{DH}	Data hold time	0	—	0	—	0	—
t_{OHZ}	Output disable to output in high-Z	0	35	0	35	0	20
t_{WHZ}	Write to output in high-Z	0	35	0	25	0	20
t_{OW}	Output active from end of write	5	—	10	—	5	—
t_{WR}	Write recovery time	0	—	0	—	0	—
t_{CW}	Chip selection to end of write	80	—	80	—	40	—

```verilog
reg [7:0] data_int;
reg [7:0] mem [2**15-1:0];
specify
   // parameters for read cycle
   specparam t_RC  = 55; // read cycle time
   specparam t_AA  = 55; // address access time
   specparam t_CE  = 55; // chip select access time
   specparam t_OE  = 35; // output enable to output valid
   specparam t_CLZ = 5;  // chip select to output in
                         // low-Z
   specparam t_OLZ = 5;  // output enable to output in
                         // low-Z
   specparam t_OH  = 5;  // output hold from address
                         // change
   specparam t_OHZ = 0;  // output disable to output in
                         // high-Z
   specparam t_CHZ = 0;  // chip deselect to output in
                         // high-Z
   // read cycle time
   $width(negedge addr, t_RC);
   // module path timing specifications
   (addr *> data) = t_AA;
   (CE_b *> data) = (t_CE, t_CE, t_CHZ);
   (OE_b *> data) = (t_OE, t_OE, t_OHZ);
   // write cycle timing checks
```

```
    // parameters for write cycle
    specparam t_WC  = 55; // write cycle time
    specparam t_AS  = 0;  // address setup time
    specparam t_AW  = 40; // address valid to end of write
    specparam t_WP  = 35; // write pulse width
    specparam t_DW  = 25; // data setup time
    specparam t_DH  = 0;  // data hold time
    specparam t_WHZ = 0;  // write to output in high-Z
    specparam t_OW  = 5;  // output active from end of
                          // write
    specparam t_WR  = 0;  // write recovery time
    specparam t_CW  = 40; // chip selection to end of
                          // write
    // write cycle time
    $width(negedge addr, t_WC);
    // address valid to end of write
    $setup (addr, posedge WE_b &&& CE_b == 0, t_AW);
    $setup (addr, posedge CE_b &&& WE_b == 0, t_AW);
    // address setup time
    $setup (addr, negedge WE_b &&& CE_b == 0, t_AS);
    $setup (addr, negedge CE_b &&& WE_b == 0, t_AS);
    // write pulse width
    $width(negedge WE_b, t_WP);
    // data setup time
    $setup (data, posedge WE_b &&& CE_b == 0, t_DW);
    $setup (data, posedge CE_b &&& WE_b == 0, t_DW);
    // data hold time
    $hold (data, posedge WE_b &&& CE_b == 0, t_DH);
    $hold (data, posedge CE_b &&& WE_b == 0, t_DH);
    // chip select to end of write
    $setup (CE_b, posedge WE_b &&& CE_b == 0, t_CW);
    $width (negedge CE_b &&& WE_b == 0, t_CW);
endspecify
// process memory access cycle
assign data = ((CE_b == 0)&& (WE_b == 1)&&(OE_b == 0))
              ? data_int: 8'bzzzz_zzzz;
always @(*) // wait for chip enable
    if (!CE_b)
        if (!WE_b)      // write cycle
            mem[addr] = data;
        else if (WE_b) // read cycle
                if (!OE_b) data_int = mem[addr];
                else data_int = 8'bzzzz_zzzz;
            else data_int = 8'bzzzz_zzzz;
    else // chip not selected, disable output
        data_int <= 8'bzzzz_zzzz;
endmodule
```

Review Questions

Q9.27 Define data setup time and data hold time.

Q9.28 What are the three constraints that must be satisfied for valid data to appear at the output data bus when reading an RAM device?

Q9.29 Define address access time, $\overline{CE}$ access time and $\overline{OE}$ access time.

Q9.30 What are the two ways that can be used to write data into RAM devices?

Q9.31 Describe the functions of $\overline{CE}$ and $\overline{OE}$ signals of RAM devices.

9.3 SHIFT REGISTERS

In addition to being used as data registers, shift registers may also shift their contents left or right. Shift registers are often required in a digital system in which they are employed to shift data or perform various data format conversions. When used as simple shift registers, they perform left or right shift operations. When used as parallel/serial format conversions, they perform one or more operations of the following four types:

- serial in, serial out (SISO);
- serial in, parallel out (SIPO);
- parallel in, serial out (PISO);
- parallel in, parallel out (PIPO).

9.3.1 Shift Registers

The basic structure of a 4-bit shift register is shown in Figure 9.16(a). An n-bit shift register generally consists of n D-type flip-flops connected in series with a common clock signal. From the timing depicted in Figure 9.16(b), the serial input data is sampled at each positive edge of the clock signal clk and shifted to the next stage per clock cycle.

The following is an n-Bit shift register with active-low asynchronous reset.

EXAMPLE 9.14 *An Example of an n-Bit Shift Register*

The output `qout` are cleared to 0 whenever the reset signal `reset_n` is active; otherwise, they are assigned the concatenation of serial input `sin` and the portion of `qout`, `qout[N-1:1]`. Hence, the result is a shift-right operation.

```
// a shift register module example
module shift_register(clk, reset_n, sin, qout);
parameter N = 4;   // default number of bits
input   clk, reset_n;
input   sin;
output reg [N-1:0] qout;
// the body of an N-bit shift register
always @(posedge clk or negedge reset_n)
    if (!reset_n) qout <= {N{1'b0}};
    else          qout <= {sin, qout[N-1:1]};
endmodule
```
■

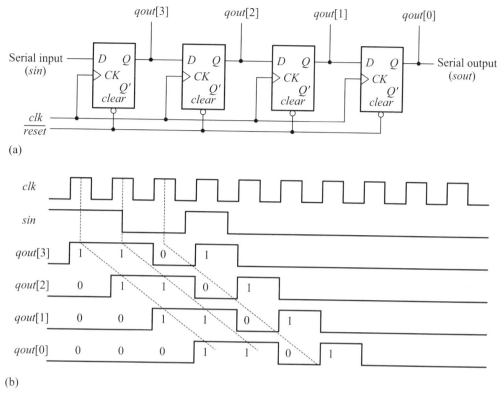

FIGURE 9.16 The logic diagram (a) and timing (b) of a 4-bit shift register

The following example gives another way of modeling a shift register by observing the operation of the shift register shown in Figure 9.16(a), where each flip-flop accepts the output value from its preceding stage in the previous cycle except the first stage which accepts the serial input data `sin`.

EXAMPLE 9.15   *A 4-Bit Shift Register as a Trivial Pipeline*

The output `qout` are cleared to 0 whenever the reset signal `reset_n` is active; otherwise, each of them is assigned the output value of its preceding stage except the first stage which accepts the serial input data `sin`. Hence, the result is a shift-right operation.

```
// a shift register as a trivial pipeline
module shift_register_pipeline(clk, reset_n, sin, qout);
input   clk, reset_n;
input   sin;
output reg [3:0] qout;
// the body of a 4-bit shift register
always @(posedge clk or negedge reset_n)
```

```
    if (!reset_n) qout <= 0;
    else begin
        qout[3] <= sin;
        qout[2] <= qout[3];
        qout[1] <= qout[2];
        qout[0] <= qout[1];
    end
endmodule
```
◼

The above example will not work properly if all nonblocking assignments are replaced with blocking assignments. Why? The reader is invited to synthesize it with both types of assignments and compare the synthesized results carefully. It is worth noting that mixed use of both blocking and nonblocking assignments in the same `always` block is usually not allowed in most synthesis tools. Another insight into this example is that it illustrates how to model a pipeline operation. In fact, a shift register may be considered as a pipeline with a trivial combinational logic between any two stages.

In addition to inputting data through the serial input, it is not uncommon to load data into a shift register in parallel. The following is such an example.

EXAMPLE 9.16   *An Example of an n-Bit Shift Register with Parallel Load*

The output `qout` are cleared to 0 whenever the reset signal `reset_n` is active; otherwise, they are loaded the external data `din` if the synchronous load control `load` is enabled or assigned the concatenation of serial input `sin` and the portion of `qout`, `qout[N-1:1]` if `load` is not enabled. Hence, the result is a shift-right register with asynchronous reset and synchronous parallel load.

```
// a shift register with parallel load
module shift_register_parallel_load(clk, load, reset_n,
    din, sin, qout);
parameter N = 4;  // default number of bits
input  sin, clk, load, reset_n;
input  [N-1:0] din;
output reg [N-1:0] qout;
// the body of an N-bit shift register
always @(posedge clk or negedge reset_n)
    if (!reset_n)  qout <= {N{1'b0}};
    else if (load) qout <= din;
    else           qout <= {sin, qout[N-1:1]};
endmodule
```
◼

9.3.2 Universal Shift Registers

As described previously, a shift-right or shift-left register can be modeled at will. In the following, we demonstrate that it is still not that difficult to model a shift register with the sense that it is universal. A shift register is said to be universal if it can carry out the following operations:

- serial in, serial out (SISO);
- serial in, parallel out (SIPO);
- parallel in, serial out (PISO);
- parallel in, parallel out (PIPO).

To perform all above four operations, the shift register must have the following capabilities:

- parallel load;
- serial in and serial out;
- shift left and shift right.

The following example gives such a universal shift register, which is modeled in behavioral style.

EXAMPLE 9.17 *An Example of an n-Bit Universal Shift Register*

Suppose that the serial data inputs for right shift and left shift are lsi and rsi, respectively. Because there are four operating modes of the shift register, two mode selection signals s1 and s0 are required. With these, the shift register is modeled with an always block contained inside a case statement to select the desired operating mode.

```
// a universal shift register module
module universal_shift_register(clk,reset_n,s1,s0,lsi,
  rsi,din,qout);
parameter N = 4;  // define the default size
input  s1, s0, lsi, rsi, clk, reset_n;
input  [N-1:0] din;
output reg [N-1:0] qout;
// shift register body
always @(posedge clk or negedge reset_n)
   if (!reset_n) qout <= {N{1'b0}};
   else case ({s1,s0})
      2'b00: ;// qout <= qout;            // no change
      2'b01: qout <= {lsi, qout[N-1:1]}; // shift right
      2'b10: qout <= {qout[N-2:0], rsi}; // shift left
      2'b11: qout <= din;                // parallel
                                         // load
   endcase
endmodule
```

Another universal shift register often encountered in digital logic textbooks is shown in Figure 9.17(a), where an explicit 4-bit 4-to-1 multiplexer is used to implement the case statement in the preceding example. The logic symbol of such a universal shift register is depicted in Figure 9.17(b). Figure 9.17(c) summarizes the operating modes of the universal shift register.

The following example models the above universal shift register using a generate-loop statement and is also a parameterized module.

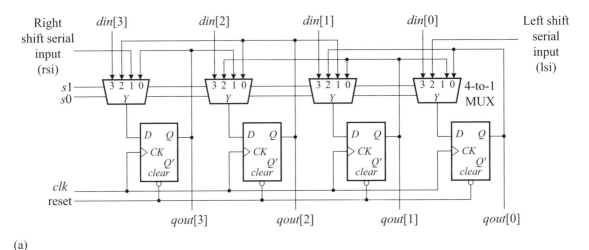

(a)

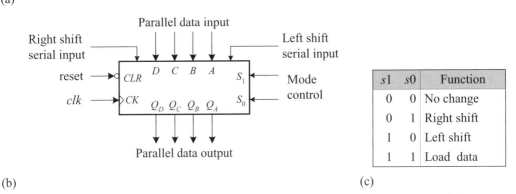

(b)

s1	s0	Function
0	0	No change
0	1	Right shift
1	0	Left shift
1	1	Load data

(c)

FIGURE 9.17 The logic diagram (a), logic symbol (b) and function table (c) of a 4-bit universal shift register

EXAMPLE 9.18 *An Example of an n-Bit Universal Shift Register Using a Generate-Loop Statement*

A continuous assignment `assign` is used to implement the 4-to-1 multiplexer and an `always` block is used to implement a *D*-type flip-flop with asynchronous reset. There are three cases needed to be separately processed. These cases include two boundary conditions, LSB and MSB, as well as the rest bits. For the LSB, it needs to take the left-shift serial input into consideration and for the MSB, it needs to take the right-shift serial input into consideration. For the rest bits, they take the output from their preceding or succeeding stage, depending on whether the shift operation is left or right.

```
// a universal shift register module using generate
// blocks
module universal_shift_register_generate
                (clk, reset_n, s1, s0, lsi, rsi, din,
                 qout);
```

```
parameter N = 4;  // define the default size
input  s1, s0, lsi, rsi, clk, reset_n;
input  [N-1:0] din;
output reg [N-1:0] qout;
wire   [N-1:0] y;
// shift register body
genvar  i;
generate for (i = 0; i < N; i = i + 1) begin: universal_
  shift_register
    if (i == 0) begin: lsb   // specify LSB
      assign y[i] = (~s1 & ~s0 & qout[i]) |
                    (~s1 & s0 & qout[i+1])|
                    ( s1 & ~s0 & lsi)     |
                    ( s1 & s0 & din[i]);
      always @(posedge clk or negedge reset_n)
          if (!reset_n) qout[i] <= 1'b0;
          else  qout[i] <= y[i]; end
    else if (i == N-1) begin: msb   // specify MSB
      assign y[i] = (~s1 & ~s0 & qout[i])   |
                    (~s1 & s0 &  rsi)        |
                    ( s1 & ~s0 & qout[i-1])  |
                    ( s1 & s0 & din[i]);
      always @(posedge clk or negedge reset_n)
          if (!reset_n) qout[i] <= 1'b0;
          else  qout[i] <= y[i]; end
    else begin: rest_bits   // specify the rest bits
      assign y[i] = (~s1 & ~s0 & qout[i])    |
                    (~s1 & s0 & qout[i+1])   |
                    ( s1 & ~s0 & qout[i-1])  |
                    ( s1 & s0 & din[i]);
      always @(posedge clk or negedge reset_n)
          if (!reset_n) qout[i] <= 1'b0;
          else  qout[i] <= y[i]; end
endgenerate // universal_shift_register
endmodule
```

Review Questions

Q9.32   What are the four possible operations when a shift register is used as a parallel/serial format conversion?

Q9.33   What does a universal shift register mean?

Q9.34   Using Verilog HDL, describe an n-bit shift register with active-high synchronous reset.

Q9.35   Using Verilog HDL, describe an n-bit shift register with active-high asynchronous reset.

9.4 COUNTERS

A counter is a device that counts the input events such as input pulses or clock pulses. There are many kinds of counters found in a variety of digital systems, depending on the attempted usage. Counters can be classified into asynchronous and synchronous in terms of whether they are in synchronism with a clock or not. In practical applications, the most common asynchronous (ripple) counters are binary counters and the most widely used synchronous counters include binary counters and BCD counters, as well as Gray counters. It is worth noting that a counter is called a *timer* when its input or clock source is from a standard or known timing source such as the system clock.

9.4.1 Ripple Counters

Like ripple-carry adders, ripple counters are the most widely known types of counters used in digital systems. An n-bit ripple binary counter consists of n T-type flip-flops connected in series; namely, the output of each flip-flop is connected to the clock input of its succeeding flip-flop. Figure 9.18(a) shows an example of a 3-bit ripple binary counter. As we can see from this figure, the essential feature of ripple counters is that each flip-flop is triggered by the output of its preceding stage, except the first stage which is triggered by an external signal, known as clock.

The following gives an example of how to model a ripple counter in behavioral style.

EXAMPLE 9.19 *An Example of a 3-Bit Ripple Counter*

This example intuitively models the ripple counter shown in Figure 9.18(a). Three `always` blocks are used, with each corresponding to a flip-flop. Due to the toggle

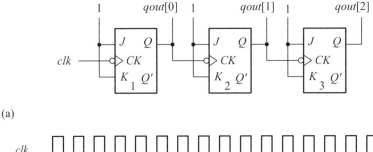

(a)

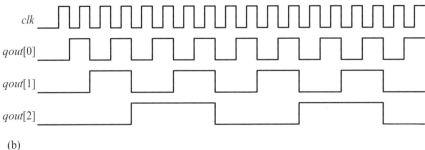

(b)

FIGURE 9.18 The logic diagram (a) and timing (b) of a 3-bit ripple binary counter

operations inherent to T-type flip-flops, the output of each flip-flop is toggled when the flip-flop is triggered.

```
// a 3-bit ripple counter module example
module ripple_counter(clk, qout);
input   clk;
output reg [2:0] qout;
// the body of the 3-bit ripple counter
always @(negedge clk)
   qout[0] <= ~qout[0];
always @(negedge qout[0])
   qout[1] <= ~qout[1];
always @(negedge qout[1])
   qout[2] <= ~qout[2];
endmodule
```

Try to synthesize it and see what happens!! The output cannot be observed from simulators due to lacking initial values of qout. In most practical applications, a useful counter often needs to be resettable such as in the following example.

EXAMPLE 9.20   *An Example of a 3-Bit Ripple Counter with Asynchronous Reset*

This example enhances the preceding one with the capability of reset. Whenever the reset signal reset_n is active, the output of the counter is cleared to 0. Otherwise, the counter functions as a normal ripple counter as described in the preceding example. The operation of each flip-flop is described by an if-else statement.

```
// a 3-bit ripple counter with enable control
module ripple_counter_enable(clk, enable, reset_n, qout);
input   clk, enable, reset_n;
output reg [2:0] qout;
// the body of the 3-bit ripple counter
always @(posedge clk or negedge reset_n)
   if (!reset_n) qout[0] <= 1'b0; else
   if (enable)   qout[0] <= ~qout[0];
always @(posedge qout[0] or negedge reset_n)
   if (!reset_n) qout[1] <= 1'b0; else
   if (enable)   qout[1] <= ~qout[1];
always @(posedge qout[1] or negedge reset_n)
   if (!reset_n) qout[2] <= 1'b0; else
   if (enable)   qout[2] <= ~qout[2];
endmodule
```

The major disadvantage of the above two examples is the lack of parameterized capability. In the following example, we demonstrate how to overcome this by using a generate-loop statement to expand the required number of stages of a parameterized ripple counter.

EXAMPLE 9.21 *An Example of an n-Bit Ripple Counter Using a Generate-Loop Statement*

As we have seen in Figure 9.18(a), except the first stage that accepts the external clock signal, all other stages accept the output from their preceding stage as their triggered signal. As a consequence, it is necessary to consider these two cases separately. The resulting module is as follows.

```
// an N-bit ripple counter using generate blocks
module ripple_counter_generate(clk, reset_n, qout);
input  clk, reset_n;
parameter N = 4; // define the size of counter
output reg [N-1:0] qout;
// the body of the N-bit ripple counter
genvar  i;
generate for (i = 0; i < N; i = i + 1) begin: ripple_
  counter
  if (i == 0) // specify LSB
    always @(negedge clk or negedge reset_n)
        if (!reset_n) qout[0] <= 1'b0;
        else          qout[0] <= ~qout[0];
  else        // specify the rest bits
    always @(negedge qout[i-1] or negedge reset_n)
        if (!reset_n) qout[i] <= 1'b0;
        else          qout[i] <= ~qout[i];
end endgenerate
endmodule
```

It is worth noting that the above three ripple counters may also be modified into down counters by simply changing the triggered means from negative-edge to positive-edge. In addition, a ripple counter can be constructed as an up/down counter under a mode control. For example, it is an up counter when the mode is 0 and a down counter when the mode is 1. Furthermore, a ripple counter is not necessary to be a binary with modulo 2^n, where n is the number of flip-flops. It may be designed to count with a modulus other than 2^n. The *modulus* of a counter is the number of states that the counter may have in its normal operation. In reality, it can be any modulus as you like. However, these issues are left to the reader as exercises. Further details about ripple counters can be found in Lin [4].

9.4.2 Synchronous Counters

As mentioned previously, synchronous counters mean that they operate in synchronism with the clock. In other words, all flip-flops of a synchronous counter change their states under the control of a clock signal, namely, at the positive or negative edge of the clock signal. Like ripple counters, synchronous counters can be designed to count with a modulus m, where m can be 2^n or any integer value small than 2^n, where n is the number of flip-flops. In addition, a synchronous counter may be either an up counter or a down counter, even an up/down counter under a mode control.

EXAMPLE 9.22 *An n-Bit Binary Counter with Synchronous Reset and Enable Control*

This example is a synchronous modulo 2^n binary counter with synchronous reset and enable. It consists of n flip-flops. In order to be cascadable, this counter produces a carry out signal cout with a duration of one clock cycle. This carry out signal cout can be used as the enable signal to the next higher module to enable its operations each time it is generated.

```
// an N-bit binary counter with synchronous
// reset and enable control
module binary_counter(clk, enable, reset, qout, cout);
parameter N = 4;
input   clk, enable, reset;
output reg [N-1:0] qout;
output cout;     // carry output
// the body of the N-bit binary counter
always @(posedge clk)
    if (reset)      qout <= 0;
    else if (enable) qout <= qout + 1;
// generate carry output
assign #2 cout = &qout; // why #2 is required ?
endmodule
```

In order to be simulated properly, a delay timing control with some constant delay is needed in the continuous assignment for generating the cout signal. Otherwise, the next higher module will not count up at all. You may justify this by simulating the above module with and without the delay timing control and then checking your results.

The following example illustrates how to model a synchronous binary up/down counter.

EXAMPLE 9.23 *An n-Bit Binary Up/Down Counter with Synchronous Reset and Enable Control (Version 1)*

This example uses a single up/down control signal to control the operations of a synchronous binary counter. The reset and enable controls of the counter are supposed to be synchronous. At each positive edge of the clock signal, the counter is cleared to 0 if the reset signal is active; otherwise, it may count up, count down or remain unchanged, depending on the values of the enable control enable and model selection upcnt. When the counter is enabled, it can count up if upcnt is 1 or down if upcnt is 0. The counter remains unchanged if the enable control enable is 0.

```
// an N-bit binary up/down counter with synchronous
// reset and enable control
```

```
module binary_up_down_counter_reset
        (clk, enable, reset, upcnt, qout, cout, bout);
parameter N = 4;
input  clk, enable, reset, upcnt;
output reg [N-1:0] qout;
output cout, bout;  // carry and borrow outputs
// the body of N-bit up/down binary counter
always @(posedge clk)
   if (reset) qout <= 0;
   else if (enable) begin
        if (upcnt) qout <= qout + 1;
        else       qout <= qout - 1;
   end
// generate carry and borrow outputs
assign #2 cout = &qout; // why #2 is required ?
assign #2 bout = ~|qout;
endmodule                                              ■
```

In order to be convenient for cascading multiple modules together to form a bigger counter, the enable control and mode selection are often recombined and separated into two mode control signals: enable count up (eup) and enable count down (edn). The following example illustrates this idea.

EXAMPLE 9.24 *An n-Bit Binary Up/Down Counter with Synchronous Reset and Enable Control (Version 2)*

This example still uses synchronous reset and enable control signals. At each positive edge of the clock signal, the counter is cleared to 0 if the reset signal is active; otherwise, it may count up, count down or remain unchanged, depending on the values of the enable count up (eup) and enable count down (edn). The counter counts up if eup is 1 and down if edn is 1. Both count enable signals eup and edn cannot be active at the same time.

```
// an N-bit up/down binary counter with synchronous
// reset and enable control
module up_dn_bin_counter(clk, reset, eup, edn, qout,
                         cout, bout);
parameter N = 4;
// enable up count (eup) and enable down count (edn)
// cannot be set to one at the same time.
input  clk, reset, eup, edn;
output reg [N-1:0] qout;
output cout, bout;
// the body of the N-bit binary counter
```

```
always @(posedge clk)
   if (reset) qout <= 0; // synchronous reset
   else if (eup)  qout <= qout + 1;
   else if (edn)  qout <= qout - 1;
assign #1 cout = ( &qout) & eup; // generate carry out
assign #1 bout = (~|qout) & edn; // generate borrow out
endmodule
```

The following example illustrates how to cascade two up/down counters to form a bigger up/down counter.

EXAMPLE 9.25 *An Example Illustrating How to Cascade Two Up/Down Counters (Version 2)*

This example instantiates two 4-bit up/down counters and cascades them together to form an 8-bit up/down counter. From this example, you can see the importance of delay timing control associated with the continuous assignments generating carry out and borrow out signals during simulation. Without them, the higher counter module cannot operate properly. However, the synthesized gate-level module can work properly. Why? Try to explain this.

```
// an example to illustrate the cascaded of two up/down
// counters.
module up_dn_bin_counter_cascaded(clk, reset,eup, edn,
   qout, cout, bout);
parameter N = 4;
input  clk, reset, eup, edn;
output [2*N-1:0] qout;
output cout, bout;
// declare internal nets for cascading both counters
wire   cout1, bout1;
// the body of the cascaded up/down counter
   up_dn_bin_counter #(4) up_dn_cnt1
                     (clk, reset,eup, edn, qout[3:0],
                      cout1, bout1);
   up_dn_bin_counter #(4) up_dn_cnt2
                     (clk, reset,cout1, bout1, qout[7:4],
                      cout, bout);
endmodule
```

It is quite often to have a counter with a modulus other than 2^n, where n is the number of flip-flops. For instance, a counter with modulus 6 or 10 is often desired in practical applications. The following example explains how to model a synchronous counter with an arbitrary modulus.

EXAMPLE 9.26 *A Modulo r Binary Counter with Synchronous Reset and Enable Control*

To construct a synchronous counter with a modulus of r, it usually constructs an n-bit binary synchronous counter with $2^n \geq r$ first and then a combinational logic is used to detect the output value of the counter. If the output value of the counter is $r - 1$, then the counter is cleared at the next clock; otherwise, the counter continues to counts up in the usual way. Consequently, by setting the appropriate value of the modulus r, a synchronous counter with an arbitrary modulus can be easily built.

```
// the body of modulo R binary counter with synchronous
// reset and enable control
module modulo_r_counter(clk, enable, reset, qout, cout);
parameter N = 4;
parameter R = 10; // BCD counter
input  clk, enable, reset;
output reg [N-1:0] qout;
output cout;       // carry output
// the body of modulo r binary counter
assign cout = (qout == R - 1);
always @(posedge clk)
   if (reset) qout <= 0;
   else begin
      if (enable) if (cout) qout <= 0;
      else qout <= qout + 1;
   end
endmodule
```

It is worthy to mention that in behavioral style, you may simply describe a modulo-r counter by using the following statement:

```
qout <= (qout + 1) % r;
```

However, the modulo operator (%) is usually not synthesizable if it is not an integer power of 2. It is instructive to justify it on your system.

Review Questions

Q9.36 Distinguish between counters and timers.

Q9.37 What does an asynchronous counter mean?

Q9.38 What does a synchronous counter mean?

Q9.39 Explain the meaning of ripple counters.

Q9.40 What are the differences between using a single statement, {cout, qout} <= qout + 1, and using two statements, qout <= qout + 1 and assign cout = &qout?

9.5 SEQUENCE GENERATORS

The important applications of shift registers are their use as sequence generators. In this section, we focus our attention on the following four circuits: pseudo-random sequence (PR-sequence) generators, cyclic-redundancy check (CRC) generators, ring counters and Johnson counters.

9.5.1 PR-Sequence Generators

The general logic diagram of a sequence generator is shown in Figure 9.19, where the outputs of all D-type flip-flops of the shift register are combined through a combinational logic circuit and the result of the combinational logic is then fed back into the input of the shift register. The sequence generator is called a *linear feedback shift register* (LFSR) if the combinational logic circuit is a network of XOR gates. An LFSR is called a *pseudo-random sequence generator* (PRSG) if it can generate a maximum-length sequence; namely, it has a period of $2^n - 1$, excluding the case of all 0s, where n is the number of flip-flops. This kind of sequence is often called a *pseudo-random sequence* (PR-sequence).

Although Figure 9.19 can also be used to generate sequences other than the PR-sequences, in the following we only focus our attention on the PR-sequences. The reader interested in the issues of generating other possible sequences can refer to Lin [4].

A PR-sequence is a polynomial code which is based on treating bit strings as a representation of a polynomial with coefficients of 0 and 1 only. Let x denote a unit delay, which corresponds to a D-type flip-flop and x^k denote k-unit delays. Then, an n-stage LFSR can be represented as the following polynomial:

$$f(x) = a_n x^n + a_{n-1} x^{n-1} + \cdots + a_1 x + a_0 \qquad (9.8)$$

where the coefficients $a_i \in \{0, 1\}$, for all $0 \leq i \leq n - 1$. Each combination of coefficients a_i of a given n corresponds to a function $f(x)$. However, not all functions $f(x)$ can produce a maximum-length sequence. In fact, only a few of them can do this. When a polynomial $f(x)$ generates a maximum-length sequence, it is called a *primitive polynomial*. Hence, a PR-sequence is generated by a primitive polynomial. Sample primitive polynomials for n from 1 to 60 are listed in Table 9.4.

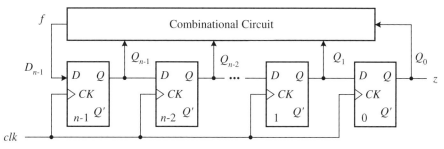

FIGURE 9.19 The block diagram of a general sequence generator

TABLE 9.4 Sample primitive polynomials for *n* from 1 to 60

n	f(x)	n	f(x)	n	f(x)
1, 2, 3, 4, 6,	$1+x+x^n$	24	$1+x+x^2+x^7+x^n$	43	$1+x+x^5+x^6+x^n$
7, 15, 22, 60		26	$1+x+x^2+x^6+x^n$	44, 50	$1+x+x^{26}+x^{27}+x^n$
5, 11, 21, 29	$1+x^2+x^n$	30	$1+x+x^2+x^{23}+x^n$	45	$1+x+x^3+x^4+x^n$
10, 17, 20, 25	$1+x^3+x^n$	32	$1+x+x^2+x^{22}+x^n$	46	$1+x+x^{20}+x^{21}+x^n$
28, 31, 41, 52		33	$1+x^{13}+x^n$	48	$1+x+x^{27}+x^{28}+x^n$
9	$1+x^4+x^n$	34	$1+x+x^{14}+x^{15}x^n$	49	$1+x^9+x^n$
23, 47	$1+x^5+x^n$	35	$1+x^2+x^n$	51, 53	$1+x+x^{15}+x^{16}+x^n$
18	$1+x^7+x^n$	36	$1+x^{11}+x^n$	54	$1+x+x^{36}+x^{37}+x^n$
8	$1+x^2+x^3+x^4+x^n$	37	$1+x^2+x^{10}+x^{12}+x^n$	55	$1+x^{24}+x^n$
12	$1+x+x^4+x^6+x^n$	38	$1+x+x^5+x^6+x^n$	56, 59	$1+x+x^{21}+x^{22}+x^n$
13	$1+x+x^3+x^4+x^n$	39	$1+x^4+x^n$	57	$1+x^7+x^n$
14, 16	$1+x^3+x^4+x^5+x^n$	40	$1+x^2+x^{19}+x^{21}+x^n$	58	$1+x^{19}+x^n$
19,27	$1+x+x^2+x^5+x^n$	42	$1+x+x^{22}+x^{23}+x^n$		

There are two standard paradigms for implementing primitive polynomials for any given n. These are known as *standard format* and *modular format*, respectively, as shown in Figure 9.20. An example of a 4-bit PR-sequence generator having a primitive polynomial, $1 + x + x^4$, and being realized in standard form is depicted in Figure 9.21.

It is quite simple to model an n-bit PR-sequence generator in standard format. The feedback XOR network can be simply computed by using a reduction operator, such as in the following example.

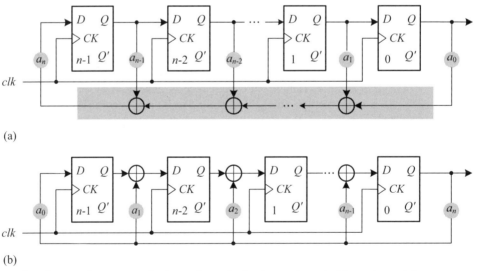

(a)

(b)

FIGURE 9.20 The two standard paradigms for implementing primitive polynomials: (a) standard format; (b) modular format

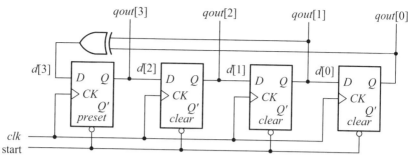

FIGURE 9.21 A 4-bit PR-sequence generator with a primitive polynomial: $1 + x + x^4$

EXAMPLE 9.27 *An n-Bit PR-Sequence Generator Module—In Standard Form*

Based on the observation of Figure 9.20(a), we can see that the basic idea of implementing the standard format is to use a reduction operator to reduce the XOR network into a single-bit result which is then fed back as the serial input to the first stage. To do this, it is very helpful to declare an $(n + 1)$-bit parameter tap to store the coefficients of the given primitive polynomial. The lowest n bits of tap and the outputs of D-type flip-flops are AND together and then uses XOR reduction operator to reduce into a single bit result. This result is fed back to the first stage as the serial input data.

```
// an N-bit pr_sequence generator module --- in standard
// form
module pr_sequence_generator (clk, qout);
parameter N = 4;   // define the default size
parameter [N:0]tap = 5'b10011;   // x^4 + x + 1
input   clk;
output reg [N-1:0] qout = 4'b0100;
wire    d;
// pseudo-random sequence generator body
assign d = ^(tap[N-1:0] & qout[N-1:0]);
always @(posedge clk)
   qout <= {d, qout[N-1:1]};
endmodule
```

The qout = 4'b0100 is used for simulation only. Without the initial value, simulators cannot calculate qout and hence we could not observe the qout values. In practical applications, a PR-sequence generator needs to be started automatically. The following example illustrates how to start a PR-sequence generator automatically.

EXAMPLE 9.28 *An n-Bit PR-Sequence Generator Module—In Standard Form*

In order to start up the circuit, an initial value other than all 0s is required. This is done in this example by using a start signal start. The circuit will start from state 4'b1000 after the start signal is applied.

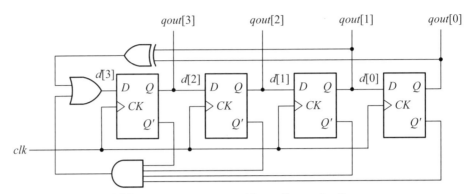

FIGURE 9.22 A 4-bit PR-sequence generator with a self-start circuit

```
// an N-bit pr_sequence generator module --- in standard
// form
module pr_sequence_generate (clk, start, qout);
parameter N = 4;   // define the default size
parameter [N:0] tap = 5'b10011;
input  clk, start;
output reg [N-1:0] qout;
wire   d;
// pseudo-random sequence generator body
assign d = ^(tap[N-1:0] & qout[N-1:0]);
always @(posedge clk or posedge start)
    if (start) qout <= {1'b1, {N-1{1'b0}}};
    else       qout <= {d, qout[N-1:1]};
endmodule                                              ■
```

Using the `start` signal to set the initial value to $4'b1000$, simulators can calculate `qout` and hence we could observe the `qout` values. Of course, the circuit will start from the state $4'b1000$ after the `start` signal is applied.

As we have mentioned previously, a PR-sequence generator will go through every possible state except the all 0s state, which will lock out the circuit forever. The simplest way to get out all 0s states is to add an all-zero detector circuit, as shown in Figure 9.22. The all-zero detector circuit is an n-input AND gate with all inputs connected to the complemented output of each D-type flip-flop of the PR-sequence generator. The output of the AND gate is combined with the output from the network of XOR gates through an OR gate and is then fed into the input of the shift register. Consequently, when all outputs of the flip-flops are 0, the output of the zero-detector is 1, which causes the shift register to start with state $100\cdots00$ at the next cycle. The following example shows how to model this self-started PR-sequence generator.

EXAMPLE 9.29 *An n-Bit PR-Sequence Generator Module—In Standard Form*

Basically, this circuit is almost the same as the preceding example, except that it adds a zero-detector circuit (AND gate) to detect the situation that all outputs of the flip-flops are 0 and generate an output value of 1. This value of 1 will transfer the shift register back to its normal loop at the next cycle.

```
// an N-bit pr_sequence generator module --- in standard
// form
module pr_sequence_generator_self_start (clk, qout);
parameter N = 4;  // define the default size
parameter [N:0]tap = 5'b10011;
input  clk;
output reg [N-1:0] qout;
wire   d;
// the body of pseudo-random sequence generator
assign d = ^(tap[N-1:0] & qout[N-1:0])|(~(|qout[N-1:0]));
always @(posedge clk)
   qout <= {d, qout[N-1:1]};
endmodule                                              ∎
```

Review Questions

Q9.41 What is an LFSR?

Q9.42 Write an *n*-bit PR-sequence generator in modular form.

Q9.43 What is a pseudo-random sequence?

Q9.44 What is a primitive polynomial?

Q9.45 Explain the function of the AND gate in the circuit of Figure 9.22.

9.5.2 CRC Generator/Detectors

Another important applications of LFSR are used to generate and check the CRC code, which is widely used in communications to ensure data integrity. Like PR-sequences, a CRC code is also a polynomial code and can also be represented as a polynomial of the *x* operator. The following are examples commonly used in communications:

$$\text{CRC-8} = x^8 + x^7 + x^6 + x^4 + x^2 + 1$$

$$\text{CRC-12} = x^{12} + x^{11} + x^3 + x^2 + x + 1$$

$$\text{CRC-16} = x^{16} + x^{15} + x^2 + 1$$

$$\text{CRC-CCITT} = x^{16} + x^{12} + x^5 + 1$$

$$\text{CRC-32a} = x^{32} + x^{30} + x^{22} + x^{15} + x^{12} + x^{11} + x^7 + x^6 + x^5 + x + 1$$

$$\text{CRC-32b} = x^{32} + x^{26} + x^{23} + x^{22} + x^{16} + x^{12} + x^{11} + x^{10} + x^8 + x^7$$
$$+ x^5 + x^4 + x^2 + x + 1$$

The CRC codes find their applications in checking data integrity in disk devices, wired and wireless communication networks. An important property of a polynomial code with *n* check bits is that it has the capability of detecting all burst errors of lengths less than or equal to *n*. Consequently, an *n*-bit CRC code can detect all burst errors of lengths less than *n* bits and all burst errors with odd numbers of bits.

The use of the CRC code to ensure data integrity is based on the following principle. At the transmitter, the message, represented as a polynomial of x and denoted as $D(x)$, to be transmitted is first divided by a CRC polynomial, called a *CRC generator*, denoted as $G(x)$ and then the remainder, denoted as $R(x)$, is appended to the message polynomial. The resulting polynomial, $T(x) = \{D(x), R(x)\}$ is transmitted to the receiver, where $\{,\}$ denotes the concatenation of two strings. At the receiver end, the received message, $T'(x)$, is divided by the same CRC polynomial, $G(x)$. If the remainder is 0, it is correct; otherwise, it has errors.

The above operations can be written as the following algorithm.

Algorithm: Computing and Appending CRC Codes

Input. The message $D(x)$ to be sent.
Output. The message containing the CRC code, namely, $T(x) = \{D(x), R(x)\}$.

1. Suppose that $G(x)$ is a degree-n polynomial. Append n 0s to the message polynomial $D(x)$. Denote the resulting polynomial as $D'(x)$.

2. Divide the resulting polynomial $D'(x)$ by $G(x)$ using modulo 2 operation.

3. Compute the message $T(x)$ to be transmitted by appending the remainder $R(x)$ obtained from the above step to the original message $D(x)$; namely, it computes $T(x) = \{D(x), R(x)\}$, which corresponds to $D'(x) - R(x)$.

Figure 9.23 gives an example showing how to compute the remainder. Here, we assume that $G(x) = x^4 + x + 1$ and the message to be transmitted is 1101101010.

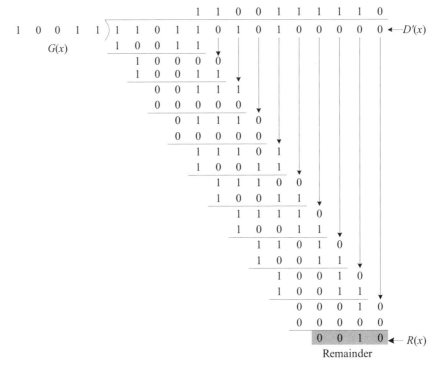

FIGURE 9.23 A numerical example of a 4-bit CRC generator

Because the degree of $G(x)$ is 4, four 0s are appended to the message polynomial. After completing the division, we obtain the remainder 0010. As a consequence, the message transmitted is 11011010100010. It is easy to check that the remainder will be 0 when this message is divided by $G(x)$ again.

The general paradigm for a CRC polynomial is basically the same as the modular format of the PR-sequence generator shown in Figure 9.20(b), except that here we need to introduce the external data into the circuit. Due to the inherent bit-serial operations of a CRC generator, the external data may be introduced into the circuit from the left-most or right-most end, as shown in Figures 9.24(a) and (b), respectively.

Figure 9.24(a) shows the first approach, where the external data is fed into the circuit from the LSB. In this approach, the input of each flip-flop receives the combined result from the output of the last stage with the input data $D(x)$ or with the output from its previous stage under the selection of the coefficients of $G(x)$. In reality, this method realizes the above algorithm directly. That is, the actual message entering into the CRC generator is a combination of the real message with n 0s. In other words, after entering the entire message into the circuit, it still requires to fill another n 0s in order to generate the final remainder (namely, the CRC code). This may cause some troubles in practical applications because it needs to wait n cycles for the generation of the CRC code after the message has been entered into the CRC generator.

To overcome the above difficulty, an alternative method is by feeding the data into the circuit from the MSB, as shown in Figure 9.24(b), where the input data $D(x)$ is combined with the output from the last stage and the result is then fed back to the preceding stages under the selection of the coefficients of $G(x)$. This circuit has an important feature that the CRC code is left in the circuit just after the message is entirely entered into the circuit. That is, it need not wait another n clock cycles for generating the CRC code. Hence, this method is most widely used in practical applications. It is worth noting that the same circuit is used for both CRC code generation and detection.

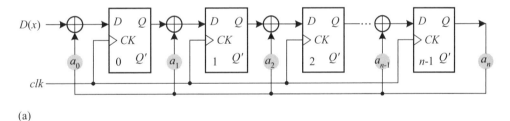

(a)

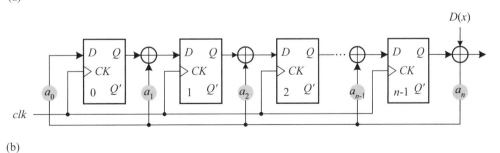

(b)

FIGURE 9.24 The general paradigms of an n-bit CRC generator: (a) feeding data from the LSB; (b) feeding data from the MSB

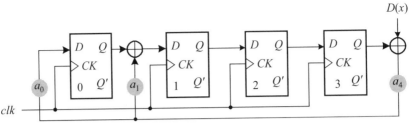

FIGURE 9.25 A 4-bit CRC generator with $G(x) = x^4 + x + 1$

An example of implementing $G(x) = x^4 + x + 1$ is shown in Figure 9.25. Because only the coefficients a_4, a_1 and a_0 are not zero, only the connections indicated by these three coefficients are needed. The following example shows how to describe an n-bit CRC generator using a generate-loop statement.

EXAMPLE 9.30 *An n-Bit CRC Generator Using a Generate-Loop Statement*

In reality, this example illustrates how to implement the modular format shown in Figure 9.20(b), except that the external data $D(x)$ needs to be included. From Figure 9.24(b), we can see that two cases are needed to be considered, the LSB and the rest bits. Suppose that d is the result from data XORing with qout[N-1]. For the case of LSB, the data d is directly input into the flip-flop when the reset signal reset_n is not active. For the case of rest bits, the input of a flip-flop is directly from the output of its preceding stage or a combined result of d with the output of its preceding stage, depending on the value of its associated coefficient of $G(x)$. It is the output from its preceding stage if the associated coefficient is 0; otherwise, it is a combined result.

```
// a parameterized CRC generator/detector module
// using generate-for statement --- feeding data from MSB
module CRC_MSB_generate(clk, reset_n, data, qout);
parameter N = 4;
parameter [N:0]tap = 5'b10011;
input   clk, reset_n, data;
output reg [N-1:0] qout;
wire d;
// CRC generator/detector body
assign d = data ^ qout[N-1];
genvar i;
generate for (i = 0; i < N ; i = i + 1) begin:
  crc_generator
    if (i == 0) // LSB
       always @(posedge clk or negedge reset_n)
          if (!reset_n) qout[i] <= 1'b0;
          else qout[i] <= d;
    else         // the rest bits
```

```
        always @(posedge clk or negedge reset_n)
            if (!reset_n) qout[i] <= 1'b0;
            else if (tap[i] == 1) qout[i] <= qout[i-1]^d;
            else qout[i] <= qout[i-1];
end endgenerate
endmodule
```

In reality, the generate-loop statement within the above module may also be replaced by the following `always` statement:

```
always @(posedge clk or negedge reset_n)
    if (!reset_n) qout <= 1'b0;
    else qout <= {qout[N-2:0], 1'b0} ^ (tap[N-1:0] &
        {N-1{d}});
```

You are encouraged to explain how this works. It is instructive to write a complete module by using the above statement and write a test bench to verify the functionality of the resulting module.

9.5.3 Ring Counters

There are two basic types of ring counters: *standard ring counter* and *twisted ring counter*. A modulo n standard ring counter is an n-stage shift register with the serial output fed back to the serial input. An essential feature of a ring counter is that it outputs a 1-out-of-n code directly from its flip-flop outputs.

An example of a 4-bit ring counter is shown in Figure 9.26, where the *qout*[3] is connected backward to the input d[0]. In addition, a start circuit is used to set the initial value of the ring counter to 1000.

EXAMPLE 9.31 *An n-Bit Ring Counter with Initial Value*

In order to start up the circuit, an initial value other than all 0s is required. This is done in this example by using the start signal start. The circuit will start from state N'b10...00 after the start signal is applied.

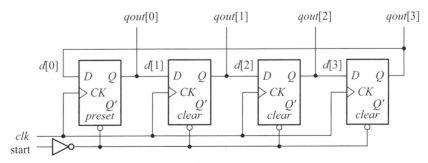

FIGURE 9.26 An example of a 4-bit ring counter

```
// a ring counter with initial value
module ring_counter(clk, start, qout);
parameter N = 4;
input   clk, start;
output reg [0:N-1] qout;
// The body of ring counter
always @(posedge clk or posedge start)
   if (start) qout <= {1'b1,{N-1{1'b0}}};
   else       qout <= {qout[N-1], qout[0:N-2]};
endmodule
```
∎

In general, a standard ring counter has n valid states and $2^n - n$ invalid states. As shown in Figure 9.27, except for a counting loop of valid states, it also includes five independent loops constructed from invalid states. Once the ring counter enters into one of these invalid-state loops, it will be locked in the loop and cannot return to its normal counting loop. To prevent the ring counter from locking inside these invalid-state loops, some extra logic circuit must be added to instruct the ring counter to return to its normal counting loop from the invalid-state loops. A widely used approach for the 4-bit standard ring counter shown in Figure 9.26 is to replace the $d[0]$ with the following switching function:

$$d[0] = (qout[0] + qout[1] + qout[2])'$$
(9.9)

It can be shown that once the ring counter enters into any invalid-state loop, it will go back to its normal counting loop after four clock cycles at most. A standard ring counter so designed is called a *self-correcting ring counter*.

9.5.4 Johnson Counters

A twisted-ring counter is also known as a *Johnson counter*, a Moebius or a *switched-tail counter*. A module $2n$ Johnson counter is an n-bit shift register with the complement

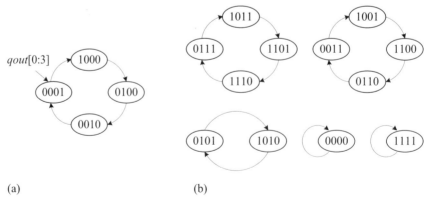

(a)   (b)

FIGURE 9.27 State diagrams of the 4-bit ring counter shown in Figure 9.26: (a) valid states; (b) invalid states

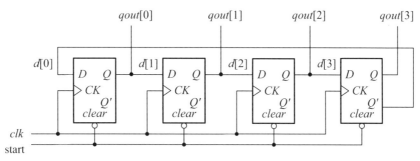

FIGURE 9.28 An example of a 4-bit Johnson counter

serial output fed back to the serial input. The essential feature of a Johnson counter is that it only needs half the number of flip-flops needed in a ring counter to achieve the same number of counting states.

An example of a 4-bit Johnson counter is shown in Figure 9.28, where the complement of the output of the last stage is fed back to the input of the first stage.

EXAMPLE 9.32   *An n-Bit Johnson Counter with Initial Value*

Like a standard ring counter, the essential structure of a Johnson counter is also an n-bit shift register, except that the complement of the output of the last stage is fed back to the first stage. In contrast to a standard ring counter, the initial value set in the Johnson counter is all 0s rather than a 1-out-of-n code.

```
// Johnson counter with initial value
module Johnson_counter(clk, start, qout);
parameter N = 4;
input  clk, start;
output reg [0:N-1] qout;
// the body of Johnson counter
always @(posedge clk or posedge start)
    if (start) qout <= {N{1'b0}};
    else       qout <= {~qout[N-1], qout[0:N-2]};
endmodule
```

An n-bit Johnson counter has $2n$ valid states and $2^n - 2n$ invalid states. Hence, a Johnson counter is subjected to the same trouble as a standard ring counter; namely, it may be locked in one of the invalid-state loops and cannot return to its normal counting loop. To make a Johnson counter self-correctable, some extra logic circuit must be added to instruct the Johnson counter to return to its normal counting loop from the invalid-state loops. A widely used approach for the 4-bit Johnson counter shown in Figure 9.28 is to replace the $d[2]$ with the following switching function:

$$d[2] = (qout[0] + qout[2])qout[1] \tag{9.10}$$

The reader is encouraged to show that the resulting Johnson counter is a self-correcting counter.

Review Questions

Q9.46   Describe the basic principle of applying a CRC code to ensure data integrity.

Q9.47   Describe how to compute the remainder of the CRC code.

Q9.48   Describe the basic structure of an n-bit ring counter.

Q9.49   Describe the basic structure of an n-bit Johnson counter.

Q9.50   What is the basic feature of an n-bit ring counter?

Q9.51   What is the meaning of a "self-correcting" counter?

Q9.52   Discuss the reasons why a counter needs to be self-correcting.

9.6   TIMING GENERATORS

A timing generator is a device that generates the timing required for specific applications. Two timing generators are widely used in designing digital systems. These are a multiphase clock signal generator and a digital monostable circuit. A multiphase clock generator can be constructed based on a ring counter, a Johnson counter or a binary counter with a decoder. The digital monostable circuit may be either retriggerable or non-retriggerable.

9.6.1   Multiphase Clock Generators

A typical multiphase clock signal is shown in Figure 9.29. Each clock phase lasts for a clock cycle and is repeated after a given number of clock cycles. From another viewpoint, an n-phase clock signal generator outputs a 1-out-of-n code at every clock cycle. This is exactly matches the feature of an n-bit ring counter. For example, Figure 9.30 displays a four-phase clock signal, which is exactly the same as the output from a 4-bit ring counter.

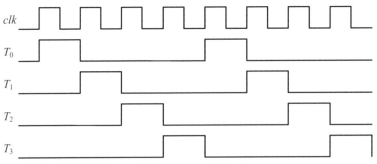

FIGURE 9.29 An example of a multiphase clock signal

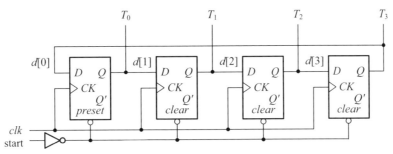

FIGURE 9.30 An example of a 4-phase clock signal generated by a 4-bit ring counter

As described above, it is quite natural to use an n-bit ring counter to generate an n-phase clock signal due to the matched feature between them.

EXAMPLE 9.33 *A Multiphase Clock Signal Generated by a Ring Counter*

This example uses an n-bit ring counter to generate an n-phase clock signal. In order to start up the counter, an initial value of N'b100...0 is set into the counter whenever the reset signal reset is active.

```
// a ring counter with initial value serve as a timing
// generator
module ring_counter_timing_generator(clk, reset, qout);
parameter N = 4;    // define the default counter size
input  clk, reset;
output reg [0:N-1] qout;
// the body of N-bit ring counter
always @(posedge clk or posedge reset)
   if (reset) qout <= {1'b1,{N-1{1'b0}}};
   else       qout <= {qout[N-1], qout[0:N-2]};
endmodule
```

Another widely used approach for generating an n-phase clock signal is to use a log_2n-bit binary counter with a log_2n-to-n decoder. An example of an 8-phase clock signal generator using the binary counter with a decoder approach is depicted in Figure 9.31.

The following example is a parameterized design for generating an n-phase clock signal using the above approach.

EXAMPLE 9.34 *A Multiphase Clock Signal Generated by Using a Binary Counter with a Decoder Approach*

There are two major components in the binary counter with a decoder approach. The first component is the m-bit binary counter, which is described by the first always block. The second component is the m-to-2^m decoder, which is modeled by the second always block.

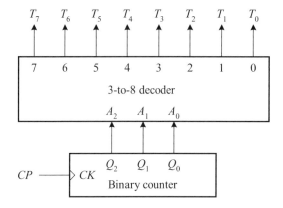

FIGURE 9.31 An example of a multiphase clock signal generated by using the binary counter with decoder approach

```
// a binary counter with decoder serve as a timing
// generator
module binary_counter_timing_generator(clk, reset, enable,
   qout);
parameter N = 8;  // define the default number of phases
parameter M = 3;  // define the bit number of binary
                  // counter
input  clk, reset, enable;
output reg [N-1:0] qout;
reg     [M-1:0] bcnt_out;
// the body of M-bit binary counter
always @(posedge clk or posedge reset)
   if (reset)  bcnt_out <= {M{1'b0}};
   else if (enable) bcnt_out <= bcnt_out + 1;
      else          bcnt_out <= bcnt_out;
// decode the output of the binary counter
always @(bcnt_out)
   qout = {{N-1{1'b0}}, 1'b1} << bcnt_out;
endmodule
```

The reader is encouraged to synthesize the above module and observe the results. It is also instructive to change the parameters and repeat the above operations.

9.6.2 Digital Monostable Circuits

Other types of basic logic circuits that are widely used in digital systems are monostable circuits. These kinds of circuit are known as "monostable" because they only have one stable state. Initially, a 'monostable' is in its stable state. Whenever it is triggered, it leaves the stable state temporarily and goes into a transient state in which it outputs a pulse signal and then returns to its stable state.

A monostable circuit can be either *retriggerable* or *non-retriggerable*. A retriggerable monostable is a circuit that could be retriggered before it ends its previous transient state and restarts another transient state. A non-retriggerable monostable circuit

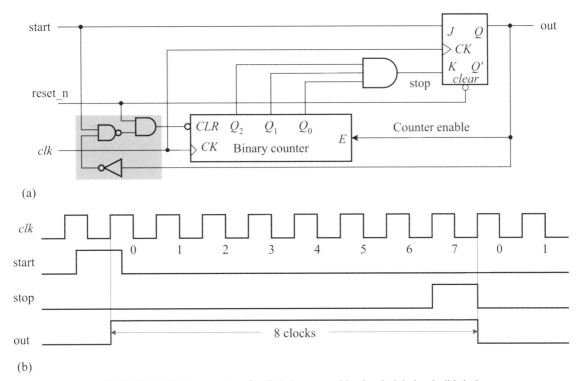

(a)

(b)

FIGURE 9.32 An example of a digital monostable circuit: (a) circuit; (b) timing

is a circuit which ignores any triggered signal before it completes its current transient state.

Although most monostable circuits are built into *RC* circuits, a digital monostable circuit is usually desired in designing modern digital systems because it facilitates a more testable and robust circuit than an *RC*-based circuit. An example of a non-retriggerable digital monostable circuit is shown in Figure 9.32. The digital monostable circuit operates as follows. Each time when the trigger signal start is active, the output out of the *JK* flip-flop rises to high at the next positive edge of the clock signal. This high-level out signal enables the binary counter to count up. When the binary counter reaches a given value (7 in the current example) set by the AND gate, the output of the *JK* flip-flop resets to 0 and the out pulse terminates. Consequently, it generates an output pulse with a width of a given number of clock cycles, namely, 8 clock cycles.

The following example gives a method to construct a parameterized non-retriggerable digital monostable circuit based on the above idea and circuit structure.

EXAMPLE 9.35   *A Non-Retriggerable Digital Monostable Example*

This example describes a parameterized non-retriggerable digital monostable with a maximum output pulse width of 2^m clock cycles. The pulse output signal out actually goes to high for *PW* clock cycles once the start signal is activated. The first always block models the binary counter and its associated logic. The continuous assignment and the second always block describe the *JK* flip-flop and its related logic circuits.

```
// a non-retriggerable digital monostable example --- the
// out goes high for 2**M cycles once start signal is
// activated.
module digital_monostable(clk, reset_n, start, out);
parameter  M = 3;   // define maximum pulse width = 2**M.
parameter PW = 5;   // define the actual pulse width
input   clk, reset_n, start;
output reg out;
reg     [M-1:0] bcnt_out;
wire    d;
// the body of binary counter
always @(posedge clk or negedge reset_n)
    if   (!reset_n || (start && ~out))
      bcnt_out <= {M{1'b0}};
    else if (out)     bcnt_out <= bcnt_out + 1;
// describe the JK flip-flop and its related circuits.
assign d = (start & ~out) | (~(bcnt_out == (PW-1)) & out);
always @(posedge clk or negedge reset_n)
    if (!reset_n) out <= 1'b0;
    else out <= d;
endmodule
```

Review Questions

Q9.53 What is a multiphase clock generator?

Q9.54 What is a digital monostable circuit?

Q9.55 Define the following terms: retriggerable and non-retriggerable.

Q9.56 Can we use a Johnson counter to construct a multiphase clock generator?

Q9.57 Modify the circuit shown in Figure 9.32 so that it can be retriggered.

SUMMARY

In this chapter, we have examined many widely used sequential modules, including flip-flops, synchronizers, a switch-debouncing circuit, registers, data registers, register files, shift registers, counters (binary, BCD, Johnson), CRC generators and detectors, clock generators, pulse generators and timing generators.

Flip-flops and latches are the basic building blocks for any sequential module. The flip-flops and latches are basically bistable devices. The fundamental difference between them is that latches have a transparent property but flip-flops do not. The transparent property is a feature that the output of the device follows the input signal when the enable control signal is active. Hence the output of a latch follows the input if its enable is active. This may cause troubles in some applications due to the ease of inducing noise into the system. Flip-flops do not have this property. They only sample the input data at some specific point, such as the positive or negative edge of the clock signal.

There are many types of flip-flops defined and implemented in discrete modules. Perhaps the fundamental application of flip-flops is to synchronize an external asynchronous signal to the system clock. These types of applications at least include synchronizers and a switch-debouncing circuit. The synchronizer is used to sample an external asynchronous signal in synchronism with the internal system clock and the switch debouncer removes the switch-bouncing effects of mechanical switches. The switch-bouncing effects are caused by the mechanical inertia of switches.

Flip-flops are also found widely as applications in constructing registers and shift registers, as well as counters. Registers usually include basic registers and register files. Shift registers are often used to perform arithmetic operations such as being divided and multiplied by 2, as well as data format conversion. Counters we have considered include both asynchronous and synchronous types. In addition, shift-register-based counters are also considered in detail.

The important applications of shift registers are for use as sequence generators. Three circuits, PR-sequence generators, ring counters and Johnson counters, are widely used in digital systems. The general scheme of a sequence generator is an n-stage LFSR in which the outputs of all D-type flip-flops of the shift register are combined through a combinational logic circuit and the result is then fed back into the input of the shift register. An LFSR is called a PR-sequence generator if it generates a maximum-length sequence, namely, it has a period of $2^n - 1$, excluding the case of all 0s, where n is the number of flip-flops. Ring counters and Johnson counters are the two basic applications of LFSR-based sequence generators.

A timing generator is a device that generates the timing required for specific applications. Two timing generators are widely used in designing most digital systems. These are multiphase clock signal generators and digital monostable circuits. A multiphase clock generator can be constructed from a ring counter, a Johnson counter or a binary counter with a decoder. For a typical multiphase clock signal, each clock phase lasts for a clock cycle and is repeated after a given clock cycles. Indeed, an n-phase clock signal generator outputs a 1-out-of-n code at every clock cycle.

The digital monostable circuit can be either retriggerable or non-retriggerable. A retriggerable monostable is a circuit that can be retriggered before it ends its previous transient state and restarts another transient state. A non-retriggerable monostable circuit is a circuit that ignores any triggered signal before it completes its current transient state. In a digital monostable circuit, when the trigger signal is active each time, the output rises to high for a given number of clock cycles.

REFERENCES

1. J. Bhasker, *A Verilog HDL Primer,* 3rd Edn, Star Galaxy Publishing, 2005.
2. M.D. Ciletti, *Modeling, Synthesis and Rapid Prototyping with the Verilog HDL,* Prentice-Hall, Upper Saddle River, NJ, USA, 1999.
3. J. F. Wakerly, *Digital Design Principles and Practices,* 3rd Edn, Prentice-Hall, Upper Saddle River, NJ, USA, 2001.
4. M.-B. Lin, *Digital System Design: Principles, Practices and ASIC Realization,* 3rd Edn, Chuan Hwa Book Company, Taipei, Taiwan, 2002.

PROBLEMS

9.1 Design a D-type flip-flop with both asynchronous clear and preset.

9.2 Design a D-type flip-flop with both synchronous clear and preset.

9.3 Consider the simple logic circuit shown in Figure 9.33.

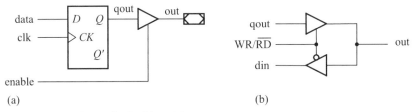

(a)   (b)

FIGURE 9.33 Figure for Problem 9.3

  (a) Write a Verilog HDL module to describe the circuit shown in Figure 9.33(a).

  (b) Replace the output tristate buffer of the circuit shown in Figure 9.33(a) with the bidirectional buffer shown in Figure 9.33(b) and re-do (a).

9.4 Suppose that $f = 50$ MHz, $a = 100$ kHz and a 74LS74 D-type flip-flop is used. The setup time of the flip-flop is 10 ns.

  (a) Calculate the $MTBF$.

  (b) What is the maximum operating frequency allowed so that the $MTBF$ can be greater than 100 years?

9.5 Suppose that $f = 50$ MHz, $a = 200$ kHz and a 74ALS74 D-type flip-flop is used. The setup time of the flip-flop is 10 ns.

  (a) Calculate the $MTBF$.

  (b) Calculate the $MTBF$ when the frequency-divided synchronizer with $N = 2$, as shown in Figure 9.9(b), is used.

9.6 Design a circuit that can generate a pulse with a width of 10 ms when a switch is pressed each time. You will have to take care of the switching-bounce problem.

9.7 Design a circuit that can produce a one-cycle pulse whenever the input x is set to high, despite how long it remains in the high state.

9.8 Write a test bench to verify the module `sram_timing_check` described in Section 9.2.4.

9.9 Suppose that a special memory with the access features shown in Figure 9.34 is required.

  (a) Model the special memory in behavioral style.

  (b) Write a test bench to verify the functionality of your module.

9.10 Consider the shift register module shown in the following:

```
module shift_register(clk, sin, qout);
input  clk;
```

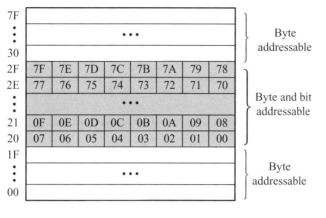

FIGURE 9.34 The layout of the special RAM under consideration in Problem 9.9

```
input  sin;  // serial data input
output reg [3:0] qout;
// the body of a 4-bit shift register
always @(posedge clk)
   begin       // using blocking assignments
      qout[0] = sin;
      qout[1] = qout[0];
      qout[2] = qout[1];
      qout[3] = qout[2];
   end
endmodule
```

   (a) Can it be operated as a 4-bit shift-left register?

   (b) Synthesize it on your own system and examine the result.

   (c) Explain the synthesized result.

   (d) Correct the statements within the module so that it can function correctly.

9.11 Design an n-bit program counter (PC) that has the following functions:

   (a) It is cleared to 0 when the asynchronous reset is asserted.

   (b) It is loaded a new value in parallel when the PCload is asserted.

   (c) It is incremented by 1 when the PCinc is asserted.

9.12 Suppose that a 4-bit programmable counter is required. The counter can be set to modulo n, where n can be set externally from 0 to 15 by the control signals $C3$ to $C0$. Write an RTL code in Verilog HDL to describe this counter and write a test bench to verify it.

9.13 Suppose that an 8-bit programmable counter is required. The counter can be set to modulo $4n$, where n can be set externally from 1 to 15 by the control signals $C3$ to $C0$. Write an RTL code in Verilog HDL to describe this counter and write a test bench to verify it.

9.14 Consider the CRC-16 circuit shown in Figure 9.35, which is found in a commercial device.

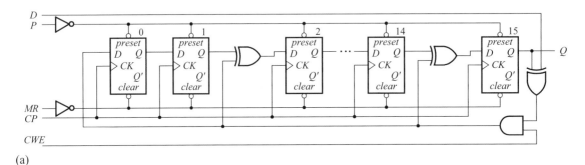

(a)

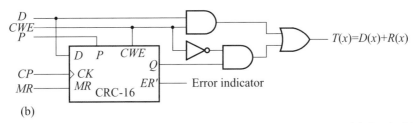

(b)

FIGURE 9.35 The block diagrams of a CRC-16 generator/detector: (a) circuit; (b) application (Problem 9.14)

(a) Write a Verilog HDL code to describe the circuit shown in Figure 9.35(a). Write a test bench to verify the function of it.

(b) Write a Verilog HDL code to describe the circuit shown in Figure 9.35(b). Write a test bench to verify the function of it.

9.15 Consider the general paradigms shown in Figures 9.24(a) and (b), respectively.

(a) Use $G(x) = x^4 + x + 1$ and message $M(x) = 1101001110$ to show that both paradigms generate the same CRC code.

(b) What are the differences between these two paradigms?

(c) If the CRC code must follow immediately after the message, give an approach to accomplishing this by using the paradigm shown in Figure 9.24(a).

9.16 Consider the general paradigms shown in Figure 9.24(a).

(a) Write a parameterized CRC generator/detector according to the paradigm shown in Figure 9.24(a) by using a generate-loop statement. Write a test bench to verify the functionality of the resulting module.

(b) Write a parameterized CRC generator/detector according to the paradigm shown in Figure 9.24(a) without using a generate-loop statement. Write a test bench to verify the functionality of the resulting module.

9.17 Implement the following three CRC generators and checkers:

(a) CRC-8 $= x^8 + x^7 + x^6 + x^4 + x^2 + 1$

(b) CRC-12 $= x^{12} + x^{11} + x^3 + x^2 + x + 1$

(c) CRC-16 $= x^{16} + x^{15} + x^2 + 1$

9.18 Implement the following two CRC generators and checkers:

(a) CRC-32a $= x^{32} + x^{30} + x^{22} + x^{15} + x^{12} + x^{11} + x^7 + x^6 + x^5 + x + 1$

(b) CRC-32b $= x^{32} + x^{26} + x^{23} + x^{22} + x^{16} + x^{12} + x^{11} + x^{10} + x^8 + x^7 + x^5 + x^4 + x^2 + x + 1$

9.19 Show that the 4-bit standard ring counter shown in Figure 9.26 is a self-correcting ring counter when its $d[0]$ function is replaced with the following switching function:

$$d[0] = (qout[0] + qout[1] + qout[2])'$$

In other words, once the ring counter enters into any invalid-state loop, it only needs to take at most four clock cycles to go back to its normal counting loop.

9.20 Show that the 4-bit Johnson counter shown in Figure 9.28 is a self-correcting Johnson counter when its $d[2]$ function is replaced with the following switching function:

$$d[2] = (qout[0] + qout[2])qout[1]$$

9.21 Using the binary counter with a decoder approach, design a timing generator with the following characteristics:

(a) The timing generator generates an eight-phase timing signal, T_i, where $0 \le i \le 7$.

(b) Each phase sustains one clock cycle except for phases T_3 and T_5.

(c) Phase T_3 sustains two clock cycles and phase T_5 sustains four clock cycles.

9.22 Design a retriggerable digital monostable circuit that has the following features:

(a) Both maximum pulse and actual pulse widths can be specified as module parameters.

(b) The output of the monostable circuit is cleared whenever the `reset_n` signal is asserted.

(c) The pulse output signal `out` goes to high for a specified clock cycles when the `start` signal is activated each time.

DESIGN OPTIONS OF DIGITAL SYSTEMS

IN THIS chapter, we examine various design options of digital systems with gate counts ranging from many tenths to several millions. These options include application-specific integrated circuits (ASICs) and field-programmable devices. ASICs are devices that must be fabricated in IC foundries and can be designed with one of the following: full-custom, cell-based and gate-array-based approaches. Field-programmable devices are the ones that can be personalized in laboratories and include programmable logic devices (PLDs), complex PLDs (CPLDs) and field-programmable gate arrays (FPGAs).

The full-custom design starts from scratch, i.e. from zero to the final product. The cell-based design incorporates a standard cell library into a synthesizable design flow so as to reduce the time to market. The gate-array-based approach is also a synthesizable design flow but targets to a prefabricated wafer known as gate arrays, in which only metal masks are left to be defined. Hence, it takes the advantages of cell-based design and speeds up the prototyping relative to full-custom design.

The PLDs are devices built on the two-level AND–OR logic structure and include three types: read-only memory (ROM), programmable logic array (PLA) and programmable array logic (PAL). CPLDs are devices that embody many PALs onto the same chip and interconnects them through a programmable interconnection network. FPGAs use configurable logic blocks (CLBs) or logic elements (LEs) to implement switching functions and can be divided into two structures, matrix type and row type, according to the arrangement of CLBs and LEs on the chip.

For a system composed of a variety of devices with different logic levels and power-supply voltages, two most important issues related to the interface between devices must be taken into account: I/O standards and voltage tolerance. An I/O standard defines a set of electrical rules that connect two devices so as to transfer information through electrical signals properly. A device is said to be voltage-tolerant if it can withstand a voltage greater than its V_{DD} on its I/O pins.

Digital System Designs and Practices   Ming-Bo Lin
© 2008 John Wiley & Sons (Asia) Pte Ltd

10.1   DESIGN OPTIONS OF DIGITAL SYSTEMS

Currently, ASICs and field-programmable devices are ubiquitous in digital systems. Although from the acronym of the word "ASIC", which is application specific integrated circuit, any integrated circuit not in a standard catalog may be called an ASIC; namely, it is a customized integrated circuit and may also include all field-programmable devices. However, in industry the word "ASIC" is usually reserved for the devices that must be produced directly from IC foundries only. In this section, we examine a variety of available options that can be used to design a digital system.

10.1.1   Hierarchical System Design

Nowadays digital system designers widely use a *hierarchical structure*, also known as a *modular design* technique, to design a digital system. The essence of a hierarchical structure is to partition the system into many smaller independent sub-structures which combine to perform the same functionality of the original system.

Basically, the design hierarchy of digital systems can be classified into four levels: system level, register transfer level, logic gate level and circuit level. The conceptual differences of these four design levels are shown in Figure 10.1.

System Level   The system level is the topmost level. In this level, the data units processed by digital systems are usually bytes or block of bytes. The processing

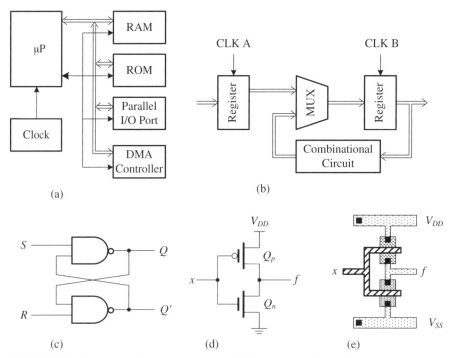

FIGURE 10.1 The design hierarchy of general digital systems: (a) system level; (b) register-transfer level; (c) logic gate level; (d) circuit level; (e) physical level

time is in the range of ms and μs. The logic devices used in this level are microprocessors/microcomputers (μPs/μCs), memory devices and peripherals such as timers, general-purpose input and output (GPIO), universal asynchronous receiver and transmitter (UART), universal serial bus (USB) controller, inter-integrated circuit (IIC) controller, and so on. In addition, a real-time operating system and C programming language are usually employed as tool chains to write the desired application softwares.

Due to the highly developed manufacturing techniques of integrated circuits, nowadays system-level design can be built on printed-circuit board (PCB) or silicon. PCB-based design consists of discrete standard components such as μPs/μCs, peripherals and memory devices. The features of this design are as follows. The fixed cost is low but the final product may cost too much to be accepted due to too many devices required. In addition, the actual size and power consumption of the products are large.

Silicon-based design can be accomplished by either system on a programmable chip (SoPC) or cell-based platform IP. SoPC is also called a *programmable system chip* (PSC). SoPC-based design consists of system cells, such as MicroBlaze and NIOS—both are 32-bit RISC CPUs—and their associated peripherals, much like the ones used in PCB-based design. The entire system can be implemented on the same FPGA chip. SoPCs will be increasingly used for designing most digital systems in the near future. Cell-based design comprises soft cells (synthesizable modules) that can be configured into an optimized system hardware. An example is Tensilica's Xtensa. For the time being, this approach is most applicable to the fields of multimedia.

Register Transfer Level   The RTL operates on bits or bytes on a time scale of 10^{-8} to 10^{-9} s. An RTL design consists of combinational logic modules and sequential logic modules. The combinational logic modules include decoders, encoders, multiplexers, demultiplexers, arithmetic circuits and magnitude comparators, among others. The sequential logic modules include registers, shift registers, counters, sequential logic modules and so on.

Gate Level   Gate level operates on bits on a time scale of 10^{-9} to 10^{-11} s. A gate-level design uses basic logic gates, flip-flops and latches to construct an RTL module.

Circuit Level   Circuit level operates on bits on a time scale of 10^{-10} to 10^{-12} s. A circuit-level design consists of MOSFETs, BJTs and MESFETs. The *physical level* is just another view of circuit level.

Review Questions

Q10.1   Describe the concept of hierarchy.

Q10.2   What are the four levels of design hierarchy?

Q10.3   What are the three major system-level design options?

Q10.4   Give the major features of SoPC-based platforms.

Q10.5   What are the major features of platform-based design?

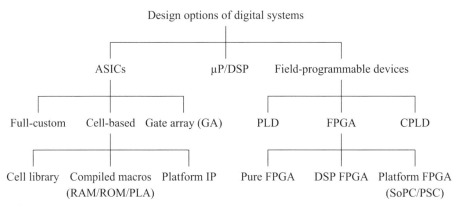

FIGURE 10.2 Design options of digital systems

10.1.2 Design Options of Digital Systems

With the maturity of very large scale integrated circuit (VLSI) technology, a variety of new options are available for designing a given digital system. These options are not only to reduce the hardware size but also to reduce the power consumption and increase the system performance considerably. As a result, by using these options, the total cost of the system is reduced and the performance of the system is increased dramatically.

Roughly speaking, the design options of digital systems can be divided into three classes: ASICs, μP/DSP and field-programmable devices, as shown in Figure 10.2.

The ASIC class contains *full-custom design*, *cell-based design* and *gate-array-based approach*. In the full-custom and cell-based approaches, the designer or design team must provide all masks required for the chip to be manufactured. The major difference between full-custom and cell-based approaches is that the former creates the design from scratch while the latter uses a standard cell library, in which the layout of each cell has been completed. In the gate-array-based approach, the wafer or chip has been processed partially by the IC foundry up to the point in which only the interconnections between cells or transistors are left for the designers to define. Hence, it is often called a *semi-custom approach*. Using gate arrays to implement a design, the designer only needs to specify the required interconnections via CAD tools so as to define or commit the specified function of the chip.

Field-programmable devices are the ones that can be personalized in laboratories and include PLDs, CPLDs and FPGAs. PLDs are devices built on two-level AND–OR logic structures and contain three types: ROM, PLA and PAL. It is worth noting that gate arrays are mask-programmable while FPGA are field-programmable. Hence, sometimes gate arrays are also known as mask-programmable gate arrays (MPGAs).

The μP/DSP approach uses general-purpose microprocessors or microcomputers, or digital-signal processing processor(s), or their combination(s) as basic components, combining with other devices such as peripherals or memories, to design the required system. The μP/DSP approach can be based on one of the following three options: *standard ICs* on a printed-circuit board (PCB), *platform-based design* and *system on a programmable chip* (SoPC).

Both platform-based and SoPC-based designs use system-level cells, such as μP/DSP, memories and peripherals, to construct the required system, much like the manner that builds a system on a printed-circuit board using standard ICs. The system cells used in both platform-based and SoPC-based designs are known as *intellectual property* (IP). In general, an intellectual property is a predesigned component that can be used in a larger design. IP can be classified into two types: *hard IP* and *soft IP*. Hard IP comes as a predesigned layout. Its block's size, performance and power consumption can be accurately measured. Soft IP comes as a synthesizable module in a hardware description language, Verilog HDL or VHDL. It has features of being flexible to new technologies but harder to characterize, larger and slower.

The essential difference between platform-based design and SoPC-based design is that the former uses a cell-based approach to build its system and the latter uses FPGA. Both designs use computer(s) as components to build systems, thereby being called *embedded systems*. In addition, these systems are founded on silicon regardless of using a cell library or FPGA devices. As a consequence, they are also called *systems-on-a-chip* (SoC). In short, an SoC is an embedded system on silicon. SoPC is also called *platform FPGA*. The design issues of both platform-based and SoPC-based approaches involve a system-level hardware and software codesign and are beyond the scope of this book.

Figure 10.3 compares the features of various digital system design options in terms of time to market and cost. The time to market and cost are represented as functions of design flexibility, (integration) density, speed, complexity, process complexity and non-recurring engineering (NRE) cost. The time to market is the time elapsed from when the design specification is produced to the point that the final product goes to the market. Roughly speaking, the NRE cost is the fixed cost needed to develop the ASIC or the field-programmable device. The fixed cost includes engineer salaries, training fees, software costs and so on. The NRE cost is amortized over all units of a given product. Consequently, for products with high NRE costs, the expected volume sold of the product should be large enough in order to lower the end-price below a point that can be acceptable by users.

Because the design flexibility, density, speed, complexity, process complexity, NRE and time to market are the highest of all design options, the full-custom design is only aimed at high-performance, high-speed and high-volume products such as CPU, memories and field-programmable devices such as PLDs and CPLDs, as well as FPGAs.

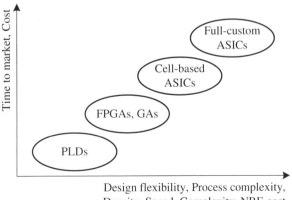

FIGURE 10.3 Comparison of various design options in terms of time to market and cost

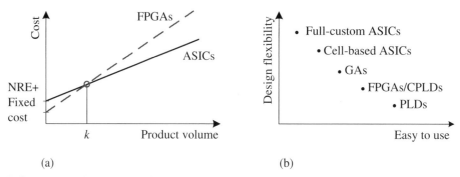

FIGURE 10.4 Comparison of various design options in terms of cost and design flexibility: (a) cost versus product volume; (b) comparison of various digital system design options

The advantages of using ASICs are as follows. Both the size and power consumption of a system are reduced. The reliability and performance of the system are increased due to the decreased number of devices used in the system.

Figure 10.4 shows the comparison of various design options of digital systems in terms of cost and design flexibility, respectively. In general, the cost of a product is a combination of NRE cost, recurring cost and fixed cost. The NRE cost includes the engineering cost and the cost of prototyping the product. Although it is only needed once during the entire life time of the given product, it is often the highest one among all three costs. The NRE cost is independent of the product volume. Recurring cost is the part that needs to be paid each time the product is produced. The amount of recurring cost is proportional to the product volume. The fixed cost is the expenses for marketing the product.

Generally speaking, the cost of a product using a given design option can be represented as a function of the product volume, as shown in Figure 10.4(a), where the intercept with the vertical coordinate is the NRE plus fixed cost of the product. The slope of the curve is the recurring cost of the product. In general, field-programming devices have a lower NRE plus fixed cost but a higher recurring cost than ASICs. For example, two examples shown in the figure are FPGAs and ASICs, respectively. The slope of FPGAs is greater than that of ASICs. Due to the lower NRE plus fixed cost of FPGAs, there must exist a crosspoint, denoted as a k point, which represents the division of product volume. Above it, the cost of the product using FPGAs is higher than using ASICs; below it, the cost of the product using FPGAs is lower than using ASICs. The same idea can be applied equally well to compare the other design options.

Figure 10.4(b) compares the relative design flexibility with the ease to use of various design options. Here, the *design flexibility* means the capability of designing a circuit with an arbitrary function. The easier to use, the less design flexibility. For example, PLDs are easiest to use but most lack design flexibility. The full-custom design is very difficult to use but has the highest design flexibility.

Review Questions

Q10.6　Distinguish the features of ASICs and field-programmable devices.

Q10.7　What are the three design options of digital systems?

Q10.8 What is the meaning of platform-based design?

Q10.9 What is the meaning of SoPC-based design?

Q10.10 What are the advantages of using ASICs to design digital systems?

Q10.11 Distinguish the features of gate arrays and FPGAs.

Q10.12 Describe the meaning of the k point in Figure 10.4(a).

10.1.3 ASIC Designs

As mentioned previously, an ASIC can be created by either full-custom, cell-based or gate-array-based approach. In this section, we will describe each of these in more detail.

10.1.3.1 Full-Custom Design   In a full-custom design, each transistor and its layout are designed carefully by the designer in order to achieve the best performance. However, although it has a very much better performance, the full-custom design has very low throughput because it needs much more time to create an ASIC. In addition, it needs much more time than other design options to prototype a design. Furthermore, due to the shorter life time of electronic products nowadays, the design time and time to market of the products need to be cut accordingly. As a result, it is apparent that using the full-custom approach cannot already satisfy those applications needed to be prototyped rapidly. To solve this problem, cell-based design, gate-array-based approach and the approach based-on field-programmable devices are increasingly popular in the market.

In summary, due to the much shorter life time of electronic products than before and increasing NRE cost of full-custom design, nowadays, full-custom design is only used when high-volume products such as CPU, memories and gate arrays, as well as field-programmable devices, are required. The rationale behind this is that the high NRE cost can be amortized over many units so that the price of each unit is low enough for realizing a cost-effective product. Please refer back to Figure 10.4(a) for gaining a more insightful explanation of this claim.

10.1.3.2 Cell-Based Design   The essence of cell-based design is that a design is composed of a set of predefined cells with layouts known as *standard cells*, as shown in Table 10.1. The layout masks of these cells have already been done in advance. The

TABLE 10.1 The basic cell types of a typical standard cell library

Standard cell types	Variations
Inverter/buffer/tristate buffer	1X, 2X, 4X, 8X, 16X
NAND/AND gate	2 ~ 8 inputs
NOR/OR gate	2 ~ 8 inputs
XOR/XNOR gate	2 ~ 8 inputs
MUX/DeMUX	2 ~ 8 inputs (inverted/noninverted output)
Encoder/Decoder	4 ~ 16 inputs (inverted/noninverted output)
Schmitt trigger circuit	Inverted/noninverted output
Latch/register/counter	D/JK (sync./async. clear/reset)
I/O pad circuits	Input/Output (tristate/bidirectional)

typical standard cell library usually includes the following:

1. *Basic logic gates*: NAND, NOR, XOR, AOI, OAI, AND, OR, buffers and inverters.

2. *Basic combinational logic modules*: decoders, encoders, priority encoders, multiplexers, demultiplexers, parity checkers, adders, subtracters and shifters.

3. *Basic sequential logic modules*: D-type flip-flops, registers, counters, timing generators and memories (ROM, RAM).

4. *System building blocks*: multipliers, arithmetic and logic unit (ALU), center-processing unit (CPU), universal asynchronous receiver and transmitter (UART), inter-integrated circuit (IIC) controller, general-purpose input and output (GPIO), universal serial bus (USB) controller and so on.

In general, cells in a standard cell library have a fixed height with power and ground separately routed at the top and bottom of cells in order to allow the cells to be abutted end-to-end and to have the power supply rails connected. The width is allowed to be variable so as to accommodate different functions. For instance, the NOT gate has only two transistors but a 2-input NAND gate has four transistors. As a result, they must have different widths when their heights are set to be equal. Figure 10.5 shows an example of using the above mentioned NAND and NOT gates to construct a D-type flip-flop. All gates have the same height (note that we have rotated the figure through 90°) and hence they can be abutted together. Both the left and right parts contain four NAND gates and one NOT gate. The left part constitutes the master latch of the D-type flip-flop and the right part constitutes the slave latch. Between them is the routing channel used for interconnecting NAND gates and NOT gates. For the multi-layer metalization process, the routing channel is virtually no longer needed.

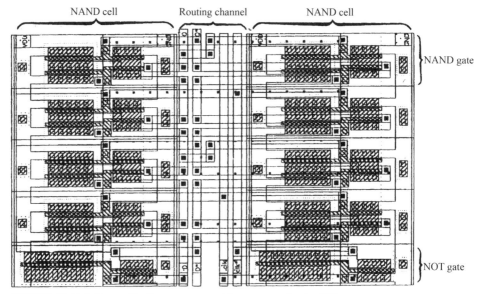

FIGURE 10.5 A D-type flip-flop constructed from a standard cell library

Compared to the full-custom approach, the cell-based approach uses predesigned cells with layouts. The size of each cell in a typical standard cell library is much larger than the one created by the full-custom approach. Therefore, an ASIC created by using a standard cell library often needs much more area and performs worse than the one created by the full-custom approach. Nonetheless, the cell-based approach can be directly embedded into the HDL design flow and thus increases the productivity of ASIC design. Therefore, the main advantage of using a standard cell library to design a system is to improve the productivity. Like the full-custom design, the prototype designed with the cell-based approach still needs to be fabricated in IC foundries.

10.1.3.3 *Gate-Array-Based Approach*   Gate arrays (GAs), also known as *uncommitted logic arrays* (ULAs), are a type of semi-custom designs of ASICs. Here, the semi-custom means that only a partial set of masks needs to be processed in IC foundries after logic functions have been committed. The basic element of gate arrays can be either NOR gates or NAND gates in CMOS technology. A particular sub-class of gate arrays is known as *sea-of-gates* (SoGs). Sea-of-gates differ from gate arrays in that the array of transistors is continuous rather than segmented.

A typical structure of CMOS gate arrays is shown in Figure 10.6, which consists of a set of pairs of CMOS FETs and routing channels. When the rows of nMOS and pMOS FETs are broken into segments with each having two or three FETs, such as the one shown in Figure 10.7, the result is the gate-array structure; otherwise, it is the sea-of-gates structure. The features of sea-of-gates are that they utilize multiple-layer metalizations and remove the routing channels. Hence, they have a higher density than gate arrays.

The basic circuit structure of gate arrays is shown in Figure 10.7, where two segments, one with two transistors and the other with three transistors, along with their equivalent circuits are depicted. Because a MOS (pMOS or nMOS) transistor has the symmetric feature that its drain and source can be interchanged without deteriorating its performance and a CMOS logic gate has the feature that the gate of an nMOS transistor must be connected to the gate of a pMOS transistor, the gates of both nMOS and pMOS

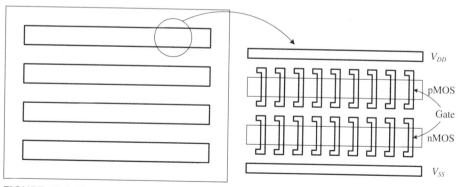

FIGURE 10.6 The basic structure of gate arrays and sea-of-gates

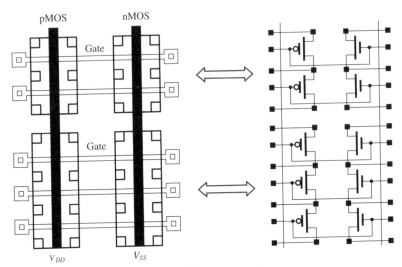

FIGURE 10.7 The basic structure of gate-array cells

are connected in pairs like the one shown in Figure 10.7 or placed closely adjacent like the one shown in Figure 10.6. In addition, all transistors are intimately placed to increase the density of the transistors.

When using the basic device to design logic gates, it is only required to connect the drain to the power supply V_{DD}, connect the source to ground GND or connect them together. The following example demonstrates how to apply this idea to separately construct a NOT gate and a NAND gate from a gate array.

EXAMPLE 10.1 *NOT Gate and NAND Gate*

Figure 10.8 is an example of using the gate array shown in Figure 10.7 to construct a NOT gate and a NAND gate.

To construct a NOT gate, one pMOS and one nMOS transistors are required. The gates of both transistors are connected together to serve as the input as they have already been done. The drains of both transistors are connected together to serve as the output. The source of nMOS is connected to V_{SS} while the source of pMOS connected to V_{DD}.

The NAND gate is constructed in a similar way to that depicted in Figure 10.8. Both gates of pMOS and nMOS transistors are connected together in pairs and serve as inputs, denoted as a and b, respectively. Two nMOS transistors are connected in series whereas two pMOS are connected in parallel, as shown in Figure 10.8. ∎

Like full-custom design, it is quite clumsy and time-consuming to start a design of an ASIC from scratch when using gate arrays. Similar to the cell-based approach, macro libraries are usually provided by the vendors to save design time. The typical macro library associated with gate arrays is similar to the standard cell library with only one major difference that it contains only the metalization masks because the transistors on gate arrays are already fabricated in advance.

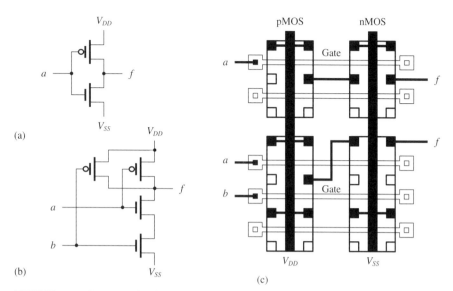

FIGURE 10.8 An example of interconnected gate-array cells: (a) NOT gate; (b) NAND gate; (c) gate-array interconnection

Like cell-based design, using gate arrays to design an ASIC also faces the same problem that some overheads of chip area are required. These overheads are due to the unused logic components associated with macros, which cannot be removed from the macro – for example, when only an 8-bit adder macro is available for designing a 6-bit adder. The area overhead is 25 % because the highest or the lowest two-bit adders are left unused but still occupy the chip area. In full-custom design, this overhead can be removed without any penalty. However, using gate arrays can save the time to market since like cell-based design it can be embedded into modern widely used HDL design flows.

In gate arrays, four to six masks are standard, independent of the final functions committed to the gate arrays. Consequently, these masks can be fabricated in advance. Later, the personalization of gate arrays is defined by the metalization masks, which are determined by the required switching functions of an ASIC. In contrast, full-custom approach needs to create every mask required in an ASIC and process-related operations. As a result, using gate arrays needs much less time to complete an ASIC design and prototyping.

Compared to cell-based approach, gate arrays have less functionality per unit area due to the overheads of interconnection and those unused prefabricated transistors. However, unlike cell-based approach which needs to process every mask, gate arrays only need to process the final metalization masks. Thus, they take less time and less cost to complete an ASIC.

Since all full-custom, cell-based and gate-array approaches need the design to be fabricated in an IC foundry, it requires a lot of time to prototype an ASIC. One technique to solve this difficulty is by using field-programmable devices, which will be described in more detail in the following sub-section.

Review Questions

Q10.13 Which approaches can be used to create an ASIC?

Q10.14 What are the features of full-custom design?

Q10.15 What are the features of cell-based design?

Q10.16 What are the features of gate arrays?

Q10.17 Define routing channel.

Q10.18 Describe the meaning of uncommitted logic array (ULA).

Q10.19 Describe the meaning of SoG.

10.1.4 Design with Field-Programmable Devices

Field-programmable devices can be divided into two types, PLD/CPLD and FPGA, according to their structures. The common feature of PLDs and CPLDs is that they use two-level AND–OR logic structure to realize (or implement) switching functions. FPGA devices combine the features of gate arrays and the on-site programmability of PLDs and CPLDs.

The programmable options of PLDs/CPLDs and FPGAs can be either *mask-programmable* or *field-programmable*. Like gate arrays, for mask-programmable devices the designer needs to provide the vendors with the designed interconnection pattern of the given device for preparing the required mask to fabricate the final ASIC. Field-programmable devices can be personalized by the designers through using appropriate *programming equipment*, also known as *programmer*. Field-programmable devices can be further divided into two types: *one-time programmable* (OTP) and *erasable*. OTP devices can only be programmed one time; erasable devices can be reprogrammed as many times as required. In summary, field-programmable devices are usually used at the start-up time of a design to gain the flexibility or in low-volume production to save the NRE cost and mask-programmable devices are used in high-volume production to reduce the cost.

10.1.4.1 *Programmable Logic Devices*   When designing a digital system, a PLD can replace many small-scale integration (SSI) and/or medium-scale integration (MSI) devices. Consequently, using PLDs to design digital systems can considerably reduce the number of interconnections, the area of the printed circuit board (PCB) and the number of connectors. The hardware cost of the resulting system is then reduced profoundly.

For convenience of the following discussion, a shorthand notation is often used to represent the PLDs, as shown in Figure 10.9. Since the current underlying technology is CMOS, the "fuse" shown in the figure is a floating-gate electrical erasable programmable read only memory (EEPROM) device or an antifuse device when the PLD is field-programmable. In mask-programmable devices, there is a MOS transistor if a "fuse" is presented; otherwise, there is no MOS transistor.

As mentioned before, PLDs can be divided into the following three categories: PLA, ROM and PAL. All of these three devices have the similar two-level AND–OR logic structure. The essential differences between these devices are the programmability

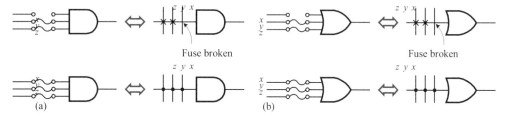

FIGURE 10.9 The shorthand notations used to describe the structures of PLDs: (a) AND gate; (b) OR gate

of AND and OR arrays, as shown in Figure 10.10. For ROM devices, the AND array generates all minterms of the inputs and hence is fixed but the OR array is programmable in order to implement the required functions. For PLA devices, both AND and OR arrays are programmable. For PAL devices, the AND array is programmable but the OR array is fixed. Nowadays, the discrete PLA devices has become obsolete. However, PLA structures accompanied with ROMs are often used in full-custom and cell-based designs to take advantage of their regular structures. Both PAL and ROM not only have commercial discrete devices but are also widely used in digital systems.

10.1.4.2 *Programmable Interconnection*   The programmable interconnection structures of field-programmable devices can be classified into the following three types: static RAM (SRAM), EEPROM (flash cell) and antifuse, as shown in Figure 10.11. The interconnection structure based on SRAM is a pass transistor or a transmission gate controlled by an SRAM cell. The basic structure of an SRAM cell is a bistable circuit, as shown in Figure 10.11(a). Once programmed, the SRAM circuit retains its state until it is reprogrammed or the power applied to it is removed. The EEPROM cell is like the cell used in flash memory devices, as shown in Figure 10.11(b).

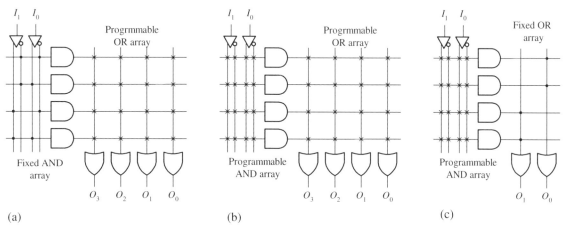

FIGURE 10.10 The basic structures of PLDs: (a) ROM structure; (b) PLA structure; (c) PAL structure

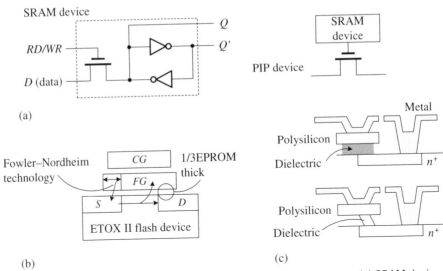

FIGURE 10.11 The basic structures of programmable interconnections: (a) SRAM device; (b) flash device; (c) antifuse device

Its basic structure is a floating-gate transistor, which can be programmed to store "1" or "0". The antifuse is a device that operates in the reverse direction of a normal fuse; namely, it has high resistance in normal condition but is changed to low resistance permanently when an appropriate voltage is applied to it.

The basic characteristics of the above mentioned three types of programmable interconnection structures are summarized in Table 10.2. From this table, we can see that the antifuse structure has a better performance since it has a lower resistance and capacitance. The other two structures have almost the same performance. However, the SRAM structure is a volatile but EEPROM is not.

10.1.4.3 Complex Programmable Logic Devices
Due to the popularity of PAL devices, along with the maturity of VLSI technology, combining many PALs with a programmable interconnection structure into the same chip is feasible. Such a device is known as a complex PLD. The PAL used in a CPLD is called a *PAL macro* or a *macrocell* for short.

TABLE 10.2 The basic features of programmable interconnections

Feature	SRAM	Flash	Antifuse
Process technology	Standard CMOS	Standard 2-level polysilicon	New type polysilicon
Programming approach	Shift register	FAMOS	Avalanche
Cell area	Very large	Large	Small
Resistance	$\approx$2 kΩ	$\approx$2 kΩ	$\approx$500 kΩ
Capacitance	$\approx$50 fF	$\approx$50 fF	$\approx$10 fF

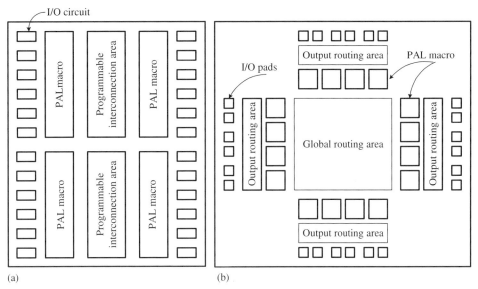

FIGURE 10.12 The basic structures of CPLDs: (a) CPLD basic structure; (b) pLSI basic structure

There are two basic types of CPLDs, which are classified according to the arrangement of PAL macros and the interconnection structure, as shown in Figures 10.12(a) and (b), respectively. The first type, as depicted in Figure 10.12(a), is most widely used in commercial CPLDs, where PAL macros are placed on both sides of the programmable interconnection. Another type of CPLD structure is depicted in Figure 10.12(b), where PAL macros are placed on all four sides and a programmable interconnection is placed at the center region called the *global routing region*. In addition, an output routing region is placed between the PAL macros and input/output blocks.

10.1.4.4 *Field-Programmable Gate Arrays*
The basic structures, known as *fabrics*, of FPGAs are composed of configurable logic blocks (CLBs) or logic elements (LEs) and interconnections, as well as input/output blocks (IOBs). The CLBs or LEs are usually used to implement combinational logic and sequential logic. Each CLB or LE is a k-input universal logic module, consisting of a look-up table or a multiplexer-tree, where k is usually set to 3 to 8. A circuit is known as a k-input *universal logic module* when it can implement any switching function with, at most, k variables. According to the arrangement of CLBs or LEs on the chip, the basic structures of FPGA devices can be divided into two types: *matrix* and *row vector*. The matrix type of FPGA is shown in Figure 10.13(a), where CLBs are placed in a two-dimensional matrix manner. Between the CLBs there are two types of interconnections, called *horizontal routing channels* and *vertical routing channels*, respectively. Figure 10.13(b) is a row-type FPGA, where the CLBs or LEs are intimately placed in a row-vector manner. The spaces between them are the routing channels.

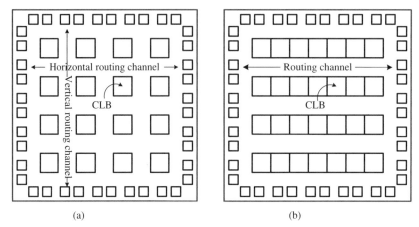

FIGURE 10.13 The basic structures of FPGAs: (a) matrix type; (b) row type

Nowadays, many FPGAs have been incorporated into specialized features for specific applications, including communications, multimedia or consumer products. These features include carry-chain adders, multipliers, shifters, block and distributed RAMs, even powerful 32-bit CPUs such as PowerPC and ARM CPU, as well as peripherals such as the Ethernet controller and USB controller.

Review Questions

Q10.20   What types can field-programmable devices be classified into?

Q10.21   What is the meaning of OTP devices?

Q10.22   Describe the basic structure of PLDs.

Q10.23   Distinguish the differences between the three types of PLDs.

Q10.24   What are the three types of programmable interconnections (PICs) commonly used in field-programmable devices?

Q10.25   Describe the feature of antifuse mechanism.

Q10.26   Describe the basic structures of CPLD devices.

Q10.27   Describe the basic structures of FPGA devices.

10.2 PLD MODELING

As described previously, PLDs are built-on the two-level AND–OR logic structure and include three types of structures, according to the programmability of AND and OR arrays. In this section, we first discuss these structures and their related devices in more detail and then use PLAs as examples to illustrate how to model these structures in Verilog HDL.

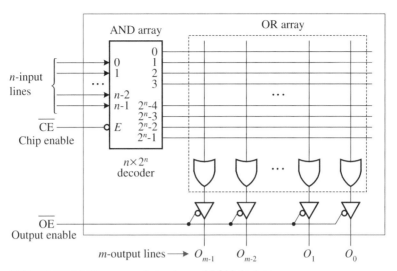

FIGURE 10.14 The general structure of ROM devices

10.2.1 ROM

As mentioned previously, the basic structure of ROM is a two-level AND–OR logic structure. A $2^n \times m$ ROM is an LSI or VLSI device that includes an $n \times 2^n$ (full) decoder and an array of m programmable 2^n-input OR gates, as shown in Figure 10.14. Each combination of input lines or variables is called an *address*. For n-input variables, there are 2^n combinations and hence addresses. Each combination of output lines is called a *word*. For a ROM with n input lines, there are 2^n words in total because there are 2^n combinations of input lines. The number of bits, namely, $2^n \times m$, contained in a ROM device is known as *capacity*. It in general uses the form $2^n \times m$ to specify the capacity of a ROM device, such as 16×4 ROM and 256×8 ROM. For commercial ROM devices, there are two control signals: chip enable ($\overline{CE}$) and output enable ($\overline{OE}$). The chip enable allows multiple ROM devices to be connected to a common bus where each is selectable by its unique address. The output enable controls the operations of the output tristate buffers. These buffers are used to either pass values of the OR array to the outputs or place the outputs to a high-impedance, depending on whether the output enable is asserted or not.

The operation of writing data into a ROM device is named as *program*. According to the programming mechanisms of memory cells, ROM devices can be divided into three types: *mask ROM*, *programmable ROM* (PROM) and *erasable programmable ROM* (EPROM). The contents of mask ROM are set by the vendor according to the pattern provided by the designer and cannot be changed again after the device is produced. A programmable ROM is an OTP device; namely, it can only be programmed one time. An erasable programmable ROM can be reprogrammable as many time as required. Currently, most EPROM devices are designed with CMOS floating-gate technology. These EPROM devices are also known as EEPROM because their contents can be erased electrically.

Input		Output	
A_1	A_0	O_1	O_0
0	0	0	1
0	1	1	0
1	0	0	0
1	1	1	1

(a)

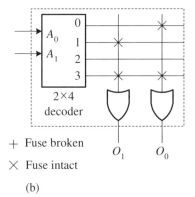

+ Fuse broken

$\times$ Fuse intact

(b)

FIGURE 10.15 An example of using a $2^2 \times 4$ ROM to implement switching functions: (A) truth table; (b) logic circuit

The more widely used EEPROM devices nowadays are known as *flash memory*. They can be used just like SRAM devices without concerning the programming process, which is automatically taken care of by an internal controller to generate all timing required for programming a specified block of cells.

Perhaps, the most common use of ROM in digital systems is to implement switching functions. As shown in Figure 10.14, the structure of ROM devices is a two-level AND–OR logic structure. The AND array can generate all minterms of inputs and the OR array can connect the required minterms to its output. In other words, it is the two-level logic circuit for multiple-output functions in standard SOP form. As a result, a $2^n \times m$ ROM can implement any m multiple-output functions with n input variables.

EXAMPLE 10.2   *ROM—Implementing Switching Functions*

Figure 10.15 shows an example of using a $2^2 \times 2$ ROM device to implement a multiple-output function. Due to the fact that the input AND array of ROM generates all minterms of input variables, it only needs to connect the minterms with value of 1 to the programmable OR array, as shown in Figure 10.15(b). It is instructive to compare both the truth table and the circuit shown in Figures 10.15(a) and (b), respectively. ∎

The following example illustrates how to model a ROM device without timing checks.

EXAMPLE 10.3   *Modeling a ROM Device*

In this example, we use a $2^2 \times 4$ ROM to implement the switching functions shown in Figure 10.15. To model a ROM device without timing checks, it only needs to declare a memory with the required number of words and the word width. In general, the contents of ROMs can be initialized by using procedural assignment statements or reading from a file through the use of system tasks: $readmemb or $readmemh. In this example, procedural assignment statements are used to initialize the contents of the ROM used to implement the switching functions. Using system tasks to fill the ROM is deferred to the PLA modeling sub-section.

```
module ROM_example(a1, a0, out1, out0);
input  a1, a0;
```

```
output out1, out0;
reg [1:0] mem[3:0];
// using procedural assignment statements to
// initialize the contents of ROM
initial begin
   mem[0] = 2'b01;
   mem[1] = 2'b10;
   mem[2] = 2'b00;
   mem[3] = 2'b11;
end
assign {out1, out0} = mem[{a1, a0}];
endmodule                                        ■
```

Review Questions

Q10.28   Describe the features of a ROM in terms of the programmability of the AND–OR logic structure.

Q10.29   Describe the following terms: address, program, capacity and word.

Q10.30   What are the types of ROM devices according to the programming mechanisms of memory cells?

Q10.31   Is the AND array of ROM devices fully or partially decoded the input?

Q10.32   How many functions can a $2^n \times m$ ROM implement?

10.2.2   PLA

The essential idea behind the PLA is based on the following observation that when using ROM to implement switching functions, a situation is often encountered that some minterms are unused, such as the minterm 2 in Figure 10.15. Consequently, these minterms are virtually wasted and can be removed. When this is the case, the AND array must be made programmable so that it only accommodates the desired minterms. The resulting structure makes both AND and OR arrays programmable. This new type of two-level AND–OR logic structure is known as *programmable logic array* (PLA).

For a typical $n \times k \times m$ PLA, as shown in Figure 10.16, there are n inputs and buffers/NOT gates, k AND gates and m OR gates, as well as m XOR gates. Between the inputs and the AND gates, there are $2n \times k$ programmable points, known as *fuses* for a historical reason that the programming points of earlier PLA devices were composed of fuses; between the AND gates and the OR gates, there are $k \times m$ programmable points. At the output XOR gates, there are m programming points, which are used to program the output as inverted or non-inverted.

The features of a PLA device are that both AND and OR arrays are programmable and all AND gates may be shared by all OR gates. As a result, when using a PLA to implement switching functions, a multiple-output minimization process is usually used to simplify the switching functions in order to obtain the shared product terms and hence reduce the required number of product terms. The following example illustrates how to implement a multiple-output switching function with a typical PLA. The multiple-output minimization process is beyond the scope of this book. The interested reader can refer to Lin [6].

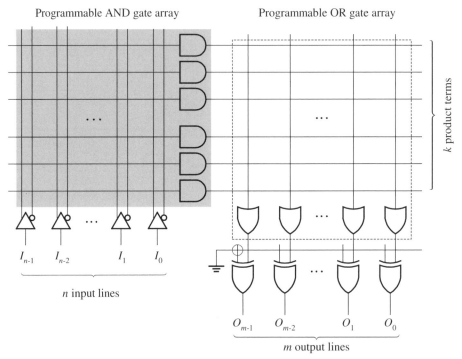

Programmable AND gate array Programmable OR gate array

k product terms

I_{n-1} I_{n-2} I_1 I_0

n input lines

O_{m-1} O_{m-2} O_1 O_0

m output lines

FIGURE 10.16 The general structure of PLA devices

EXAMPLE 10.4 *PLA—Implementing Switching Functions*

Figure 10.17 shows that three switching functions are implemented by using a PLA device. These three switching functions are as follows:

$$f_1(w, x, y, z) = \Sigma(2, 3, 5, 7, 8, 9, 10, 11, 13, 15)$$
$$f_2(w, x, y, z) = \Sigma(2, 3, 5, 6, 7, 10, 11, 14, 15)$$
$$f_3(w, x, y, z) = \Sigma(6, 7, 8, 9, 13, 14, 15) \tag{10.1}$$

Simplifying by using a multiple-output minimization process, such as the one introduced in Lin [6], we obtain the following simplified switching expressions:

$$f_1(w, x, y, z) = x'y + wx'y' + w'xz + wxz$$
$$f_2(w, x, y, z) = w'xz + y$$
$$f_3(w, x, y, z) = xy + wx'y' + wxz \tag{10.2}$$

These three switching functions can be represented as the PLA programming table shown in Figure 10.17(a). The product term $wx'y'$ and wxz are shared by functions f_1 and f_3, and $w'xz$ is shared by functions f_1 and f_2. There are six product terms, numbered from 0 to 5. The function f_1 consists of the product terms 2, 3, 4 and 5. The function f_2 consists of the product terms 0 and 4. The function f_3 consists of the product terms 1, 3 and 5. The logic circuit that realizes the programming table shown in Figure 10.17(a) is depicted in Figure 10.17(b). ■

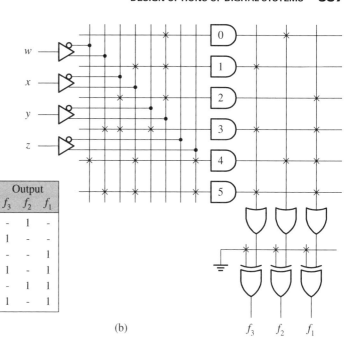

Product	Input				Output		
term	w	x	y	z	f_3	f_2	f_1
y 0	-	-	1	-	-	1	-
xy 1	-	1	1	-	1	-	-
$x'y$ 2	-	0	1	-	-	-	1
$wx'y'$ 3	1	0	0	-	1	-	1
$w'xz$ 4	0	1	-	1	-	1	1
wxz 5	1	1	-	1	1	-	1

(a)   (b)

FIGURE 10.17 An example to explain how a PLA is used to implement switching functions: (a) PLA programming table; (b) logic circuit

Review Questions

Q10.33 Describe the features of a PLA device in terms of the programmability of AND–OR logic structure.

Q10.34 How many programmable points can an $n \times k \times m$ PLA have?

Q10.35 Can a product term be shared among multiple switching functions in a PLA device?

Q10.36 Is the AND array of PLAs fully or partially decoded the input combinations?

10.2.3 PAL

It is shown that most switching functions in practical use contain less than five product terms. Consequently, it is unnecessary to allow all product terms to be connected to all OR gates. Based on this reason, the third variation of two-level AND–OR logic structure, known as *programmable array logic* (PAL), is obtained. In PALs, AND arrays are programmable and OR arrays are connected to a fixed number of AND gates. The resulting structure is that each OR gate along with its associated AND array comprises a logic module suitable for implementing a switching function with limited product terms. In general, the number of product terms ranges from five to eight.

Based on the above description, when implementing a multiple-output function with a PAL device, it is sufficient to simplify each output function independently because there are no AND gates that can be shared by different OR gates.

Nowadays, there are a lot of discrete PAL devices on the market. These PAL devices have the standard features as listed in Table 10.3. They can generally be

TABLE 10.3 The standard features of PAL devices

PAL devices	Package pins	AND-gate inputs	Primary inputs	Bidirectional I/Os	Registered outputs	Combinational outputs
PAL16L8	20	16	10	6	0	2
PAL16R4	20	16	8	4	4	0
PAL16R6	20	16	8	2	6	0
PAL16R8	20	16	8	0	8	0
PAL20L8	24	20	14	6	0	2
PAL20R4	24	20	12	4	4	0
PAL20R6	24	20	12	2	6	0
PAL20R8	24	20	12	0	8	0

grouped into two types: PAL16xx and PAL20xx series. The former has 16 inputs for each AND gate and the latter has 20 inputs for each AND gate. The next character "L" or "R" distinguishes combinational logic from registered output type. The last digit denotes the number of outputs. An example of a PAL device, 16R8, is shown in Figure 10.18.

The important features of PALs are as follows. First, each OR gate associated with an AND array forms a two-level AND–OR logic structure. This AND–OR structure can be used to implement a switching function. Second, since there are no shared product terms in the PAL devices, a single-output minimization process is sufficient for the purpose of minimizing switching functions. The following example illustrates how to implement a multiple-output switching function with a typical PAL.

EXAMPLE 10.5   *PAL—Implementing Switching Functions*

Figure 10.19 shows that four switching functions are implemented by using a PAL device. These four switching functions are as follows:

$$f_1(w, x, y, z) = \Sigma(2, 12, 13)$$

$$f_2(w, x, y, z) = \Sigma(7, 8, 9, 10, 11, 12, 13, 14, 15)$$

$$f_3(w, x, y, z) = \Sigma(0, 2, 3, 4, 5, 6, 7, 8, 10, 11, 15)$$

$$f_4(w, x, y, z) = \Sigma(1, 2, 8, 12, 13) \tag{10.3}$$

Simplifying them by using a Karnaugh map or another minimization process, we obtain the following simplified switching expressions:

$$f_1(w, x, y, z) = wxy' + w'x'yz'$$

$$f_2(w, x, y, z) = w + xyz$$

$$f_3(w, x, y, z) = w'x + yz + x'z'$$

$$f_4(w, x, y, z) = wxy' + w'x'yz' + wy'z' + w'x'y'z \tag{10.4}$$

The resulting PAL programming table and logic diagram are shown in Figures 10.19(a) and (b), respectively. ■

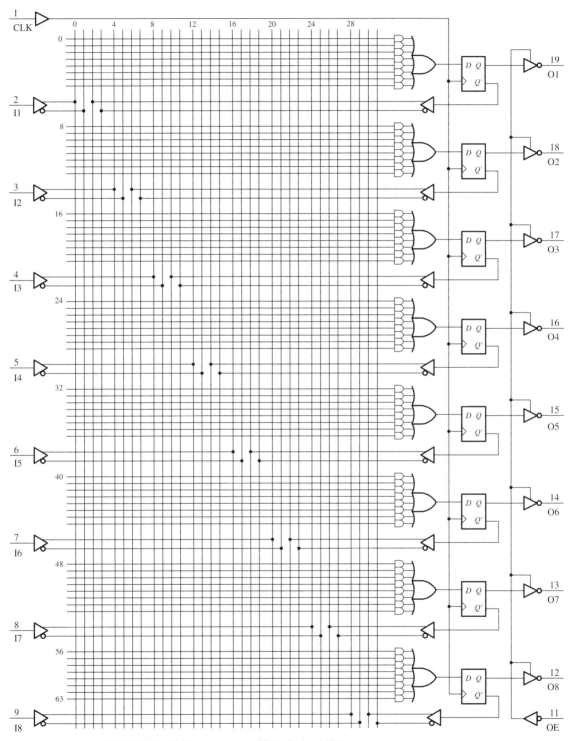

FIGURE 10.18 The structure of PAL device 16R8

Product term		Input				Output function
		w	x	y	z	
wxy'	16	1	1	0	-	f_1
$w'x'yz'$	17	0	0	1	0	
w	24	1	-	-	-	f_2
xyz	25	-	1	1	1	
$w'x$	32	0	1	-	-	
yz	33	-	-	1	1	f_3
$x'z'$	34	-	0	-	0	
wxy'	40	1	1	0	-	
$w'x'yz'$	41	0	0	1	0	f_4
$wy'z'$	42	1	-	0	0	
$w'x'y'z$	43	0	0	0	1	

(a)

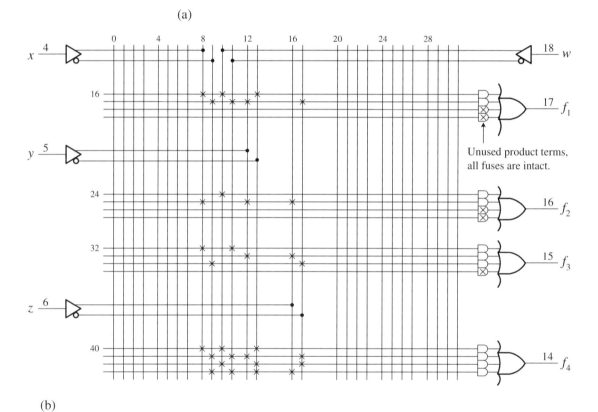

(b)

FIGURE 10.19 An example to explain how a PAL is used to implement switching functions:
(a) PAL programming table; (b) logic circuit

Review Questions

Q10.37 Describe the features of a PAL in terms of the programmability of AND–OR logic structure.

Q10.38 Can a product term be shared among multiple switching functions in a PAL device?

Q10.39 Is the AND array of PALs fully or partially decoded the input combinations?

10.2.4 PLA Modeling

As we have discussed, the structure of any PLD device can be one of the following types:

1. *Sum of product* (SOP). An SOP expression comprises an AND plane (array) and an OR plane (array). It can also be equivalently represented by a NAND–NAND structure through using DeMorgan's law.

2. *Product of sum* (POS): A POS expression consists of an OR plane and an AND plane. It can also be equivalently represented by a NOR–NOR structure through using DeMorgan's law.

An AND plane or an OR plane is often called a *personality array* (or memory) for historical reasons. In the following, we will focus our attention on the modeling of PLAs because they are most widely used in HDL-based design. However, the same technique can be applied equally well to PALs.

In Verilog HDL, a PLA device is modeled by a group of system tasks. Through an appropriate combination of two system tasks, any type of PLA device can be modeled. These system tasks can be either asynchronous or synchronous. For asynchronous system tasks, the evaluation is executed whenever any input changes. For synchronous system tasks, the evaluation time and the update of outputs can be controlled. However, regardless of which kind of system tasks are used, the outputs are updated without any delay. The system tasks of PLA modeling are shown in Table 10.4.

The general form of PLA modeling system tasks is as follows:

```
$array_type$logic$format (memory_type, input_terms,
   output_terms);
```

TABLE 10.4 The PLA modeling tasks in Verilog HDL

Array format		Plane format	
Asynchronous	Synchronous	Asynchronous	Synchronous
$async$and$array	$sync$and$array	$async$and$plane	$sync$and$plane
$async$nand$array	$sync$nand$array	$async$nand$plane	$sync$nand$plane
$async$or$array	$sync$or$array	$async$or$plane	$sync$or$plane
$async$nor$array	$sync$nor$array	$async$nor$plane	$sync$nor$plane

where `array_type` can be either `sync` (synchronous) or `async` (asynchronous); the `logic` can be any of the following: `and`, `or`, `nand` and `nor`; the `format` can be either `array` or `plane`. As a consequence, there are sixteen system tasks that can be obtained through these combinations, as shown in Table 10.4.

Two formats of logic array personality are as follows:

1. The array format uses only 1 or 0 to denote whether the input value is taken or not.

2. The plane format (complies with Espresso) uses:

   (a) 1 or 0 to denote that the true or complement input value is taken;

   (b) x to denote that the worst-case input value is taken;

   (c) ? or z to denote don't care.

The following is an example of how to use a PLA modeling system task.

EXAMPLE 10.6 *Declarations of Asynchronous and Synchronous Array Forms*

This example demonstrates the use of a PLA modeling system task. The memory array is declared as an array of three 7-bit vectors since there are seven input terms and three output terms. The personality of AND plane is defined by the contents of `mem`, which can be loaded through `$readmemb` or `$readmemh` system task, or by using procedural assignments.

```
wire a1, a2, a3, a4, a5, a6, a7;
reg  b1, b2, b3;
wire [1:7] awire;
reg  [1:3] breg;
reg  [1:7] mem [1:3]; // define personality memory
   // asynchronous AND plane
   $async$and$array(mem, {a1, a2, a3, a4, a5, a6, a7},
     {b1, b2, b3});
   // or using the following statement
   $async$and$array(mem, awire, breg);
   // synchronous AND plane
   forever @(posedge clock)
      $sync$and$array(mem, {a1, a2, a3, a4, a5, a6, a7},
         {b1, b2, b3});
```

From the above example, it is easy to see that logic array personality is declared as a vector array of `reg` variable with the vector size equal to the number of input terms and the array size equal to the number of output terms. The AND array personality can be loaded into memory using either `$readmemb` or `$readmemh` system task, or the procedural assignment statements. The personality can be changed dynamically during simulation simply by changing the contents of the memory.

The following example addresses how to define the personality of the AND array from a file.

EXAMPLE 10.7 *Defining the Personality of and Array from a File*

In this example, only one AND plane is considered. The contents of the personality of and array mem are read from the file `array.dat` by using $readmemb system task.

```
// an example of the usage of the array format.
module async_array(a1, a2, a3, a4, a5, a6, a7, b1, b2,
  b3);
input  a1, a2, a3, a4, a5, a6, a7 ;
output b1, b2, b3;
reg [1:7] mem [1:3]; // memory declaration for array
                     // personality
reg b1, b2, b3;
initial begin
   // setup the personality from the file array.dat
   $readmemb ("array.dat", mem);
   // setup an asynchronous logic array with the input
   // and output terms expressed as concatenations
   $async$and$array(mem, {a1, a2, a3, a4, a5, a6, a7},
      {b1, b2, b3});
end
endmodule                                             ■
```

The output functions are as follows:

```
b1 = a1 & a2
b2 = a3 & a4 & a5
b3 = a5 & a6 & a7
```

Hence, the personality of AND array is defined as follows:

```
1100000
0011100
0000111
```

   The following example demonstrates how to define the personality of the AND array by using procedural assignments.

EXAMPLE 10.8 *Defining the Personality of and Array Using Procedural Assignments*

In this example, only one AND plane is considered. The contents of the personality of AND array mem are assigned by using procedural assignments. Three assignments are required in this example since there are three output terms.

```
module pla(a0, a1, a2, a3, a4, a5, a6, a7, b0, b1, b2);
input  a0, a1, a2, a3, a4, a5, a6, a7;
```

```
output b0, b1, b2;
reg    b0, b1, b2;
reg [7:0] mem [0:2];
// an example of the usage of array format
initial begin    // using procedural assignment statements.
   mem[0] = 8'b11001100;
   mem[1] = 8'b00110011;
   mem[2] = 8'b00001111;
   $async$and$array(mem, {a0,a1,a2,a3,a4,a5,a6,a7}, {b0,
     b1,b2});
   end
endmodule
```

The following shows the results of simulation:

```
A = 11001100 -> B = 100
A = 00110011 -> B = 010
A = 00001111 -> B = 001
A = 10101010 -> B = 000
A = 01010101 -> B = 000
A = 11000000 -> B = 000
A = 00111111 -> B = 011
```

Figure 10.20 shows that two switching functions are implemented by using a PLA device. The two switching functions are functions of three input variables: x, y and z. After minimization, these two switching functions can be represented as the PLA programming table shown in Figure 10.20(a). The product term xz is shared by the two functions. There are four product terms, numbered from 0 to 3. The function f_1 consists of the product terms 0, 1 and 3. The function f_2 consists of the product terms 1 and 2. The programming table shown in Figure 10.20(a) can be represented by a shorthand notation, named as a *PLA symbolic diagram*, as shown in Figure 10.20(b). Figure 10.20(c) is the logic circuit that realizes the programming table shown in Figure 10.20(a).

In the following, we give two examples to address the modeling of this PLA device in the array and plane formats, respectively. First, we consider the case of using array format.

EXAMPLE 10.9   *A PLA Modeling Example by Using the Array Format*

As mentioned previously, in the personality array of array format, only two values of 1 and 0 can be used to denote whether the input variable is taken or not. Hence, we need to produce the complement signals from the inputs by using continuous assignments. The results is shown in the following module.

```
module pla_example(input x, y, z, output reg f1, f2);
// an example of the usage of the array format.
reg  p0, p1, p2, p3;        // internal minterms
reg  [0:5] mem_and[0:3];
reg  [0:3] mem_or[1:2];
```

Product		Input			Output	
		x	y	z	f_2	f_1
xy'	0	1	0	-	-	1
xz	1	1	-	1	1	1
$y'z$	2	-	0	1	1	-
$x'yz'$	3	0	1	0	-	1

(a)

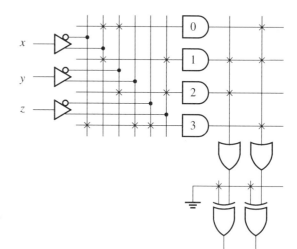

(c)

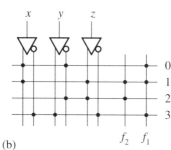

(b)

FIGURE 10.20 An example illustrating how to model a PLA: (a) PLA programming table; (b) PLA symbolic diagram; (c) logic circuit

```
wire x_n, y_n, z_n;
assign x_n = ~x, y_n = ~y, z_n = ~z;
initial begin: pla_model_array
   mem_and[0] = 6'b100100;   // Define AND plane
   mem_and[1] = 6'b100010;
   mem_and[2] = 6'b000110;
   mem_and[3] = 6'b011001;
   $async$and$array(mem_and, {x, x_n, y, y_n, z, z_n},
      {p0, p1, p2, p3});
   mem_or[1] = 4'b1101;      // Define OR plane
   mem_or[2] = 4'b0110;
   $async$or$array(mem_or, {p0, p1, p2, p3}, {f1, f2});
end
endmodule
```

The simulation results are as follows:

```
# 0 ns   x x x x x
# 5 ns   0 0 0 0 0
# 10 ns  0 0 1 0 1
# 15 ns  0 1 0 1 0
# 20 ns  0 1 1 0 0
# 25 ns  1 0 0 1 0
# 30 ns  1 0 1 1 1
# 35 ns  1 1 0 0 0
# 40 ns  1 1 1 1 1
```

The following example demonstrates how to model a PLA using the plane format.

EXAMPLE 10.10 *A PLA Modeling Example by Using the Plane Format*

As mentioned previously, in the personality array of plane format, values 1 and 0 denote the true and complement of the inputs, respectively. In addition, the ? and z are used to denote the don't care. Hence, the AND plane personality memory only needs three bits and four words because there are three input variables and four product terms. The contents of the AND plane personality memory directly correspond to the programming table shown in Figure 10.20(a). Similarly, the OR plane personality memory needs four bits and two words becasue there are four product terms and two switching functions. The contents of the OR plane personality memory can also be obtained from the programming table of Figure 10.20(a). The results are shown in the following module.

```
module pla_example(input x, y, z, output reg f1, f2);
reg  p0, p1, p2, p3;        // internal minterms
reg  [0:2] mem_and[0:3];    // AND plane personality matrix
reg  [0:3] mem_or[1:2];     // OR plane personality matrix
// an example of the usage of the plane format.
initial begin: pla_model_plane
   $async$and$plane(mem_and, {x, y, z}, {p0, p1, p2, p3});
   mem_and[0] = 3'b10?;     // Define AND plane
   mem_and[1] = 3'b1?1;
   mem_and[2] = 3'b?01;
   mem_and[3] = 3'b010;
   $async$or$plane(mem_or, {p0, p1, p2, p3}, {f1, f2});
   mem_or[1] = 4'b11?1;     // Define OR plane
   mem_or[2] = 4'b?11?;
end
endmodule
```
∎

Review Questions

Q10.40 Describe the general form of PLA modeling system tasks.

Q10.41 Describe the meaning of personality array.

Q10.42 Describe the differences between synchronous and asynchronous PLAs.

Q10.43 Describe the features of array format used in modeling PLAs.

Q10.44 Describe the features of plane format used in modeling PLAs.

10.3 CPLD

In general, CPLDs comprise three types of constituent components: *function block*, *interconnect* and *input/output block*. Function blocks are used to implement the desired logic functions; interconnect provides paths for communicating between function blocks and IOBs; IOBs are I/O pins associated with the required circuits and can program the I/O pins into input, output or bidirectional operations as needed. IOBs also provide other features such as low-power or high-speed connections. In this section, we describe two examples of CPLDs, the Xilinx XC9500 family and the Altera MAX7000 family, based on the above three types of components. Both CPLDs are EEPROM-based devices and are *in-system programmable* (ISP).

10.3.1 XC9500 Family

As we have introduced previously, a CPLD device can usually replace many PAL devices. For instance, the smallest member of the XC9500 family, XC9536, has 36 macrocells, and the largest one, XC95288, has 288 macrocells. Each macrocell corresponds to an AND–OR logic structure in PAL devices.

The basic structure of the XC9500 CPLD family is shown in Figure 10.21. It can be divided into three major parts: function blocks, switch matrix and IOBs. Each function block contains 18 macrocells, where each macrocell is responsible to implement a logic function with 54 input variables. The switch matrix connects all inputs and outputs from function blocks to any function block. IOBs provide buffers for inputs and outputs.

Function Block   The structure of each function block is depicted in Figure 10.22, where each function block consists of 18 independent macrocells. Each macrocell

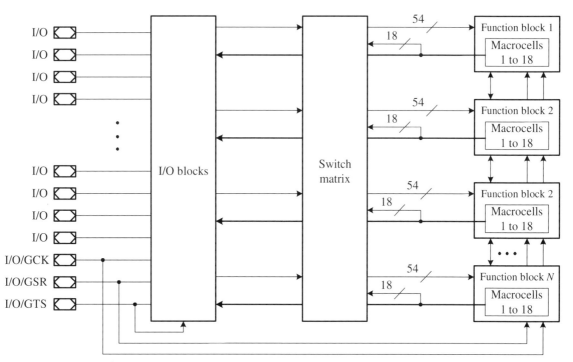

FIGURE 10.21 Basic structure of the XC9500XL family. Reproduced by permission of Xilinx, Inc.

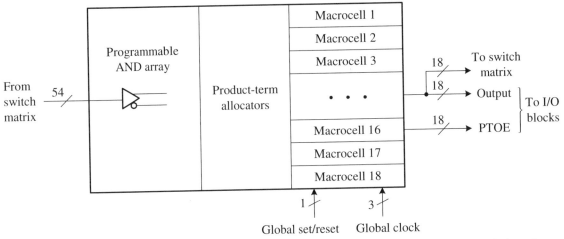

FIGURE 10.22 The function block of the XC9500XL family. Reproduced by permission of Xilinx, Inc.

contains 5 product terms and is responsible to implement a combinational logic or sequential logic function. All 18 outputs from a function block can serve as the inputs to both switch matrix and IOBs at the same time. In addition, there are global clock, set and reset signals that can be used to control the programmable flip-flop within function blocks.

Due to the fact that each macrocell is a form of AND–OR logic structure, the switching function realized by each macrocell is in SOP form. Each function block accepts 54 input signals from and outputs 18 signals to the switch matrix, through which it connects to IOBs and other macrocells. Although each macrocell only has 5 product terms, each function block has 90 product terms. These product terms may be partially or totally switched into a macrocell in the same function block.

Each macrocell in a function block has the structure as shown in Figure 10.23. The basic 5 product terms act as an AND array that can be directly used as an input source of OR and XOR gates to implement the desired function or as clock, clock enable or set/reset control signals for the programmable flip-flop within the macrocell, and the output enable signal to control the output buffer within its associated IOB.

The product-term allocator determines the actual usage of product terms. It can accept the product terms from other macrocells, only use basic product terms or allocate the unused product terms to other macrocells. The programmable flip-flop within each macrocell can be programmed as either a D-type or a T-type flip-flop to fit the actual requirement. Of course, it is bypassed when realizing a combinational logic. Each flip-flop can be set or cleared asynchronously through its asynchronous set and reset control inputs, which can be either the global set/reset control signal or the product-term set and product-term reset control signals. When the power starts up, flip-flops are set to the default value of 0 unless they are set to the values specified by users. The clock input of the flip-flop may be programmed as positive-edge or negative-edge triggered. The clock signal can be obtained either from the global clock signal or from product-term clock, as shown in Figure 10.23. The use of a global clock signal achieves the fastest

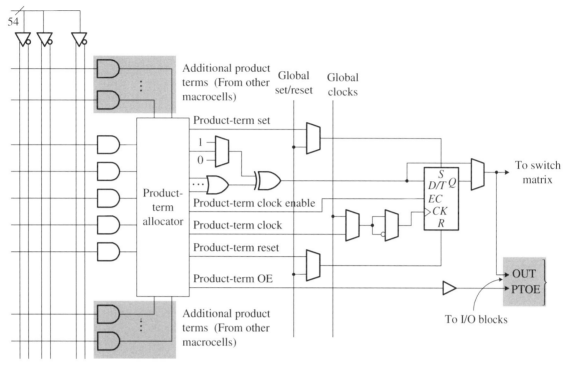

FIGURE 10.23 The macrocell of the function block of the XC9500XL family. Reproduced by permission of Xilinx, Inc.

clock-to-output performance while the use of a product-term clock has the feature that the flip-flop can be clocked by signals from macrocells or I/O pins.

Switch Matrix   The switch matrix functions as an interconnection network used to provide paths from IOBs to function blocks as well as from function blocks to function blocks, as shown in Figure 10.24. Recall that each function block can have at most 54 inputs.

I/O Block (IOB)   The IOBs are used to route the signals either from the outputs of function blocks to I/O pins or from I/O pins to the inputs of function blocks. The structure of IOBs is depicted in Figure 10.25. Each IOB contains an input buffer, an output driver, an output enbale select multiplexer, a pull-resistor and a programmable ground.

The input buffer is compatible to 5-V CMOS, 5-V TTL, 3.3-V CMOS, and 2.5-V CMOS. In order to provide a constant input threshold voltage and avoid the threshold varying with the I/O power-supply voltage (V_{CCIO}), an internal +3.3-V power-supply voltage (V_{CCINT}) is used.

All output drivers can be either tolerant to 3.3-V CMOS and 5-V TTL when $V_{CCIO} = 3.3$ V or tolerant to 2.5-V CMOS when $V_{CCIO} = 2.5$ V. In addition, each output driver has a slew-rate control circuit which controls both rise time and fall time of the output signal. There are four options for the output enable control. These are

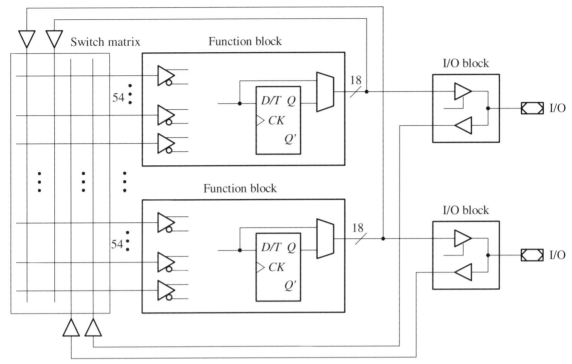

FIGURE 10.24 The switch matrix of the XC9500XL family. Reproduced by permission of Xilinx, Inc.

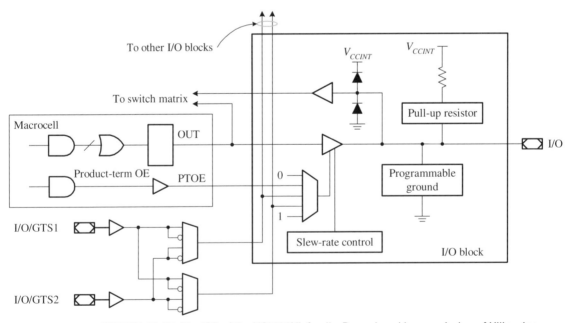

FIGURE 10.25 The IOB of the XC9500XL family. Reproduced by permission of Xilinx, Inc.

always "1", always "0", the product-term output enable (PTOE) signal generated from the macrocell, and global output enable signal (GTS). In the XC9500 family, there are two GTS signals when the devices have 72 or less macros and four GTS signals when the devices have 144 or more macros. Finally, each IOB also has the capability of programmable ground to ground the unused I/O pins so as to reduce power dissipation and switching noise.

Review Questions

Q10.45 Describe the basic structure of the XC9500 CPLD family.

Q10.46 Describe the structure of function blocks.

Q10.47 Describe the structure of macrocells.

Q10.48 Describe the structure of switch matrix.

Q10.49 Describe the structure of IOBs.

10.3.2 MAX7000 Family

Like the XC9500 family, the MAX7000 family is also an EEPROM-based CPLD with the capability of in-system programmability (ISP). The basic structure of the MAX7000 family is shown in Figure 10.26, which also includes four dedicated inputs (clock, clear and two output enable signals) that can be used as general-purpose inputs or as high-speed, global control signals for each macrocell and I/O pin. As we have mentioned before, the structure of a field-programmable device consists of three basic components: function block, interconnect and IOB. Using the terminology of the MAX7000 family, these three components are logic array blocks (LABs), programmable interconnect array (PIA) and I/O control blocks (IOBs), respectively.

Logic Array Blocks (LABs) Each device in the MAX7000 family consists of many LABs, each having 16 macrocells. These LABs are linked together through a global bus known as a PIA, which is fed by all dedicated inputs, I/O pins and macrocells. Each LAB is fed by two kinds of signals. The first one is 36 signals from the PIA that are used for general logic inputs; the second one is the global control signals used for controlling flip-flop functions. For MAX 7000E and MAX 7000S devices, there are direct input paths from I/O pins to the flip-flops in order to provide fast setup times.

Figure 10.27 shows the structure of each macrocell, which consists of three basic blocks: a logic array, a product-term select matrix and a programmable flip-flop. Each macrocell is responsible to implement a sequential or a combinatorial logic function.

Each macrocell contains five product terms. Through the product-term select matrix, these product terms can be allocated for use as either primary logic inputs to the OR and XOR gates to implement combinatorial functions or as the control signals, clear, preset, and clock and clock enable control signals, for the macrocell flip-flop.

The programmable flip-flop in each macrocell can be programmed as D-type, T-type, JK or SR flip-flop. The flip-flop is bypassed for combinational logic. The clock of each programmable flip-flop can be either from the global clock signal or from a

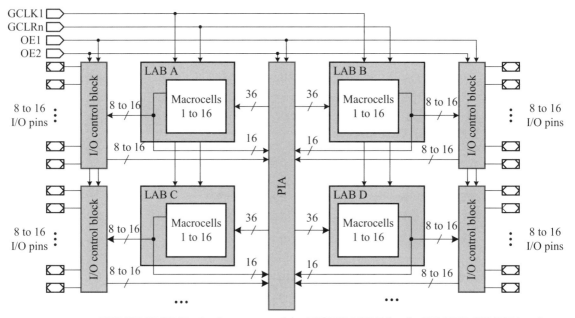

FIGURE 10.26 The basic structure of the MAX7000 CPLD family: EPM7032, EPM7064 and EPM7096. Reproduced by permission of Altera International Ltd

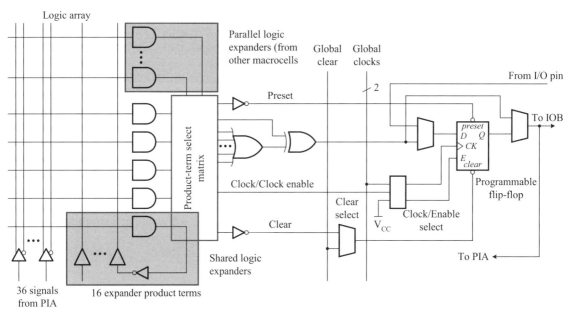

FIGURE 10.27 The macrocell structure of the MAX 7000 CPLD family: EPM7032, EPM7064 and EPM7096. Reproduced by permission of Altera International Ltd

product term of logic array. The use of a global clock signal achieves the fastest clock-to-output performance while the use of a product-term clock has the feature that the flip-flop can be clocked by signals from macrocells or I/O pins. The global clock signal is available from a dedicated clock pin, GCLK1, as shown in Figure 10.26.

Each flip-flop also supports asynchronous preset and clear functions. As shown in Figure 10.27, the product-term select matrix allocates product terms to control these operations. In addition, the active-low dedicated global clear pin (GCLRn) can be used to drive the clear function of each flip-flop individually, as shown in Figure 10.26. Each flip-flop in the macrocells is set to a low state when the power is on.

Programmable Interconnect Array (PIA)   As mentioned previously, a CPLD is virtually a set of PALs connected by a programmable interconnect. For the MAX7000 CPLD family, the interconnect between LABs is through the programmable intercon-nect array (PIA). The PIA is a programmable path that connects any signal source to any destination on the device, as shown in Figure 10.26. All dedicated inputs, macrocell outputs and I/O pins feed the PIA, which makes the signals available throughout the entire device. However, only the signals required by each LAB are actually routed from the PIA into the LAB.

I/O Control Block (IOB)   The I/O control block (IOB) allows each I/O pin to be separately configured for input, output or bidirectional operation, as shown in Figure 10.28. Each I/O pin associates with a tristate buffer which is disabled or enabled continuously by its enable control being connected to ground, or V_{CC}, or controlled by one of the active-low global output enable signals, OE1 and OE2, through a 4-to-1 multiplexer. In MAX 7000E and MAX 7000S devices, the IOB has six global output enable signals, driven by the true or complement of two output enable signals, a sub-set of the I/O pins, or a sub-set of the I/O macrocells.

The output of the tristate buffer is in a high-impedance and the I/O pin associated with the buffer can be used as a dedicated input when its enable control is connected to ground; the output of the tristate buffer is enabled when its control is connected to V_{CC}. Due to the fact that macrocell and pin feedback are independent, when an I/O pin is configured as an input, the associated macrocell can still be used for buried logic, namely, implementing a logic function without an I/O pin.

Expanders   Although most logic functions can be implemented with the five product terms available in each macrocell, the more complex logic functions require

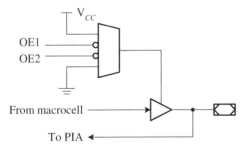

FIGURE 10.28 The structure of the IOB of the MAX 7000 CPLD family: EPM7032, EPM7064 and EPM7096. Reproduced by permission of Altera International Ltd

additional product terms. To facilitate this requirement, the simplest solution is to use another macrocell to supply the required logic gates. One major disadvantage of this is that the propagation delay is long due to many additional gates needed to be passed, as shown in Figure 10.27. Another approach like that used in most commercial CPLDs is to provide a sharable mechanism within the structure. The sharable mechanism allows additional product terms from neighboring macrocells to be allocated together with the macrocell for implementing the desired logic function. This sharable mechanism helps ensure that logic is synthesized with the fewest possible logic gates and multiplexers to obtain the fastest possible speed.

In the MAX7000 CPLD family, the sharable mechanism is supported by two kinds of expander product terms, known as *expanders*. These are *shareable expanders* and *parallel expanders*. Shareable expanders are inverted product terms that are fed back into the logic array and parallel expanders are product terms borrowed from neighboring macrocells, as depicted in Figure 10.27.

Shareable Expanders Each LAB has 16 shareable expanders, with one extracted from each macrocell. These sharable product terms feed back into the logic array with inverted outputs, as shown in Figure 10.29. Each shareable product term can be used and shared by any or all macrocells in the LAB to build complex logic functions.

Parallel Expanders Parallel expanders are unused product terms that can be allocated to a neighboring macrocell to implement complex logic functions. For the MAX7000 CPLD family, parallel expanders allow up to 20 product terms to directly

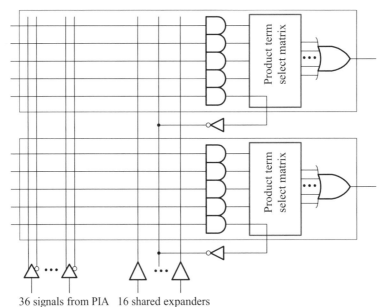

FIGURE 10.29 The structure of the sharable expanders of the Max 7000 CPLD family.
Reproduced by permission of Altera International Ltd

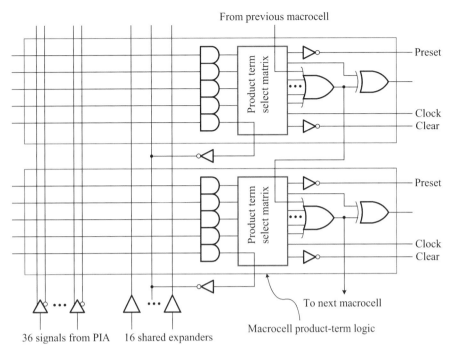

FIGURE 10.30 The structure of the parallel expanders of the Max 7000 CPLD family. Reproduced by permission of Altera International Ltd

feed the macrocell OR logic. Five of these product terms are supported by the macrocell and the rest fifteen parallel expanders (namely, three neighboring macrocells) are provided by neighboring macrocells in the LAB, as shown in Figure 10.30.

The 16 macrocells in each LAB is divided into two groups: macrocells 1 through 8 and 9 through 16. These two groups of 8 macrocells form two chains and can lend or borrow parallel expanders from neighboring mcrocells with one constraint that a macrocell can only borrow parallel expanders from lower-numbered macrocells. For instance, macrocell 6 can borrow parallel expanders from macrocell 5, from macrocells 5 and 4, or from macrocells 5, 4, and 3. Within each group, the highest-numbered macrocell can only borrow parallel expanders but the lowest-numbered macrocell can only lend them. If a macrocell requires 16 product terms, the five dedicated product terms within the macrocell, along with three sets of parallel expanders, are allocated; the first two sets include five product terms and the third set includes one product term.

Review Questions

Q10.50 Describe the basic structure of the MAX7000 CPLD family.

Q10.51 Describe the structure of LABs.

Q10.52 Describe the structure of IOBs.

Q10.53 Describe the structure of sharable expanders.

Q10.54 Describe the structure of parallel expanders.

10.4 FPGA

Aside from CPLDs, FPGAs are another kind of large-scale field-programmable devices. The basic difference between FPGAs and CPLDs is that the configurable logic block (CLB) or loigic element (LE) of the former has less function than the latter but the total number of CLBs/LEs of the former is much larger than the latter. For FPGAs, CLBs/LEs are dispersed on the entire chip and interconnect is prewired between the CLBs/LEs to configure the CLBs/LEs into the required logic circuit. The CLBs/LEs are usually composed of one or more k-input (k is usually set to 3 to 8) universal logic module known as look-up table (LUT) and a programmable output-stage logic circuit. Each LUT can be either a truth table or a multiplexer and the output-stage logic circuit consists of a multiplexer and a flip-flop. As a result, each CLB/LE can implement one or more combinational logic or sequential logic functions.

10.4.1 Xilinx FPGA Devices

Like CPLDs, FPGAs also comprise three basic components: function blocks (namely, CLBs), programmable interconnect and IOBs. CLBs provide the functional elements for constructing the required logic and IOBs provide the interface between I/O pins and internal signals. The programmable interconnect provides required interconnections between various CLBs and IOBs. In this section, we introduce the basic structure of the Xilinx XC4000XL family FPGA devices.

10.4.1.1 XC4000XL Family   The basic structure of the XC4000XL family is shown in Figure 10.31. The CLBs are arranged into a two-dimensional array and the interconnect is placed in between the CLBs. Each CLB contains two 4-input universal logic modules known as look-up tables (LUTs), with each having a D-type flop-flop. As a consequence, each LUT is responsible to implement a 4-variable combinational logic or a sequential logic coupled with its D-type flop-flop. The combination of two LUTs can implement a 5-variable combinational logic or a sequential logic coupled with their D-type flop-flops.

Configurable Logic Block (CLB)   The CLBs of the XC4000XL family have the structure as shown in Figure 10.32. Each CLB consists of two D-type flip-flops, two 4-input LUTs (F) and (G) and one 3-input LUT (H), as well as many multiplexers. The LUTs are also known as *logic functions*. For XC4000X, the two D-type flip-flops can also be programmed as latches.

Each LUT (F and G) has four independent inputs (F1 to F4 and G1 to G4) and is capable of implementing any switching function of four inputs at most. Because the function is implemented as look-up table, the propagation delay is independent of the function implemented. The LUT H can implement any switching function of its three inputs. Two of these inputs can optionally be the outputs from LUTs F and G or come from outside of the CLB (H2, H0). The third input must come from outside of the block (H1). With the help of LUT H, each CLB can implement certain functions of up to nine variables, such as parity check or its equivalent function.

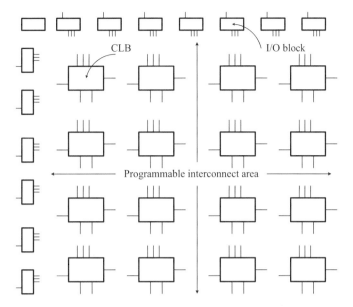

FIGURE 10.31 A corner of the basic structure of the XC4000XL family. Reproduced by permission of Xilink, Inc.

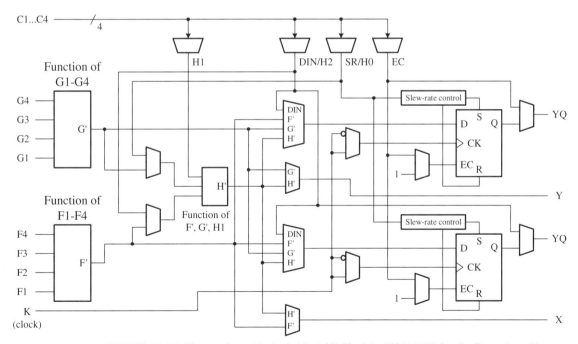

FIGURE 10.32 The configurable logic block (CLB) of the XC4000XL family. Reproduced by permission of Xilinx, Inc.

Both D-type flip-flops in each CLB can be used independently with LUTs and can directly accept the data from DIN and H1. These two edge-triggered D-type flip-flops have a common clock (K) and a clock enable (EC) inputs. Either or both clock inputs (K and EC) can also be permanently enabled. Each D-type flip-flop can be triggered at either the positive or negative edge of its clock signal. Although the clock pin is shared by both D-type flip-flops, the clock is individually invertible for each D-type flip-flop. The data input of each D-type flip-flop comes from one of four sources: direct input (C2), F$'$, G$'$ or H$'$.

The slew-rate control input of a D-type flip-flop can be configured as either set or reset, as shown in Figure 10.32. The set/reset state can be specified and disabled separately for both D-type flip-flops within each CLB. In addition, a separate global set/reset (GSR) input, through a dedicated distribution network, sets or clears each D-type flip-flop during power-up, re-configuration or when a dedicated reset net is driven active.

Both F and G LUTs of each CLB contain dedicated carry generation logic for the generation of carry and borrow signals. This extra output is passed onto the LUT in the adjacent CLB and can be expanded into any length. Through this routing-independent carry chain, fast adders, subtractors, accumulators, comparators and counters can be constructed.

A single CLB can also be configured as either a 16×2, 32×1 or 16×1 bit array of memory. The memory can be configured as either edge-triggered (synchronous) or dual-port RAM. Edge-triggered RAM simplifies system timing while dual-port (such as one read and one write) RAM doubles the effective throughput of FIFO applications. The following example illustrates this kind of applications.

EXAMPLE 10.11 *Single-Port RAM: the Data Port is Bidirectional*

The simplest synchronous memory constructed from CLBs is the one with separate write and read ports. This is a little difficult if we would like to combine both ports into a bidirectional port. To do this, it is necessary to use a tristate buffers at the read port before it can be combined with the write port. The continuous assignment shown in the following module demonstrates how this idea is realized.

```
// an n-word single-port ram
module single_port_ram(clk, addr, write, data);
parameter M = 6;  // number of address bits
parameter N = 64; // number of words, n = 2**m
parameter W = 8;  // number of bits in a word
input clk, write;
inout [W-1:0] data;
input [M-1:0] addr;
reg   [W-1:0] ram [N-1:0];
// the body of the n-word single-port ram
assign data = (!write) ? ram[addr] : {W{1'bz}};
always @(posedge clk)
   if (write) ram[addr] <= data;
endmodule
```

It is not easy to realize a bidirectional two-port memory using CLBs because the arbitration of two write operations cannot be resolved automatically by synthesis tools. Fortunately, such general multiple-port memory is not so often needed in digital systems. In contrast, most applications only need a restricted-type multiple-port memory. For instance, a dual-port memory with one write and one read port is usually used as the FIFO in digital systems, and a three-port memory with one write and two read ports is often used as the register file when designing datapaths. Details of a three-port register file can be found in Section 9.2.2.

Programmable Interconnect The XC4000XL family has the programmable interconnect structure shown in Figure 10.33. It mainly consists of interconnection lines and programmable switch matrices (PSMs). The interconnection lines can be categorized into the following types according to the relative lengths of their segments: single-length lines, double-length lines, quad lines and long lines.

Single-length lines are vertical and horizontal lines used to connect two switch matrices, as shown in Figure 10.33(a). Double-length lines are also vertical and horizontal lines used to connect two switch matrices but they go through two CLBs before entering into a switch matrix, as shown in Figrue 10.33(b). As a consequence, they are so named. Quad lines are four times as long as the single-length lines and run past four CLBs before entering a switch matrix. They are interconnected through switch matrices. The long lines cross the entire chip without passing through switch matrices and thus have a short transport delay. They are intended to be used for high fanout, timing-critical signal nets. Two horizontal long lines per CLB can be driven by tristate or open-drain drivers (TBUFs). As a result, they can implement unidirectional or bidirectional buses, wide multiplexers or wired-AND functions.

In addition to the above mentioned types of interconnect, the XC4000X family also provides eight interconnect tracks between the CLB array and the pad ring.

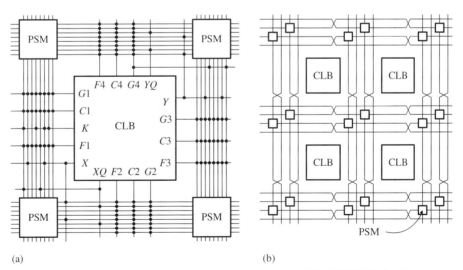

(a)   (b)

FIGURE 10.33 The programmable interconnect structure of the XC4000XL family: (a) single-length lines; (b) double-length lines. Reproduced by permission of Xilinx, Inc.

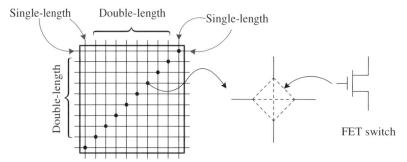

FIGURE 10.34 The programmable-switch matrix (PSM) structure of the XC4000XL family. Reproduced by permission of Xilinx, Inc.

These routing tracks can be broken every eight CLBs, namely, sixteen IOBs, by a programmable buffer that also functions as a splitter switch. Hence, they are named as octal lines. Moreover, the XC4000X family has direct connections to allow fast data flow between adjacent CLBs, and between IOBs and CLBs.

The programmable switch matrix (PSM) has the structure shown in Figure 10.34. Each crosspoint is composed of a switch circuit that contains six CMOS pass transistors; that is, it forms a fully connected switch in all directions. As a result, the horizontal and vertical single- and double-length lines intersect PSM can establish the connections between them. For example, a single-length signal entering on the left side of the PSM can be routed to a single-length line on the top, right or bottom side, or any combination of them, if multiple branches are required. Likewise, a double-length signal can be routed to a double-length line on any or all of the other three edges of the PSM.

Input/Output Block (IOB)   The IOBs provide the interface between external I/O pins and the internal logic. Each IOB controls one I/O pin and can be configured for input, output or bidirectional operations. For XC4000XL devices, inputs are both TTL and 3.3-V CMOS compatible and outputs are pulled to the 3.3-V positive supply. In addition, all I/Os on the XC4000XL devices are fully 5-V tolerant even though the V_{CC} is 3.3 V. The structure of the IOB is depicted in Figure 10.35, which consists of many multiplexers, buffers and D-type flip-flops.

The output D-type flip-flop can only be used as a flip-flop and the input D-type flip-flop can be programmed into, either as a flip-flop or as a latch. Recall that the basic difference between a flip-flop and a latch is that the flip-flop does not have the transparent property but the latch does. Both input and output D-type flip-flops have their own clock signals but with a common clock enable input. Each flip-flop can be programmed into positive-edge or negative-edge, triggered through a 2-to-1 multiplexer at the clock input of the flip-flop.

The input signal is connected to an input D-type flip-flop that can be programmed as either an edge-triggered flip-flop or a level-sensitive latch. The input signal can be delayed a few ns before entering into the input D-type flip-flop so as to compensate the hold time requirement of the external signal due to the delay of the clock signal. Signals I_1 and I_2 are either from the I/O pin or from the output of the input D-type flip-flop/latch.

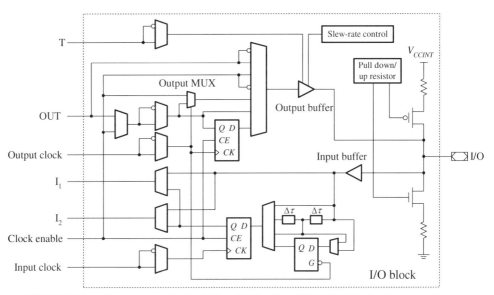

FIGURE 10.35 The input and output block (IOB) of the XC4000XL family. Reproduced by permission of Xilinx, Inc.

The output signal can be from the OUT end or from the output of the output D-type flip-flop. Besides output enable control, there is a slew-rate control used to control the slew rate of the output buffer. The function of the slew rate control is to control the magnitude of the current sunk or sourced by the output buffer. The slew rate of the buffer is reduced in order to minimize the power-bus transients when switching noncritical signals and increased to promote the performance when switching critical signals. In general, the unused I/O pins are pulled up through pull-up resistors or grounded in order to reduce power consumption.

Design Approach Like cell-based design, a macro library is usually provided by vendors for users to easily use the underlying FPGA devices to design their systems. A typical macro library is shown in Table 10.5. Macros may be one of the following: *soft macro*, *hard macro* and *permanent macro*. Soft macros only describe logic and interconnections without containing the information of partition and routing. Hard macros describe where the elements are placed; namely, they contain the complete information about partition, placement and routing. Permanent macros are the ones that reside on the chip permanently; namely, they are already manufactured along with the FPGA device. For this reason, permanent macros are often named as *hardwired macros*.

10.4.1.2 Spartan Family The XC3000, XC4000, Spartan and Spartan/XL use the same structure which consists of an array of CLBs interdispersed with switch matrixes surrounded by IOBs. Both the Spartan and Spartan/XL families use the same structure of CLBs and IOBs as the XC4000 family but have higher performance and are targeted for high-volume applications. These devices support only distributed memory,

TABLE 10.5 The macro library example of using FPGA devices

Soft macros		Hard macros	
Accumulators	Adder/Subtractors	Accumulators	Adders
Counters	Comparators	Counters	Comparators
Data/shift registers	Dividers	Data/shift registers	Decoders
Flip-Flops/latches	Encoders/decoders	RAM	Dividers
FSMs	Gates/buffers		Priority encoders
RAMs/ROMs	Logical shifters		Logical shifters
	Multiplexers		Multiplexers
	Muitipliers		Multipliers
	Parity checkers		Parity checkers
	Tristate buffers		

which is constituted of CLBs and thus whose use will reduce the number of CLBs that could be used for logic.

Spartan can also accommodate embedded soft cores such as Microblaze (a 32-bit RISC CPU soft core provided by Xilinx) and many useful peripherals. The on-chip distributed, dual-port, synchronous RAM (Select RAM) can be used to implement FIFOs, shift registers and scratch memories.

10.4.1.3 Spartan II and 3 Families
Starting with Spartan II, a configurable block memory is added in addition to distributed memory. Furthermore, delay-locked loops and multiple I/O standards are supported.

In Spartan II and 3, each CLB contains four logic cells, which is organized as a pair of slices. Each logic cell has a four-input LUT, carry and control logic, and a D-type flip-flop much similar to the structure of the XC4000 family. The CLB also contains logic for configuring functions of five or six inputs. As a result, each CLB can implement four logic functions of four inputs, two logic functions of five inputs or one logic function of six inputs.

The IOBs of Spartan II and 3 provide programmable, bidirectional interface between an I/O pin and the FPGA's internal logic. More precisely, they are programmable to support the reference, output voltage and termination voltages of a variety of high-speed memory and bus standards. Each IOB has three registers, which can be used as either D-type edge-triggered flip-flops or D-type latches. The first register can be utilized to store the signal that controls the programmable output buffer. The second register can be used to store the signal from the internal logic, which of course can alternatively bypass the register through a 2-to-1 multiplexer. The third register can be employed to store the signal coming from the I/O pin. Like the XC4000 family, this signal also passes through a programmable delay element used to compensate the pad-to-pad hold time requirement.

10.4.1.4 Virtex Family
The Virtex family is the leading edge of Xilinx technology, which incorporates embedded block memories, hardwired multipliers, even PowerPC CPU, in addition to FPGA fabrics. The multipliers are mainly used in DSP-related applications.

Review Questions

Q10.55 What are the three basic constituent components of a CPLD or FPGA?

Q10.56 Describe the basic structure of the XC4000XL family.

Q10.57 Describe the structure of CLBs of the XC4000XL family.

Q10.58 Describe the structure of the programmable interconnect of the XC4000XL family.

Q10.59 Describe the structure of the PSM of the XC4000XL family.

Q10.60 Describe the structure of the IOB of the XC4000XL family.

10.4.2 Altera FPGA Devices

Altera APEX 20K devices combine the features of function blocks (LUT-based logic) used in FPGA and macrocells (product-term logic) used in PLDs with an enhanced memory structure. The signal interconnections between function blocks and I/O pins are provided by a series of continuous row and column channels that run the entire length and width of the device. This interconnect is known as *FastTrack interconnect* in Altera literature.

As we have mentioned before, the structure of a field-programmable device consists of three constituent components: function block, interconnect and IOB. Using the terminology of the APEX 20K family devices, these are MegaLAB, FastTrack interconnect and I/O element (IOE), respectively, as shown in Figure 10.36.

MegaLABs are employed to implement logic functions, which may be either combinational or sequential, or implement various types of memory blocks. Edge MegaLABs can also be driven by I/O pins via the local interconnect. The FastTrack interconnect routes signals between MegaLAB structures and I/O pins that are fed by IOEs located at the end of each row and each column of the FastTrack interconnect. Each IOE contains a bidirectional I/O buffer and a flip-flop which can be utilized as either an input or output flip-flop to feed input, output or bidirectional signals. In addition, IOEs provide a variety of features, such as JTAG boundary-scan test (BST), slew-rate control, tristate buffers and support for various I/O standards.

APEX 20K devices provide two dedicated clock pins and APEX 20KE devices provide two additional dedicated clock pins, for a total of four dedicated clock pins. The dedicated clock pins can also feed logic. These signals use dedicated routing channels in order to provide low-delay and low-skew clock signals. In addition, there are four global signals that use dedicated inputs and can also be driven by internal logic, thereby providing a logic circuit for clock divider or internally generated asynchronous clear signals.

Structure of MegaLAB The structure of MegaLAB is shown in Figure 10.37, which contains a group of logic array blocks (LABs), one embedded system block (ESB) and local interconnect, as well as a MegaLAB interconnect. Each LAB consists of 10 logic elements (LEs) with each containing carry and cascade chains, LAB control signals and the local interconnect. The ESB can be configured as a block of macrocells or implement various types of memory blocks on an ESB-by-ESB basis. The MegaLAB interconnect routes signals within the MegaLAB structure.

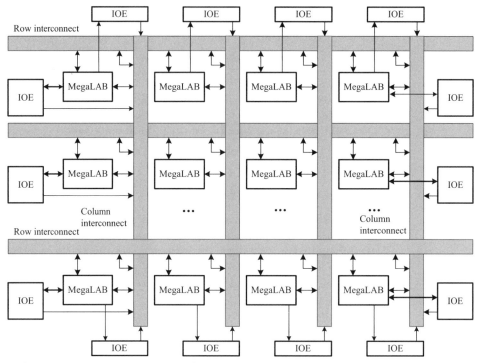

FIGURE 10.36 The basic structure of the Altera APEX 20K family. Reproduced by permission of Altera International Ltd

Dedicated logic is also enclosed in each LAB for driving various control signals to its LEs and ESBs. These control signals include clock, clock enable, asynchronous preset, asynchronous clear, asynchronous load, synchronous clear and synchronous load as well. A maximum of six control signals can be utilized at a time. Each LAB can use two clocks and two clock enable signals.

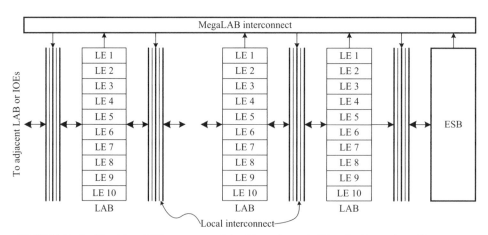

FIGURE 10.37 The MegaLAB structure of the Altera Apex 20K family. Reproduced by permission of Altera International Ltd

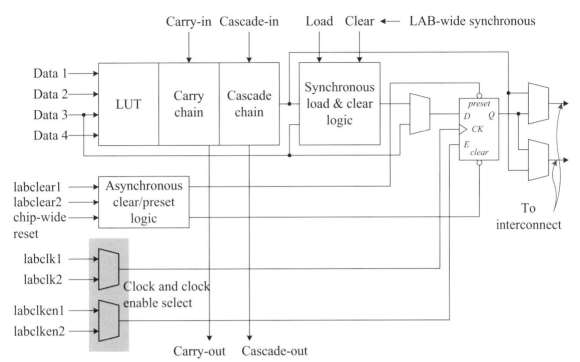

FIGURE 10.38 The basic structure of logic elements of the Altera Apex 20K family. Reproduced by permission of Altera International Ltd

Structure of Logic Element   Each LE has the structure shown in Figure 10.38, which consists of a four-input LUT, carry and cascade chains, LAB control signals, a programmable D-type flip-flop and the local interconnect. The four-input LUT is a universal logic module that can implement any four-variable function. The local interconnect routes signals between LEs in the same or adjacent LABs, IOEs or ESBs. Each LE can drive 29 other LEs through the local interconnect and has two outputs that drive the local interconnect, MegaLAB interconnect and FastTrack interconnect routing structures. Each output can be independently set to the output from LUT or D-type flip-flop.

The programmable D-type flip-flop in each LE can be configured into D-type, T-type, JK or SR flip-flop. The clock and clear control signals of the flip-flop can be from any of the following sources: global signals, general-purpose I/O pins and any internal logic. For combinational functions, the D-type flip-flop is bypassed and the output of the four-input LUT directly routes to the outputs of the LE.

In order to provide high-speed arithmetic functions, the APEX 20 structure provides two dedicated datapaths, *carry chains* and *cascade chains*, through which adjacent LEs can be connected without using local interconnect paths. A carry chain supports arithmetic functions such as counters and adders and a cascade chain implements wide-input functions such as equality comparators. Carry and cascade chains connect LEs 1 through 10 in an LAB and all LABs in the same MegaLAB structure.

With the carry chain, the carry-in signal from a lower-order bit drives forward into the higher-order bit through the carry chain, and feeds into both the LUT and the next portion of the carry chain so as to allow to implement counters, adders, and comparators of arbitrary width. With the cascade chain, adjacent LUTs can compute portions of a function in parallel. The cascade chain can use a logical AND or logical OR to connect the outputs of adjacent LEs. Each LE may contribute four inputs to the effective width of a function.

The LE of APEX 20K devices can operate in one of the following three modes: *normal mode*, *arithmetic mode* and *counter mode*. In normal mode, four data inputs from the LAB local interconnect and the carry-in are input to a four-input LUT. The LUT output can be combined with the cascade-in signal to form a cascade chain through the cascade-out signal. In arithmetic mode, the four-input LE is split into two 3-input LUTs, with one LUT computing a three-input function and the other generating a carry output. In the counter mode, two three-input LUTs are employed. One generates the counter data and the other generates the carry out.

Embedded System Block (ESB)   On an ESB-by-ESB basis, the ESB can be configured to either act as a set of macrocells or to implement various types of memory blocks, including ROM, FIFO, CAM and dual-port RAM. In macrocell mode, each ESB contains a macrocell array of 16 macrocells, with each containing two product terms and a programmable *D*-type flip-flop. As shown in Figure 10.39, the macrocell array is fed by 32 inputs from the adjacent local interconnect and nine outputs from the macrocell array feed back into the local interconnect. As a result, the macrocell array can be driven by the MegaLAB interconnect or the adjacent LAB. Dedicated clock pins, global signals and additional inputs from the local interconnect drive the control signals of the macrocell array. The function of each macrocell is much similar to that of the MAX7000 family. In addition, parallel expanders are provided in the macrocell mode.

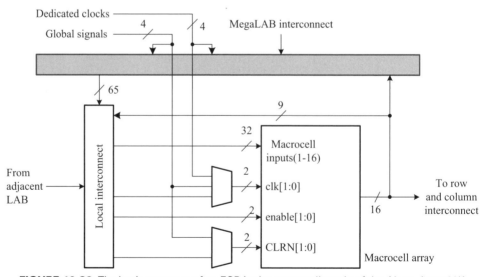

FIGURE 10.39 The basic structure of an ESB in the macrocell mode of the Altera Apex 20K family. Reproduced by permission of Altera International Ltd

In the memory block mode, each ESB can be configured into any of the following sizes: 128×16, 256×8, 512×4, 1024×2, or 2048×1. By combining multiple ESBs, a larger memory block can be created. For example, two 128×16 RAM blocks can be combined to form a 128×32 RAM block. In addition, the ESB includes input and output registers, which can support simultaneous reads and writes at two different clock frequencies. As a result, the ESB may offer a variety of dual-port modes, such as read/write clock mode, input/output clock mode or bidirectional mode.

10.4.2.1 Platform FPGA

Altera Stratix devices are platform-based FPGAs. The features that distinguish it from APEX 20K are as follows. First, three types of hardwired memory blocks are provided on chip. Second, digital-signal processing (DSP) blocks are embedded onto the chip for implementing DSP-related applications.

The basic structure of Stratix devices includes the following major components, LABs, digital-signal processing (DSP) blocks, memory blocks, input and output elements (IOEs) and interconnect, as shown in Figure 10.40. The interconnect consists of a series of column and row wires of varying length to route signals between logic array blocks (LABs), memory block structures and DSP blocks.

The logic array consists of LABs, with each containing 10 LEs. The structure and function of LEs are much similar to that of APEX 20K. LABs are grouped into

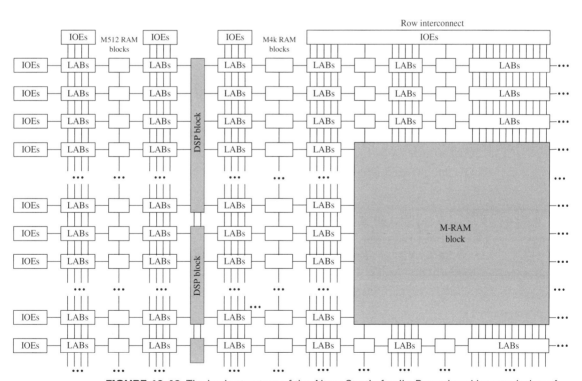

FIGURE 10.40 The basic structure of the Altera Stratix family. Reproduced by permission of Altera International Ltd

rows and columns across the device. There are three types of memory blocks, M512-RAM blocks, M4K-RAM blocks and M-RAM block, as shown in Figure 10.40. M512-RAM blocks are simple dual-port memory blocks with 512 bits plus parity (576 bits). These blocks support dedicated simple dual-port or single-port memory up to 18-bit wide and can be used in dual-port memory, shift registers and FIFO buffers. M512 blocks are grouped into columns across the device in between certain LABs.

M4K-RAM blocks are true dual-port memory blocks with 4 kb (k bits) plus parity (4608 bits). These blocks provide dedicated true dual-port, simple dual-port or single-port memory up to 36-bit wide and can be used in true dual-port memory and other embedded memory applications. Like M512-RAM blocks, M4K-RAM blocks are also grouped into columns across the device in between certain LABs.

M-RAM blocks are true dual-port memory blocks with 512 kb plus parity (589 824 bits). These blocks provide dedicated true dual-port, simple dual-port or single-port memory up to 144-bit wide. Several M-RAM blocks are located individually or in pairs within the device, as shown in Figure 10.40.

Each digital signal processing (DSP) block can implement up to eight 9×9-bit multipliers, four 18×18-bit multipliers or one 36×36-bit multiplier with add or subtract features. These DSP blocks also contain 18-bit input shift registers for digital signal processing applications, including FIR and infinite impulse response (IIR) filters. DSP blocks are grouped into two columns in each device, as shown in Figure 10.40.

Review Questions

Q10.61   Describe the basic structure of the APEX20K FPGA family.

Q10.62   Describe the structure of logic elements (LEs).

Q10.63   Describe the structure of MegaLAB.

Q10.64   Describe the structure of an embedded system block (ESB).

Q10.65   Describe the structure of an IOB.

Q10.66   What are the features of Stratix devices?

10.5   PRACTICAL ISSUES

Due to the evolution of VLSI technology, a variety of devices, including ASICs and field-programmable devices, may co-exist in the same system. These devices have different logic levels and power-supply voltages. Consequently, the interface between them becomes an important issue. The two most important issues related to the interface between two devices are then considered in this section: *I/O standards* and *voltage tolerance*. An I/O standard defines a set of electrical rules that connect two devices in order to transfer information through electrical signals. A device is said to be voltage-tolerant when it can withstand a voltage greater than its V_{DD} on its I/O pins. Both input and output stages have voltage-tolerant problems. The *input voltage tolerance* needs to be considered when an input of a logic circuit with a low power-supply voltage is driven by an output from a logic circuit with a high power-supply voltage. The *output voltage tolerance* needs to be considered when two or more outputs of logic circuits with different power-supply voltages are connected together.

10.5.1 I/O Standards

Due to the fast evolution of VLSI techniques, the feature size of MOS transistors has been reduced from several micrometers down to tenths of nanometers (50 nm and beyond). This evolution has led the operating voltage of chips to move toward lower voltages. Nowadays, it is not uncommon that many operating voltages co-exist in the same PCB system, even in the same chip, where the core logic is operated at a low voltage, say 2.5 V, but the I/O circuits are operated at a higher voltage, say 3.3 V, for being compatible to other older-generation devices in the system. For the easy of interfacing chips with different operating voltages, an IC industry standards group, called the *Joint Electron Device Engineering Council* (JEDEC), selected three standard power-supply voltages for logic circuits. These standard voltages are 3.3 V $\pm$ 0.3 V, 2.5 V $\pm$ 0.2 V and 1.8 V $\pm$ 0.15 V. The logic levels associated with these standard voltages are listed in Table 10.6.

The JEDEC standard for 3.3-V logic actually contains two sets of logic levels: low-voltage CMOS (LVCMOS) and low-voltage TTL (LVTTL). LVCMOS levels are used in CMOS applications where the outputs have light loads (currents), usually less than 100 μA, so that V_{OL} and V_{OH} are maintained within 0.2 V of the power-supply rails, namely, $V_{OL} = 0.2$ V and $V_{OH} = 3.1$ V. LVTTL levels are used in TTL applications where the outputs have significant loads (currents). The $V_{OL} = 0.4$ V and $V_{OH} = 2.4$ V, which are compatible with traditional 5-V TTL levels.

Although a lot of I/O standards are defined and used in various application environments, they can be classified into the following three categories: chip-to-chip interface, backplane interface and high-speed memory interface. In the following, we only focus on the two most widely used chip-to-chip interfaces: low-voltage TTL (LVTTL) and low-voltage differential signal (LVDS). The interested reader may refer to Granberg [5] for details of these I/O standards.

Figure 10.41(a) shows the single-ended signal transfer circuit and its related logic levels. The voltage swing between the two levels is 1.2 V. Figure 10.41(b) gives an illustration of how a differential signal can be transferred between a transmitter and a receiver. The use of LVDS allows us to transmit data at the rate of hundreds (>400) and even thousands of megabits per second (Mbps). The low swing and current-mode

TABLE 10.6 The most widely used standard logic levels

Voltage	5-V CMOS (V)	5-V TTL (V)	LVTTL (V)	2.5-V CMOS (V)	1.8-CMOS (V)
V_{CC}	5.0	5.0	3.3	2.5	1.8
V_{OH}	4.44	2.4	2.4	2.0	1.45
V_{IH}	3.5	2.0	2.0	1.7	1.2
V_{IL}	1.5	0.8	0.8	0.7	0.65
V_{OL}	0.5	0.4	0.4	0.4	0.45

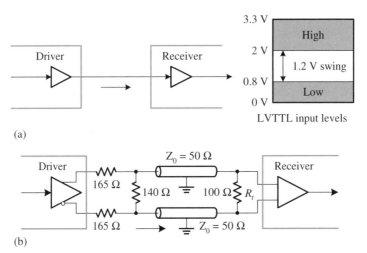

(a)

(b)

FIGURE 10.41 LVTTL signal versus LVDS signal: (a) single-ended data transfer; (b) differential signal data transfer (LVDS)

driver outputs create low noise and provide very low power dissipation. The LVDS drivers are capable of transmitting a signal over cable up to 10 to 15 m. As shown in Figure 10.41(b), the output voltage swing is ± 350 mV when the value of the terminal resistor R_t is 100 Ω. The receiver threshold voltage is ± 100 mV and can accommodate 1.2-V common mode input voltage.

10.5.2 Voltage Tolerance

A device is said to be voltage tolerant when it can withstand a voltage greater than its V_{DD} on its I/O pins. When the device can receive a voltage greater than its V_{DD} on its input pin, it is said to have input voltage tolerance. When the device can drive a load with a power-supply voltage greater than its V_{DD} on its output pin, it is said to have output voltage tolerance. A device is voltage compliance if it can receive and transmit signals from and to a device with a higher power-supply voltage than itself.

Because the interface between two devices separately operating 5-V TTL and 3.3-V LVTTL logic levels are usually encountered in many practical applications, in the following we only consider the input-tolerant and output-tolerant problems between these two voltages.

Input Voltage Tolerance   As we have shown in Table 10.6, even though both LVTTL and TTL families have the same logic levels, it probably could not work properly if we ignore the potential voltage-tolerant problem between them. In order to illustrate this, let us consider the circuit shown in Figure 10.42(a), where a standard dual-diode input protection network is used to protect the input buffer from being damaged by electrostatic discharge (ESD). Due to the use of the clamped diodes $D1$ and $D2$, the input voltage will be clamped in between −0.7 V and +0.7 + V_{DD}, where 0.7 V is the turn-on voltage of either diode. A problem will arise when the input is from the output

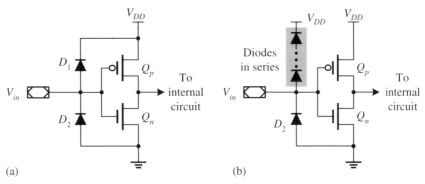

FIGURE 10.42 The typical input stage of CMOS logic circuits: (a) non 5-V tolerant input; (b) 5-V tolerant input

of a CMOS device operating at 5 V and V_{DD} is 3.3 V. From Table 10.6, the V_{OH} is greater than 4.44 V, which makes the diode $D1$ turn on and a low-resistance is formed between the input and power-supply rail V_{DD}.

A simple way to solve the above problem is to cascade several diodes, but still keep the $D2$ for clamping the negative voltage. The resulting circuit is shown in Figure 10.42(b). The number of diodes cascaded is determined by the voltage difference between the input and the supply voltage. As a result, the inputs of the device are 5-V tolerant.

Output Voltage Tolerance When the outputs of two logic circuits with separate power-supply voltages of 3.3 and 5.0 V are connected together, there may exist the 5-V output tolerant problem. Consider the circuit shown in Figure 10.43(a), where two output stages with different power-supply voltages are connected together. Under normal conditions without taking the output voltage at Y into consideration, both nMOS transistors Q_{1n} and Q_{2n}, and pMOS transistor Q_{1p}, are off. However,

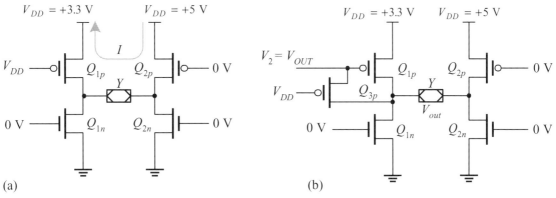

FIGURE 10.43 The typical output stage of CMOS logic circuits: (a) non 5-V tolerant output; (b) 5-V tolerant output

the voltage at node Y is 5 V due to the fact that the pMOS transistor Q_{2p} is actually on because it has 0-V gate voltage. Hence, the pMOS transistor Q_{1p} is indeed turned on by the high voltage appearing at the output Y, which is 5 V and greater than the 3.3 V at the gate by at least an amount that can turn on the transistor. As a result, there exists a low-resistance path between two power-supply rails and a small direct current is passing through both pMOS transistors. This current will consume power and generate heat on the I/O circuits and even degrade the performance of the devices.

Figure 10.43(b) shows a solution of the problem encountered in Figure 10.43(a) by adding a pMOS transistor Q_{3p} to the output buffer of the device with low power-supply voltage. This transistor is connected across the gate and drain of pMOS transistor Q_{1p} and its gate is connected to V_{DD} (3.3 V) so that it can be turned on and make the gate and drain of pMOS transistor Q_{1p} at the same potential when the output Y is at high voltage of 5 V. As a result, the pMOS transistor Q_{1p} remains off. In a commercial 3.3-V LVTTL device, when its outputs can be directly connected with the outputs of a 5-V TTL device, the outputs of the device are 5-V tolerant.

Review Questions

Q10.67 Explain the meaning of voltage tolerance.

Q10.68 Explain the meaning of voltage compliance.

Q10.69 Describe the features of LVDS.

Q10.70 Can an LVTTL device be driven by a standard TTL device?

Q10.71 Can both the outputs of an LVTTL device and a standard TTL device be connected together?

SUMMARY

In this chapter, we have examined a hierarchical structure and various design options of digital systems. The hierarchical structure, also known as a modular design technique, is often used to design a digital system. The essence of a hierarchical structure is to partition the system into many smaller independent sub-structures which combine to perform the same functionality of the original system.

The design options of digital systems include field-programmable devices and ASICs. Field-programmable devices are the ones that can be programmed in laboratories and include PLDs, CPLDs and FPGAs. The PLDs are devices built on the two-level AND–OR logic structure. The PLDs contain three types: ROM, PLA and PAL. PLDs are used in low-end applications.

The CPLD and FPGA devices are used in more complex digital systems and dominate the current market of digital systems. The basic functional structure of CPLDs is the same as that of PLDs; namely, it is still a two-level AND–OR logic structure. A CPLD virtually embeds many PLDs onto the same chip and interconnects them through a programmable interconnection network. As a result, a complex digital system may be designed by using a single CPLD. FPGAs have a quite different structure to CPLDs. They use configurable logic blocks (CLBs) or logic elements (LEs) to implement switching functions and can be roughly divided into two structures: matrix type and row type, according to the arrangement of CLBs and LEs on the chip.

ASICs mean those devices needed to be manufactured in IC foundries. An ASIC may be designed with either full-custom, cell-based or gate-array-based approaches. The full-custom design starts from scratch, namely, from zero to the final product. The cell-based design incorporates a standard cell library into a synthesizable design flow so that it can reduce both the design time and the time to market considerably. However, the standard cell library still needs to be designed in full-custom approach and the final prototypes using it need to be fabricated by IC foundries. Gate arrays are the wafers with prefabricated transistors and leave the metalization masks undefined. Design with gate arrays is still a synthesizable design flow by incorporating into a macro library supplied by the vendor of gate arrays. Consequently, compared to the full-custom approach it not only can shorten the design time like the cell-based approach but can also speed up the prototyping of the design. In addition, the time to market of a product designed with the gate-array-based approach is shorter than that with cell-based and full-custom approaches since gate arrays are prefabricated wafers and hence save a lot of time-consuming steps.

Due to the evolution of VLSI technology, it is not uncommon that a variety of devices, including ASICs and field-programmable devices, co-exist in the same system. Because of being manufactured with different process technologies, these devices have different logic levels and power-supply voltages. Consequently, the interface between them becomes an important issue. The two most important issues related to the interface between two devices are I/O standards and voltage tolerance. An I/O standard defines a set of electrical rules that connect two devices in order to transfer information through electrical signals. A device is said to be voltage-tolerant if it can withstand a voltage greater than its V_{DD} on its I/O pins.

REFERENCES

1. Altera Corporation, *MAX 7000 Programmable Logic Device Family,* ver. 6.7, 2005 (http://www.altera.com).
2. Altera Corporation, *APEX 20K Programmable Logic Device Family,* ver. 5.1, 2004.
3. Altera Corporation, *Stratix Device Handbook,* Vol. 1, 2006.
4. M.D. Ciletti, *Modeling, Synthesis and Rapid Prototyping with the Verilog HDL,* Prentice-Hall, Upper Saddle River, NJ, USA, 1999.
5. T. Granberg, *Handbook of Digital Techniques for High-Speed Design,* Prentice-Hall, Upper Saddle River, NJ, USA, 2004.
6. M.-B. Lin, *Digital System Design: Principles, Practices and ASIC Realization,* 3rd Edn, Chuan Hwa Book Company, Taipei, Taiwan, 2002.
7. J.F. Wakerly, *Digital Design Principles and Practices,* 3rd Edn, Prentice-Hall, Upper Saddle River, NJ, USA, 2001.
8. W. Wolf, *FPGA-Based System Design,* Prentice-Hall, Upper Saddle River, NJ, USA, 2004.
9. Xilinx Corporation, *FastFLASH XC9500XV: High-Performance, Low-Cost CPLD Family,* v1.1, San Jose, CA, USA, 2000 (http://www.xilinx.com).

10. Xilinx Corporation, *XC4000E and XC4000X Series Field Programmable Gate Arrays,* v1.6, San Jose, CA, USA, 1999.

11. Xilinx Corporation, *Spartan-3 FPGA Family: Complete Data Sheet,* San Jose, CA, USA, 2005.

12. Xilinx Corporation, *Virtex-II Platform FPGAs: Complete Data Sheet,* v3.4, San Jose, CA, USA, 2005.

PROBLEMS

10.1 Consider the switching functions shown in Figure 10.15.

   (a) Use a PLA device to implement the switching functions.

   (b) Write a Verilog HDL module to describe the PLA device you used.

10.2 Consider the switching functions shown in Figure 10.15.

   (a) Use a PAL device to implement the switching functions.

   (b) Write a Verilog HDL module to describe the PAL device you used.

10.3 Consider the switching functions shown in Figure 10.17.

   (a) Use a ROM device to implement the switching functions.

   (b) Write a Verilog HDL module to describe the ROM device you used.

10.4 Consider the switching functions shown in Figure 10.17.

   (a) Use a PAL device to implement the switching functions.

   (b) Write a Verilog HDL module to describe the PAL device you used.

10.5 Consider the switching functions shown in Figure 10.19.

   (a) Use a ROM device to implement the switching functions.

   (b) Write a Verilog HDL module to describe the ROM device you used.

10.6 Consider the switching functions shown in Figure 10.19.

   (a) Use a PLA device to implement the switching functions.

   (b) Write a Verilog HDL module to describe the PLA device you used.

10.7 Consider the switching functions shown in Figure 10.20.

   (a) Use a ROM device to implement the switching functions.

   (b) Write a Verilog HDL module to describe the ROM device you used.

10.8 Consider the switching functions shown in Figure 10.20.

   (a) Use a PAL device to implement the switching functions.

   (b) Write a Verilog HDL module to describe the PAL device you used.

10.9 Use the XC9500 CPLD family to implement the Booth multiplier: style2 step 3 in Section 11.2.

   (a) Observe how many macrocells are used.

   (b) Observe the longest propagation time; namely, find the minimum clock period.

10.10 Use the MAX7000 CPLD family to implement the Booth multiplier: style2 step 3 in Section 11.2.

(a) Observe how many macrocells are used.

(b) Observe the longest propagation time; namely, find the minimum clock period.

10.11 Use the XC4000 FPGA family to implement the Booth multiplier: style2 step 3 in Section 11.2.

(a) Observe how many LUTs are used.

(b) Observe the longest propagation time; namely, find the minimum clock period.

10.12 Use the APEX20K FPGA family to implement the Booth multiplier: style2 step 3 in Section 11.2.

(a) Observe how many LEs are used.

(b) Observe the longest propagation time; namely, find the minimum clock period.

10.13 Use the XC9500 CPLD family to implement the Booth array multiplier in Section 11.3.

(a) Observe how many macrocells are used.

(b) Observe the longest propagation time; namely, find the critical path.

10.14 Use the MAX7000 CPLD family to implement the Booth array multiplier in Section 11.3.

(a) Observe how many macrocells are used.

(b) Observe the longest propagation time; namely, find the critical path.

10.15 Use the XC4000 FPGA family to implement the Booth array multiplier in Section 11.3.

(a) Observe how many LUTs are used.

(b) Observe the longest propagation time; namely, find the critical path.

10.16 Use the APEX20K FPGA family to implement the Booth array multiplier in Section 11.3.

(a) Observe how many LEs are used.

(b) Observe the longest propagation time; namely, find the critical path.

SYSTEM DESIGN METHODOLOGY

IN THE previous chapters, we have dealt with various combinational and sequential logic modules widely used in digital systems. In this chapter, we will introduce two useful techniques by which a system can be designed. These techniques include the finite-state machine (FSM) and register-transfer level (RTL) design approaches. The former may be described by using a state diagram or an algorithmic state machine (ASM) chart; the latter may be described by an ASM chart. The ASM chart is also known as a state machine (SM) chart and only composed of three blocks: state block, decision block and conditional output block. An important feature of ASM charts is that it precisely defines the operations on every cycle and clearly shows the flow of control from state to state. In this chapter, we are concerned with how to model both the state diagram and ASM chart in Verilog HDL.

An RTL design may be realized by either coding an ASM chart directly or using the datapath and controller (DP+CU) paradigm. For simple systems, the datapath and controller of a design can be derived from its ASM chart directly. The datapath portion corresponds to the registers and function units in the ASM chart, and the controller portion corresponds to the generation of control signals needed in the datapath. In order to illuminate this technique, we outline a three-step paradigm by which the datapath and controller of a design could be derived from its ASM chart easily. For complex systems, their datapaths and controllers are often derived from specifications in a state-of-the-art manner. An example of displaying a four-digit data on a commercial dot-matrix liquid-crystal display (LCD) module is used to illustrate this approach.

11.1 FINITE-STATE MACHINE

A finite-state machine is usually used to model a circuit with memory such as a sequential circuit or the controller in a small digital system. In a sequential circuit, the output is determined not only by the present inputs but also by the past history. Hence, it is a circuit with memory. In this section, we focus our attention on finite-state machines and their related issues often encountered when designing digital systems.

11.1.1 Types of Sequential Circuits

A finite-state machine (FSM) $\mathcal{M}$ is a quintuple $\mathcal{M} = (\mathcal{I}, \mathcal{O}, \mathcal{S}, \delta, \lambda)$, where $\mathcal{I}, \mathcal{O}$ and $\mathcal{S}$ are finite, nonempty sets of input, output and state symbols, respectively; δ and λ are the state transition function and output function, respectively. Here, a symbol is a combination of input, output or state variables.

The state transition function δ is defined as follows:

$$\delta : \mathcal{I} \times \mathcal{S} \to \mathcal{S} \tag{11.1}$$

which means that the δ function is determined by both input and state symbols.

The output function λ is defined as follows:

$$\lambda : \mathcal{I} \times \mathcal{S} \to \mathcal{O} \quad \text{(Mealy machine)} \tag{11.2}$$

$$\lambda : \mathcal{I} \to \mathcal{O} \quad \quad \text{(Moore machine)} \tag{11.3}$$

where in a Mealy machine, the output function is determined by both the present input and state, whereas in a Moore machine the output function is only determined by the present state. Illustrations of both Mealy and Moore machines are portrayed in Figure 11.1.

The design issues of an FSM used in digital systems are to compute both the state transition and output functions from input and (present) states. In other words, we need to design both combinational logic circuits of the state transition function and output

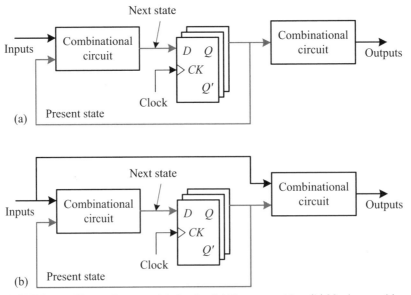

FIGURE 11.1 Types of sequential circuits: (a) Moore machine; (b) Mealey machine

TABLE 11.1 Some state encoding examples

State	Binary	Gray	One hot
A	00	00	1000
B	01	01	0100
C	10	11	0010
D	11	10	0001

function from the present input values and present state according to the behavioral description of the digital system.

In order to design a sequential circuit based on the FSM approach, it is necessary to derive a state diagram (or state table) from the circuit specification. Then, the states are encoded into an appropriate code from which the output and state transition functions can be derived. The most common FSM state encoding options include the following:

1. *One-hot encoding* sets one of the state registers to 1 at each state; namely, it uses the 1-out-of-n coding scheme.

2. *Binary encoding* assigns binary code to state registers according to the state sequence.

3. *Gray encoding* assigns a Gray code to state registers according to the state sequence.

4. *Random encoding* assigns a random code to state registers for each state.

One-hot encoding is usually used in FPGA-based design due to a lot of registers available for use in theses devices. An example of state encoding using binary, Gray and one-hot is shown in Table 11.1.

11.1.2 FSM Modeling Styles

Now that we know the definition and circuit types of FSMs, the next issue naturally arising is how to implement (or realize) an FSM in Verilog HDL. Two major implementation issues of an FSM are the declaration and update of state register, and the computation of both next-state and output functions. State register can be declared as one register such as `state` and two registers such as `present_state` and `next_state`. One-register style is harder to follow and model. Two-register style is easier to follow and model. Output and next state functions can be computed by one of the following approaches or their combinations: continuous assignment, function and `always` block.

11.1.2.1 Modeling Styles   As mentioned above, to model an FSM explicitly two styles can be used depending on whether one or two state registers are used. As a consequence, in the following we divide the modeling styles of an FSM into two types according to whether the state register is declared as one register or two registers.

Style 1: One State Register   In this style, only one state register is declared. The description can be divided into three parts.

Part 1: *Initialize and update the state register.* Usually, an `always` block is used for this purpose. The next state is computed by a function `next_state()`.

```
always @(posedge clk or negedge reset_n)
   if (!reset_n) state <= A;
   else state <= next_state (state, x) ;
```

Part 2: *Compute next state using function.* It usually uses a function to compute the next state function from both current inputs and present state.

```
function next_state (input present_state, x)
   case (present_state )
   ...
   endcase
endfunction
```

Part 3: *Compute output function.* The output function can be computed by using either an `always` block or continuous assignments. When using an `always` block, a `case` statement is often employed inside the `always` block to compute the output function associated with each state.

```
always @(state or x)
   case (state)... endcase or
continuous assignments
```

Style 2: Two State Registers   In this style, two state registers, `present_state` and `next_state`, are declared. The description can be divided into three parts.

Part 1: *Initialize and update the state register.* Usually, an `always` block is used for this purpose. Inside the `always` block, an `if-else` statement is used to select whether the state assigned to the `present_state` is the start state or next state.

```
always @(posedge clk or negedge reset_n)
if (!reset_n) present_state <= A;
else present_state <= next_state;
```

Part 2: *Compute next state.* It usually uses an `always` block to compute the next state function from both current input and present state. Sometimes continuous assignments are used instead.

```
always @(present_state or x)
   case (present_state )... endcase or
continuous assignments
```

Part 3: *Compute output function.* The output function can be computed by using either an `always` block or continuous assignments. When using an `always` block, a `case`

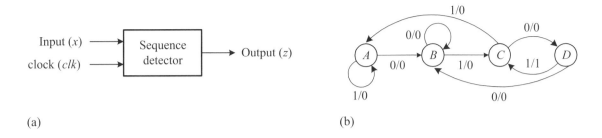

(a)   (b)

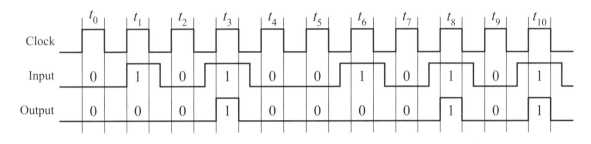

(c)

FIGURE 11.2 An example of a sequence 0101 detector: (a) block diagram; (b) state diagram; (c) timing chart

statement is often employed inside the `always` block to compute the output function associated with each state.

```
always @(present_state or x)
    case (present_state ) ... endcase or
continuous assignments
```

In the rest of this section, we use examples to illustrate how to describe an FSM in Verilog HDL using the above modeling styles.

Problem Description   Suppose that a sequential circuit has an input x and an output z. The input x receives a sequence of 1 and 0. The output z outputs 1 whenever the circuit detects a pattern of 0101 in its input x. An instance of this 0101 sequence detector is shown in Figure 11.2. The state diagram of the sequence 0101 detector is shown in Figure 11.2(b), which can be derived by using the approach described in most digital logic textbooks, such as Gajski [3].

The following example models the 0101 sequence detector in behavioral style using one state register.

EXAMPLE 11.1 *Behavioral Description of a 0101 Sequence Detector—Style 1a*

This example declares a state register `state` and uses an `always` block to update the state register. The next state is computed by a function `fsm_next_state`, which uses a `case` statement containing `if-else` statements to encode the state diagram into RTL code. The output function is simply computed by a continuous assignment.

```
// behavioral description of 0101 sequence detector ---
// style 1a
// Mealy machine example ---
// using only state register and one always block.
module sequence_detector_mealy(clk, reset_n, x, z);
input  x, clk, reset_n;
output wire z;
// declare state register
reg  [1:0] state;  // for both ps and ns
localparam A = 2'b00, B = 2'b01, C = 2'b10, D = 2'b11;
// part 1: update state register
always @(posedge clk or negedge reset_n) begin
   if (!reset_n) state <= A;
   else state <= fsm_next_state(state, x);
end
// part 2: compute the next state function
function [1:0] fsm_next_state (input [1:0] present_state,
input x);
reg [1:0] next_state;
begin
   case (present_state)
      A: if (x) next_state = A;  else next_state = B;
      B: if (x) next_state = C;  else next_state = B;
      C: if (x) next_state = A;  else next_state = D;
      D: if (x) next_state = C;  else next_state = B;
   endcase
   fsm_next_state = next_state;
end endfunction
// part 3: evaluate the output function
assign z = (x == 1 && state == D);
endmodule
```

The following is another example of using style 1. It uses two `always` blocks, one for updating state register and the other for computing the output function.

EXAMPLE 11.2 *Behavioral Description of a 0101 Sequence Detector— Style 1b*

This example declares a state register `state` and uses an `always` block to update the state register. The next state is computed by a function `fsm_next_state`. The output function is simply computed by another `always` block. Both next state and output functions are separately computed by using a `case` statement and its related `if-else` statements.

```
// behavioral description of 0101 sequence detector:
// style 1b
// Mealy machine example ---
// using only one state register and two always blocks.
module sequence_detector_mealy(clk, reset_n, x, z);
input  x, clk, reset_n;
output reg z;
// declare state register
reg  [1:0] state;  // for both ps and ns
localparam A = 2'b00, B = 2'b01, C = 2'b10, D = 2'b11;
// part 1: update state register
always @(posedge clk or negedge reset_n) begin
   if (!reset_n) state <= A;
   else state <= fsm_next_state(state, x);
end
// part 2: compute the next state function
function [1:0] fsm_next_state (input [1:0] present_state,
   input x);
reg [1:0] next_state;
begin
   case (present_state)
      A: if (x) next_state = A;  else next_state = B;
      B: if (x) next_state = C;  else next_state = B;
      C: if (x) next_state = A;  else next_state = D;
      D: if (x) next_state = C;  else next_state = B;
   endcase
   fsm_next_state = next_state;
end endfunction
// part 3: evaluate output function z
always @(state or x)
   case (state)
      A: if (x) z = 1'b0; else z = 1'b0;
      B: if (x) z = 1'b0; else z = 1'b0;
      C: if (x) z = 1'b0; else z = 1'b0;
      D: if (x) z = 1'b1; else z = 1'b0;
   endcase
endmodule
```

■

The following is an example of using style 2. It uses three `always` blocks, for updating the state register, computing the next state function and computing the output function.

EXAMPLE 11.3 *Behavioral Description of a 0101 Sequence Detector— Style 2*

This example declares two state registers `present_state` and `next_state` and uses an `always` block to update the state register. The next state is computed by another `always` block. The output function is simply computed by the third `always` block. Both next state and output functions are separately computed by using a `case` statement and its related `if-else` statements.

```
// behavioral description of 0101 sequence detector:
// style 2
// Mealy machine example ---
// using two state registers and three always blocks.
module sequence_detector_mealy(clk, reset_n, x, z);
input  x, clk, reset_n;
output z;
// declare state registers: present state and next state
reg [1:0] present_state, next_state;
reg z;
localparam A = 2'b00, B = 2'b01, C = 2'b10, D = 2'b11;
// part 1: Initialize to state A and update state register
always @(posedge clk or negedge reset_n)
   if (!reset_n) present_state <= A;
   else present_state <= next_state; // update present
                                           state
// part 2: compute the next state function
always @(present_state or x)
   case (present_state)
      A: if (x) next_state = A; else next_state = B;
      B: if (x) next_state = C; else next_state = B;
      C: if (x) next_state = A; else next_state = D;
      D: if (x) next_state = C; else next_state = B;
   endcase
// part 3: evaluate output function z
always @(present_state or x)
   case (present_state)
      A: if (x) z = 1'b0; else z = 1'b0;
      B: if (x) z = 1'b0; else z = 1'b0;
      C: if (x) z = 1'b0; else z = 1'b0;
      D: if (x) z = 1'b1; else z = 1'b0;
   endcase
endmodule                                               ■
```

In this simple example of a 0101 sequence detector, both modeling styles yield the same result, as shown in Figure 11.3, regardless of whether one state register or two state registers are used. In addition, the declaration of using two registers, `present_state` and `next_state`, does not mean the synthesized result will have two state registers. It only provides a convenient way to write the RTL code.

Guidelines for writing finite state machines that may help in optimizing the logic are listed as follows.

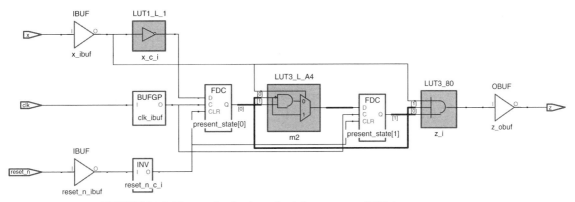

FIGURE 11.3 The synthesized result of the sequence 0101 detector

Coding Styles

1. State name should be described by using the parameters, `parameter` or `localparam`.

2. Combinational logic for computing the next state should be separated from the state registers, i.e. using its own `always` block or a function.

3. The combinational logic of the next state function should be implemented with a `case` statement.

11.1.3 Implicit versus Explicit FSM

Essentially, we have two basic approaches to realizing an FSM circuit in Verilog HDL: *implicit FSM* and *explicit FSM*.

Implicit FSM   The essential idea of an implicit FSM is that states are not declared explicitly. The synthesis tools infer the state from the activity within a cyclic (such as `always @(posedge clk)...`) behavior. It is so hard to write a source code in this approach clearly and correctly. In addition, not every synthesis tool supports this type of FSM. As a result, it is often not suggested to use this method to realize an FSM circuit.

Explicit FSM   The essential idea of an explicit FSM is that states are declared explicitly. It is rather easy to write a source code in this approach clearly and correctly and every synthesis tool supports this type of FSM. Almost all FSMs are written in this type. Therefore, an explicit FSM is usually called an FSM for short.

Due to the difficulty of writing a correct implicit FSM program, we only give an example to illustrate this modeling style and will not further discuss it here.

EXAMPLE 11.4 *An Implicit FSM Example*

An implicit FSM means that we do not declare the state register explicitly. The FSM behavior is automatically inferred from the source code by synthesis tools. It is

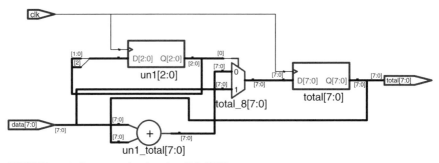

FIGURE 11.4 An example of an implicit FSM

suggested that an implicit FSM be used only when the data flow is in a linear fashion, such as in this example.

```
// an implicit FSM example
module sum_3data_implicit(clk, data, total);
parameter n = 8;
input   clk;
input   [n-1:0] data;
output  [n-1:0] total;
reg     [n-1:0] total;
// we do not declare state registers.
always begin
   @(posedge clk) total <= data;
   @(posedge clk) total <= total + data;
   @(posedge clk) total <= total + data;
end
endmodule
```

The synthesized result is shown in Figure 11.4. It is worth noting that an implicit FSM is seldom used in digital system designs except something like the one shown in the preceding example. The following example uses the explicit style to rewrite the above example.

EXAMPLE 11.5 *An Explicit FSM Example*

An explicit FSM means that we need to declare the state register explicitly. That is, we need to schedule the sequential operations of the circuit. For example, we use an FSM with three states to arrange the operations of the preceding example.

```
// an explicit FSM example
module sum_3data_explicit(clk, reset, data, total);
parameter n = 8;
input   clk, reset;
input   [n-1:0] data;
output  [n-1:0] total;
```

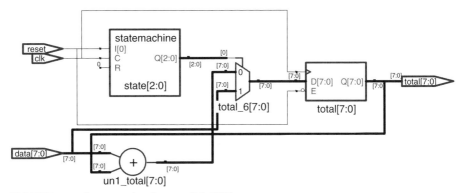

FIGURE 11.5 An example of an explicit FSM

```
reg     [n-1:0] total;
reg     [1:0]   state;
localparam A = 2'b00, B =2'b01, C = 2'b10;
// the FSM used to schedule the sequential operations
always @(posedge clk)
   if (reset) state <= A; else
   case (state)
      A: begin total <= data; state <= B; end
      B: begin total <= total + data; state <= C; end
      C: begin total <= total + data; state <= A; end
   endcase
endmodule
```

The synthesized result is shown in Figure 11.5. Although both implicit and explicit FSMs have the same synthesized results for the above two examples, it is better to use an explicit FSM in practical digital systems.

Coding Styles

1. It should avoid mixing positive and negative edge-triggered flip flops.

2. It is good practice to use an explicit FSM instead of an implicit FSM.

3. It should isolate a state-machine from other logic.

4. It should avoid multiple clocks within a block.

Review Questions

Q11.1 Define a finite-state machine.

Q11.2 What are the two major types when using the FSM to model a sequential circuit?

Q11.3 What are the differences between a Moore machine and a Mealy machine?

Q11.4 Explain what an explicit FSM is.

Q11.5 Explain what an implicit FSM is.

11.2 RTL DESIGN

The register-transfer level (RTL) design is often used for designing a large system at the algorithmic level. The rationale behind the RTL design is that any digital system can be considered as a system of interconnected registers being placed between two registers for a combinational logic to perform the desired functions. As a consequence, the digital system operates in RTL in such a way that data from the output of registers are transferred to other registers, through which a combinational logic is used to change the property of the data.

The RTL design has the following important features. First, a register-transfer design is a sequential machine. Second, a register-transfer design is structural. A complex combinations of state machines may not be easily described solely by a large state diagram or ASM chart. Third, an RTL design concentrates on functionality, not details of logic design. Today, two widely used design approaches at register-transfer level (RTL) are an *ASM chart*, as well as the *datapath and controller* (usually called the DP+CU approach for short). In this section, we will address these two approaches in sequence.

11.2.1 ASM Chart

The ASM chart, sometimes called a *state machine* (SM) chart, is often used to describe a digital design at the algorithmic level. Two essential features of ASM charts are as follows. First, they specify RTL operations since they define what happens on every cycle of operation. Second, they show clearly the flow of control from state to state.

An ASM chart is composed of three blocks, *state block*, *decision block* and *conditional output block*, as shown in Figure 11.6. A state block, as shown in Figure 11.6(a), specifies a machine state and a set of unconditional RTL operations associated with the state. It may execute as many actions as you want and all actions in a state block occur in parallel. Each state, along with its related operations, occupies a clock

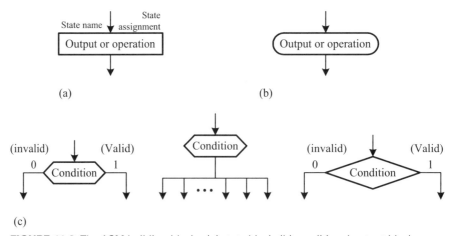

FIGURE 11.6 The ASM building blocks: (a) state block; (b) conditional output block; (c) decision block

period. In an ASM chart, state blocks are executed sequentially. In addition, a register may be assigned to only once in a state block. This is known as a *single-assignment rule*.

A conditional output block, as shown in Figure 11.6(b), describes the RTL operations that are executed under conditions specified by one or more decision blocks. The input of a conditional output block must be from the output of a decision block or the other conditional output block. A conditional output block can only evaluate the present state or primary input value on a present cycle.

A decision block, as shown in Figure 11.6(c), describes the condition under which an ASM will execute specific actions and select the next state based on the value of the primary input or present state. It can be drawn in either two-way selection or multi-way selection.

An ASM block contains exactly one state block, together with the possible decision blocks and conditional output blocks associated with that state. Each ASM block describes the operations executed in one state. An ASM block has the following two features. First, it has exactly one entrance path and one or more exit paths. Second, it contains one state block and a serial-parallel network of decision blocks and conditional output blocks. The following example gives two instances of valid ASM blocks.

EXAMPLE 11.6   *Examples of Valid ASM Blocks*

The ASM blocks shown in Figure 11.7(a) and (b) are both valid. Figure 11.7(a) shows a serial test but Figure 11.7(a) displays a parallel test. In Figure 11.7(a), condition x_1 is first tested and then condition x_2. Conditional output block $c \leftarrow 0$ is performed whenever $x_1 = 0$ and conditional output block $d \leftarrow 5$ is performed whenever $x_2 = 1$. The reader can check that all four combinations of x_1 and x_2 are consistent without any conflictions. In Figure 11.7(b), conditions x_1 and x_2 are tested independently and in parallel. Conditional output block $c \leftarrow 0$ is performed whenever $x_1 = 0$ and conditional output block $d \leftarrow 5$ is performed whenever $x_2 = 1$. It is worth noting that although both ASM blocks seem to be different, they indeed perform the same function. ■

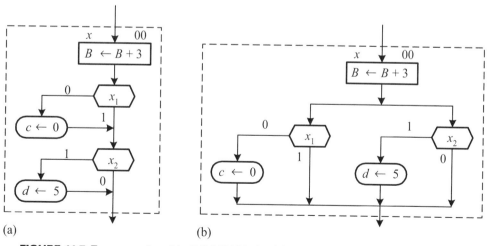

(a)   (b)

FIGURE 11.7 Two examples of (valid) ASM blocks: (a) serial test; (b) parallel test

The basic constructing rules for ASM charts are as follows:

Rule 1. Each state and its associated set of conditions must define a unique next state.

Rule 2. Every path of the network of conditions must terminate at a next state.

Rule 3. There cannot exist any loop in the network of conditions.

The following gives examples of invalid ASM blocks.

EXAMPLE 11.7 *Examples of Invalid ASM Blocks*

For example, the ASM block shown in Figure 11.8(a) is invalid because the next state is undefined. This can be seen from the figure as when x_1 and x_2 are both 0, the next state will be states B and C, respectively, not a unique state. Hence, it violates **rule 1**. Figure 11.8(b) shows another invalid ASM block. There is a loop around the decision blocks; namely, a loop is formed when both x_1 and x_2 are equal to 1 and conditional output block $c \leftarrow 0$. Consequently, it violates **rule 3.** ∎

An ASM chart is more powerful than a Moore or a Mealy machine. A restricted form of an ASM chart is often used to describe a Moore or Mealy machine. For example, when describing a Moore (sequential) machine, only state and decision blocks are required. In this case, the state blocks require assignments associated with it because in Moore machines the outputs are associated with states. When describing a Mealy (sequential) machine, it requires conditional output blocks (i.e. outputs) in addition to state and decision blocks. In this case, the state blocks do not require any assignments associated with it because in Mealey machines the outputs are associated with transitions.

An example of the equivalence between a state diagram of a Mealy machine and an ASM chart is shown in Figure 11.9. Each state of the state diagram corresponds to

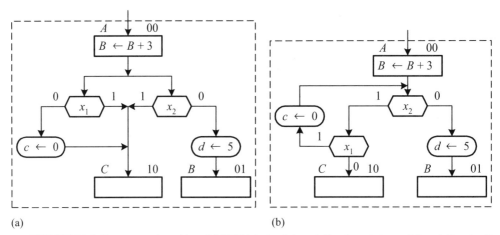

(a)   (b)

FIGURE 11.8 Two examples of invalid ASM blocks: (a) undefined next stage; (b) undefined exit path

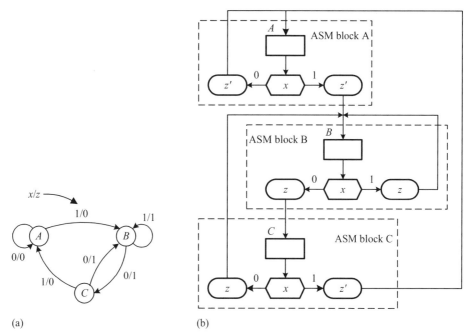

(a) (b)

FIGURE 11.9 An example of the equivalence between (a) a state diagram and (b) an ASM chart

an ASM block, as indicated in the figure with dashed-line blocks. For example, state A in the state diagram tests the input signal x and outputs 0, regardless of what the value of x is. The next state of state A is A if the input signal x is 0 and is state B if the input signal x is 1. The other two ASM blocks corresponding to states B and C, respectively, can be derived in a similar way. The complete ASM chart is depicted in Figure 11.9(b).

Review Questions

Q11.6 Describe the rationale behind the RTL design.

Q11.7 What are the major features of ASM charts?

Q11.8 What are the three building blocks of ASM charts?

Q11.9 When would you use conditional output blocks in an ASM chart?

Q11.10 When using an ASM chart to describe a Moore machine, what kinds of building blocks need to be used?

Q11.11 When using an ASM chart to describe a Mealy machine, what kinds of building blocks need to be used?

11.2.2 ASM Modeling Styles

Much like an FSM, an ASM chart can be modeled in either of two styles depending on whether one or two state registers are used. As a consequence, in the following we divide the modeling styles of an ASM chart into two types according to whether the state register is declared as one register or two registers.

11.2.2.1 Style 1a: One State Register and One `always` Block (not a Good Style!!!) In this modeling style, only one state register and one `always` block are used to describe the ASM chart.

Part 1: *Initialize, compute, update the state register and execute RTL operations.* This usually uses an `always` block to initialize, compute and update the state register, as well as execute RTL operations. An `if-else` statement is used to separate the initial operations from the rest of the operations in the ASM chart.

```
always @(posedge clk or negedge start_n)
   if (!start_n) begin
      state <= A;
         ...  // set initial values
   end
   else begin
      case     // perform RTL operations and determine
               // next state
         ...;
      endcase
   end
```

11.2.2.2 Style 1b: One State Register and Two `always` Blocks In this modeling style, one state register and two `always` blocks are used to describe the ASM chart. It can be divided into two parts.

Part 1: *Initialize, compute and update the state register.* This usually uses an `always` block to compute the next state function from both the current input and present state. An `if-else` statement is used to select whether the state assigned to the `state` is the start state or next state.

```
always @(posedge clk or negedge start_n)
   if (!start_n) state <= A;
   else  state <= ...;
```

Part 2: *Execute RTL operations.* This usually uses an `always` block coupled with a `case` statement to perform the desired RTL operations. Sometimes, continuous assignments are used instead.

```
always @(posedge clk or negedge start_n)
   if (!start_n) // initialization
   else case (state )
      ...
   endcase or
continuous assignments
```

11.2.2.3 Style 2: Two State Registers and Three `always` Blocks (the Best Style !!!!) In this modeling style, two state registers and three `always` blocks are used to describe the ASM chart. This style is much like the style 2 of FSM modeling.

Part 1: *Initialize and update the state register.* Usually, an `always` block is used for this purpose. Inside the `always` block, an `if-else` statement is used to select whether the state assigned to the `present_state` is the start state or next state.

```
always @(posedge clk or negedge start_n)
    if (!start_n) present_state <= A;
    else present_state <= next_state;
```

Part 2: *Compute next state.* This usually uses an `always` block to compute the next state function from both the current input and present state. Sometimes continuous assignments are used instead.

```
always @(present_state or x)
    case (present_state )
      . . .
    endcase  or
continuous assignments
```

Part 3: *Execute RTL operations.* The RTL operations can be computed by using either an `always` block or continuous assignments. Inside the `always` block, a `case` statement is usually employed to compute the RTL operations associated with each state.

```
always @(posedge clk or negedge start_n)
    if (!start_n) // initialization
    else case (present_state )
      . . .
    endcase or
continuous assignments
```

11.2.2.4 An ASM Example—Booth Algorithm

Problem Specification   Assume that two inputs are $X = x_{n-1}x_{n-2}\cdots x_2x_1x_0$ and $Y = y_{n-1}y_{n-2}\cdots y_2y_1y_0$. Then, the Booth (multiplication) algorithm can be described as follows. At the ith step, where $0 \leq i \leq n-1$, one of the following operations is performed according to the value of x_ix_{i-1}:

- Add 0 to the partial product P if $x_ix_{i-1} = 00$.
- Add Y to the partial product P if $x_ix_{i-1} = 01$.
- Subtract Y from the partial product P if $x_ix_{i-1} = 10$.
- Add 0 to the partial product P if $x_ix_{i-1} = 11$.

In order to derive the ASM chart for the Booth algorithm, we suppose that the multiplicand Y and multiplier X are stored in the registers *mcand* and *mp*, respectively.

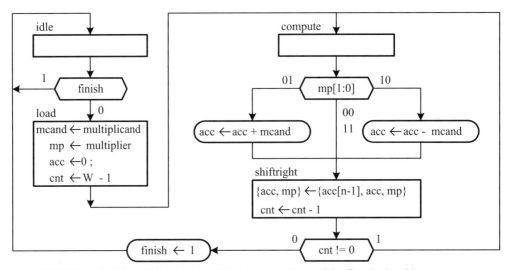

FIGURE 11.10 The ASM chart describing the operations of the Booth algorithm

The register mp has an extra bit in the right-most position to accommodate the $x_{-1} = 0$ for the 0th step. The multiplicand is added to or subtracted from acc at each step when the values of the lowest two bits of mp are 01 and 10, respectively. The result in acc, concatenated with the content of mp, is then arithmetically shifted right by one bit position. The acc, concatenated with the left-most n bits of register mp, forms a $2n$-bit partial product register. To control the desired iterations of the above operations, a counter is used for this purpose. Based on this description, an ASM chart is derived, as shown in Figure 11.10.

In order to further illustrate the detailed operations of the Booth algorithm, a numerical example is given in Figure 11.11. The reader is encouraged to trace each step of the numerical example along with the ASM chart shown in Figure 11.10.

The following gives an example of modeling the Booth algorithm described in the ASM chart shown in Figure 11.10 in style 1a.

EXAMPLE 11.8 *Booth Algorithm: Style 1a*

When using style 1a, only one `always` block is used to initialize, compute and update the state register, as well as execute RTL operations. An `if-else` statement is used to separate the initial operations from the rest of the operations in the ASM chart. When the `start_n` signal is active, all registers, `cnt`, `mp`, `acc` and `mcand`, as well as the flag `finish`, are cleared. Otherwise, the system enters into its normal operations. Each ASM block is modeled by using a case item selected by the `state` variable, which is used as the conditional expression of the `case` statement.

```
// Booth multiplier: style 1a --- realizing directly its
// ASM
module booth_step1(clk,start_n,multiplier,multiplicand,
  product,finish);
parameter W = 8; // default word size
parameter N = 3; // N = log2(W)
```

	acc	mp	mp(0)	cnt	
mp = 77H	0 0 0 0 0 0 0 0	0 1 1 1 0 1 1 1	0	7	acc ← acc - mcand
mcand = 13H	- 0 0 0 1 0 0 1 1				
	1 1 1 0 1 1 0 1				
	1 1 1 1 0 1 1 0	1 0 1 1 1 0 1 1	1	7	right shift acc:mp
	1 1 1 1 1 0 1 1	0 1 0 1 1 1 0 1	1	6	right shift acc:mp
	1 1 1 1 1 1 0 1	1 0 1 0 1 1 1 0	1	5	right shift acc:mp
	+ 0 0 0 1 0 0 1 1				acc ← acc +mcand
	0 0 0 1 0 0 0 0				
	0 0 0 0 1 0 0 0	0 1 0 1 0 1 1 1	0	4	right shift acc:mp
	- 0 0 0 1 0 0 1 1				
	1 1 1 1 0 1 0 1				
	1 1 1 1 1 0 1 0	1 0 1 0 1 0 1 1	1	3	right shift acc:mp
	1 1 1 1 1 1 0 1	0 1 0 1 0 1 0 1	1	2	right shift acc:mp
	1 1 1 1 1 1 1 0	1 0 1 0 1 0 1 0	1	1	right shift acc:mp
	+ 0 0 0 1 0 0 1 1				acc ← acc +mcand
	0 0 0 1 0 0 0 1				
08D5H →	0 0 0 0 1 0 0 0	1 1 0 1 0 1 0 1	0	0	right shift acc:mp

FIGURE 11.11 A numerical example illustrating the operations of the Booth algorithm

```
input   clk, start_n;
input   [W-1:0]   multiplier, multiplicand;
output  [2*W-1:0] product;
wire    [2*W-1:0] product;
reg     [W-1:0]   mcand, acc;
reg     [N-1:0]   cnt;
reg     [W:0]     mp;      // one extra bit
reg     [1:0]     state;
output reg        finish;  // finish flag
localparam idle = 2'b00, load = 2'b01,
           compute = 2'b10, shift_right = 2'b11;
// the body of the W-bit booth multiplier
always @(posedge clk or negedge start_n)
   if (!start_n) begin
      state <= idle; finish <= 0;
      cnt <= 0; mp <= 0; acc <= 0; mcand <= 0;
   end
   else begin
      case (state)
         idle: if (!finish) state <= load;
               else state <= idle;
         load: begin
            mp  <= {multiplier,1'b0};
            mcand <= multiplicand;
```

```
                    acc <= 0; cnt <= W - 1;
                    state <= compute; end
                compute: begin
                    case (mp[1:0])
                        2'b01: acc <= acc + mcand;
                        2'b10: acc <= acc - mcand;
                        default: ; // do nothing
                    endcase
                    state <= shift_right; end
                shift_right: begin
                    {acc, mp} <= {acc[W-1], acc, mp[W:1]};
                    cnt <= cnt - 1; if (cnt == 0) finish <= 1;
                    if (cnt == 0) state <= idle;
                    else state <= compute; end
            endcase
        end
assign product = {acc, mp[W:1]};
endmodule
```

The following is an example of modeling the Booth algorithm described in the ASM chart given in Figure 11.10 in style 1b.

EXAMPLE 11.9 *Booth Multiplier: Style 1b*

When using style 1b, one state register and two `always` blocks are used to describe the Booth algorithm. It can be divided into two parts. The first `always` block coupled with a function `fsm_next_state` initializes, computes and updates the state register. The second `always` block performs the desired RTL operations.

```
//Booth multiplier: style 1b --- using two always blocks
module booth_step1(clk,start_n,multiplier,multiplicand,
                   product,finish);
parameter W = 8; // default word size
parameter N = 3; // N = log2(W)
input  clk, start_n;
input  [W-1:0]   multiplier, multiplicand;
output [2*W-1:0] product;
wire   [2*W-1:0] product;
reg    [W-1:0]   mcand, acc;
reg    [N:0]     cnt;
reg    [W:0]     mp;      // one extra bit
reg    [1:0]     state;   // only declare one state
                          // register
output reg       finish;  // finish flag
localparam idle = 2'b00,    load = 2'b01,
           compute = 2'b10, shift_right = 2'b11;
```

```verilog
// the body of the W-bit booth multiplier
// part 1: Initialize, determine next state,
// and update the state
always @(posedge clk or negedge start_n)
   if (!start_n) state <= idle;
   else  state <= fsm_next_state(state, cnt, finish);
// function to compute next state.
function [1:0] fsm_next_state;
input [1:0] fsm_present_state;
input [N:0] fsm_cnt;
input fsm_finish;
reg  [1:0] next_state;
begin
   case (fsm_present_state)
      idle:    if (!fsm_finish) next_state = load;
               else next_state = idle;
      load:    next_state = compute;
      compute: next_state = shift_right;
      shift_right: if (fsm_cnt == 0) next_state = idle;
                   else next_state = compute;
   endcase
   fsm_next_state = next_state;
end endfunction
// part 2: execute RTL operations
always @(posedge clk or negedge start_n) begin
   if (!start_n) finish <= 0;
   case (state)
      idle: ;
      load: begin
         mp <= {multiplier,1'b0};
         mcand <= multiplicand;
         acc <= 0; cnt <= W - 1;    end
      compute: begin
         case (mp[1:0])
            2'b01: acc <= acc + mcand;
            2'b10: acc <= acc - mcand;
            default: ; // do nothing
         endcase end
      shift_right: begin
         {acc, mp} <= {acc[W-1], acc, mp[W:1]};
         cnt <= cnt - 1;
         if (cnt == 0) finish <= 1; end
   endcase
end
assign product = {acc, mp[W:1]};
endmodule
```

Like the case of an FSM, an ASM chart can also be described by using two state registers and three `always` blocks. This style is much like the style 2 of the FSM modeling.

EXAMPLE 11.10 *Booth Multiplier: Style 2*

In this style, three `always` blocks are used. The first `always` block initializes and updates the state register. The second `always` block computes the next state from the present state, current input as well as the condition computed. The third `always` block performs the desired RTL operations.

```
//Booth multiplier: style 2 --- using three always blocks
module booth_style2_step1(clk, start_n, multiplier,
                          multiplicand, product, finish);
parameter W = 8; // default word size
parameter N = 3; // n = log2(w)
input   clk, start_n;
input   [W-1:0]   multiplier, multiplicand;
output  [2*W-1:0] product;
wire    [2*W-1:0] product;
reg     [W-1:0]   mcand, acc;
reg     [N-1:0]   cnt;
reg     [W:0]     mp;  // one extra bit
reg     [1:0]     ps, ns;
output reg        finish;  // finish flag
localparam idle = 2'b00,    load = 2'b01,
           compute = 2'b10, shift_right = 2'b11;
// the body of the W-bit booth multiplier
// part 1: Initialize and update state registers
always @(posedge clk or negedge start_n)
   if (!start_n) ps <= idle;
   else  ps <= ns;
// part 2: compute next state
always @(*)
   case (ps)
      idle:    if (!finish) ns = load; else ns = idle;
      load:    ns = compute;
      compute: ns = shift_right;
      shift_right: if (cnt == 0) ns = idle;
                   else ns = compute;
   endcase
// part 3: execute RTL operations
always @(posedge clk or negedge start_n) begin
   if (!start_n) finish <= 0;
   case (ps)
      idle: ;  // do nothing
```

```
        load: begin
          mp <= {multiplier,1'b0};
          mcand <= multiplicand;
          acc   <= 0;
          cnt   <= W - 1; end
        compute: begin
          case (mp[1:0])
            2'b01: acc <= acc + mcand;
            2'b10: acc <= acc - mcand;
            default: ; // do nothing
          endcase end
        shift_right: begin
          {acc, mp} <= {acc[W-1], acc, mp[W:1]};
          cnt <= cnt - 1;
          if (cnt == 0) finish <= 1; end
    endcase
end
assign product = {acc, mp[W:1]};
endmodule
```

Almost all of the above three modeling styles produce the same synthesized results. The synthesized result of the preceding example is given in Figure 11.12. The simulation results of the Booth algorithm are shown in Figure 11.13.

11.2.3 Datapath and Controller Design

In general, a digital system can be considered as a system composed of three major parts, *datapath*, *memory* and *controller*, as shown in Figure 11.14. Datapath performs all operations desired in the system. Memory temporarily stores intermediate data used and generated by the datapath unit. Controller controls and schedules all operations performed by the datapath unit.

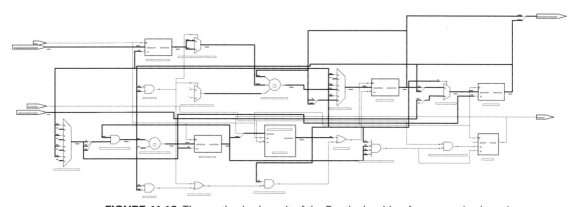

FIGURE 11.12 The synthesized result of the Booth algorithm from a synthesis tool

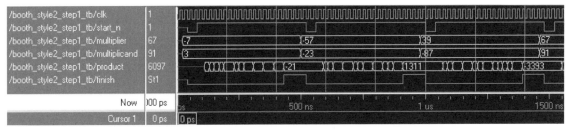

FIGURE 11.13 The simulation results of the Booth multiplier algorithm

A datapath is a logical and a physical structure, which usually comprises arithmetic units such as adder/subtractor, multiplier, shifter, comparator, registers, ALU and some other function units. Memory may be one or more of the following, RAM, CAM, ROM, FIFO (buffer), shift registers and registers, depending on the actual requirements. Controller is a finite state machine, being described by a state diagram or an ASM chart and implemented by using PLA, ROM or random logic. In addition, an interconnect network is often used to pass data among the above three portions. The most widely used interconnects are switches, arbiters, buses and multiplexers. In large digital systems, such as CPUs, the most important problems in datapath design are memory (sometimes registers) and interconnects because they severely affect the system performance and cost.

Sometimes, the datapath and controller (DP+CU) approach is also called the *control-point style*. For simple systems, the datapath and controller of a design can be derived from its ASM chart easily. The datapath portion corresponds to the registers and function units in the ASM chart and the controller portion corresponds to the needed control signals in the ASM chart. For complex systems, the datapath and controller of a design are often derived from the specifications in a state-of-the-art manner.

11.2.3.1 Three-Step Paradigm Transforming an ASM chart into a datapath and controller architecture generally follows three steps:

1. *Model the design* (usually described by an ASM chart) using any modeling style described above as a single module.

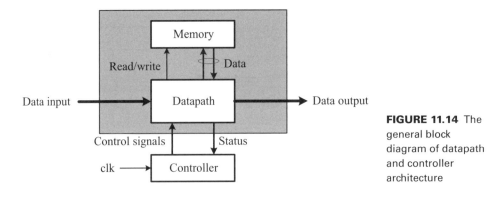

FIGURE 11.14 The general block diagram of datapath and controller architecture

2. *Extract the datapath* from the module and construct it as an independent module. The datapath module is then instantiated by the original module, which now contains only the portion of control signal generation.

3. *Extract control-unit module and construct top module.* Rewrite the portion of the control signal generation as an independent module in parallel to the datapath module. Add a top module that instantiates both datapath and control-unit modules.

The following three examples illustrate the above three-step paradigm in sequence by using a very simple example of summing an input number in a number of times that is determined by the number of 1s in another input number. Indeed, this problem can be done in one cycle by using a combinational logic.

EXAMPLE 11.11 *Step 1: Modeling the Problem Using a Single Verilog HDL Module*

Suppose that a bit counter `cnt` is used to control the iterations of addition, where a sum `total` summarizes the specified input number `data_b` under the control of another input `data_a`. The `data_b` is added to `total` when the bit value of the input number `data_a[cnt]` is 1; otherwise, the `data_b` does not need to be added to `total` at all. In order to implement this idea, an FSM with two states is used. In the IDLE state, the counter `cnt` is loaded by the bit numbers $n - 1$. In the ADDITION state, the input number `data_a` is examined bit-by-bit in order to determine whether we need to add the input number `data_b` to `total` or not. Additionally, in this state, the counter is decremented by 1 and the next state remains in this state until the counter `cnt` reaches 0. Then, the operation has been completed.

```
// a simple example to illustrate the three-step design
// paradigm
module three_steps_step1(clk, start_n, data_a, data_b,
                         total, finish);
parameter N = 8;    // default word size
parameter M = 3;    // M = log_2(N)
input    clk, start_n;
input    [N-1:0] data_a;
input    [N-1:0] data_b;
output   reg [N-1:0] total;
reg      next_state, present_state;
reg      [M-1:0] cnt;
output reg  finish;  // finish flag
localparam IDLE = 1'b0, ADDITION = 1'b1;
// part 1: initialize and update state register
always @(posedge clk or negedge start_n)
   if   (!start_n) present_state <= IDLE;
   else present_state <= next_state;
// part 2: compute next state
```

```
always @(present_state or finish or cnt)
   case (present_state)
      IDLE:     if (!finish)  next_state = ADDITION;
                else next_state = IDLE;
      ADDITION: if (cnt == 0) next_state = IDLE;
                else next_state = ADDITION;
   endcase
// part 3: compute RTL operations
always @(posedge clk or negedge start_n) begin
   if (!start_n) begin finish <= 0; total <= 0; end
   else case (present_state)
      IDLE: cnt <= N - 1;
      ADDITION: begin
         if (data_a[cnt] == 1) total <= total + data_b;
         cnt <= cnt - 1;
         if (cnt == 0) finish <= 1; end
   endcase end
endmodule
```
■

The second step is to extract the datapath operations from the Verilog HDL description. The following example explains how to do this.

EXAMPLE 11.12 *Step 2: Extract the Datapath Portion from the Module and Form an Independent Module*

The datapath structure is extracted from part 3. In this example, only one register `total` is used to accumulate the sum of the input number and two operations are performed: clear and load the accumulated sum. Hence, the datapath is quite simple and is only composed of an accumulator, as shown in the datapath module. What is left in part 3 are those signals without relating to the datapath, and the control signals for the datapath. In this example, only one control signal is required, namely, `acc_load`, which loads the accumulator.

```
// a simple example to illustrate the three-step design
// paradigm
module three_steps_step2(clk, start_n, data_a, data_b,
                         total, finish);
parameter N = 8; // default word size
parameter M = 3; // M = log_2(N)
input  clk, start_n;
input  [N-1:0] data_a;
input  [N-1:0] data_b;
output wire [N-1:0] total;
reg    next_state, present_state;
reg    [M-1:0] cnt;
wire   load;
```

```
output reg  finish;  // finish flag
localparam IDLE = 1'b0, ADDITION = 1'b1;
// datapath --- consisting of a single accumulator.
datapath  #(8) dp (.clk(clk), .acc_reset_n(start_n),
                   .acc_load(load), .acc_data_b(data_b),
                   .acc_total(total));
// part 1: initialize and update state register
always @(posedge clk or negedge start_n)
   if  (!start_n) present_state <= IDLE;
   else present_state <= next_state;
// part 2: compute next state
always @(present_state or finish or cnt)
   case (present_state)
      IDLE:     if (!finish)  next_state = ADDITION;
                else next_state = IDLE;
      ADDITION: if (cnt == 0) next_state = IDLE;
                else next_state = ADDITION;
   endcase
// part 3: execute RTL operations
always @(posedge clk or negedge start_n) begin
   if  (!start_n) finish <= 0;
   else case (present_state)
      IDLE: cnt <= N - 1;
      ADDITION: begin
         cnt <= cnt -1;
         if (cnt == 0) finish <= 1; end
   endcase end
assign load = (present_state == ADDITION ) &&
             (data_a [cnt] == 1);
endmodule
// define the datapath module
module datapath (clk, acc_reset_n, acc_load, acc_data_b,
                 acc_total);
parameter N = 8; // default word size
input  clk, acc_reset_n, acc_load;
input  [N-1:0] acc_data_b;
output reg [N-1:0] acc_total;
always @(posedge clk or negedge acc_reset_n)
   if (!acc_reset_n) acc_total <= 0;
   else if (acc_load) acc_total <= acc_total + acc_data_b;
endmodule
```

The third step is to rewrite the module as a control module and construct another module, called the top module, to embody both datapath and control modules as its instances.

EXAMPLE 11.13 *Step 3: Extract the Control-Unit Module and Construct the Top Module*

This part is simply to rewrite the module, except for the instance of the datapath as a control module. In order to make the notation more clear, in this example, we prefix cu_ to all ports of the control module so that they can be distinguished from the top module. Of course, it is not necessary to do this. However, it is good practice to make the signals distinguishable.

```
// a simple example to illustrate the three-step design
// paradigm
module three_steps_step3(clk, start_n, data_a, data_b,
                         total, finish);
parameter N = 8; // word size
parameter M = 3; // M = log_2(N)
input   clk, start_n;
input   [N-1:0] data_a;
input   [N-1:0] data_b;
output wire [N-1:0] total;
wire    load;
output wire finish;  // finish flag
localparam IDLE = 1'b0, ADDITION = 1'b1;
// datapath --- consisting of a single accumulator.
datapath  #(8) dp (.clk(clk), .acc_reset_n(start_n),
                .acc_load(load), .acc_data_b(data_b),
                .acc_total(total));
controller cu (.cu_clk(clk), .cu_reset_n(start_n),
            .cu_data_a(data_a), .cu_load(load),
            .cu_finish(finish));
endmodule
// the controller module
module controller (cu_clk, cu_reset_n, cu_data_a,
                   cu_load, cu_finish);
parameter N = 8;  // default word size
parameter M = 3;  // M = log_2(N)
input cu_clk, cu_reset_n;
input [N-1:0] cu_data_a;
reg   next_state, present_state;
reg   [M-1:0] cnt;
output wire cu_load;
output reg  cu_finish;  // finish flag
localparam IDLE = 1'b0, ADDITION = 1'b1;
// part 1: initialize and update state register
always @(posedge cu_clk or negedge cu_reset_n)
   if (!cu_reset_n) present_state <= IDLE;
   else present_state <= next_state;
```

```
// part 2: compute next state
always' @(present_state or cu_finish or cnt)
   case (present_state)
      IDLE:      if (!cu_finish)  next_state = ADDITION;
                 else next_state = IDLE;
      ADDITION: if (cnt == 0) next_state = IDLE;
                 else next_state = ADDITION;
   endcase
// part 3: execute RTL operations
always @(posedge cu_clk or negedge cu_reset_n) begin
   if (!cu_reset_n) cu_finish <= 0;
   case (present_state)
      IDLE: cnt <= N - 1;
      ADDITION: begin
         cnt <= cnt - 1;
         if (cnt == 0) cu_finish <= 1; end
   endcase
end
assign cu_load = (present_state == ADDITION ) &&
                 (cu_data_a[cnt] == 1);
endmodule
// define the datapath module
module datapath (clk, acc_reset_n, acc_load,
                 acc_data_b, acc_total);
parameter N = 8; // default word size
input  clk, acc_reset_n, acc_load;
input  [N-1:0] acc_data_b;
output reg [N-1:0] acc_total;
always @(posedge clk or negedge acc_reset_n)
  if (!acc_reset_n)  acc_total <= 0;
  else if (acc_load) acc_total <= acc_total + acc_data_b;
endmodule                                               ■
```

In summary, the datapath and controller architecture generally not only has a simple architecture but also consumes less area than the ASM chart approach. In the rest of this section, we give another more complex example—the Booth algorithm.

11.2.3.2 Booth Algorithm Example The datapath of the Booth algorithm can be derived from the ASM chart given in Figure 11.10 and is shown in Figure 11.15. The datapath has an *n*-bit adder and subtractor, *add_sub*, one accumulator, *acc*, and two registers, *mp* and *mcand*. The accumulator *acc* and register *mp* constitute a shift register and serve as the partial product register.

The controller generates all control signals desired in the datapath. It reads the lowest two bits *mp*[1 : 0] and generates all control signals accordingly. In addition, a reset signal *start_n* is input to both datapath and controller to reset all modules to their

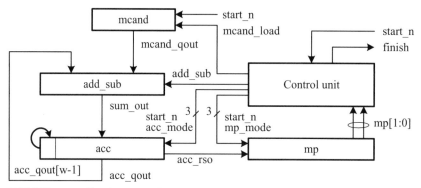

FIGURE 11.15 The datapath of the Booth algorithm derived from the ASM chart given in Figure 11.10

initial states. The control signal *mcand_load* loads the multiplicand into the register *mcand*. The control signal *acc_sub* determines whether the operation is addition or subtraction. The control signal *acc_mode* is a two-bit mode selection signal, which controls the operations of the accumulator. There are three modes associated with the accumulator *acc*: 00 does nothing, 10 loads the accumulator, and 01 shifts right the accumulator. Similarly, the control signal *mp_mode* is also a two-bit mode selection signal, which controls the operations of the register *mp*. The register *mp* has three modes: 00 does nothing, 10 loads register *mp*, and 01 shifts right the register *mp*.

The following further uses the Booth algorithm as an example to illustrate how to transfer an ASM chart or a Verilog HDL description into a datapath and controller architecture.

EXAMPLE 11.14   *Booth Multiplier: Style 2–Step 2*

As described before, the second step of the three-step paradigm is to extract the datapath structure from the RTL operations. The datapath has the structure shown in Figure 11.15, along with related control signals.

```
// Booth multiplier --- style 2 : step 2
module booth_style2_step2(clk, start_n, multiplier,
                          multiplicand, product, finish);
parameter W = 8; // default word size
parameter N = 3; // N = log2(W)
input   clk, start_n;
output reg finish;
input   [W-1:0] multiplier, multiplicand;
output wire [2*W-1:0] product;
reg    [1:0] acc_mode, mp_mode;
reg    mcand_load, add_sub;
wire   [1:0] mp_qb10;
localparam idle = 2'b00, load = 2'b01,
           compute = 2'b10, shift_right = 2'b11;
```

```
// the body of the W-bit booth multiplier
// datapath part --- structural description
booth_datapath #(W)  bdatapath(.clk(clk),
                               .start_n(start_n),
                               .multiplier(multiplier),
                               .multiplicand
                                  (multiplicand),
                               .mp_qb10(mp_qb10),
                               .add_sub(add_sub),
                               .acc_mode(acc_mode),
                               .mp_mode(mp_mode),
                               .mcand_load(mcand_load),
                               .product(product));
// control part --- behavioral description
// local variables declaration
reg [1:0] ns, ps;
reg [N-1:0] cnt;
// part 1: initialize and update state registers
always @(posedge clk or negedge start_n)
   if (!start_n) ps <= idle;
   else  ps <= ns;
// part 2: compute next state
always @(*)
   case (ps)
      idle:    if (!finish) ns = load; else ns = idle;
      load:    ns = compute;
      compute: ns = shift_right;
      shift_right: if (cnt == 0) ns = idle; else ns =
         compute;
   endcase
// part 3: execute RTL operations
// define finish signal
always @(posedge clk or negedge start_n) begin
   if (!start_n) finish <= 0; else
   if ((ps == shift_right) && (cnt == 0)) finish <= 1;
      end
// define counter block
always @(posedge clk or negedge start_n)
   if (!start_n) cnt <= 0;
   else if (ps == load) cnt <= W - 1 ;
       else if (ps == shift_right) cnt <= cnt - 1;
// generate control signals. all signals are initially
// set to their initial values and then set to
// be active accordingly.
always @(*) begin begin
   mcand_load = 1'b0; mp_mode = 2'b00;
```

```
        acc_mode = 2'b00;   add_sub = 1'b0;      end
      case (ps)
        idle: ; // do nothing
        load: begin
          mcand_load = 1'b1;   mp_mode = 2'b10; end
        compute: case (mp_qb10)
          2'b01: begin
            add_sub = 1'b0;
            acc_mode = 2'b10; end // addition
          2'b10: begin
            add_sub = 1'b1;
            acc_mode = 2'b10; end // subtraction
          default: ; endcase
        shift_right: begin
          acc_mode = 2'b01;
          mp_mode  = 2'b01;   end // shift right one bit
                                  // position
        endcase
end // end of always block
endmodule
//
//  Booth datapath is described in a structural way
module booth_datapath(clk, start_n, multiplier,
                      multiplicand, mp_qb10, add_sub,
                      acc_mode, mp_mode, mcand_load,
                      product);
parameter W = 8; // word size
input  clk, start_n;
input  [W-1:0] multiplier, multiplicand;
output wire [2*W-1:0] product;
wire   [W-1:0] acc_qout, mcand_qout, sum_out;
wire   [W:0]   mp_qout;
input  [1:0]   acc_mode, mp_mode;
input  mcand_load, add_sub;
wire   acc_rso, mp_rso;
output wire [1:0] mp_qb10;
// instantiate data path components
addsub    #(W)    addsub(.a(acc_qout), .b(mcand_qout),
                          .mode(add_sub), .sum(sum_out));
shift_reg #(W)    acc(.clk(clk), .reset_n(start_n),
                      .mode(acc_mode), .data(sum_out),
                      .lsi(acc_qout[W-1]), .qout(acc_qout),
                      .rso(acc_rso));
shift_reg #(W+1) mp(.clk(clk), .reset_n(start_n),
                      .mode(mp_mode), .data({multiplier,
                        1'b0}), .lsi(acc_rso),
                      .qout(mp_qout), .rso(mp_rso));
```

```verilog
register  #(W)   mcand(.clk(clk), .reset_n(start_n),
                      .load(mcand_load), .data
                         (multiplicand),
                      .qout(mcand_qout));
assign product = {acc_qout, mp_qout[W:1]};
assign mp_qb10 = mp_qout[1:0];
endmodule
// shift register module
module shift_reg(clk, reset_n, mode, data, lsi, qout,
                rso);
parameter N = 8;
input  clk, reset_n, lsi;
input  [1:0]  mode;
input  [N-1:0] data;
output reg [N-1:0] qout;
output rso;
wire   rso;
// shift register body
assign rso = qout[0];
always @(posedge clk or negedge reset_n)
   if (!reset_n) qout   <= {N{1'b0}};
   else case (mode)
      2'b00: ;  // No change
      2'b01: qout <= {lsi, qout[N-1:1]}; // shift right
      2'b10: qout <= data;  // parallel load
      default:; // no operation
   endcase
endmodule
// register module
module register(clk, reset_n, load, data, qout);
parameter N = 8;
input  clk, load, reset_n;
input  [N-1:0] data;
output reg [N-1:0] qout;
// register body
always @(posedge clk or negedge reset_n)
   if (!reset_n) qout   <= {N{1'b0}};
   else if (load) qout <= data;
endmodule
// addition and subtraction module
module addsub(a, b, mode, sum);
parameter N = 8;
input  [N-1:0] a, b;
input  mode; // define addition or subtraction
output reg [N-1:0]   sum;
// adder-subtractor body
always @(a or b or mode)
```

```
   if   (mode) sum = a - b; // mode = 1 --> subtraction
   else sum = a + b;        // mode = 0 --> addition
endmodule
```

The third step is to rewrite the module as a control module and construct another module, called the top module, to embody both datapath and control modules as its instances.

EXAMPLE 11.15   *Booth Multiplier: Style 2–Step 3*

This part is simply to rewrite the module except for the instance of the datapath as a control module.

```
// Booth multiplier: datapath and control approach
module booth_style2_step3(clk, start_n, multiplier,
                          multiplicand, product, finish);
parameter W = 8; // default word size
parameter N = 3; // N = log2(W)
input  clk, start_n;
output finish;
input  [W-1:0] multiplier, multiplicand;
output wire [2*W-1:0] product;
wire   [1:0] acc_mode, mp_mode;
wire   mcand_load, add_sub;
wire   [1:0] mp_qb10;
// the body of the W-bit booth multiplier
// both datapath and controller are described in
// structural ways.
booth_datapath #(W)  bdatapath(.clk(clk),
                               .start_n(start_n),
                               .multiplier(multiplier),
                               .multiplicand
                                  (multiplicand),
                               .mp_qb10(mp_qb10),
                               .add_sub(add_sub),
                               .acc_mode(acc_mode),
                               .mp_mode(mp_mode),
                               .mcand_load(mcand_load),
                               .product(product));
booth_control #(W,N) bcontrol(.clk(clk),
                               .start_n(start_n),
                               .mp_qout(mp_qb10),
                               .add_sub(add_sub),
                               .acc_mode(acc_mode),
                               .mp_mode(mp_mode),
                               .mcand_load(mcand_load),
                               .finish(finish));
endmodule
```

```verilog
//  Booth datapath is described in structural ways.
module booth_datapath(clk, start_n, multiplier,
                      multiplicand, mp_qb10, add_sub,
                      acc_mode, mp_mode, mcand_load,
                      product);
parameter W = 8; // word size
input  clk, start_n;
input  [W-1:0] multiplier, multiplicand;
output wire [2*W-1:0] product;
wire   [W-1:0] acc_qout, mcand_qout, sum_out;
wire   [W:0]   mp_qout;
input  [1:0]   acc_mode, mp_mode;
input  mcand_load, add_sub;
wire   acc_rso, mp_rso;
output wire [1:0] mp_qb10;
// instantiate data path components
addsub    #(W)    addsub(.a(acc_qout), .b(mcand_qout),
                         .mode(add_sub), .sum(sum_out));
shift_reg #(W)    acc(.clk(clk), .reset_n(start_n), .mode
                     (acc_mode), .data(sum_out), .lsi
                     (acc_qout[W-1]), .qout(acc_qout),
                     .rso(acc_rso));
shift_reg #(W+1) mp(.clk(clk), .reset_n(start_n), .mode
                     (mp_mode), .data({multiplier, 1'b0}),
                     .lsi(acc_rso), .qout(mp_qout), .rso
                     (mp_rso));
register  #(W)    mcand(.clk(clk), .reset_n(start_n),
                        .load(mcand_load), .data
                        (multiplicand), .qout(mcand_qout));
assign product = {acc_qout, mp_qout[W:1]};
assign mp_qb10 = mp_qout[1:0];
endmodule
//
// control part --- behavioral description
module booth_control(clk, start_n, mp_qout, add_sub,
                     acc_mode, mp_mode, mcand_load,
                     finish);
parameter W = 8;
parameter N = 3;
input  clk, start_n;
input  [1:0] mp_qout;
output reg add_sub, mcand_load;
output reg [1:0] acc_mode, mp_mode;
output reg finish;   // finish flag
localparam idle = 2'b00, load = 2'b01,
           compute = 2'b10, shift_right = 2'b11;
// Local variables declaration
```

```verilog
reg [1:0] ns, ps;
reg [N-1:0] cnt;
// part 1: initialize and update state registers
always @(posedge clk or negedge start_n)
   if (!start_n) ps <= idle;
   else  ps <= ns;
// part 2: compute next state
always @(*)
   case (ps)
      idle:    if (!finish) ns = load; else ns = idle;
      load:    ns = compute;
      compute: ns = shift_right;
      shift_right: if (cnt == 0) ns = idle; else ns =
         compute;
   endcase
// part 3: execute RTL operations
// define finish signal
always @(posedge clk or negedge start_n)begin
  if (!start_n) finish <= 0; else
  if ((ps == shift_right) && (cnt == 0)) finish <= 1; end
// define counter block
always @(posedge clk or negedge start_n)
   if (!start_n) cnt <= 0;
   else if (ps == load) cnt <= W - 1 ;
      else if (ps == shift_right) cnt <= cnt - 1;
// generate control signals. All signals are initially
// set to their initial values and then set
// to be active accordingly.
always @(*) begin begin
   mcand_load = 1'b0;  mp_mode = 2'b00;
   acc_mode = 2'b00;   add_sub = 1'b0;    end
   case (ps)
      idle: ; // do nothing
      load: begin
         mcand_load = 1'b1;  mp_mode = 2'b10; end
      compute: case (mp_qout[1:0])
         2'b01: begin
            add_sub = 1'b0;
            acc_mode = 2'b10; end // addition
         2'b10: begin
            add_sub = 1'b1;
            acc_mode = 2'b10; end // subtraction
         default: ; endcase
      shift_right: begin
         acc_mode = 2'b01;
         mp_mode  = 2'b01;  end // shift right one bit
```

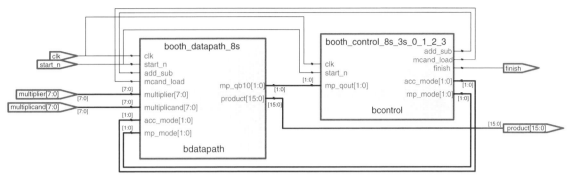

FIGURE 11.16 The synthesized result of the Booth's algorithm based on the DP+CP architecture

```
          // position
    endcase
end // end of always block
endmodule
```

The synthesized result is given in Figure 11.16. From this figure, it is apparent that there are two modules: datapath and control. You can see the details of each module by dissolving it. As described previously, the datapath and controller approach in general consumes less area than the ASM chart approach. The reader is encouraged to justify this by comparing various modeling approaches described in this section. The simulation results are shown in Figure 11.17.

Review Questions

Q11.12 What are the three basic components in datapath and controller architecture?

Q11.13 Describe how to derive the datapath from an ASM chart?

Q11.14 Describe how to derive the controller from an ASM chart?

Q11.15 Describe the operations of the three-step paradigm.

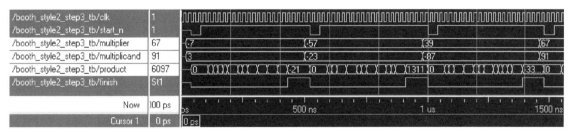

FIGURE 11.17 The simulation results of the Booth algorithm

11.3 RTL IMPLEMENTATION OPTIONS

There are many options that can be used to implement an RTL design. The rationale behind these options is a tradeoff among performance (throughput), space (area) and time (operating frequency). The most common options that are widely used in practical systems are, *single cycle*, *multiple cycle* and *pipeline*, as shown in Figure 11.18. In this section, we discuss these options in more detail.

11.3.1 Single-Cycle Structure

Generally speaking, a single-cycle logic structure only uses combinational logic to realize the desired functions as shown in Figure 11.18(a). This may require a quite long propagation time to finish a computation of the desired functions. A simple example of using the single-cycle structure is an *n*-bit adder, which has appeared many times in this book.

The following is another simple example of using the single-cycle structure to compute an arithmetic expression, which includes an addition and a subtraction.

EXAMPLE 11.16 *A Single-Cycle Example*

This example computes an arithmetic expression, $total = data\_c - (data\_a + data\_b)$, in one cycle. The propagation delay is one addition and one subtraction.

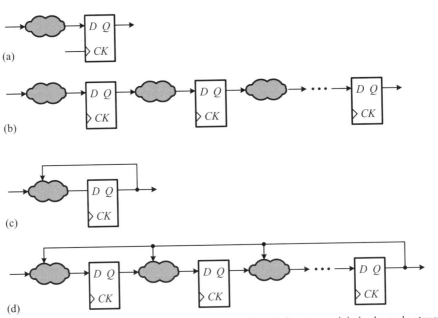

FIGURE 11.18 The RTL implementation options of digital systems: (a) single-cycle structure; (b) linear multiple-cycle structure; (c) nonlinear single-stage multiple-cycle structure; (d) nonlinear multiple-stage multiple-cycle structure

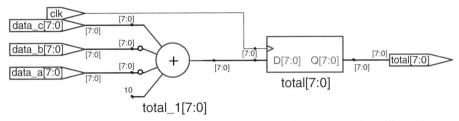

FIGURE 11.19 The single-cycle structure of the logic used to compute the arithmetic expression: *total = data_c − (data_a + data_b)*

```
// a single-cycle example
module single_cycle_example(clk, data_a, data_b, data_c,
                            total);
parameter N = 8;
input   clk;
input   [N-1:0] data_a, data_b, data_c;
output reg [N-1:0] total;
// compute total = data_c - (data_a + data_b)
always @(posedge clk) begin
   total <= data_c - (data_a + data_b);
end
endmodule                                              ■
```

The synthesized result of the above single-cycle structure is given in Figure 11.19. From this figure, we can see that all input data are summed up together using addition/subtraction units. The resulting values are then stored in register `total`.

11.3.2 Multiple-Cycle Structure

A multiple-cycle structure executes the desired functions in consecutive clock cycles. This may be further classified into two basic structures: *linear structure* and *nonlinear structure*. The linear structure simply performs the desired functions without sharing resources, as shown in Figure 11.18(b). The nonlinear (feedback or feed-forward) structure performs the required functions with sharing resources by using a feedback or feed-forward connection. It can be either single stage or multiple stage, as shown in Figures 11.18(c) and (d), respectively.

The following example uses the multiple-cycle structure to re-do the preceding example.

EXAMPLE 11.17   *A Multiple-Cycle Example*

At the first positive edge of the clock signal, input data `data_a` is latched into a register `qout_a`. The next clock cycle computes the addition of `qout_a` and `data_b`. The result is stored into the register `qout_b`. The third clock cycle computes the final result, which is stored in the register `total`. As a result, a complete computation of the arithmetic expression requires three clock cycles.

```
// a multiple cycle example
module multiple_cycle_example(clk, data_a, data_b, data_c,
                                          total);
parameter N = 8;
input  clk;
input  [N-1:0] data_a, data_b, data_c;
output reg [N-1:0] total;
reg    [N-1:0] qout_a, qout_b;
// compute total = data_c - (data_a + data_b)
always @(posedge clk) begin
   qout_a <= data_a;
   @(posedge clk)
      qout_b <= qout_a + data_b;
   @(posedge clk)
      total <= data_c - qout_b;
end
endmodule
```

The synthesized result of the above example is shown in Figure 11.20. From this figure, we can see that a controller is inferred from the module to schedule the operations that compute the arithmetic expression. Indeed, this is an example of implicit FSM. See Section 11.1.3 for details of the implicit FSM.

A nonlinear single-stage multiple-cycle structure can generally be transformed into a linear multiple-stage multiple-cycle structure by duplicating both combinational logic and register the number of times corresponding to the required cycles, and then cascading together all stages in a linear manner. The major advantage of using nonlinear single-stage multiple-cycle structure is that it can save hardware resources because it explores the features of resource sharing. However, using the linear multiple-stage multiple-cycle structure has a feature that it is ready to pipeline the operations it performs.

11.3.3 Pipeline Structure

A pipeline structure is also a multiple-cycle multiple-stage structure in which new data can be fed into the structure at every clock cycle. Hence, it may output a result per clock cycle after the pipeline is fully filled.

The general block diagram of a pipeline system is depicted in Figure 11.21. From this figure, we can see that a pipeline is composed of several stages, with each stage

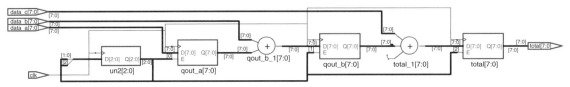

FIGURE 11.20 The multiple-cycle structure of the logic used to compute the arithmetic expression: *total = data_c − (data_a + data_b)*

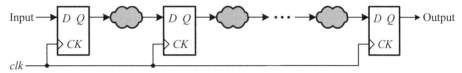

FIGURE 11.21 The general block diagram of a pipeline system

consisting of a combinational logic circuit and a register, called the *pipeline register*. The data in the pipeline are forwarded to the next stage through computation of the combinational logic between them in every clock cycle.

In order to measure the performance of a pipeline, two metrics are often used: *pipeline latency* and *pipeline clock period*. Pipeline latency is the number of cycles required between the presentation of an input value and the appearance of its associated output. Sometimes a *pipeline depth* is used to denote the number of stages in a pipeline. The pipeline clock, f_{pipe}, is set by the slowest stage and is usually larger than the frequency used in the original cascaded system.

For the ith-stage, the smallest allowable clock period, T_i, is determined by the following condition:

$$T_i = t_q + t_{setup} + t_{d,i} - t_{skew,i+1} \tag{11.4}$$

where the first two terms, t_q and t_{setup}, are the propagation delay and set up time of flip-flops, respectively. The propagation delay of a flip-flop is sometimes also called clock-to-Q delay. The third term, $t_{d,i}$, is the propagation delay of the ith-stage combinational logic. The final term, $t_{skew,i+1}$ is the clock skew. Based on these factors, the pipeline clock period for an m-stage pipeline is chosen to be:

$$T_{pipe} = max\{T_i | i = 1, 2, \ldots, m\}. \tag{11.5}$$

The preceding multiple-cycle example can be computed by using a pipeline structure, such as in the following example.

EXAMPLE 11.18   *A Simple Pipeline Example—not a Good One*

As we have described in Section 9.3.1, a pipeline structure is like a shift register in which the output of a register is connected to the input of another register in a linear structure and all registers are clocked by a common clock signal. Indeed, a shift register is virtually a trivial pipeline system in which no nontrivial combinational logic is placed between any two registers. Based on these observations, a pipeline system can then be easily modeled by using the same skill of the shift register, namely, using nonblocking assignments. Consequently, the preceding arithmetic expression can then be rewritten as three nonblocking assignments as follows.

```
// a simple pipeline example --- not a good one
module simple_pipeline(clk, data_a, data_b, data_c,
                       total);
parameter N = 8;
input  clk;
input  [N-1:0] data_a, data_b, data_c;
```

```
output reg [N-1:0] total;
reg    [N-1:0] qout_a, qout_b;
// compute total = data_c - (data_a + data_b)
always @(posedge clk) begin
   qout_a <= data_a;
   qout_b <= qout_a + data_b;
   total  <= data_c - qout_b;
end
endmodule
```
■

The synthesized result of the above simple pipeline module is shown in Figure 11.22. You are encouraged to synthesize this module on your system and see the features and problems of it. Indeed, the above module is an incorrect implementation of a pipeline system. A correct pipeline implementation of the preceding arithmetic expression computation is given in the following example.

EXAMPLE 11.19   *A Pipeline Example*

In order to perform a correct pipeline operation, the data in the system must be forwarded in a hop-by-hop manner; namely, they must be forwarded to their next stage at the next clock cycle. Based on this idea, the preceding example is modified as follows. The reader is encouraged to compare both `always` statements and see the differences between them.

```
// a simple pipeline example
module pipeline_example(clk, data_a, data_b, data_c,
                        total);
parameter N = 8;
input  clk;
input  [N-1:0] data_a, data_b, data_c;
output reg [N-1:0] total;
reg    [N-1:0] qout_a, qout_b, qout_c, qout_d, qout_e;
// compute total = data_c - (data_a + data_b)
always @(posedge clk) begin
   qout_a <= data_a; qout_b <= data_b; qout_c <= data_c;
   qout_d <= qout_a + qout_b;
   qout_e <= qout_c;
   total  <= qout_e - qout_d;
end
endmodule
```
■

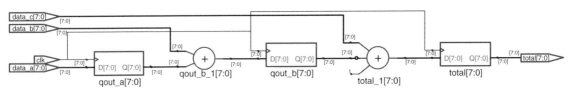

FIGURE 11.22 The synthesized result of the simple pipeline

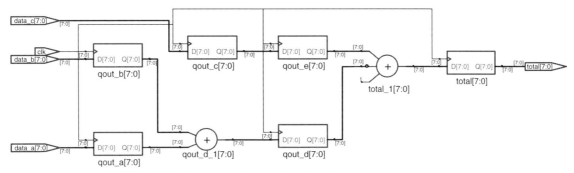

FIGURE 11.23 The synthesized result of the simple pipeline

The synthesized result of the above pipeline module is shown in Figure 11.23. It is easy to justify that the result is indeed a pipeline structure, exactly like the one shown in Figure 11.21.

It is not so difficult to observe the relationship between pipeline depth and pipeline clock period, as well as pipeline latency. As shown in Figure 11.24, as pipeline depth increases the pipeline clock period decreases but the pipeline latency increases accordingly. The pipeline latency increases linearly with pipeline depth because it is the number of clock cycles that an input has to be elapsed before it can reach the output.

11.3.4 FSM versus Iterative Logic

An iterative logic can be considered as an FSM being expanded in time. At each time step, the combinational logic part of the FSM is duplicated and the memory element part is ignored, as shown in Figure 11.25. Figure 11.25(a) is the general structure of an FSM, where the next state function is stored in the state register. The combinational logic portion of the FSM is shown in Figure 11.25(b). When an FSM is expanded in time, a 1-dimensional iterative logic is obtained, as illustrated in Figure 11.25(c),

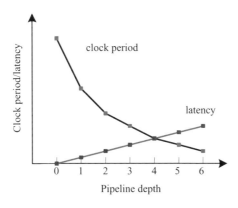

FIGURE 11.24 The clock period and latency versus pipeline depth

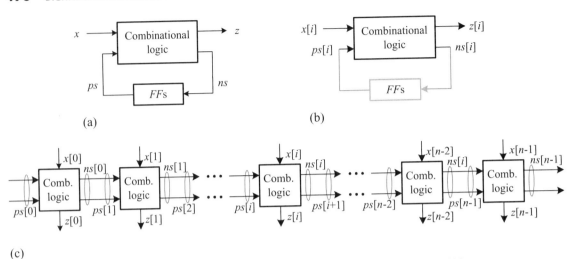

FIGURE 11.25 The relationship between FSM and iterative logic: (a) FSM structure; (b) combinational logic portion; (c) 1-D iterative logic

where each cell corresponds to the combinational logic portion of the FSM in Figure 11.25(a).

In theory, any sequential logic circuit implemented by an FSM can be implemented by iterative logic by duplicating and cascading the combinational logic portion of the FSM in a number of times equal to the length of the input sequence. The following example illustrates this idea through using iterative logic to reimplement the sequence detector discussed in Section 11.1.2.

The implementation of an 0101 sequence detector using iterative logic is given in Figure 11.26. As can be seen from Figure 11.26(a), each cell receives present state and input, and produces next state and output. The function of each cell is described by a state table exactly the same as that of its FSM. Indeed, the design of the ith cell is exactly the same as that of an FSM. Hence, we can derive the transition as well as output tables from the state table. With the help of the Karnaugh maps shown in Figure 11.26(d), we obtain the state transition and output functions and have the combinational logic depicted in 11.26(e). An example of a 6-bit 0101 sequence detector implemented by iterative logic is given in Figure 11.27.

In the following, we rewrite the FSM example discussed in Section 11.1.2 using the technique of iterative logic.

EXAMPLE 11.20 *An Iterative Logic Version of the 0101 Sequence Detector*

Due to the inherent structure features of iterative logic, it is convenient to model the iterative logic using a generate-loop statement. The description of the 0101 sequence detector can be divided into two kinds of cells: LSB and the rest bits. The LSB cell receives the initial state, namely, A, and input x, and then produces output z and next state. The rest-bit cells have almost the same operations of the LSB except that they receive the next state from their previous cells instead of the initial state A. The definition of each cell contains only the computation of next state and output functions.

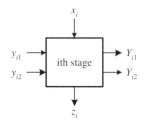

(a)

PS	x_i NS, z_i 0	1
A	B,0	A,0
B	B,0	C,0
C	D,0	A,0
D	B,0	C,1

(b)

y_{i1} y_{i2}	x_i Y_{i1} Y_{i2} 0	1	x_i z_i 0	1
00	01	00	0	0
01	01	10	0	0
10	11	00	0	0
11	01	10	0	1

(c)

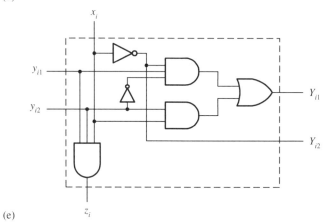

(d)

(e)

FIGURE 11.26 The iterative logic implementation of an 0101 sequence detector:
(a) ith stage block diagram; (b) ith stage state table; (c) transition table and output table;
(d) Karnaugh map; (e) logic circuit

```
// behavioral description of 0101 sequence detector
// an iterative logic version
module sequence_detector_iterative (x, z);
parameter N = 6; // define the size of 0101 sequence
                 // detector
input  [N-1:0] x;
output wire [N-1:0] z;
// Local declaration
```

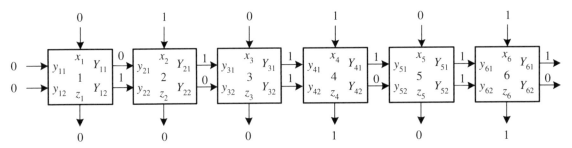

FIGURE 11.27 An example of a 6-bit 0101 sequence detector implemented by iterative logic

```
wire  [1:0]next_state [N-1:0];
localparam A = 2'b00, B = 2'b01, C = 2'b10, D = 2'b11;
// using generate block to produce an N-bit 0101 sequence
// detector
genvar i;
generate for(i = 0; i < N; i = i + 1) begin: detector_0101
   if (i == 0) begin: LSB   // describe LSB cell
      basic_cell bc (x[i], A, z[i], next_state[i]); end
   else begin: rest_bits    // describe the rest bits
      basic_cell bc (x[i], next_state[i-1], z[i],
                     next_state[i]);
   end
end endgenerate
endmodule
// define the basic cell
module basic_cell (x, present_state, output_z,
                   next_state);
input x;
input [1:0] present_state;
output reg [1:0] next_state;
output reg output_z;
localparam A = 2'b00, B = 2'b01, C = 2'b10, D = 2'b11;
// determine the next state
always @(present_state or x)
   case (present_state)
      A: if (x) next_state = A; else next_state = B;
      B: if (x) next_state = C; else next_state = B;
      C: if (x) next_state = A; else next_state = D;
      D: if (x) next_state = C; else next_state = B;
   endcase
// determine the output value
always @(present_state or x)
   case (present_state)
```

```
        A: if (x) output_z = 1'b0; else output_z = 1'b0;
        B: if (x) output_z = 1'b0; else output_z = 1'b0;
        C: if (x) output_z = 1'b0; else output_z = 1'b0;
        D: if (x) output_z = 1'b1; else output_z = 1'b0;
    endcase
endmodule                                                      ∎
```

The above example is an instance of one-dimensional iterative logic. However, there are many circuits of two-dimensional iterative logic in practice. The following Booth array multiplier is such an example.

11.3.4.1 *Booth Array Multiplier*   We have described in Section 11.2.2 how to implement the Booth algorithm as a sequential circuit, namely, using a multiple-cycle structure. In the following, we will show how to implement it by using a 2-*D* iterative logic structure, often called an *array structure*. Remember that an iterative logic usually means a single-cycle structure.

Actually, an array structure for the Booth algorithm is simply to mimic the operations of the multiple-cycle structure described previously. Recall that at each step of the Booth algorithm, two bits (from LSB to MSB) of the multiplier are examined and taken an appropriate operation accordingly. Then, the partial product is shifted right by one bit position, which corresponds to shifting the multiplicand one bit left. Based on these observations, a 4×4 Booth array multiplier can be derived and is shown in Figure 11.28(d). It requires 4×4 complementer and subtractor (CAS) units and 4 controllers (CTRLs). The function and logic circuit of CAS are given in Figure 11.28(a) and (c), respectively. The CTRL controls the operations of CAS and has the logic circuit shown in Figure 11.28(b).

The interested reader is easy to verify that the Booth array multiplier can work properly. In addition, it is not difficult to construct an $n \times n$ Booth array multiplier from the one shown in Figure 11.28(d). In the following, we show how to model an $m \times n$ Booth array multiplier with generate-loop statements.

EXAMPLE 11.21   *An $m \times n$ Booth Array Multiplier Using Generate-Loop Statements*

This module consists of two parts: CTRL units and multiplication array. The CTRL units are generated according to the logic circuit shown in Figure 11.28(b) except that for the LSB which can be further simplified. The multiplication array is also divided into two cases: the first row and the rest rows. In the first row, the operand P_{in} of CAS is zero and in the rest rows, they are from their previous rows. The other operand, y_i, of all rows is the multiplicand. As shown in Figure 11.28(d), the P_{in} of the last CAS cell of all rows except the first row, is from its previous cell at the same row. Based on these observations, the multiplication array can then be easily modeled with generate-loop statements.

x_i	x_{i-1}	add_sub_sel	add_sub_en	Operation
0	0	ϕ	0	Shift only
0	1	0	1	Add and shift
1	0	1	1	Subtract and shift
1	1	ϕ	0	Shift only

(a)

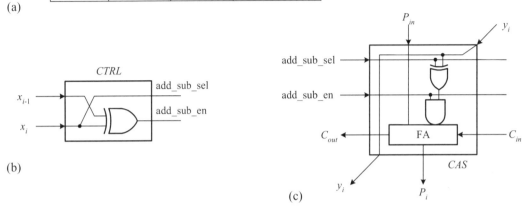

(b)

(c)

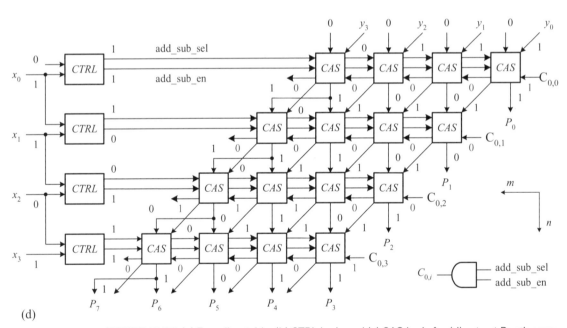

(d)

FIGURE 11.28 (a) Encoding table, (b) CTRL logic and (c) CAS logic for (d) a 4 × 4 Booth array multiplier

```verilog
// an M-by-N radix-2 Booth array multiplier
module booth_array_multiplier(x, y, product);
// inputs and outputs
parameter M = 4;
parameter N = 2;
input   [M-1:0] y;   // multiplicand
input   [N-1:0] x;   // multiplier
output [M+N-1:0] product;
// Internal wires
wire    sum   [M-1:0][N-1:0]; // internal nets
wire    carry [M-1:0][N-1:0];
wire    add_sub_en[N-1:0], add_sub_sel[N-1:0];
genvar i, j;
generate for (i = 0; i < N; i = i + 1)
begin: booth_encoder
   // generate add_sub_en and add_sub_sel control signals
   if (i == 0) begin
    assign add_sub_en[i] = x[0];
      assign add_sub_sel[i] = x[0]; end
   else begin
     assign add_sub_en[i] = x[i]^x[i-1];
  assign add_sub_sel[i] = x[i]; end
end endgenerate
   // generate the multiplication array
generate for (i = 0; i < N; i = i + 1) begin: booth_array
  if (i == 0) begin  // describe the first row
     for (j = 0; j < M; j = j + 1) begin: first_row
        if (j == 0) // the LSB of first row.
           assign {carry[j][i],sum[j][i]}=(add_sub_en[i]&
                 (add_sub_sel[i]^y[j]))+
                 (add_sub_sel[i]& add_sub_en[i]);
        else        // the rest bits of first row
           assign {carry[j][i],sum[j][i]}=(add_sub_en[i]&
                 (add_sub_sel[i]^y[j]))+ carry[j-1][i];
     end
  end
  else begin: rest_rows // describe the rest rows
     for (j = 0; j < M; j = j + 1) begin: rest_rows
        if (j == 0) // describe LSB of each row
           assign {carry[j][i],sum[j][i]}=(add_sub_en[i]&
                 (add_sub_sel[i]^y[j]))+sum[j+1][i-1]+
                 (add_sub_sel[i]& add_sub_en[i]);
        else if (j == M - 1) // describe MSB of each row
           assign {carry[j][i],sum[j][i]}=(add_sub_en[i]&
                 (add_sub_sel[i]^y[j]))+sum[j][i-1]+
                 carry[j-1][i];
```

```
          else      // describe the rest bits of each row
            assign {carry[j][i],sum[j][i]}=(add_sub_en[i]&
                    (add_sub_sel[i]^y[j]))+sum[j+1][i-1]+
                    carry[j-1][i];
        end
      end
end endgenerate
// generate product bits
generate for (i = 0; i < N ; i = i + 1) begin: product_
    lower_part
      assign product[i] = sum[0][i];
end endgenerate
generate for (i = 1; i < M ; i = i + 1) begin: product_
    upper_part
      assign product[N-1+i] = sum[i][N-1];
end endgenerate
assign product[M+N-1] = sum[M-1][N-1];
endmodule
```

■

In summary, a design can be usually implemented by using either a multiple-cycle or a single-cycle structure, depending on the performance required and the hardware allowed. Up to now, we have described several design examples with both single-cycle and multiple-cycle implementations. For the case of an 0101 sequence detector, when it is implemented using the multiple-cycle structure, only one 2-bit register and one combinational logic circuit are required. It iterates n times when the length of the input sequence is n bits. As a consequence, the total running time is $n \times (t_{comb} + t_q + t_{setup})$. When the sequence detector is implemented by using iterative logic (namely, single-cycle structure), n copies of combinational logic circuits, but no flip-flop, are required. The total running time in this case is $n \times t_{comb}$. As a result, iterative logic runs faster at the expense of more hardware.

Similarly, for the case of an $m \times n$ Booth algorithm, when implemented with the multiple-cycle structure, it requires three registers for multiplicand, multiplier and partial product, respectively. In addition, an m-bit adder is used to add the multiplicand to partial product. The total running time is at least $2nt_{clk}$, where t_{clk} is determined by the propagation delay of the m-bit adder, as well as the t_q and t_{setup} of the registers used. The constant factor 2 counts both the compute and shift-right steps of the ASM chart shown in Figure 11.10. However, when implemented with iterative logic (namely, single-cycle structure), it requires $m \times n$ CAS units and n CTRL units. The total running time is $[2(m-1) + n]t_{CAS}$, where t_{CAS} is the propagation delay of the CAS cell. Of course, this is one more example of a tradeoff between performance and the cost of hardware.

Review Questions

Q11.16 Describe the basic feature of a single-cycle structure.

Q11.17 Describe the basic feature of a multiple-cycle structure.

Q11.18 Describe the basic feature of a pipeline structure.

Q11.19   How would you determine the pipeline clock period?

Q11.20   Describe the basic feature of iterative logic.

11.4 A CASE STUDY: LIQUID-CRYSTAL DISPLAYS

The liquid-crystal display (LCD) is another widely used display in addition to seven-segment LED displays. It has an ultra-low power consumption of the order of microwatts, compared to the order of milliwatts for LEDs. In this section, we first consider the basic principles of LCDs and then use a commercial reflective field-effect LCD module as an example to illustrate how to interface to and utilize an LCD to display information.

11.4.1 Principles of LCDs

A variety of LCDs have been constructed in the past decades. The types of major interest are field-effect and dynamic scattering devices. Both LCDs can be operated in the *reflective mode* or the *transmissive mode*. In the following, we only consider the field-effect devices because they are the most widely used today. The discussion will include both reflective and transmissive modes.

A field-effect liquid crystal is a material that flows like a liquid and the individual molecules have a rodlike appearance. A useful feature of a liquid crystal is that its optical characteristics can be influenced by an externally applied electric field. Under normal conditions (namely, with no applied bias), all of the rodlike molecules are aligned in a spiral way such that the light passing through is shifted by 90°. However, when a bias is applied, the rodlike molecules align themselves with the field and the light passes directly through without a 90° shift.

Based on the above property of field-effect liquid crystals, two widely used types of displays are constructed according to whether the light source is internally or externally provided. The liquid-crystal display (LCD) is in the reflective mode when an external light source is used and in the transmissive mode when an internal light source is provided.

11.4.1.1 Reflective Field-Effect LCDs   The basic structure of a reflective field-effect LCD is shown in Figure 11.29, which consists of six planes: *vertical light polarizer*, *frontplane electrodes*, *liquid crystal*, *backplane electrodes*, *horizontal light polarizer* and *reflector* as well.

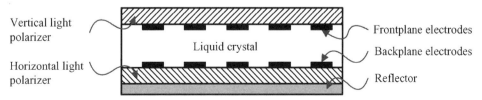

FIGURE 11.29 The structure of a typical reflective field-effect LCD device

As described above, the property of a liquid crystal is that the light passing through it is shifted by $90°$ when no bias is applied and without a $90°$ shift when a bias is applied. As a result, when no bias is applied to the liquid crystal, the vertically polarized light passing through the vertical light polarizer is shifted by $90°$ and transferred into horizontally polarized light. This horizontally polarized light encounters a horizontal light polarizer and passes through to the reflector, where it is reflected back into the liquid crystal, bent back to the other vertical polarization and returned to the observer. A transparent area results on the liquid crystal.

However, when a bias is applied, the rodlike molecules align themselves with the field and the light passes directly through without a $90°$ shift. The vertically incident light cannot pass through the horizontal light polarizer and reach the reflector. A dark area results on the liquid crystal. Based on this observation, by designing the frontplane and backplane electrodes into a desired pattern, the resulting device is an LCD.

11.4.1.2 Transmissive Field-Effect LCDs
In fact, the degree of polarization of a liquid crystal can be controlled by the amount of applied bias. In other words, we can control how much light passes through the liquid crystal by controlling the magnitude of the applied bias. Based on this idea, an LCD with gray level can be built. In addition, by properly using a color filter array and controlling the active pixels (picture element), a color display results.

In order to display an image, the basic pixel cells are arranged in a 2-D array. Each pixel cell controls the amount of light passed through it. By grouping three basic cells and activating separately each individual cell with a variable bias, a color image dot may result if a color filter is used appropriately. To achieve this and provide high-integration density, the bias for the basic cell is applied by using an active element, usually a *thin-film transistor* (TFT). The active element allows the transfer of the signal at each pixel to the liquid crystal cell to be controlled by a proper timing signal. Due to the fact that active elements are used and the pixel cells are arranged into an array, the resulting LCD is known as an *active-matrix LCD* or a *TFT LCD panel*. Active-matrix LCDs are usually used in TVs, computer systems, instruments, consumer products, cell phones, and so on.

Figure 11.30 shows the basic structure of an active-matrix LCD device [4]. Like the reflective LCD, the active-matrix LCD device is still composed of a polarizer, electrodes, and liquid crystal. However, an internal light source known as *backlight* is also used. Nowadays, the dominant backlighting technology for LCDs is the *cold cathode fluorescent lamp* (CCFL), which uses a low-voltage DC to high-voltage AC converter as the driver. This driver consumes the largest amount of power in the display system.

One major feature distinguishing it from the above mentioned reflective LCD is that each pixel cell of the active-matrix LCD is controlled in such a way that the amount of light passed through it is determined by the amount of the bias applied to it. In order to achieve this, both electrodes of pixel cell are made of *indium tin oxide* (ITO), which is a transparent conducting material.

In the rest of this section, we consider the most common dot-matrix LCD modules, which are based on the reflective field-effect LCD and often find their applications in

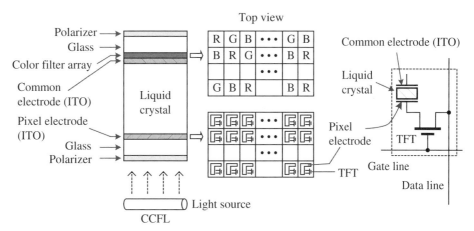

FIGURE 11.30 The structure of a typical active-matrix LCD device [4]

various consumer products except the large-scale displays. These LCDs are usually made into various patterns, such as seven-segment digit and dot-matrices, in order to display digits and characters.

11.4.2 Commercial Dot-Matrix LCD Modules

A widely used commercial dot-matrix LCD module is the one that uses a Hitachi 44780 or its equivalence as a controller, as shown in Figure 11.31. The controller contains a command register, an address counter (AC), a character generator (CG) ROM, a character generator (CG) RAM, a display data (DD) RAM and a busy flag (BF).

The CG ROM is accessed by ASCII code and can generate 192 character patterns, including 160 5×7 character patterns and 32 5×10 character patterns. The CG RAM allows the user to create up eight custom 5×7 characters plus a 5×8 cursor. The display data (DD) RAM can buffer up to 80 or even more bytes, depending on the types of modules. The command register stores the command to be executed. The address counter (AC) stores the address of the CG ROM or DD RAM. Due to only one address counter being available, only either the CG ROM or the DD RAM can be accessed at any time. The busy flag (BF) is used to indicate whether the LCD module is working a command or not.

The standard electrical interface of the module is shown in Figure 11.31(a). The pin assignments and associated functions are listed in Figure 11.31(b). The functions of pins are as follows. $DB7$ to $DB0$ is the eight-bit data bus. In 4-bit mode, only $DB7$ to $DB4$ are required and each command or data are sent twice, one nibble each time. The upper nibble is transferred first and then the lower nibble follows. This mode is useful for interfacing with a 4-bit microcomputer, which often find applications in low-end consumer products. The other three control signals, enable (E), register select (RS) and

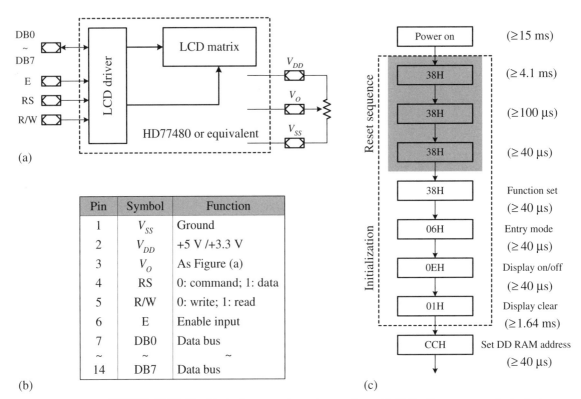

(a)

(b)

(c)

FIGURE 11.31 The block diagram, pin assignments and initialization process of a typical commercial reflected LCD device: (a) block diagram; (b) pin assignments and functions; (c) initialization process

read/write (R/W), are used to control the access of the module. The signal E is to enable the read or write cycle. The RS selects the register to be accessed. When $RS = 1$, data buffer is selected; when $RS = 0$, command register is selected. The R/W controls the access mode. When $R/W = 1$, the access mode is read; when $R/W = 0$, the access mode is write.

In order to access the LCD module properly, the timing relationship of the above interface signals must be applied in accordance with the timing specification, as shown in Figure 11.32. Both read and write timing of the LCD module are shown in Figure 11.32. In read cycle, both RS and R/W must be stable for a period of time, t_{AS}, before the rising edge of E and must remain stable for a period of time, t_{AH}, after the falling edge of E. The data $DB7$ to $DB0$ appears at the data bus within a period of time, t_{DDR}, after the rising edge of E and remain for a period of time, t_{DHR}, after the falling edge of E. The pulse width of E must be at least t_{WEP} and the period of E must be at least t_{CYC}. Detailed values of the various parameters are listed in Figure 11.32(c).

In write cycle, both RS and R/W have exactly the same timing relationship as in read cycle. The data on the data bus is latched into the module by the falling edge of E. The data $DB7$ to $DB0$ to be written must be stable at the data bus for a period of time,

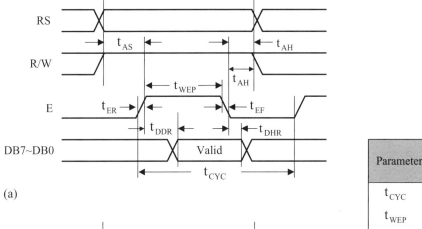

(a)

(b)

(c)

Parameter	Value	
	Minimum	Maximum
t_{CYC}	450 ns	
t_{WEP}	220 ns	
t_{ER}		25 ns
t_{EF}		25 ns
t_{AS}	40 ns	
t_{AH}	10 ns	
t_{DDR}		120 ns
t_{DS}	60 ns	
t_{DH}	10 ns	
t_{DHR}	20 ns	

FIGURE 11.32 The access timing of a typical LCD module: (a) read cycle; (b) write cycle; (c) parameters

t_{DS}, before the falling edge of E and must remain for a period of time, t_{DH}, after the falling edge of E. Detailed values of the various parameters are listed in Figure 11.32(c).

In the LCD module, there are two registers: *command register* and *data register*. These two registers are selected by the *RS* signal, as described before. Another register is called the address counter (AC), which specifies an address when accessing CG RAM or DD RAM.

There are many commands associated with the LCD module for using to control and access the module, as shown in Table 11.2. In the following, we describe each command in greater detail.

- **Clear display.** Clear all display memory and return the cursor to the home position.

- **Cursor home.** Return the cursor to the home position.

- **Entry mode set.** Set the entry mode and the cursor automatically moves to the right when incremented by one or to the left when decremented by one. The I/D bit sets the address counter (AC) to increment ($I/D = 1$) or decrement

TABLE 11.2 The command summary of typical commercial dot-matrix LCD devices

RS	RW	DB7	DB6	DB5	DB4	DB3	DB2	DB1	DB0	Function
Clear display										Execution time: 82 μs ~1.64 ms
0	0	0	0	0	0	0	0	0	1	Clear all display memory and position the cursor to home location (i.e. 0)
Return home										Execution time: 40 μs ~1.6 ms
0	0	0	0	0	0	0	0	1	*	Return the cursor to home location (i.e. 0). DD RAM contents remain unchanged
Entry mode set										Execution time: 40 μs ~1.64 ms
0	0	0	0	0	0	0	1	I/D	S	I/D = 1: Increment; I/D = 0: Decrement. S = 1: Display shift; S= 0: No display shift
Display and cursor on/off control										Execution time: 40 μs
0	0	0	0	0	0	1	D	C	B	D = 1: Display ON; D = 0: Display OFF C = 1: Cursor ON; C = 0: Cursor OFF B = 1: Blink OB; B = 0: Blink OFF
Cursor or display shift										Execution time: 40 μs
0	0	0	0	0	1	S/C	R/L	*	*	S/C = 1: Display shift; S/C = 0: Cursor movement. R/L = 1: Right shift; R/L = 0: Left shift. DD RAM remains unchanged
Function set										Execution time: 40 μs
0	0	0	0	1	DL	N	F	*	*	DL = 1: 8 bits; DL = 0: 4 bits N= 1: 2 lines; F = 0: 5 * 7 dot matrix
CG RAM address set (CG RAM: Character generator RAM)										Execution time: 40 μs
0	0	0	1	A_{CG} (CG RAM address)						
DD RAM address set (DD RAM: Display data RAM)										Execution time: 40 μs
0	0	1	A_{DD} (DD RAM address)							
Read busy flag and address										Execution time: 1 μs
0	1	BF	A_C (Address conter for DD and CG RAM address)							
Data write to CG or DD RAM										Execution time: 40 μs
1	0	Data to be written								
Data read from CG or DD RAM										Execution time: 40 μs
1	1	Data read								

($I/D = 0$) by one when writing or reading a character code to or from DD RAM or CG RAM. After each data write to DD RAM, the entire display can be shifted either right or left ($S = 1$). When $S = 1$ and $I/D = 1$, the display shifts one position to the left; when $S = 1$ and $I/D = 0$, the display shifts one position to the right.

- **Display and cursor on/off control.** This command controls the display and cursor operations. It includes three control bits: D, C, and B. D bit controls whether the display is turned on ($D = 1$) or off ($D = 0$). The display data remain unchanged when the display is turned off. C bit controls whether the cursor is displayed ($C = 1$) or not displayed ($C = 0$). B bit controls whether the cursor is blinking ($B = 1$) or not ($B = 0$).

- **Cursor or display shift.** This command moves the cursor or shifts the display without changing the DD RAM contents. It consists of two bits: S/C and R/L. The four operation modes are as follows:

  1. $S/C \ R/L = 00$. The cursor position is shifted to the left (the AC decrements one).
  2. $S/C \ R/L = 01$. The cursor position is shifted to the right (the AC increments one).
  3. $S/C \ R/L = 10$. The entire display is shifted to the left with the cursor.
  4. $S/C \ R/L = 11$. The entire display is shifted to the right with the cursor.

- **Function set.** This command sets the bus width and the number of display lines, as well as the character font. It consists of three bits: DL, N, and F. When $DL = 1$, the data bus width is set to 8 bits; when $DL = 0$, the data bus width is set to 4 bits ($DB7$ to $DB4$). The upper nibble is transferred first, then the lower nibble follows. N specifies the number of display lines. When $N = 1$, the display is two lines. F sets the character font. When $F = 0$, the character is a 5×7 dot-matrix.

- **CG RAM address set.** This command sets the CG RAM address to the AC before the CG RAM can be accessed.

- **DD RAM address set.** This command sets the DD RAM address to the AC before the DD RAM can be accessed.

- **Busy flag/address read.** This command reads the busy flag and the contents of AC. The busy flag indicates whether the controller is working on a current command ($BF = 1$) or not ($BF = 0$). When $BF = 1$, the module is working internally and cannot accept a new command. Therefore, it is necessary to make sure $BF = 0$ before writing the next command.

- **Data write to CG RAM or DD RAM.** Write data to either CG RAM or DD RAM, which is set by the CG RAM address set command or DD RAM address set command before this command is executed.

- **Data read from CG RAM or DD RAM.** Read data from either CG RAM or DD RAM, which is set by the CG RAM address set command or DD RAM address set command before this command is executed.

11.4.2.1 Initialization Process   When the power is turned on and before any access to the LCD module can proceed, an initialization process has to be carried out. The initialization process is a sequence of commands used to set-up the modes of the

LCD module, as shown in Figure 11.31(c). Details of the initialization sequence are repeated here as follows.

Procedure: Initialization of LCD module

Begin

1. Wait for at least 15 ms after V_{DD} has arisen to 4.5 V.
2. Write command 8'h38 and wait for at least 4.1 ms
3. Write command 8'h38 and wait for at least 100 μs
4. Write command 8'h38 and wait for at least 40 μs
5. Write the following commands in sequence:

```
function set:     8'b0011_NFxx
entry mode:       8'b0000_0110
display on/off:   8'b0000_1110
clear display:    8'b0000_0001
```

End

11.4.3 Datapath Design

Now that we know the details of the widely used commercial LCD module, our focus in the rest of this section is on the design of a hardware module to interface the commercial LCD module.

For simplicity, suppose that we want to design an LCD driver that displays four decimal digits on the LCD module starting at the address of 8'hcc (8'hcc to 8'hcf.) The digits may come from various sources such as the output of a four-digit decimal counter, or other data sources. A clock signal with a period of 1 ms is employed in all modules in order to simplify the design. After understanding this hardware module, interested readers can easily modify the underlying hardware module to their own applications.

As described before, the essential operations of using an LCD module for displaying something are to write commands and data through the electrical interface of the module properly. Due to many constants used in the initialization process, we may collect these constants coupled with the starting address together with the four-digit inputs as a unit called the datapath, because these data are required to be sent to the LCD module through the data bus.

The next problem needed to be considered is how to schedule the constants and the four-digit inputs to be sent to the LCD module in such a way that all data must be written into the LCD module in accordance with the timing constraints of the write cycle depicted in Figure 11.32(b). This problem is solved by a circuit called the controller. As a result, the LCD driver is composed of two parts, datapath and controller, as shown in Figure 11.33. In this figure, we also show the details of the interface between the LCD driver realized in FPGA and the LCD module.

As mentioned above, the datapath of the LCD driver contains only the constants used in the initialization process and the starting address of DD RAM, as well as the

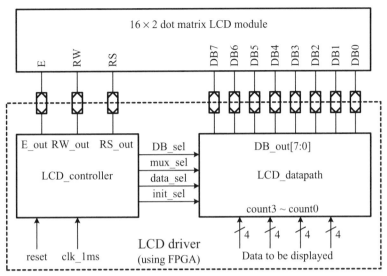

FIGURE 11.33 Block diagram of a four-digit LCD driver

four-digit data to be displayed. However, only one constant or piece of data can be routed to the data bus at a time. Consequently, a multiplexer tree is used to route these data one by one to the data bus $DB7$ to $DB0$ in a proper way scheduled by the controller. Details of the datapath of the LCD driver are depicted in Figure 11.34.

The datapath consists of two 4-to-1 multiplexers and two 2-to-1 multiplexers. The initialization multiplexer is a 6-bit 4-to-1 multiplexer since the highest two bits of all initialization constants are 0s. As a result, they can be concatenated at the output

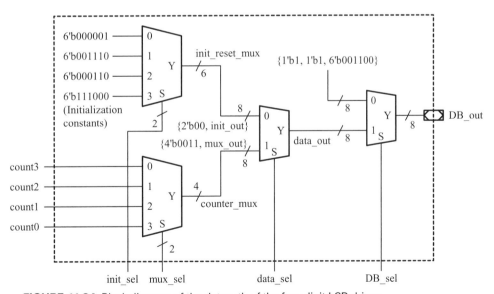

FIGURE 11.34 Block diagram of the datapath of the four-digit LCD driver

of the multiplexer. Similarly, the multiplexer used for routing the four-digit inputs is a 4-bit 4-to-1 multiplexer. At the output of this multiplexer, a 4-bit constant 4'b0011 is prefixed to convert the input decimal digits into their ASCII codes. The 8-bit 2-to-1 data select multiplexer chooses either the initialization constants or the input data to send to the LCD module. Another 8-bit 2-to-1 multiplexer is used to route another constant, starting address, to the LCD module. As a result, the datapath has four sets of control signals that are under the control of the controller.

In the following, we give an example of modeling the datapath described in Figure 11.34.

EXAMPLE 11.22   *Datapath of a Four-Digit LCD Driver*

For simplicity, in this example we use case statements to describe various multiplexers used in the datapath shown in Figure 11.34. Each case statement implements a multiplexer.

```
// LCD datapath --- a multiplexer used to select
// appropriate data for displaying on a two-line
// dot-matrix LCD panel
module LCD_datapath (mux_sel, init_sel, data_sel, DB_sel,
                     count3, count2, count1, count0,
                     DB_out);
input  data_sel, DB_sel;
input  [3:0] count3, count2, count1, count0;
input  [1:0] mux_sel, init_sel;
output reg [7:0] DB_out;
reg    [3:0] counter_mux;
reg    [5:0] init_reset_mux;
reg    [7:0] data_out;
// counter_mux --- multiplexing output the counter.
always @(*)
   case (mux_sel)
      2'b00: counter_mux = count3;
      2'b01: counter_mux = count2;
      2'b10: counter_mux = count1;
      2'b11: counter_mux = count0;
   endcase
// initialization and reset process
always @(*) // multiplexer init_reset_mux
   case (init_sel)
      2'b00: init_reset_mux = 6'b000001; // 01H --
                                         // 00_000001
      2'b01: init_reset_mux = 6'b001110; // 0EH --
                                         // 00_001110
      2'b10: init_reset_mux = 6'b000110; // 06H --
                                         // 00_000110
```

```
              2'b11: init_reset_mux = 6'b111000; // 38H --
                                                 // 00_111000
      endcase
   always @(*) // selecting the initialization or normal
               // data
      case (data_sel)
         1'b0: data_out = {2'b00, init_reset_mux};
         1'b1: data_out = {4'b0011, counter_mux};
      endcase
   always @(*)
      case (DB_sel)
         1'b0: DB_out = {1'b1, 1'b1, 6'b001100}; // set
                                                 // address at CCH
         1'b1: DB_out = data_out;
      endcase
   endmodule
```

11.4.4 Controller Design

As described before, the controller has to schedule the constants or input data sent to the LCD module properly and generate a correct timing relationship between various control signals according to the specification of write cycle, shown in Figure 11.32(b). Consequently, the controller is divided into three modules, main controller, LCD initialization and refresh, as well as write cycle, as shown in Figure 11.35.

The main controller module determines when to initialize and when to refresh the LCD module. The LCD initialization and refresh module performs the initialization process and refresh operations of the LCD module. The write cycle module generates the required timing of write cycle. In the design of a controller, we assume that a clock signal with a period of 1 ms is used. This assumption not only simplifies the design considerably but also conforms to the timing specification of write cycle. In the following, we describe each module in more detail.

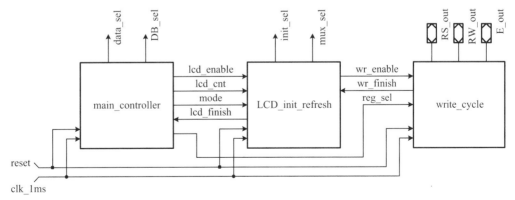

FIGURE 11.35 The controller of a four-digit LCD system

The following example describes the interface between the three modules of the controller of the LCD driver. Actually, it is the top module of the controller of the four-digit LCD system.

EXAMPLE 11.23 *Controller of a Four-Digit LCD Driver*

This is the top module of the controller of the LCD driver under consideration. It instantiates three lower-level modules, `main_controller` and `LCD_init_refresh`, as well as `write_cycle`.

```
//LCD driver --- initialize and write data to LCD module.
module LCD_controller(clk_1ms, reset, mux_sel, init_sel,
                      data_sel, DB_sel, RS_out, RW_out,
                      E_out);
input   clk_1ms, reset;
output wire data_sel, DB_sel, RS_out, RW_out, E_out;
output wire [1:0] init_sel, mux_sel;
wire mode, lcd_enable, lcd_finish, reg_sel, wr_enable,
     wr_finish;
wire [1:0] lcd_cnt;
// instantiate various module associated with lcd_driver_
// test
main_controller  main_contl  (clk_1ms, reset, lcd_finish,
                              lcd_enable, lcd_cnt, mode,
                              reg_sel, data_sel, DB_sel);
LCD_init_refresh LCD_wr_data (clk_1ms, reset, lcd_enable,
                              wr_finish, lcd_cnt, mode,
                              mux_sel, init_sel,
                              lcd_finish, wr_enable);
write_cycle      write_cycle (clk_1ms, reset, wr_enable,
                              reg_sel, wr_finish, RS_out,
                              RW_out, E_out);
endmodule
```

Write Cycle Module   The write cycle module generates the required timing of write cycle conforming to the timing specification shown in Figure 11.32(b). According to the timing specification, it is easy to divide the timing into four states: outputs *RS* and *R/W* signals, activates the *E* signal for two states, and then deactivates the *E* signal. More details of write cycle are shown in Figure 11.36 in the form of an ASM chart.

Because the write cycle module is required for writing each piece of data or constant to the LCD module, a mechanism somewhat like a subroutine call and return has to be made such that the write cycle module can be enabled and notifies that its assigned operations have been completed. This mechanism is called an *enable and finish control* mechanism and contains two control signals: `wr_enable` and `wr_finish`. The `wr_enable` enables the operations of the write cycle module. When the write cycle module completes its assigned operations, it asserts a `wr_finish` signal to notify the

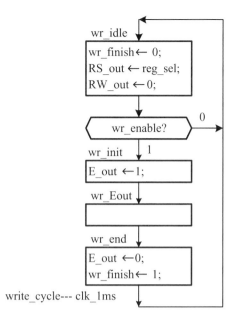

FIGURE 11.36 The ASM for the write-cycle operations of the four-digit LCD driver

enabling module. In the running example, this is the `LCD_init_refresh` module, which will be described next. In reality, the enable and finish control mechanism can also be considered as a form of handshaking protocol. The operations of the handshaking mechanism will be described in greater detail in Section 15.2.2.

In the following, we give an example to model the ASM chart shown in Figure 11.36.

EXAMPLE 11.24 *Write Cycle of the Controller*

This module generates the required timing for writing a piece of data or a constant to the LCD module. The detailed operations of the module are exactly the same as those described in the ASM chart shown in Figure 11.36.

```
// write timing generator -- a four-state FSM
// it is activated by wr_enable and signaled by wr_finish
// its operation is completed
module write_cycle(clk_1ms, reset, wr_enable, reg_sel,
                   wr_finish, RS_out, RW_out, E_out);
input  clk_1ms, reg_sel, reset, wr_enable;
output reg wr_finish, RS_out, E_out;
output wire RW_out;
reg [1:0] wr_state;
// define local parameters
localparam wr_idle = 2'b00, wr_init = 2'b01,
           wr_Eout = 2'b10, wr_end  = 2'b11;
// in write cycle RW_out is always 0.
assign RW_out = 1'b0;
```

```
always @(posedge clk_1ms or posedge reset)
    if (reset) begin  RS_out <= 1'b0;  wr_finish <= 1'b0;
        E_out <= 1'b0; wr_state <= wr_idle; end
    else case (wr_state)
        wr_idle: begin
            if (wr_enable) wr_state <= wr_init;
            else wr_state <= wr_idle;
            RS_out <= reg_sel;
            wr_finish <= 0; end
        wr_init: begin wr_state <= wr_Eout; E_out <= 1'b1;
            end
        wr_Eout: wr_state <= wr_end;
        wr_end:   begin
            wr_state <= wr_idle; E_out <= 1'b0;
            wr_finish <= 1; end
    endcase
endmodule
```

LCD Initialization and Refresh Module   Due to the same operation of both initialization process and data refresh, we combine both operations into a single module, denoted as the LCD initialization and refresh module, `LCD_init_refresh`. The goal of the `LCD_init_refresh` module is to schedule the initialization constants and the input data to be sent to the LCD module through the data bus. Hence, it has to generate the desired multiplexer selection signals `mux_sel`, `init_sel`, as shown in Figure 11.37. After this, it has to enable the write cycle module to generate the required write cycle control signals. In addition, the `LCD_init_refresh` module is enabled in either the initialization process or the refresh process. The initialization process is only executed once when the reset signal is activated, whereas the refresh process is executed repeatedly to refresh the input data onto the LCD module. In the same way that the write cycle module is enabled by the LCD initialization and refresh module, the LCD initialization and refresh module is enabled by the main controller module, which determines when to initialize the LCD module and when to refresh the LCD module with the input data. The control mechanism used in this module is `lcd_enable` and `lcd_finish`. The detailed operations of this module are represented as the ASM chart shown in Figure 11.37.

The following is an example of representing the ASM chart as a Verilog HDL module.

EXAMPLE 11.25   *LCD Initialization and Refresh of the Controller*

This module accepts the `lcd_enable`, `lcd_cnt` and `mode` control signals from the main controller module, `main_controller`, and generates the required control signals `init_sel` or `mux_sel` to the LCD module. In addition, it enables the write cycle module, `write_cycle`, to generate the required timing for writing initialization constant or data onto the LCD module. The detailed operations of the module are exactly the same as those described in the ASM chart shown in Figure 11.37.

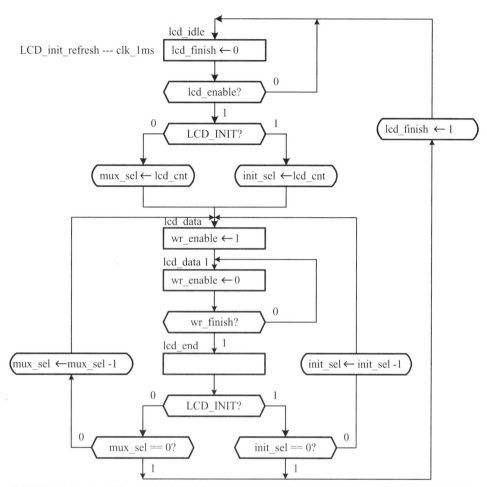

FIGURE 11.37 The ASM for the LCD initialization and refresh operations of the four-digit LCD driver

```
// LCD module initialization and refresh--- a four-state
// FSM
module LCD_init_refresh(clk_1ms, reset, lcd_enable,
                        wr_finish, lcd_cnt, mode, mux_sel,
                        init_sel, lcd_finish, wr_enable);
input  clk_1ms, reset, mode, lcd_enable, wr_finish;
input  [1:0] lcd_cnt;
output reg lcd_finish,wr_enable;
output reg [1:0] mux_sel, init_sel;
reg [1:0] state;
localparam LCD_INIT = 0;
localparam lcd_idle  = 2'b00, lcd_data = 2'b01,
           lcd_data1 = 2'b10, lcd_end  = 2'b11;
```

```
always @(posedge clk_1ms or posedge reset)
   if (reset) begin  mux_sel <= 0; init_sel <= 0;
      state <= lcd_idle; wr_enable <= 0;
      lcd_finish <= 0;  end
   else case (state)
      lcd_idle: begin
         if (lcd_enable) begin
            state <= lcd_data;
            if (mode == LCD_INIT) init_sel <= lcd_cnt;
            else mux_sel  <= lcd_cnt; end
         else state <= lcd_idle;
         lcd_finish <= 0; end
      lcd_data: begin
         wr_enable <= 1; state <= lcd_data1; end
      lcd_data1:  begin
         wr_enable <= 0;
         if (wr_finish == 1) state <= lcd_end;
         else state <= lcd_data1; end
      lcd_end:
         if (mode == LCD_INIT) begin
            if (init_sel == 0) begin
               lcd_finish <= 1; state <= lcd_idle; end
            else begin
               init_sel <= init_sel - 1;
               state <= lcd_data; end end
         else begin
            if (mux_sel == 0) begin
               lcd_finish <= 1; state <= lcd_idle; end
            else begin
               mux_sel <= mux_sel - 1;
               state <= lcd_data; end end
   endcase
endmodule                                              ∎
```

Main Controller Module The main controller module determines when to initialize the LCD module and when to refresh the LCD module with the input data and then routes the required data to the LCD module through the right-most two 8-bit 2-to-1 multiplexers, as shown in Figure 11.34. It enables the LCD initialization and refresh module through the enable and finish control mechanism: lcd_enable and lcd_finish. The detailed operations of the main controller module are represented as the ASM chart shown in Figure 11.38.

The ASM chart can be divided into three parts, with each containing two states. The first part performs the initialization process, which enables the LCD_init_refresh module to write the initialization constants into the LCD

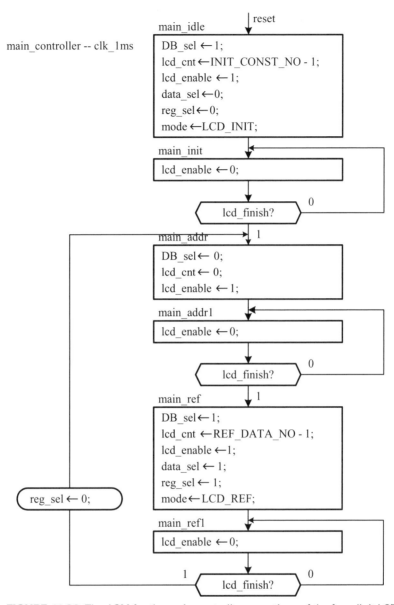

FIGURE 11.38 The ASM for the main controller operations of the four-digit LCD system

module. The second part sets the starting address of DD RAM. This address must be set each time when the refresh operation is executed so that the data may be displayed on the LCD module at the same place. The third part carries out the refresh operations. The following example describes the detailed operations of the main controller module.

EXAMPLE 11.26 *Main Controller of the Controller*

This module generates both data_sel and DB_sel control signals for the datapath shown in Figure 11.34 in order to route initialization constants, data to be displayed or starting address to the LCD module. It also generates lcd_enable, lcd_cnt and mode signals to control the appropriate operations of the LCD_init_refresh module. The detailed operations of the module are exactly the same as those described in the ASM chart shown in Figure 11.38.

```verilog
// main controller --- schedule initialization and
// refresh
module main_controller(clk_1ms, reset, lcd_finish,
                       lcd_enable, lcd_cnt, mode, reg_sel,
                       data_sel, DB_sel);
input  clk_1ms, reset, lcd_finish;
output reg lcd_enable, mode, reg_sel, data_sel, DB_sel;
output reg [1:0] lcd_cnt;
reg [2:0] main_state;
localparam INIT_CONST_NO = 4,
           REF_DATA_NO   = 4;
localparam LCD_INIT = 0, LCD_REF = 1;
// main-loop operation --- a seven-state fsm.
localparam main_idle  = 3'b000, main_init = 3'b001,
        main_addr = 3'b010, main_addr1 = 3'b011,
           main_ref  = 3'b100, main_ref1 = 3'b101;
always @(posedge clk_1ms or posedge reset)
   if (reset) begin
      DB_sel <= 1'b1; lcd_enable <= 0; lcd_cnt <= 0;
      data_sel <= 1'b0; mode <= LCD_INIT;
      reg_sel <= 1'b0;  main_state <= main_idle;  end
   else case (main_state)
      main_idle: begin // begin initialization
          lcd_cnt <= INIT_CONST_NO - 1; DB_sel <= 1'b1;
          lcd_enable <= 1; data_sel <= 1'b0;
          mode <= LCD_INIT;
          main_state <= main_init; reg_sel <= 1'b0; end
      main_init: begin // inializing ...
          lcd_enable <= 0;
          if (lcd_finish) main_state <= main_addr;
          else main_state <= main_init; end
      main_addr:  begin // set up starting address
          lcd_enable <= 1;  DB_sel <= 1'b0;
          lcd_cnt <= 0;  // send only one data
          main_state <= main_addr1; end
      main_addr1: begin
          lcd_enable <= 0;
```

```
                if (lcd_finish) main_state <= main_ref;
                else main_state <= main_addr1; end
          main_ref:  begin   // begin to refresh
                lcd_cnt <= REF_DATA_NO - 1; DB_sel <= 1'b1;
                lcd_enable <= 1;  data_sel <= 1'b1;
                mode <= LCD_REF;
                main_state <= main_ref1; reg_sel <= 1'b1; end
          main_ref1: begin
                lcd_enable <= 0;
                if (lcd_finish) begin
                    main_state <= main_addr; reg_sel <= 1'b0; end
                else main_state <= main_ref1; end
       endcase
   endmodule
```

Review Questions

Q11.21 Describe the basic principles of reflective field-effect LCDs.

Q11.22 What is the function of CG ROM? Can we access the CG ROM directly?

Q11.23 What is the function of DD RAM?

Q11.24 Can we access both DD RAM and CG RAM at the same time?

Q11.25 Describe the write cycle of the LCD module we have introduced.

Q11.26 Describe the initialization process of the LCD module we have introduced.

Q11.27 Describe the basic process of using an LCD module to display characters.

Q11.28 What are the basic components of the controller of the LCD module?

SUMMARY

In this chapter, we have introduced two major methods: the FSM-based approach and the RTL design, for designing digital systems. The FSM-based approach is employed in designing small systems, whereas the RTL design is used for designing large systems.

A finite-state machine is usually used to model a circuit with memory, such as a sequential circuit or a controller in digital systems. Finite-state machines may be described by using either state diagrams or ASM charts. Two widely used finite-state machine models, the Mealy machine and Moore machine, are introduced to model sequential circuits. In the Mealy machine, the output function is determined by both the present input and state, whereas in the Moore machine the output function is solely determined by the present state.

The RTL design is based on the observation that any digital system can be considered as a system of interconnected registers being placed between two registers, a combinational logic, to perform the required functions. As a consequence, the digital system operates in RTL in such a way that data from the output of registers are transferred to other registers through which between them a combinational logic is used to change the property of the data.

Two widely used RTL designs are the ASM chart, as well as the datapath and controller approach. The ASM chart, also known as a state machine (SM) chart, is often used to describe a digital design at the algorithmic level. An ASM chart is only composed of three blocks: state block, decision block and conditional output block. It is much like a flow chart but incorporates timing into consideration. ASM charts have the following features. They specify RTL operations since they precisely define the operation on every cycle and clearly show the flow of control from state to state.

The datapath and controller approach is also sometimes called the control-point style, which can be easily derived from the ASM chart of the design. The datapath part corresponds to the registers and function units in the ASM chart and the controller part corresponds to the control signals in the ASM chart. In this chapter, we outlined a three-step paradigm by which a datapath and controller architecture can be easily derived from the ASM chart of the RTL design.

A state-of-the-art approach could also be used to derive the datapath and controller architecture of an RTL design. To illustrate this technique, an example of displaying data on an LCD module is given. In this example, an LCD driver module which displays four-digit data on a commercial dot-matrix LCD module is designed by partitioning it into datapath and controller in a heuristic way. Then, ASM charts are applied to describe each part of the controller.

There are many options that can be used to implement an RTL design. The rationale behind these options is a tradeoff among performance (throughput), space (area) and time (operating frequency). The most common options that are widely used in practical systems are: single-cycle structure, multiple-cycle structure and pipeline. For a single-cycle structure, combinational logic is used to perform the desired computation. This structure often runs faster than the other two structures at the expense of a tremendous amount of hardware. The multiple-cycle structure uses much less hardware than the single-cycle structure but has a worse performance. The pipeline structure is often used in the case of a stream input to speed up the total throughput at the expense of both hardware and data latency.

REFERENCES

1. M.G. Arnold, *Verilog Digital Computer Design: Algorithms into Hardware,* Prentice-Hall, Upper Saddle River, NJ, USA, 1999.
2. M.D. Ciletti, *Modeling, Synthesis and Rapid Prototyping with the Verilog HDL,* Prentice-Hall, Upper Saddle River, NJ, USA, 1999.
3. D.D. Gajski, *Principles of Digital Design,* Prentice-Hall, Upper Saddle River, NJ, USA, 1997.
4. H.-C. Lee, *Introduction to Color Imaging Science,* Cambridge University Press, New York, NY, USA, 2005.
5. M.-B. Lin, *Digital System Design: Principles, Practices and ASIC Realization,* 3rd Edn, Chuan Hwa Book Company, Taipei, Taiwan, 2002.
6. M.-B. Lin, *Basic Principles and Applications of Microprocessors: MCS-51 Embedded Microcomputer System, Software and Hardware,* 2nd Edn, Chuan Hwa Book Company, Taipei, Taiwan, 2003.
7. C.H. Roth, Jr, *Fundamentals of Logic Design,* 4th Edn, PWS, Boston, MA, USA, 1992.
8. W. Wolf, *FPGA-Based System Design,* Prentice-Hall, Upper Saddle River, NJ, USA, 2004.

PROBLEMS

11.1 A 4-bit ECC memory. The rationale behind the Hamming code is to insert k parity bits into an m-bit message to form a $(k + m)$-bit codeword. The parity bits are placed at the positions of 2^i, where $0 \leq i < k$. In order to correct single-bit error, including the m-bit message and k parity bits, m and k must satisfy the equation: $2^k \geq m + k + 1$. For instance, when the message is 4 bits, at least three parity bits are required in order to satisfy the inequality equation and they are placed at positions of $2^0 = 1$, $2^1 = 2$ and $2^2 = 4$, respectively. As a result, the new codeword is $p_0 p_1 m_0 p_2 m_1 m_2 m_3$, where p_0 is at position 1 and m_3 is at position 7. When writing an m-bit message to a memory, the parity bits are computed by a criterion that each parity bit combined with some message bits must form an even parity; namely, all bits indicated by digit 1 in each row of Table 11.3 have to form an even parity. For instance, p_0 combined with m_0, m_1, and m_3 have to form an even parity. These parity bits along with the m-bit message, called a codeword, are written into the memory. The parity property is checked whenever the codeword is read out from memory and the result called a syndrome is represented as $c_2 c_1 c_0$. If $c_2 c_1 c_0 = 0$, the codeword is error free; otherwise, the bit indicated by $c_2 c_1 c_0$ is erroneous.

(a) Design a logic circuit to implement the above parity check generator, as well as the error check and correction.

(b) Describe your design in Verilog HDL as an ECC memory module.

(c) Write a test bench to verify the functionality of your ECC memory module.

11.2 An 8-bit ECC memory. This problem is an extension of the above 4-bit ECC memory. The positional numbers of the Hamming (12, 8) code are shown in Table 11.4.

(a) Design a logic circuit to implement the parity check generator, as well as the error check and correction.

(b) Describe your design in Verilog HDL as an ECC memory module.

(c) Write a test bench to verify the functionality of your ECC memory module.

11.3 An 8-bit SECMED memory. In order to provide the capability of multiple-bit error detection in addition to single-bit error correction, an extra parity check bit has to be added. The resulting parity check generation is listed as Table 11.5. Each data bit is checked by three parity bits. Each parity bit, c_i, along with its associated data bits (marked by "x") in the same row of the table, forms an even

TABLE 11.3 The positional numbers of the Hamming (7, 4) code

	1	2	3	4	5	6	7
c_n	p_0	p_1	m_0	p_2	m_1	m_2	m_3
c_0	1	0	1	0	1	0	1
c_1	0	1	1	0	0	1	1
c_2	0	0	0	1	1	1	1

TABLE 11.4 The positional numbers of the Hamming (12, 8) code

	1	2	3	4	5	6	7	8	9	10	11	12
c_n	p_0	p_1	m_0	p_2	m_1	m_2	m_3	p_3	m_4	m_5	m_6	m_7
c_0	1	0	1	0	1	0	1	0	1	0	1	0
c_1	0	1	1	0	0	1	1	0	0	1	1	0
c_2	0	0	0	1	1	1	1	0	0	0	0	1
c_3	0	0	0	0	0	0	0	1	1	1	1	1

parity. For instance, c_0, and data bits d_3, d_4, d_6 and d_7 form an even parity. In order to locate a possible single-bit error, it is necessary to determine whether the parity property of the codeword stored in memory is still valid or not. This is achieved by calculating the syndrome (i.e. the combined parity values of all rows) and the result is denoted as new c_i, This new c_i is then used to locate the single error bit and correct the erroneous bit accordingly. The relationship between syndrome and bit location is shown in Table 11.6.

(a) Design a logic circuit to implement the parity check generator, as well as the error check and correction.

(b) Describe your design in Verilog HDL as an ECC memory module.

(c) Write a test bench to verify the functionality of your ECC memory module.

11.4 RTL implementation options. Consider an n-bit adder which can be implemented by either a one-cycle or multiple-cycle structure.

(a) Suppose that both inputs x and y are stored in each individual n-bit register; the output *sum* is also stored in an n-bit register. Design a circuit using only one 1-bit full adder to add these two inputs and generate the sum. Write a test bench to verify the functionality of the module.

(b) Using iterative logic circuit, re-design the above module. Write a test bench to verify the functionality of this module.

11.5 An LCD module. Modify the LCD display system described in Section 11.4 so that it can display eight digits starting from address 8'hC8.

(a) Draw the datapath module.

(b) Construct the ASM chart for the controller.

(c) Write a test bench to verify the functionality of your system.

TABLE 11.5 The parity check generation for an SECMED code

c_n	d_0	d_1	d_2	d_3	d_4	d_5	d_6	d_7
c_0				×	×		×	×
c_1		×	×		×	×		×
c_2	×		×	×		×	×	
c_3	×	×				×	×	×
c_4	×	×	×	×	×			

TABLE 11.6 Locating a single error in the SECMED code shown in Table 11.5

c_n	No error	c_0	c_1	c_2	c_3	c_4	d_0	d_1	d_2	d_3	d_4	d_5	d_6	d_7
c_0	1	0	1	1	1	1	1	1	1	0	0	1	0	0
c_1	1	1	0	1	1	1	1	0	0	1	0	0	1	0
c_2	1	1	1	0	1	1	0	1	0	0	1	0	0	1
c_3	1	1	1	1	1	1	0	0	1	1	1	0	0	0
c_4	1	1	1	1	0	0	0	0	0	0	0	1	1	1

11.6 Synchronization between two different clock domains. This problem is to explore the synchronization between two different clock domains. In order to examine this, suppose that the `write_cycle` shown in Figure 11.36 uses a clock signal with a period of 50 μs rather than 1 ms. Redesign the `write_cycle` module and write a test bench to make sure that the resulting function of the LCD display system is still correct.

11.7 A stop watch. Suppose that a stop watch with a resolution of 10 ms and a maximum count of 99 s is desired. The stop watch has a start/stop button that starts and stops the operations of the stop watch. In addition, a clear button is facilitated to reset the stop watch. Design this stop watch module and write a test bench to verify the functionality of it.

11.8 A clock. Suppose that a clock with a 24-h display format is desired. The clock may display hours, minutes and seconds, and it has three buttons which are employed to separately set its hours, minutes and seconds. Assume that the master clock available is a 4-MHz clock signal. Design this clock module and write a test bench to verify the functionality of it.

11.9 A timer. Suppose that a timer with a maximum duration of 24 h is desired. The timer may display hours, minutes and seconds, and it has three buttons which are employed to separately set its hours, minutes and seconds. The timer outputs an alarm signal for 30 s once it counts down to 0. In addition, the timer can be stopped and cleared by another two buttons: stop and clear, respectively. Assume that the master clock available is a 4-MHz clock signal. Design this timer module and write a test bench to verify the functionality of it.

11.10 An FIFO. Suppose that a 16-bit FIFO with a depth of 16 words is desired. The front end (read port) and rear end (write port) have a separate set of control signals: clock and access enable (write or read). In addition, two flags, empty and full, are used to indicate the status of the FIFO. Design this FIFO module and write a test bench to verify the functionality of it.

11.11 A stack. Suppose that a 16-bit stack with a depth of 16 words is desired. The top end (read and write port) has the following control signals: clock, push and pop. In addition, two flags, empty and full, are used to indicate the status of the stack. Design this stack module and write a test bench to verify the functionality of it.

11.12 A 16-key keypad. Suppose that a 16-key keypad is desired. The layout of the keypad is shown in Figure 11.39.

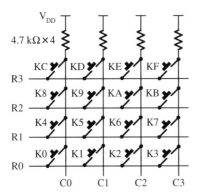

FIGURE 11.39 The layout of a typical 16-key keypad

(a) Design a scanning circuit to scan the keypad in sequence. In addition, the switch debouncing problem must be taken into account. Draw the circuit in register-transfer level.

(b) Describe the keypad scanning circuit in Verilog HDL.

(c) Write a test bench to verify the functionality of the module you designed.

11.13 A programmable lock. Suppose that a programmable lock with the capability of a password verification is desired. The length of the password is limited to 6 digits and the initial password is all 0s. In order to change the password, the following procedure is followed. First, the key denoted as the change-password is pressed and then enter the old password and the new password twice. After finishing the entering of the password each time, you need to press the key denoted as end to notify the lock that the password has completely been entered. To open the lock, just enter the password and the lock is open (sends an open signal) if the password entered is exactly matched with the one stored. Otherwise, it does nothing. Design this module and write a test bench to verify the functionality of it.

11.14 A keypad plus LED module. By combining the LED module described in Section 8.6.1 and the keypad module from problem 11.12, use an available FPGA device to construct a keypad plus LED module system. You may limit the maximum number of digits that may be entered and design the system accordingly.

11.15 A keypad plus LCD module. By combining the LCD module described in Section 11.4 and the keypad module from problem 11.12, use an available FPGA device to construct a keypad plus LCD module system. You may limit the maximum number of digits that may be entered and design the system accordingly.

11.16 An FPGA-based calculator. Use an available FPGA device to design a simple but complete calculator. The calculator may perform the basic arithmetic operations: $+$, $-$, $*$ and $/$, just according to the entry sequence without the capability of parentheses. By combining the 16-key keypad and LCD display system modules, use an available FPGA device to construct the real system.

SYNTHESIS

THE SUCCESS of logic synthesis has dramatically cut the design time and pushed HDLs into the forefront of digital designs. Accompanying this success, massive consumer's products are produced such as MP3 players and DVD players, and have changed people's daily lives.

In this chapter, we deal with the principles of logic synthesis and the general architecture of synthesis tools. In general, synthesis can be divided into logic synthesis and high-level synthesis. The former transforms an RTL representation into gate-level netlists while the latter transforms a high-level representation into an RTL representation. The general architecture of synthesis tools is divided into two parts: the front end and the back end. The front end consists of two phases: parsing and elaboration. The back end contains three phases: analysis, optimization and netlist generation.

In order to make good use of synthesis tools, we need to provide the design environment and design constraints along with the RTL code and technology library. The design environment provides the process parameters, I/O port attributes and statistical wireload models for the synthesis tools to synthesize a design. The design constraints provide the clock related parameters, input and output delays and timing exceptions.

Finally, we give some guidelines about how to write a good Verilog HDL code such that it can be acceptable by most logic synthesis tools and can achieve the best compilation times and synthesis results. These guidelines also include clock signals, reset signals and how to partition a design.

12.1 DESIGN FLOW OF ASICs AND FPGA-BASED SYSTEMS

Quite often, it is required to follow a design flow when designing an ASIC or an FPGA-based system. A *design flow* is a set of procedures that allows designers to progress from a specification for an ASIC or FPGA-based system to the final chip or

FPGA implementation in an efficient and error-free way. In this section, we describe the general design flow of designing ASICs and FPGA-based systems.

12.1.1 The General Design Flow

Recall that the RTL design is a mixed style combining both dataflow and behavioral styles with the constraint that the resulting description can be acceptable by synthesis tools. The general design flow of an ASIC and FPGA-based design is shown in Figure 12.1.

From this figure, we know that the design flow can be divided into two major parts: *front end* and *back end*. The front end contains three phases, starting from product requirement, behavioral/RTL description and ending with RTL synthesis, which generates a gate-level netlist. The back end also contains three phases, starting from the structural description of gate-level netlist, physical synthesis and ending with physical specification. In other words, the RTL synthesis is at the heart of the front-end part and the physical synthesis is the essential component of the back-end part.

12.1.1.1 RTL Synthesis Flow The general RTL design flow is shown in Figure 12.2. The RTL design flow starts with product requirement, which is converted into a *design specification*. The specification is then described with RTL behavioral style in Verilog HDL or VHDL. The results are then verified by using a set of test benches written by HDL. This verifying process is called *RTL functional verification*.

The design is synthesized by a logic synthesizer after its function has been verified correctly. This process is denoted as *RTL synthesis* or *logic synthesis*. The function of the logic synthesizer is to convert an RTL description to generic gates and registers and then optimize the logic to improve speed and area. In addition, finite state machine decomposition, data path optimization and power optimization may also be performed

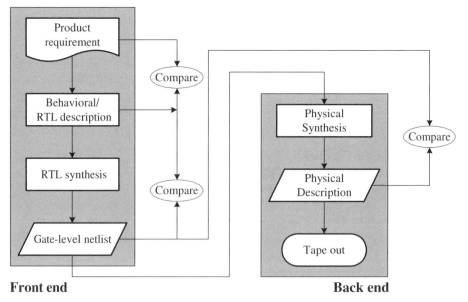

Front end   **Back end**

FIGURE 12.1 The general design flow of an ASIC and FPGA-based design

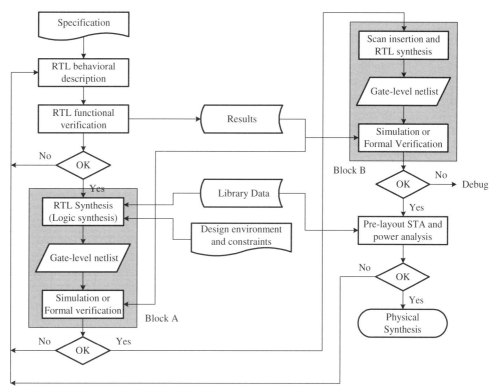

FIGURE 12.2 The general flow of RTL synthesis

at this stage. In general, a logic synthesizer accepts three inputs, *RTL code*, *technology library* and *design environment and constraints*, and generates a gate-level netlist.

After a gate-level netlist is generated, it is necessary to re-run the test benches used in the stage of RTL functional verification to check if they produce exactly the same output for the behavioral and structural descriptions or to perform an RTL versus gate equivalence checking to compare the logical equivalence of the two descriptions.

The next three steps often used in ASIC (namely, cell-based design), but not in an FPGA-based design are shown in the shaded block B, which incorporates the *scan-chain logic insertion*, re-synthesis and verification. In fact, this block may be combined together with block A. The scan-chain (or test logic) insertion step is to insert or modify logic and registers to aid in manufacturing tests. Automatic test pattern generation (ATPG) and built-in self-test (BIST) are usually used in most modern ASIC designs. The details of these topics are addressed in Chapter 16.

The final stage of RTL design flow is the pre-layout static timing analysis (STA) and power consumption analysis. Static timing analysis checks the temporal requirements of the design. We describe the STA in greater detail in Section 13.5. Power analysis estimates the power consumption of the circuit. The power consumption depends on the activity factors of the gates. Power analysis can be performed for a particular set of test vectors by running the simulator and evaluating the total capacitances switched at each clock transition of each node.

12.1.1.2 *Physical Synthesis Flow*

As mentioned before, the second part of the design flow of an ASIC or FPGA-based system is the physical synthesis. Regardless of ASIC or an FPGA-based system, physical synthesis can be further divided into two major stages: *placement* and *routing*, as shown in Figure 12.3. Because of this, physical synthesis is usually called *place and route* (PAR) in CAD tools.

In the placement stage, logic cells (standard cells or building blocks) are placed at fixed positions in order to minimize the total area and wire length. In other words, placement defines the location of logic cells (modules) on a chip and sets aside space for the interconnect to each logic cell (module). This stage, in general, is a mixture of three operations: *partitioning*, *floorplanning* and *placement*.

Partitioning partitions the circuit into parts such that the sizes of the components (modules) are within prescribed ranges and the number of connections between components is minimized. Floorplanning determines the appropriate (relative) location of each module in a rectangular chip area. Placement finds the best position of each module on the chip such that the total chip area is minimized or the total length of the wires is minimized. Of course, not all CAD tools have their placement divided into the above three sub-steps. Some CAD tools combine the above three sub-steps into one big step, simply called placement, depending on what kinds of algorithms are used.

After placement, a clock tree is inserted in the design. In this step a clock tree is generated and routed coupled with the required buffers. A clock tree is often placed before the main logic placement and routing is completed in order to minimize the clock skew. This step is not necessary in an FPGA-based design, in which a clock distribution network is already fixed on the chip.

The next big stage is known as route, which is used to complete the connections of signal nets among the cell modules placed by placement. This stage is further divided

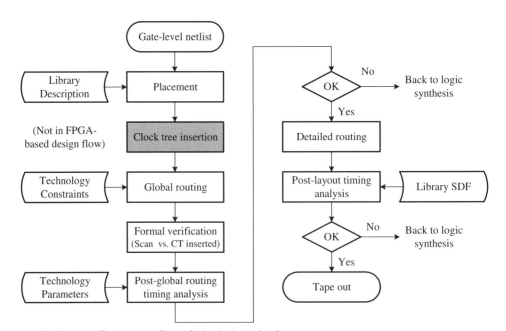

FIGURE 12.3 The general flow of physical synthesis

into two sub-stages: *global routing* and *detailed routing*. Global routing decomposes a large routing problem into small and manageable sub-problems (detailed routing) by finding a rough path for each net in order to reduce chip size, shorten the total length of the wires, and evenly distribute the congestion over the routing area. Detailed routing carries out the actual connections of signal nets among the modules.

After global and detailed routing, the post-global routing and post-layout static timing analysis are performed, separately. These timing analyses re-run the timing analysis with the actual routing loads placed on the gates to check whether the timing constraints are still valid or not.

Review Questions

Q12.1 What is a design flow?

Q12.2 Which phases are included in the front-end part of the general design flow?

Q12.3 Why is the front-end part often called logic synthesis?

Q12.4 Which phases are included in the back-end part of the general design flow?

Q12.5 Why is the back-end part often called physical synthesis?

Q12.6 What are the two major stages of physical synthesis?

12.1.2 Timing-Driven Placement

A problem with the PAR and timing analysis described previously is that the timing information can only be obtained after the layout has been completed. As a result, in order to satisfy the post-layout timing requirement, it may go back to the logic synthesis and starts there again, as illustrated in Figure 12.3. Even more, this process may be repeated several times. One technique to solve this is by incorporating the timing analysis into the placement stage so that the critical path can be placed into the layout with priority. This technique is called *timing-driven placement*. In this section, we will introduce a simple timing-driven placement based on the concept of *slack time*.

Before describing the meaning of slack time, we need to define the *arrival time* and *required time*, as shown in Figure 12.4. Assume that d_i, for $0 \leq i \leq m$, are net

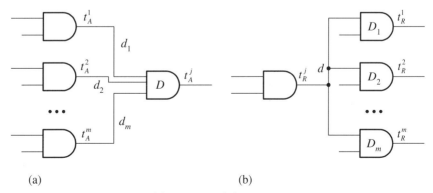

(a)　　　　　　　　　　　　　　　　　　(b)

FIGURE 12.4 Definitions of (a) arrival and (b) required times

delays and D is the delay of the cell containing net j. Arrival time is defined as:

$$t_A^j = max_{i=1}^m (t_A^i + d_i) + D$$

Assume that D_i, for $0 \leq i \leq m$, are cell delays and d is the delay of the net that connects the cell to its fanout cells. The required time is defined as:

$$t_R^j = max_{i=1}^m (t_R^i - D_i) - d$$

The arrival time at the output of a gate is known only when the arrival time at every input of that gate has been computed.

Assume that t_A and t_R are the arrival time and the required time of a signal at a given net, respectively. The slack time is defined as:

$$t_{slack} = t_R - t_A$$

The features of slack time are as follows. First, the paths with the smallest slacks are called *timing critical paths* or simply, *critical paths*. Second, the negative value of the slack means that the associated paths are the critical paths. As a result, the slack time can be used to analyze the critical paths of a design and this is often the method used in timing analysis tools.

Another use of slack time is found in the timing-driven placement algorithm. In the following, we only give a simple version of this kind of algorithms, which is called a *zero-slack algorithm*. However, it gives us a flavor of how a timing-driven placement works.

In order to describe the zero-slack algorithm [11], suppose that the signal-arriving time at each primary input and the time a signal is required at a primary output are known. Then, the algorithm can be described as follows.

Algorithm: Zero-slack algorithm

Begin
    1. repeat
        1.1 Compute all slacks;
        1.2 Find the minimum positive slack;
        1.3 Find a path with all slacks equal to the minimum slack;
        1.4 Distribute the slacks along the path segment;
      until (there exists no positive slack);
End // end of the algorithm

The following example demonstrates the operations of a zero-slack algorithm.

EXAMPLE 12.1 *An Example of a Zero-Slack Algorithm*

This example illustrates the operations of a zero-slack algorithm. As shown in Figure 12.5(a), the arrival time of all primary inputs and the required time of all primary

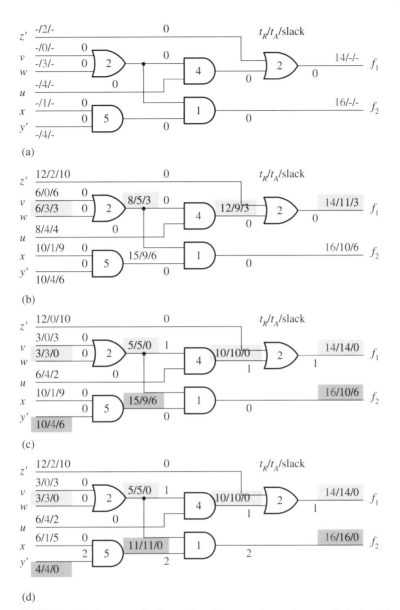

FIGURE 12.5 An example illustrating the operations of a zero-slack algorithm: (a) original; (b) step 1; (c) step 2; (d) step 3

outputs are known in advance. In addition, all gates have their propagation delays as shown in the figure. The purpose of the zero-slack algorithm is to compute and distribute the slack time evenly in the interconnect along with the path from a primary input to a primary output.

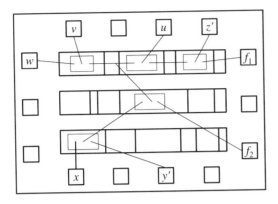

FIGURE 12.6 The final placement of the results of Figure 12.5

According to the definitions of both arrival time and required time, the arrival time and required time at each node, along with each possible path between a primary input and a primary output, can be easily computed. The arrival time of each node is computed from a primary input to a primary output and the required time is computed from a primary output backward to a primary input. Once both arrival time and required time are computed, the slack time of each node can then be easily determined. The result is as shown in Figure 12.5(b). The minimum slack time of the resulting is 3, which is distributed evenly in each net from the primary output backward to the primary input, as shown in Figure 12.5(c), coupled with the re-computed arrival, required and slack times of each node.

The same process is applied to the remaining portion of the figure. The final result is shown in Figure 12.5(d). This result is used as a reference to place the gates, as depicted in Figure 12.6. The nets with a higher slack time may use longer wires and the nets with a lower slack time use shorter wires. ∎

Review Questions

Q12.7 Define arrival time.

Q12.8 Define required time.

Q12.9 Define the slack time.

Q12.10 What are the features of slack time?

Q12.11 How can the slack time be used in the timing-driven placement?

12.2 DESIGN ENVIRONMENT AND CONSTRAINTS

The design environment provides the process parameters and I/O port attributes (both FPGA-based and cell-based designs), as well as statistical wire-load models (cell-based design only) to the synthesis tools in order to synthesize a design. The design constraints provide the clock related parameters, input and output delays and timing exceptions. Both design environment and constraints [1] are required for a design to be synthesized. They must be provided along with the RTL code and technology library

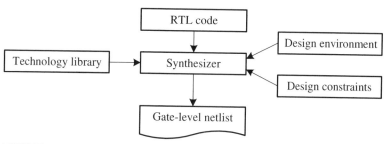

FIGURE 12.7 The general design environment of synthesis tools

to the synthesis tools, as shown in Figure 12.7. In this section, we introduce the general design environment and design constraints.

12.2.1 Design Environment

After a design has been partitioned, coded and verified, the next step is to specify the design environment and constraints before proceeding to the synthesis process. The design environment is a process that specifies the I/O port attributes and process parameters, as well as statistical wire-load models to be used with the design.

A general design environment of a digital system is shown in Figure 12.8. From this figure, we can see that the design environment includes the following:

1. *The process parameters* include technology library and operating conditions.

2. *I/O port attributes* contain drive strength of input port, capacitive loading of output port and design rule constraints.

3. *Statistical wire-load model* provides a wire-load model for processing the pre-layout static timing analysis.

The technology libraries include both the worst-case and the best-case libraries. Operating conditions describe the process of technology, operating voltage and operating temperature. Operating conditions can be worst-case, typical case or best-case. The worst-case condition is used during the pre-layout synthesis phase to optimize the design for maximum setup-time and the best-case condition is commonly used to fix the hold-time violations. The typical case is mostly ignored because it is already covered by the analyses at worst and best cases.

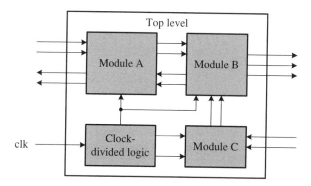

FIGURE 12.8 The general design environment of digital systems

In order to proceed the pre-layout static timing analysis, a wire-load model must be provided so that the timing analysis tool can use it to calculate the interconnect delay. The wire-load model usually models net delays as a function of capacitive loading.

For input and output ports, I/O attributes must be provided. These I/O attributes include the drive strength at the input ports as well as the capacitive loading on the output ports of the blocks. In addition, the design rule constraints, such as maximum transition, fanout and capacitance, must also be provided to the synthesis tool for synthesizing an optimized result in terms of area, power or both.

12.2.2 Design Constraints

Another portion besides design environment that must be provided to the synthesis tool is design constraints. The design constraints include clock, input delay and output delay.

There are several factors that are combined together to specify a clock signal. These are:

- *Period*—defines the clock period.
- *Duty cycle*—defines the clock duty cycle.
- *Transition time*—defines the rise time and fall time of the clock.
- *Skew*—defines the clock network delay and hence the clock skew.

The constraints of *input* and *output delays* (some tools use *offset-in* and *offset-out*) specify an amount of time relative to a reference edge of a clock signal. Input delay constraint specifies the input arrival time of a signal related to the clock, as shown in Figure 12.9(a). It is used as the setup/hold time requirements for input signals. Output delay constraint specifies the time it takes for the data to be available before the clock edge at the output port, as shown in Figure 12.9(b). Note that the sum of input delay and offset-in is the clock period. The same situation is applied to the output delay and offset-out.

The advanced design constraints include *false path* setting and *multicycle path* setting. The false path and multicycle path problems will be discussed in more detail

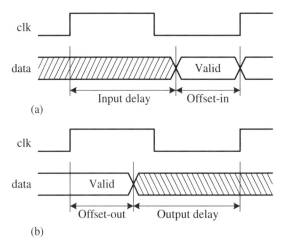

(a)

(b)

FIGURE 12.9 The general definitions of (a) input delay (offset-in) and (b) output delay (offset-out) constraints

in the next chapter. False path setting is used to instruct the synthesis tools to ignore a particular path, which is a false path, for timing optimization. Multicycle path setting is used to inform the synthesis tools regarding the number of clock cycles that a particular path requires in order to reach its endpoint. Most synthesis tools always assume that all paths are single cycle paths by default.

The other common design constraints are *minimum* and *maximum delay* specifications, as well as *paths grouping*. The minimum and maximum delay specifications can only be applied to the combinational logic blocks. Minimum delay specification defines the minimum delay required in terms of time units for a particular path. Maximum delay specification defines the maximum delay required in terms of time units for a particular path. Paths grouping is used to bundle together timing critical paths in a design for cost function calculations.

12.2.3 Optimization

Design space exploration is the process of analyzing the design for speed and area to achieve the fastest logic with minimum area on condition that the HDL code is frozen. The objective of optimization is to minimize the area but maintain the target timing requirement through synthesis and optimization. The factors that can be traded off are among area, delay and constraints. In general, the area increases considerably with tightening constraints whereas the actual delay of the logic decreases with tightening constrains. The actual delay of the logic increases with relaxed constrains.

The following are the most commonly used approaches to optimizing a design: *compiling*, *flattening*, *structuring* and *removing hierarchy*. Compiling the design performs the actual mapping of the HDL code to gate from the specified target library. A variety of optimization levels can be performed when the actual mapping is carried out.

The flattening technique is used to reduce logic with intermediate variables and parentheses to a pure two-level sum-of-product (SOP) representation. Hence, the resulting logic has few logic levels between the input and output.

Structuring is used to extract the shared logic from designs containing regular structured logic such as a carry-look-ahead (CLA) adder. The result has an impact on the total delay of logic. The structuring technique can be performed on the basis of either timing (default) optimization or Boolean optimization. Boolean optimization is often used to reduce area but may increase the timing of the resulting logic. Therefore, it is more suitable for non-critical timing circuits such as random logic and finite-state machine structures.

Although hierarchy is an effective way to design a large system, many designers often create unnecessary hierarchy for unknown reasons. In addition, synthesis tools maintain the original hierarchy of the design by default. Consequently, the resulting design is often not optimized. To overcome this, a technique known as removing hierarchy is usually used to allow synthesis tools to optimize the design across the hierarchy, namely, crossing the module boundaries.

The clock network is one of the most important issues in an ASIC design. It is also one of the hardest operations to be performed. A *big buffer approach* is only for use in features of 0.35 μm and above. For very deep sub-micron (VDSM) technologies, whose features are below 0.25 μm, *clock tree synthesis* (CTS) is a dominant method since it has minimal clock latency and skew compared to the big buffer approach.

A final comment is that synthesis tools try to optimize the design for timing by default. Designs that are non-timing critical but area intensive should be optimized for area.

Review Questions

Q12.12 What are usually specified in the design environment?

Q12.13 What are usually specified in design constraints?

Q12.14 What does design space exploration mean?

Q12.15 What does flattering mean?

Q12.16 Explain why removing hierarchy can help optimize the design.

Q12.17 What does structuring mean?

12.3 LOGIC SYNTHESIS

As mentioned previously, a synthesis tool (called a *design compiler* or *logic synthesizer*) usually accepts an RTL code, technology library and design environment, as well as design constraints, and generates a gate-level netlist as output.

12.3.1 Architecture of Logic Synthesizers

The general architecture of synthesis tools is depicted in Figure 12.10. It consists of two major parts: *front end* and *back end*. The front end consists of two phases, *parsing* and *elaboration*, while the back end contains three phases, *analysis/translation*, *logic synthesis* (*logic optimization*) and *netlist generation*.

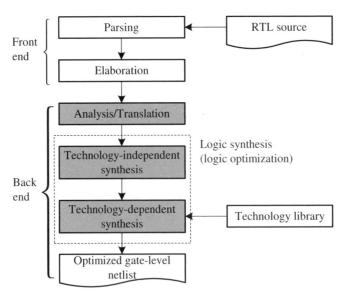

FIGURE 12.10 The general architecture of logic synthesis tools

In the parsing phase, the parser responses to check the syntax of the source code and creates internal components to be used by the next phase. In the elaboration phase, the elaborator constructs a representation of the input circuit by connecting the internal components, unrolling loops, expanding generate-loops, setting up parameters passing for tasks and functions, and so on. The end result from an elaborator is a complete description of the input circuit, which can then input to the back end for generating the final gate-level netlist.

The analysis/translation phase of the back end prepares for technology-independent logic synthesis, including managing the design hierarchy, extracting the finite-state machine (FSM), exploring resource sharing, and so on. The logic synthesis (often called *logic optimization*) is the heart of synthesis tools. It creates a new gate network which computes the functions specified by a set of Boolean functions, one per primary output. Synthesizing a gate network needs to balance a number of different concerns, which can be categorized into two classes: *functional metric* such as fanin, fanout and others, as well as *non-functional metric*, such as area, power and delay.

Logic synthesis is usually divided into two phases: *technology-independent* synthesis and *technology-dependent* synthesis. The latter takes constraints into account. The process of translating from technology-independent to the technology-dependent gate network is called *library binding*.

The general operations involved in technology-independent synthesis are classified into two classes: *restructuring operations* and *node minimization*. The former includes operations that modify the structure of the Boolean network by introducing new nodes, eliminating others, and by adding and removing edges as well; the latter includes operations that simplify the logic equations associated with nodes.

12.3.1.1 Restructuring Operations

Restructuring operations [10] include *decomposition*, *extraction*, *factorization*, *substitution* and *collapsing*.

Decomposition Decomposition is the process that expresses a given logic function in terms of a number of new functions. Decomposition replaces a divisor of a function by a new literal. For example:

$$f = wyz + xyz + uv$$

Let $g = yz$ and $h = w + x$ and then function f can be represented as a function of both g and h and is as follows:

$$f = gh + uv.$$

where the common factor yz is extracted from the first two terms of function f and defined as a new function g, and another new function h is also defined as well; the function f is then represented as the product of functions g and h plus the original term uv of function f. Decomposition is an essential step in logic optimization for FPGAs.

Extraction Extraction replaces a common divisor of two functions by a new literal. Extraction is related to decomposition but operates on a number of given functions. With extraction, the given functions are expressed in terms of newly created intermediate functions and variables. For example:

$$f = xyz + uw$$

$$g = xyz + uv$$

The product term xyz in functions f and g can be extracted and denoted by a new function, say h. Functions f and g can then be represented by using the newly created function h as follows:

$$h = xyz$$

$$f = h + uw$$

$$g = h + uv$$

Factorization   Factorization is usually based on Boolean division to extract common factors (new nodes) and then used to restructure the network. Factorization generally takes three steps. First, it generates all potential common factors and estimates how many literals can be saved for each when substituted into the network, then it chooses which factors to substitute into the network, and finally it restructures the network by adding the new factors and modifying the other functions to use these factors. For example:

$$f = wyz + xyz + uv$$

can be factored as:

$$f = yz(w + x) + uv$$

For the situation that delay is the major consideration, the factorization of a sub-network is dedicated to reduce the number of function nodes through which delay-critical signal must pass.

Substitution   Substitution, also called *resubstitution*, replaces a divisor of a function with an existing literal and then resubstitutes the literal into the original function. For example, let $g = yz$ and then:

$$f = wyz + xyz + uv$$

can be rewritten as follows:

$$f = gw + gx + uv$$

Collapsing   Collapsing, also called *elimination* or *flattening*, is the inverse operation of substitution. It removes a literal from a function by replacing it with its corresponding function. For example:

$$g = yz$$

$$f = g(w + x) + uv$$

By substituting function g into f, we obtain:

$$f = wyz + xyz + uv$$

12.3.1.2 *Node Minimization*

Node minimization [10], also called *simplification*, attempts to reduce the complexity of a given Boolean network by using Boolean minimization techniques on its nodes. The complexity is measured by the number of literals. There are three don't care sets that are often used to simplify logic functions. These are *output don't care*, *satisfiability don't care* and *observability don't care*. The output don't care occurs when the function value is not specified for some input values. The satisfiability don't care means that an intermediate variable value is inconsistent with its function inputs. The observability don't care occurs when an intermediate variable value does not affect the primary outputs of the Boolean network.

An example of satisfiability don't care is as follows. If node i of the network carries the Boolean function $f(a, b)$, where $a = xy$, $b = (x + y)z$ and x, y, z are primary inputs of the Boolean network, then $a(xy)' + a'(xy)$ and $b[(x + y)z]' + b'[(x + y)z]$ are satisfiability don't cares. In other words, the satisfiability don't cares represent combinations of variables of the Boolean network that can never occur because of the structure of the network itself.

Node minimization has been proven to be very effective for a wide variety of cases and is very often the only Boolean operation performed during the course of a network optimization.

Review Questions

Q12.18   What are the functions of parser of logic synthesizers?

Q12.19   What are the functions of elaborator of logic synthesizers?

Q12.20   What does library binding mean?

Q12.21   Describe the operation of decomposition.

Q12.22   Describe the operation of extraction.

Q12.23   Describe the operation of factorization.

Q12.24   Describe the operation of substitution.

Q12.25   Describe the operation of collapsing.

Q12.26   Describe the function of technology-independent synthesis.

12.3.2 Two-Level Logic Synthesis

In this section, we begin with definitions of several related terminologies that are often used in logic synthesis and then briefly discuss two well-known two-level logic synthesis approaches.

As we know, a logic or Boolean variable x takes on one of two values 0 and 1. The complement of the variable x is denoted by x'. Both x and x' are referred to as *literals*. A Boolean function $f : \{0, 1\}^n \longrightarrow \{0, 1\}$ is a binary function of logic variables. It is convenient to represent the n-dimensional Boolean space by an n-dimensional hypercube of 2^n nodes. In general, the set of nodes on the n-dimensional hypercube can be divided into the following three sets. The sets of nodes of the hypercube with function values 1 is referred to as the *on-set*, the set of nodes with function values 0 is referred to as the *off-set*, and the set of nodes with unspecified function values is referred to as the *don't-care-set* or *dc-set*.

A *product term* is a product of literals. A *cube* of a logic function f is a product term whose on-set does not have vertices in the off-set of f. A *minterm* is a cube in which all variables are assigned a value 0 or 1. In other words, a minterm is a single on-set node of an n-dimensional hypercube. A product term p is denoted as an *implicant* of f if for every input combination such that $p = 1$ then $f = 1$ also. Hence, a cube is an implicant of f. A *prime implicant* is an implicant of f such that if any variable is removed from p, then the resulting product term does not imply f again. A *cover* is an expression which contains all minterms. Note that a cover is not necessary minimal. It can be shown that the minimal sum (minimal cover) is the sum of prime implicants.

A Boolean network [10] is a standard technology-independent model for a logic circuit. A Boolean network $B = (V, E)$ is a directed acyclic graph, where each of the nodes $v \in V$ represents a primary input, primary output or function, and each of the edges $e \in E$ denotes the relationship between nodes. Each node is referred to as a variable name. An example of a Boolean graph is shown in Figure 12.11. There is an edge from node j to node i if the function represented by node i depends explicitly on the function represented by j. Node j is said to be a fanin of node i and node i is said to be a fanout of node j. There are two sets of special nodes. Input nodes with no incoming edges represent primary inputs and output nodes with no outgoing edges represent primary outputs.

Now that we have defined the various related terms, we can describe the two-level logic optimization methods. The most well known two-level logic optimization method is the Quine–McCluskey algorithm, which has two major steps:

1. Generates all prime implicants.

2. Finds the minimum-cost cover of all minterms by prime implicants.

A major disadvantage of this method is its high time and space complexities. The detailed operations of the Quine–McCluskey algorithm can be found in Lin [8].

Another widely used two-level logic optimization method is the Espresso [2] algorithm. The Espresso algorithm starts with a given cover and repeatedly reduces it

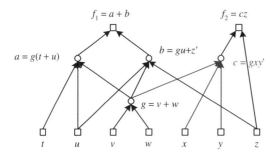

FIGURE 12.11 An example of a Boolean network

in size using an expand-irredundant-reduce loop. There are three basic steps:

1. *Expansion.* Expansion is a process that makes each cube as large as possible without covering a node in the off-set.

2. *Making irredundant.* This step is to throw out smaller (i.e. redundant) cubes so that there are no redundant cubes.

3. *Reduction.* Reduction is to reduce the size of cubes in the cover, namely, reducing the number of literals.

The rationale behind the Espresso algorithm is to throw out the smaller cubes covered by larger cubes. The following example illuminates how the Espresso algorithm works.

EXAMPLE 12.2 *The Operations of the Espresso Algorithm*

As mentioned above, the basic operations of the Espresso algorithm are expansion, making irredundant and reduction. Suppose that the initial cover is $f(x, y, z) = xy + x'y'z + (x'yz + xy'z')$. As depicted in Figure 12.12(a), the initial hypercube contains three on-set nodes and two don't care nodes. After expanding the cubes, the result is shown in Figure 12.12(b). There are four cubes in total. The next step is to make irredundant cubes. The result is shown in Figure 12.12(c). The final reduced expression is $f(x, y, z) = xy + x'z$, as shown in Figure 12.12(d). ■

Review Questions

Q12.27 What is a cube?

Q12.28 What is a minterm?

Q12.29 What is a prime implicant?

Q12.30 Describe the operations of the Espresso algorithm.

12.3.3 Multilevel Logic Synthesis

The motivation of multilevel logic synthesis is that there are many functions that are too expensive, in terms of area and propagation time, to implement in two-level logic. For example, consider the following two logic functions:

$$f(w, x, y, z) = wx + xy + xz$$

$$g(w, x, y, t) = wx' + x'y + x't \qquad (12.1)$$

which contain 12 literals and need eight gates, including two three-input gates and six two-input gates. However, the hardware requirement is reduced considerably when using three-level logic structure by factoring both functions and extracting the common factor $(w + y)$:

$$f(w, x, y, z) = wx + xy + xz$$

$$= x(w + y) + xz$$

$$= x \cdot k + xz$$

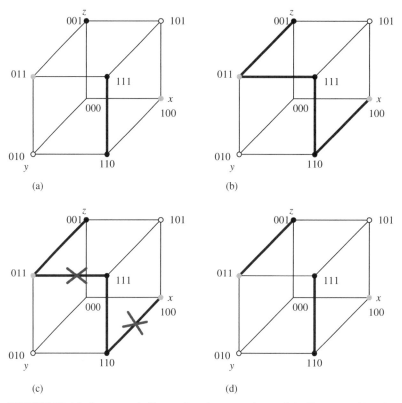

FIGURE 12.12 An example illustrating the operations of the Expresso Algorithm: (a) start; (b) expansion; (c) make irredundant; (d) reduction

$$g(w, x, y, t) = wx' + x'y + x't$$
$$= x'(w + y) + x't$$
$$= x' \cdot k + x't$$
$$k(w, y) = w + y$$

The result has 10 literals and needs only seven two-input gates.

The general operations involved in multilevel synthesis are minimizing two-level logic functions, finding common sub-expressions, substituting one expression into another and factoring single functions. In the following, we introduce a systematic approach to synthesizing a multilevel logic from two-level logic functions. This approach is named as the *kernel approach*. Before going into an in-depth discussion of the approach, we need to define some terminology first.

The divisors of f are defined as the set:

$$D(f) = \{g | f/g \neq \phi\}$$

The primary divisors of an expression f are defined as the set:

$$P(f) = \{f/c \,|\, c \text{ is a cube}\}$$

For example:

$$f = wxy + wxzt$$

and then:

$$f/w = xy + xzt$$

is a primary divisor. Every divisor of f is contained in a primary divisor. If g divides f, then $g \subseteq p \in P(f)$; g is termed *cube-free* if the only cube dividing g evenly is 1. We say g divides f evenly if the remainder is ϕ. In other words, an expression is cube-free if no cube divides the expression evenly. For example: $xy + z$ is cube-free; $xy + xz$ and xyz are not cube-free.

The *kernels* of an expression f are the cube-free primary divisors of f. The kernels of f are defined as:

$$K(f) = \{k/k \in P(f), k \text{ is cube-free}\}$$

For example, consider the logic function $f = wxy + wxzt$; then $f/w = xy + xzt$ is a primary divisor but is not cube-free since x is a factor: $f/w = x(y + zt)$. A cube c used to obtain the kernel $k = f/c$ is called a *cokernel* of k. For example, $f/wx = y + zt$ is a kernel and wx is the cokernel. The cokernel of a kernel is not unique in general.

EXAMPLE 12.3 *Kernel and Cokernel*

Consider the following switching function:

$$f(w, x, y, z) = xz + yz + wxy$$

There are 7 literals. To find the cokernels and kernels, we proceed the following divisions:

$$f/w = xy$$
$$f/x = z + wy$$
$$f/y = z + wx$$
$$f/z = x + y$$

where w is not a cokernel because vernal xy is not cube-free, which has x or y as a factor. Consequently, the cokernel set is $\{x, y, z\}$ and the kernel set is $\{z + wy, z + wx, x + y\}$. ∎

The cokernels of a switching function can be found by using a *cokernel table*, in which the headers of columns and rows list all nontrivial product terms one by one, as shown in Figure 12.13. The entries in the table are the intersection of the column and row

	twy	txy	uwy	uxy	vwy	vxy
twy	*					
txy	ty	*				
uwy	wy	y	*			
uxy	y	xy	uy	*		
vwy	wy	y	wy	y	*	
vxy	y	xy	y	xy	vy	*

	ty	uy	vy	wy	xy
ty	*				
uy	y	*			
vy	y	y	*		
wy	y	y	y	*	
xy	y	y	y	y	*

FIGURE 12.13 An example of multilevel logic synthesis

product terms. All entries except the ϕ in the cokernel table are cokernels. If all entries in the cokernel table are single literals or ϕ, then we have done the job. Otherwise, the entries not in single-literal form are collected and another cokernel table is used to find the cokernels again. This process is repeated until all entries in the cokernel table are ϕ or literals. The union of all cokernels found in all cokernel tables is the cokernels of the switching function on hand. The following example manifests this approach to finding cokernels.

EXAMPLE 12.4   *Multilevel Logic Synthesis*

Consider the following switching function:

$$f(t, u, v, w, x, y, z) = twy + txy + uwy + uxy + vwy + vxy + z$$

There are 19 literals. To find the cokernels, we can use the tables shown in Figure 12.13. From the first cokernel table, we obtain the cokernel set $\{ty, uy, vy, wy, xy, y\}$. Due to the fact that not all entries are single literals or ϕ, another cokernel table is constructed by the set $\{ty, uy, vy, wy, xy\}$, from which the cokernel y is obtained. Combining the above two cokernel sets, the resulting cokernel set is $\{ty, uy, vy, wy, xy, y\}$. The kernels corresponding to the cokernel set are calculated as follows:

$$f/ty = w + x = K_1$$
$$f/uy = w + x = K_1$$
$$f/vy = w + x = K_1$$
$$f/wy = t + u + v = K_2$$
$$f/xy = t + u + v = K_2$$
$$f/y = tw + tx + uw + ux + vw + vx$$
$$= t(w + x) + u(w + x) + v(w + x) = (w + x)(t + u + v) = K_3 = K_1 K_2$$

The function f can then be reduced to:

$$f = K_3 y + z = (w + x)(t + u + v)y + z$$

There are only 7 literals. The resulting logic diagram is shown in Figure 12.14. ■

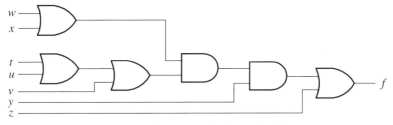

FIGURE 12.14 An example of multilevel logic synthesis

The kernel approach can also be used to decompose multiple-output switching functions. The rationale behind this is as follows. The objective of multilevel logic synthesis is to find common divisors of two (or more) functions f and g. Two switching functions f and g have a non-trivial common divisor d ($d \neq$ cube), if and only if, there exist kernels $k_f \in K(f)$ and $k_g \in K(g)$ such that $k_f \cap k_g$ is non-trivial, i.e. not a cube. Hence, we can use kernels of f and g to extract common divisors.

EXAMPLE 12.5 *A Multiple-Output Example*

Consider the following three switching functions:

$$f_1(t, u, v, w) = tv + tw + uv + uw$$
$$f_2(v, w, x, y) = vxy' + wxy'$$
$$f_3(u, v, w, x, y, z) = uv + uw + z' \tag{12.2}$$

There are 19 literals. The cokernel set is as follows:

$$C_{f_1} = \{t, u, v, w\}$$
$$C_{f_2} = \{xy'\}$$
$$C_{f_3} = \{u\} \tag{12.3}$$

The kernel set is obtained by dividing each switching function by its corresponding cokernels, one by one:

$$K_{f_1} = \{v + w, t + u\}$$
$$K_{f_2} = \{v + w\}$$
$$K_{f_3} = \{v + w\} \tag{12.4}$$

From the kernel set, it is easy to see that the common divisor is $v + w$. The switching functions can then be represented as functions of the common divisor:

$$f_1(t, u, v, w) = g(t + u)$$
$$f_2(v, w, x, y) = gxy'$$
$$f_3(u, v, w, x, y, z) = gu + z' \tag{12.5}$$

where $g = v + w$. The resulting logic diagram is shown in Figure 12.15. ∎

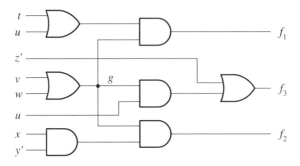

FIGURE 12.15 An example of multilevel logic synthesis

Review Questions

Q12.31 Describe the meaning of the term cube-free.

Q12.32 Define divisor and prime divisor.

Q12.33 Define kernel and cokernel.

12.3.4 Technology-Dependent Synthesis

The technology-dependent synthesis (or simply *technology mapping*) problem is the optimization problem of finding a minimum cost covering of the Boolean network by choosing from the collection of primitive logic elements in the target library. We can optimize for both area and delay during technology-dependent synthesis. In the following, we introduce two simple approaches for LUT-based FPGA architectures.

In the following discussion, suppose that each LUT has 4 inputs at most, as in the most widely used FPGA architectures. For LUT-based FPGA architecture, a two-step approach [10, 12] can be used. In the first step, the technology-independent network is decomposed into nodes with no nodes more than k inputs, where k is determined by how many number of inputs each LUT can have, which is 4 in our running example. Using this idea, the number of nodes is reduced. In the second step, the number of nodes is reduced by combining some of them, taking into account the special features of LUTs. For instance, consider the gate network shown in Figure 12.16(a) which requires four 4-input LUTs. However, if we pay more attention to it, we can find that it only requires three 4-input LUTs. In general, the mapping result depends on how much effort is used to search the netlist.

Another technology-dependent synthesis approach is the FlowMap [4, 12] method. The FlowMap method is a delay-optimized technology-mapping algorithm for LUT-based FPGA fabrics. The essential idea of a FlowMap algorithm is to break the network into LUT-sized blocks by using an algorithm that finds a minimum-height k-feasible cut in the network. It uses network flow algorithms to optimally find the minimum-height cut and uses heuristics to maximize the amount of logic fit into each cut in order to reduce the number of LUTs required.

An example for explaining the idea of the FlowMap method is depicted in Figure 12.17. Assume that 4-input LUTs are available for implementing the network. The method progresses from the primary output f backward to the primary inputs. It calculates the cut at each gate level until a cut can fit into an LUT. It removes the portion and processes the rest of the network repeatedly until it reaches the primary inputs. As a result, three LUTs are required to implement the network.

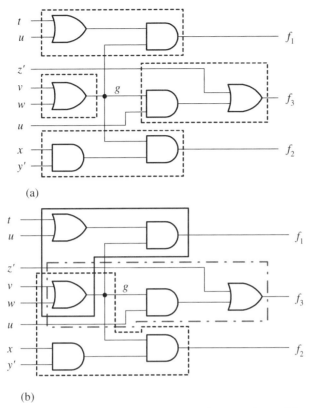

(a)

(b)

FIGURE 12.16 An example illustrating the operations of the two-step approach: (a) Mapping I requires four LUTs; (b) Mapping II only requires three LUTs

Review Questions

Q12.34 Describe the function of technology-dependent mapping.

Q12.35 Describe the operations of the FlowMap method.

Q12.36 Describe the operations of the two-step approach.

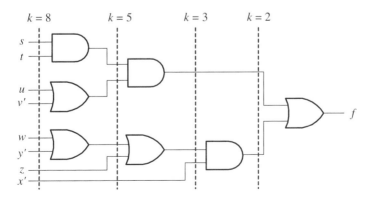

FIGURE 12.17 An example illustrating the operations of the FlowMap method

Q12.37 Apply the two-step approach to the circuit of Figure 12.17.

Q12.38 Apply the FlowMap method to the circuit of Figure 12.16(a).

12.4 LANGUAGE STRUCTURE SYNTHESIS

From the viewpoint of users, a synthesis tool at least performs the following critical tasks: to detect and eliminate redundant logic and combinational feedback loops as well, to exploit don't-care conditions, to detect unused states and collapse equivalent states, to make state assignments and to synthesize optimal multilevel logic subject to constraints.

In this section, we address the behavioral operations of synthesis tools from the viewpoint of Verilog HDL programmers. In other words, we describe how the synthesis tools deal with the language structures, including assignment statement, `if-else` statement, `case` statement, loop statements, `always` statement and memory synthesis approaches.

12.4.1 Synthesis of Assignment Statements

The analysis/translation step extracts the logical functions from a language structure. The most straightforward language structures in Verilog HDL are assignment statements, including continuous and procedural assignments. As introduced in Chapter 3, a continuous assignment is basically an expression, which then comprises operands and operators. Almost all operators in Verilog HDL are synthesizable. The case equality, arithmetic shifts, exponent and module operators are generally the exceptions. Operators defined in Verilog HDL are listed in Table 12.1. The operands can be `wire` and `tri` net data types, `reg` variable data type and parameter, as well as functions.

TABLE 12.1 The synthesizable operators of Verilog HDL by most synthesis tools

Arithmetic	Bitwise	Reduction	Relational
+: add	~: NOT	& : AND	>: greater than
−: subtract	&: AND	|: OR	<: less than
*: multiply	|: OR	~ &: NAND	>=: greater than or equal
/: divide	^ : XOR	~| : NOR	<=: less than or equal
%: modulus	~^, ^~ : XNOR	^ : XOR	
** : exponent		~^, ^~ : XNOR	**Equality**
	Case equality		==: equality
Shift		**Logical**	!=: inequality
<<: left shift	===: equality	&& : AND	
>>: right shift	!==: inequality	||: OR	**Miscellaneous**
<<<: arithmetic left shift		! : NOT	{ , }: concatenation
>>>: arithmetic right shift			{const_expr{ }}:replication
			? : : conditional

Module instances, primitive gate instances and tasks are synthesizable with that the timing constructs are ignored. Procedural statements `always`, `if-else`, `case`, `casex` and `casez` are all synthesizable but `initial` statement is not supported. Procedural blocks `begin-end`, `named blocks`, as well as `disable` statement, are also synthesizable.

Loop statements, `for`, `while` and `forever`, are synthesizable except that `while` and `forever` statements must contain timing control `@(posedge clk)` or `@(negedge clk)`.

12.4.2 Synthesis of Selection Statements

As we have described, both `if-else` and `case` statements are used to perform two-way or multiple-way selection. An `if-else` statement is usually synthesized into a 2-to-1 multiplexer. A nested `if-else` statement is synthesized into a priority-encoded, cascaded combination of multiplexers.

For some applications, it is quite often to use only the `if` part without the `else` part. When this is the situation, the result will be different, depending on whether the logic under description is combinational or sequential. For combinational logic, we need to specify a complete `if-else` statement; otherwise, a latch will be inferred. For sequential logic, we need not specify a complete `if-else` statement; otherwise, we will get a notice removing redundant expression from synthesis tools.

For example, the following statement will infer a latch because the value of `y` is only specified when the control signal `enable` is true. The synthesis tools will by default assume that the value of variable `y` is unchanged otherwise. As a result, a latch is inferred in order to keep the value of `y` unchanged.

```
always @(enable or data)
   if (enable) y = data;   //infer a latch
```

However, the situation is different when the `if-else` statement is used to describe a sequential logic, such as the following example.

```
always @(posedge clk)
   if (enable) y <= data;
   else y <= y;   // a redundant expression
```

A warning message of "a redundant expression" will be obtained when the `else` part is specified because for sequential logic the value of a variable is unchanged by default unless it is changed explicitly. As a consequence, the synthesis tool will remove the redundant expression in the optimization phase.

The following example demonstrates how an incomplete `if-else` statement will infer a latch by the synthesis tools.

EXAMPLE 12.6   *Latch Inference—an Incomplete* `if-else` *Statement*

This example illustrates that a latch is inferred by synthesis tools due to lack of the `else` part when describing a combinational logic.

```
// an example of latch inference
module latch_infer_if(enable, data, y);
input   enable, data;
output y;
reg     y;
// the body of testing program.
always @(enable or data)
   // due to lack of else part, synthesizer infer a latch
   // for y.
   if (enable) y = data;
endmodule
```

The synthesized result of the above example of the incomplete `if-else` statement is shown in Figure 12.18.

Coding Styles

1. *It should avoid using any latches in a design.*

2. *It is good practice to assign outputs for all input conditions to avoid inferred latches. For example:*

```
always @(enable or data)
   y = 1'b0;  // initialize y to its initial value.
   if (enable) y = data;
```

As mentioned in Section 4.4.2, a `case` statement is a nested `if-else` statement. The `case` statement infers a multiplexer and the consideration of using a complete or an incomplete specified statement is the same as that of an `if-else` statement. As we have introduced in the previous chapters, `case` statements are often used to model multiplexers. The following is an example of a latch inferred due to an incomplete `case` statement.

EXAMPLE 12.7   *Latch Inference—An Incomplete* `case` *Statement*

This example illustrates the situation of latch inference due to an incomplete `case` statement. The example intends to model a 3-to-1 multiplexer; hence, three case items

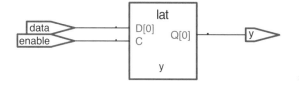

FIGURE 12.18 The synthesized result of an incomplete `if-else` statement

are naturally used to select and route the input data to the output of the multiplexer. However, it leaves a case without being specified, which is then the reason to be inferred a latch. The `default` case item is usually used to resolve this.

```verilog
// an example of latch inference
module latch_infer_case(select, data, y);
input  [1:0] select;
input  [2:0] data;
output reg y;
// the body of 3-to-1 MUX
always @(select or data)
   case (select)
      2'b00: y = data[select];
      2'b01: y = data[select];
      2'b10: y = data[select];
// the following statement is used to avoid inferring a
// latch
//    default: y = 1'bz;
   endcase
endmodule
```

The synthesized result of an incomplete `case` statement is shown in Figure 12.19. Readers are encouraged to synthesize the above example with and without the `default` case item, respectively, and see the difference between them.

12.4.3 Delay Values

The synthesis tools ignore the delay values appearing in RTL descriptions because the ultimate delays of the logic network will be determined by the actual delays of the gates used to implement the gate-level netlist. However, the delay values are often used to mimic the delays of logic gates or the period of a clock signal during simulation.

The following example demonstrates how synthesis tools ignore the delay values and generate a meaningless logic circuit.

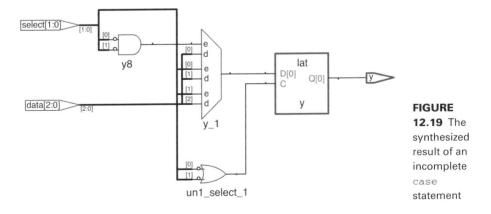

FIGURE 12.19 The synthesized result of an incomplete `case` statement

EXAMPLE 12.8   *Ignored Delay Values—A Not Synthesizable Version*

This example intends to model a four-phase clock generator. The phase 0 starts at 0 time unit from the positive edge of the clock signal. Each phase lasts for 5 time units. Although this program may generate the desired timing during the simulation time, it is not synthesizable because all delay values are ignored by the synthesis tools. The synthesized result is only the last statement left; namely, it assigns 4'b0001 to phase_out after logic optimization.

```verilog
// a four phase clock example--- generated incorrect
// hardware
module four_phase_clock_wrong(clk, phase_out);
input  clk;
output reg [3:0] phase_out;  // phase output
// the body of the four phase clock
always @(posedge clk) begin
   phase_out <= #0  4'b0001;
   phase_out <= #5  4'b0010;
   phase_out <= #10 4'b0100;
   phase_out <= #15 4'b1000;
   phase_out <= #20 4'b0001;
   end
endmodule
```

Owing to ignoring the delay values associated with statements by the synthesis tools, in those cases that need timing control they should use event timing control rather than delay timing control so that the results can be synthesizable.

The following is an example of using event timing control for generating a four-phase clock signal by modeling the circuit with a way similar to ring counter structure.

EXAMPLE 12.9   *A Four-Phase-Clock Generator—A Correct Version*

This example uses a ring-counter-like structure to model a four-phase clock generator. At each positive edge of the clock signal clk, output signals phase_out change one time; namely, the value 1 is shifted to the next higher significant bit. The vector phase_out returns to its original status after four clock cycles. As a result, it generates the desired four-phase clock signals.

```verilog
// a four phase clock example --- a synthesizable version
module four_phase_clock_correct(clk, phase_out);
input  clk;
output reg [3:0] phase_out;  // phase output
// the body of the four phase clock
always @(posedge clk)
   case (phase_out)
      4'b0000: phase_out <= 4'b0001;
```

```
        4'b0001: phase_out <= 4'b0010;
        4'b0010: phase_out <= 4'b0100;
        4'b0100: phase_out <= 4'b1000;
        default: phase_out <= 4'b0000;
     endcase
endmodule
```

The synthesized result of the above example is shown in Figure 12.20—it is indeed a four-phase clock generator based on a ring counter structure. Readers are suggested to compare it with the multiphase timing generators described in Section 9.6.1.

12.4.4 Synthesis of Positive and Negative Signals

For sequential logic circuits, it is often required that a positive- or negative-edge clock signal be used as an event that triggers a set of operations to be performed in sequence. There are no constraints on the use of posedge and negedge. Most synthesis tools support the mixed use of two or more different edge-triggered signals but cannot accept the mixed use of edge-triggered signals with level-sensitive signals in the same sensitivity list of an always block. The following example give some insight into this idea.

EXAMPLE 12.10 *An Example of Mixed use of Edge-Triggered and Level-Sensitive Signals*

Although it is so convenient to use both edge-triggered signals and level-sensitive signals in the same sensitivity list of an always block, it is considered to be a bad coding style because this kind of statement is often not acceptable by synthesis tools. The always block uses both edge-triggered signal clk and level-sensitive signal reset.

```
// an example of mixed use of posedge and level signals.
// the result cannot be synthesized. try it on your own
// system !!
module DFF_bad (clk, reset, d, q);
input  clk, reset, d;
output reg q;
// the body of DFF
```

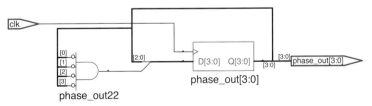

FIGURE 12.20 The synthesized result of a four-phase clock

```
always @(posedge clk or reset)
   if (reset) q <= 1'b0;
   else       q <= d;
endmodule
```

When we synthesize the above example, an error message is generally obtained from the synthesis tool: "Error: Can't mix posedge/negedge use with plain signal references". Another similar situation is to mix the use of posedge and negedge signals such as in the following example. However, this is synthesizable.

EXAMPLE 12.11 *An Example of Mixed use of Posedge and Negedge Signals*

In this example, a positive-edge event of clock signal clk is used to trigger the sampling operation of a *D*-type flip-flop and a negative-edge event of reset signal reset_n to reset the flip-flop. This is a common way that is often used to model a *D*-type flip-flop with an asynchronous reset.

```
//an example of mixed use of posedge and negedge signals.
//try it in your system !!
module DFF_good(clk, reset_n, d, q);
input  clk, reset_n, d;
output reg q;
//the body of DFF
always @(posedge clk or negedge reset_n)
   if (!reset_n) q <= 1'b0;
   else          q <= d;
endmodule
```

Please refer to Section 9.1.1 for details of how to model a *D*-type flip-flop with an asynchronous or a synchronous reset control.

12.4.5 Synthesis of Loop Statements

As we have introduced in Section 4.5, loop statements contain for, while, repeat and forever. However, not all of these statements can be synthesizable. Loop statements for, while and forever are synthesizable except that while and forever must contain timing control @(posedge clk) or @(negedge clk). The loop statement repeat is generally not synthesizable.

To synthesize a for loop statement, the elaborator unrolls the for loop and then proceeds with the analysis/translation and logic optimization. The following example demonstrates this idea.

EXAMPLE 12.12 *An Example Illustrating the Loop Statements*

This example is to add two *n*-bit operands together and produce an $(n + 1)$-bit sum. The default value of *n* is set to 4. At the elaboration phase, the always block is expanded into the following statements:

```
{co, sum[0]} = x[0] + y[0] + c_in;
{co, sum[1]} = x[1] + y[1] + co;
{co, sum[2]} = x[2] + y[2] + co;
{c_out, sum[3]} = x[3] + y[3] + co;
```

where each statement corresponds to a full adder. The four full adders are then cascaded together as a 4-bit ripple-carry adder.

```
// an N-bit adder using for loop.
module nbit_adder_for( x, y, c_in, sum, c_out);
parameter N = 4;
input   [N-1:0] x, y;     // declare as a 4-bit array
input   c_in;
output reg [N-1:0] sum; // declare as a 4-bit array
output reg c_out;
reg     co;
integer i;
// specify the function of an N-bit adder using for loop.
always @(x or y or c_in) begin
    co = c_in;
    for (i = 0; i < N; i = i + 1)
        {co, sum[i]} = x[i] + y[i] + co;
    c_out = co;
end
endmodule
```

The synthesized result of the above 4-bit adder using for-loop statement is shown in Figure 12.21.

The following example gives another application of a for-loop statement. Here, we would like to add a number data_b to the sum total under the control of input data_a. The number of times that data_b is added to the sum total is equal to the number of 1-bit of the input number data_a.

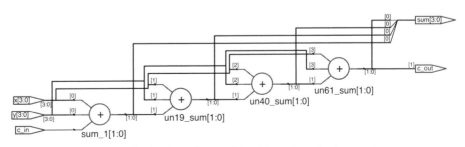

FIGURE 12.21 The synthesized result of a 4-bit adder using a for-loop statement

EXAMPLE 12.13 *An Incorrect Version of a Single-Cycle Multiple-Iteration Example*

This example would like to compute the sum total at each positive edge of the clock signal. The sum total initially sets to 0 by the reset reset_n signal. Then at each incoming positive edge clock signal clk, a for loop is used to detect the bit value of input data data_a from LSB to MSB in sequence. If the bit is 1 then data_b is added to total; otherwise, it does nothing. However, it does not work as what we have expected. Actually, it adds data_b to total at every positive edge of the clock signal whenever not all bits of data_a are zeros.

```
// a single-cycle multiple-iteration example --- an
// incorrect version. please try to correct it!
module multiple_iteration_example_a(clk, reset_n, data_a,
    data_b, total);
parameter N = 8;
parameter M = 4;
input   clk, reset_n;
input   [M-1:0] data_a;
input   [N-1:0] data_b;
output  [N-1:0] total;
reg     [N-1:0] total;
integer i;
// what does the following statement do?
always @(posedge clk or negedge reset_n) begin
    if  (!reset_n) total <= 0; else
    for (i = 0; i < M; i = i + 1)
        if (data_a[i] == 1) total <= total + data_b;
end
endmodule
```
■

The synthesized result of module multiple_iteration_example_a is shown in Figure 12.22. This is certainly not the result that we have expected. Can you explain the

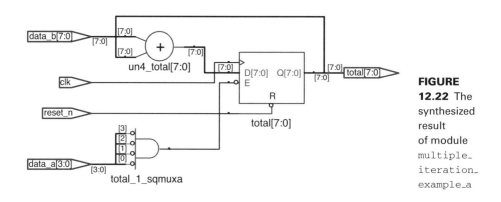

FIGURE 12.22 The synthesized result of module multiple_iteration_example_a

reason why the synthesis tool produces such a result? Moreover, what would happen if we replace all nonblocking assignments with blocking assignments within the `always` block. Try it on your own system to confirm whatever your answer is.

12.4.6 Memory and Register Files

As we have discussed in Chapter 11, memory and registers or register files often play important roles in designing a digital system. The most commonly used memory modules in digital system design based on synthesis flow are static random-access memories (SRAMs), called RAMs for short. The memory modules have a variety of types, such as RAM, register file, first-in first-out (FIFO) buffer and dual-port RAM.

Memory modules can be constructed in many ways. One way is to use flip-flops or latches. This approach is independent of any synthesis software and type of target system, cell-based or FPGA-based. It is easy to use but inefficient in terms of area because, in general, a flip-flop may take up 10 to 20 times the area of a 6-transistor static RAM cell. Another way is by using standard components supplied by cell library or FPGA vendors. Of course, they depend on the process technology. The third way is to use the RAM compiler supplied by most cell-library vendors. This may be the most area-efficient approach because the size of the memory module can be generally customized to fit into the actual requirement.

Another kind of memory module often used in digital system design is the register file. Register files are usually generated by using a synthesis directive or hand instantiation RAM.

In summary, a flip-flop- or latch-based RAM or register is only applied to the cases in which the size required is small. For the cases where a large size of RAM module or register is required, it is better to use RAM compilers or standard components supplied by the vendors.

Review Questions

Q12.39 Explain why the following statement will infer a latch.

```
always @(enable or data)
    if (enable) y = data;  //infer a latch
```

Q12.40 Explain why the `else` statement in the following is a redundant expression.

```
always @(posedge clk)
    if (enable) y <= data;
    else y <= y;   // a redundant expression
```

12.5 CODING GUIDELINES

Successful synthesis strongly depends on proper partitioning of the design together with a good HDL coding style. A good coding style not only results in reduction of the chip area (namely, hardware cost) and aids in top-level timing but also produces faster logic. A frequent obstacle to writing HDL code is the software mind-set. *You should think in hardware* when designing a hardware module. In this section, we introduce some coding guidelines to help the reader write a code that achieves the best synthesis results and reduction in compile time [1, 7].

12.5.1 Guidelines for Clocks

Clock signals are the heart of any digital system. For a digital system to be worked properly, the clock signals must be applied very carefully. In this sub-section, we discuss several issues related to clock signals. These are using a single global clock, avoiding using gated clocks, avoiding mixed use of both positive and negative edge-triggered flip-flops in the same design and avoiding using internally generated clock signals.

Using a Single Global Clock   In designing a digital system, it should use a single global clock as the preferred clock structure and use positive edge-triggered flip flops as the only memory components, as shown in Figure 12.23(a). For ASIC design, it is necessary to avoid instantiating clock buffers in RTL code manually because clock buffers are normally inserted after synthesis as part of the physical synthesis, as we have mentioned earlier (see Figure 12.3).

Avoiding Using Gated Clocks   We should avoid coding gated clocks in an RTL code because of the following reasons. First, clock gating circuits tend to be timing dependent and technology specific. Second, gated clocks may cause clock skews of local clock signals which then may cause hold time violations. Third, gated clocks limit the testability because the logic clocked by a gated clock cannot be made part of a scan chain. A multiplexer may be used to bypass the gated clock in the test mode if gated clocks are required for some reasons.

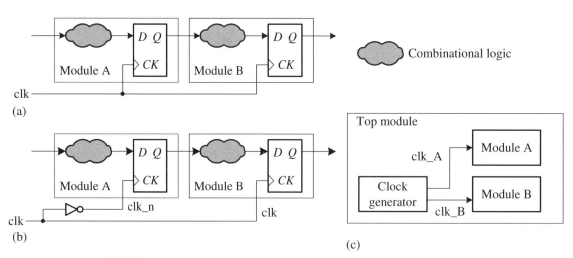

FIGURE 12.23 The clocking schemes for general digital systems: (a) an ideal clock scheme; (b) an example of using both positive and negative edge-triggered flip-flops; (c) using a sequence clock module at the top level

Avoiding Mixed use of Both Positive and Negative Edge-Triggered Flip-Flops
Although the mixed use of both positive and negative edge-triggered flip-flops in a design is so convenient, it is considered as a bad coding style nowadays because it needs to tackle the stringent timing requirements. However, if this is indeed needed due to performance reasons, the following two problems must be dealt with caution. First, the duty cycle of the clock now becomes a critical issue in the timing analysis. Second, most scan-based testing methodologies require separate handling of positive and negative edge-triggered flip-flops. A good example of using both positive and negative edge-triggered flip-flops is depicted in Figure 12.23(b).

Avoiding Using Internally Generated Clock Signals   The final concern about the issues of clock signals is to avoid using internally generated clocks in the design as much as possible. Internally generated clocks, at least, have the following two drawbacks. First, they make it more difficult to constrain the design for synthesis. Second, they limit the testability because the logic driven by the internal clock cannot be made part of a scan chain much like the case of gated clocks. When an internally generated clock, reset or gated clock is required, it is good practice to keep the clock and/or reset generation circuit as a separate module at the top level of the design, as shown in Figure 12.23(c).

12.5.2 Guidelines for Resets

The reset signal plays an important role in any digital system because it initializes the system to its known status by clearing all flip-flops. From the viewpoint of designers, the basic design issues related to reset signals are asynchronous versus synchronous, an internal or external power-on reset, as well as hard versus soft reset when more than one reset signals are used.

Asynchronous Versus Synchronous Reset   Both asynchronous and synchronous reset signal generations have their features. Asynchronous reset does not require a free-running clock and does not affect flip-flop data timing. FPGAs provide global reset control signals for all flip-flops associated with them. For a cell-based design, even though all flip-flops have asynchronous reset control signals, it is still harder to implement the system reset circuit since reset is a special signal like clock, which requires a tree of buffers to be inserted at place and route. Moreover, it makes both static timing analysis (or cycle-based simulation) and the automatic scan-chain (test structure) insertion more difficult. In contrast, synchronous reset is easy to synthesize because it is just another synchronous signal to the input. However, it requires a free-running clock, in particular at power-up, for reset to occur. In addition, all cell-based or FPGA-based flip-flops do not support this kind of reset mechanism.

The basic coding styles for both asynchronous and synchronous reset are as follows:

```
// asynchronous reset
always @(posedge clk or posedge reset)
    if  (reset) ...
    else ...
```

```
// synchronous reset
always @(posedge clk)
   if (reset) ...
   else ...
```

Avoid Internally Generated Conditional Resets   The reset signal may be generated internally or externally when power is turned on. However, it is necessary to avoid internally generated conditional resets if possible. When a conditional reset is required, it is necessary to create a separate signal for the reset signal and isolate the conditional reset logic in a separate logic block in order to improve the synthesis result and make the code more readable. For instance, in the following `always` block, the `timer_load_clear` is a conditional reset used to clear the `timer_load` flip-flop to 0 whenever it is activated.

```
always @(posedge gate or negedge reset_n or posedge
  timer_load_clear)
    if (!reset_n || timer_load_clear)
        timer_load <= 1'b0;
    else timer_load <= 1'b1;
```

The better coding style is to combine both the system reset signals `!reset_n` with the conditional reset signal `timer_load_clear` into one separate block, say `timer_load_reset`, and then apply to the flip-flop, as shown in the following:

```
assign timer_load_reset = !reset_n || timer_load_clear;
always @(posedge gate or posedge timer_load_reset)
    if (timer_load_reset) timer_load <= 1'b0;
    else timer_load <= 1'b1;
```

More detailed examples concerning this topic can be found in Section 15.4.

12.5.3 Partitioning for Synthesis

Due to the increasing complexity of digital systems, it is necessary to partition a design into many smaller modules so that they can be handled easily. In addition, the purposes of a partition are to obtain a faster compile time and a better synthesis result, as well as the capability to use simpler synthesis strategies to meet timing requirements. In the following, we give some guidelines to carry out a partition.

Register All Outputs   In order to make the output drive strengths and input delays predictable, all outputs need to be registered, as shown in Figure 12.24. In this case, there are no combinational logic being placed between the register and the output port (pad).

Keep Related Logic within the Same Module   When partitioning a design into smaller modules, it needs to keep in mind that all related combinational logic should

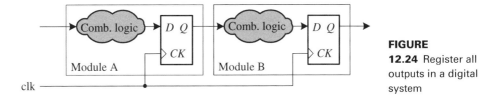

FIGURE
12.24 Register all outputs in a digital system

be put into the same module as possibly as you can. For example, as shown in Figure 12.25(a), the combinational logic *A* and *B* are displaced in two separate modules. It is better to combine them into one module, as shown in Figure 12.25(b). Here, the resulting two modules are also output registered.

Separating Structural Logic from Random Logic   Another guideline for partitioning a design is to separate structural logic from random logic. When partitioning a design, the following features must be kept in mind at all times: to limit a reasonable block size, to partition the top level and to separate I/O pads from boundary scan as well as core logic. In addition, it is necessary to remember that it should not add glue-logic at the top level. Moreover, it should avoid using asynchronous logic in a design.

Synthesis Tools Tend to Maintain the Original Hierarchy   As we have mentioned previously, synthesis tools tend to maintain the original hierarchy of a design and only optimize the codes within the same module. This feature has two implications. First, the codes in a design that have different design goals should be separated into different modules so that they can be optimized separately. For example, for a module with the critical path logic, speed optimization is applied to optimize the logic. For a module with the noncritical path logic, the area optimization is applied to optimize the logic.

Second, in order for synthesis tools to consider resource sharing, all relevant resources need to be in the same module. For example, the two adders shown in

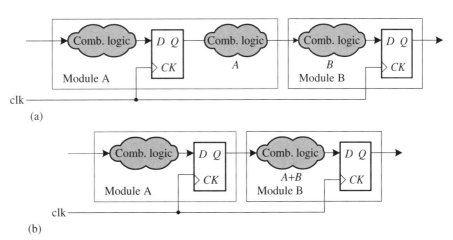

FIGURE 12.25 Keep all related combinational logic circuits together: (a) bad style; (b) good style

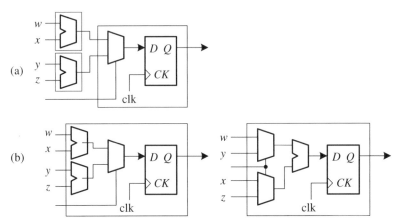

FIGURE 12.26 Partition for resource sharing: (a) resources in different modules cannot be shared; (b) resources in the same module can be shared

Figure 12.26(a) cannot be shared because they are in separate modules. However, the two adders shown in Figure 12.26(b) can be shared because they are in the same module.

SUMMARY

In this chapter, we have dealt with the principles of logic synthesis and introduced how to write a good Verilog HDL code such that it can be acceptable by most logic synthesis tools. In addition, we gave some comments about the constructs that can and cannot be synthesizable.

In general, synthesis can be divided into logic synthesis and high-level synthesis. The former transforms an RTL representation into gate-level netlists while the latter transforms a high-level representation into an RTL representation. At present, logic synthesis is the most commonly used approach when designing general digital systems and high-level synthesis is only successfully applied in specific domains for which intensive computation is required, such as in digital signal processing (DSP) applications.

In order to synthesize a design, we need to provide the design environment and design constraints along with the RTL code and technology library to the synthesis tools. The design environment includes the process parameters, I/O port attributes and statistical wire-load models. The design constraints include the clock related parameters, input and output delays and timing exceptions. Note that both design environment and design constraints are required for a design to be synthesized.

The general architecture of synthesis tools is divided into two parts: the front end and the back end. The front end consists of two phases, parsing and elaboration, while the back end contains three phases: analysis/translation, optimization and netlist generation.

From the viewpoint of users, a synthesis tool, at least, performs the following critical tasks: to detect and eliminate redundant logic and combinational feedback loops as well, to exploit don't-care conditions, to detect unused states and collapse equivalent states, to make state assignments and to synthesize optimal multilevel logic subject to constraints.

It is good practice to use a single global clock as the preferred clock structure and positive edge-triggered flip flops as the only sequential memories. For ASIC design, it is necessary to avoid hand instantiating clock buffers in the RTL code because clock buffers are normally inserted after synthesis as part of the physical synthesis.

The reset signal plays an important role in any digital system because it initializes the system to its known status; namely, it clears all flip-flops. From the viewpoint of designers, the basic design issues of reset signals are: asynchronous versus synchronous, an internal or external power-on reset, as well as hard versus soft reset when more than one reset signals are used.

REFERENCES

1. H. Bhatnagar, *Advanced ASIC Chip Synthesis: Using Synopses Design Compiler and Prime Time,* Kluwer Academic Publishers, Boston, MA, USA, 1999.
2. R.K. Brayton, C. McMullen, G.D. Hachtel and A. Sangiovanni-Vincentelli, *Logic Minimization Algorithms for VLSI Synthesis,* Kluwer Academic Publishers, Norwell, MA, USA, 1984.
3. M.D. Ciletti, *Modeling, Synthesis and Rapid Prototyping with the Verilog HDL,* Prentice-Hall, Upper Saddle River, NJ, USA, 1999.
4. J. Cong and Y. Ding, "FlowMap: an optimal technology mapping algorithm for delay optimization in lookup-table based FPGA designs", *IEEE Transactions on Computer-Aided Design of Integrated Circuits and Systems,* **13**(1), 1–12, 1994.
5. S.H. Gerez, *Algorithms for VLSI Design Automation,* John Wiley & Sons, Inc., New York, NY, USA, 1999.
6. IEEE 1364-2001 Standard, *IEEE Standard Verilog Hardware Description Language,* 2001.
7. M. Keating and P. Bricaud, *Reuse Methodology Manual: For System-on-a-Chip Designs,* Kluwer Academic Publishers, Boston, MA, USA, 2002.
8. M.-Bo. Lin, *Digital System Design: Principles, Practices and ASIC Realization,* 3rd Edn, Chuan Hwa Book Company, Taipei, Taiwan, 2002.
9. S. Palnitkar, *Verilog HDL: A Guide to Digital Design and Synthesis,* 2nd Edn, SunSoft Press, 2003.
10. A. Sangiovanni-Vincentelli, A. El Gamal and J. Rose, "Synthesis methods for field programmable gate arrays", *Proceedings of the IEEE,* **81**(7), 1057–1083, 1993.
11. M. Sarrafzadeh and C.K. Wong, *An Introduction to VLSI Physical Design,* McGraw-Hill, New York, NY, USA, 1996.
12. W. Wolf, *FPGA-Based System Design,* Prentice-Hall, Upper Saddle River, NJ, USA, 2004.

PROBLEMS

12.1 Suppose that the switching functions f, g, and h are as follows:

$$f(v, w, x, y, z) = vxy + wxy + z$$

$$g(v, w, x, y, z) = v + wx$$

$$h(v, w, x, y, z) = v + w$$

Find the quotient functions of f/g and f/h.

12.2 Find the kernel and cokernel sets of the following switching expression:

$$f(s, t, u, v, w, x, y, z) = tsu + tsv + wz + xz + yz + xy$$

12.3 Find the kernel and cokernel sets of the following switching expressions:

(a) $f(t, u, v, w, x, y, z) = twy + txy + uwy + uxy + vwy + vxy + z$

(b) $g(t, u, v, w, x, y, z) = tx + ty + uvx + uwx + uvy + uwy + z$

12.4 Use the kernel approach to implement the following multilevel logic circuits and calculate the number of literals of each multiple-output switching function:

(a) $f_1(v, w, x, y, z) = vx + vy + vz$
$f_2(v, w, x, y, z) = wx + wy + wz$

(b) $f_1(u, v, w, x, y, z) = uy + uz + vy + vz$
$f_2(u, v, w, x, y, z) = uy + uz + wy + wz$
$f_3(u, v, w, x, y, z) = vy + vz + xy + xz$

12.5 Use the kernel approach to implement the following multilevel logic circuits and calculate the number of literals of each multiple-output switching function:

(a) $f_1(v, w, x, y, z) = vw' + v'w$
$f_2(v, w, x, y, z) = wz + v'z + v'xy + wxy$

(b) $f_1(u, v, w, x, y, z) = v + w$
$f_2(u, v, w, x, y, z) = vx + vy + wx + wy + z$

12.6 Use the kernel approach to implement the following multilevel logic circuits and calculate the number of literals of each multiple-output switching function:

$$f_1(t, u, v, w, x, y, z) = tuvwz + tuvxz + tuvyz$$

$$f_2(t, u, v, w, x, y, z) = tuvwz + tuwxz + tuwyz$$

12.7 Use the simple two-step technology mapping technique to map Problem 12.4(a) into 4-input LUTs.

(a) How many LUTs are required for the original switching expressions?

(b) How many LUTs are required for the results after being implemented by using the kernel approach?

12.8 Use the simple two-step technology mapping technique to map Problem 12.4(b) into 4-input LUTs.

(a) How many LUTs are required for the original switching expressions?

(b) How many LUTs are required for the results after being implemented by using the kernel approach?

12.9 Use the simple two-step technology mapping technique to map Problem 12.5(a) into 4-input LUTs.

(a) How many LUTs are required for the original switching expressions?

(b) How many LUTs are required for the results after being implemented by using the kernel approach?

12.10 Use the simple two-step technology mapping technique to map Problem 12.5(b) into 4-input LUTs.

(a) How many LUTs are required for the original switching expressions?

(b) How many LUTs are required for the results after being implemented by using the kernel approach?

12.11 Use the simple two-step technology mapping technique to map Problem 12.6 into 4-input LUTs.

  (a) How many LUTs are required for the original switching expressions?

  (b) How many LUTs are required for the results after being implemented by using the kernel approach?

12.12 If we want to use four inverters cascaded together in some applications, write a Verilog HDL to model it. Synthesize your design with an available FPGA device and see what happens.

12.13 A simple frequency doubler uses XOR gates to extract the edge information from an input clock signal, such as the one shown in Figure 12.27.

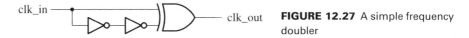

clk_in   clk_out   **FIGURE 12.27** A simple frequency doubler

  (a) Describe the circuit shown in Figure 12.27 in Verilog HDL.

  (b) Synthesize your design with an available FPGA device and see what happens.

12.14 Synthesize the following two arithmetic expressions and see what happens.

$$f = w + x + y + z$$
$$g = (w + x) + (y + z)$$

12.15 For every cycle, we want to compute the following for loop and then store the result at the next positive edge of the clock signal clk:

```
for (i = 0; i < m; i = i + 1)
   if (data_a[i] == 1) then total = total + data_b;
```

  (a) Explain why the `multiple_iteration_example_a` described in Section 12.4.5 cannot correctly compute the desired results. Of course, you may synthesize it by using synthesis tools and examine the synthesized result carefully. Perhaps this may help you to explore the reason why it cannot work properly.

  (b) Of course, if we change the nonblocking operators into blocking ones, the results will be correct. Please explain why?

  (c) Assume that we want to insist using the coding style set in this book, namely, still using a nonblocking assignment. Rewrite or modify the code so that it can work properly.

CHAPTER *13*

VERIFICATION

VERIFICATION IS a necessary process that makes sure a design can meet its specifications. Due to the inherent features of timing relationship between various components of the hardware module in question, the verification process needs to assure that the design works not only correctly in function but also correctly in timing. As a consequence, the verification of a design can be divided into two types: functional verification and timing verification. Functional verification only considers whether the logic function of the design meets the specifications and can be done either by simulation or by formal proof. Timing verification considers whether the design meets the timing constraints and can be performed through dynamic timing simulation or static timing analysis (STA).

In order to use dynamic timing simulation to verify the timing constraints of a design, two essential components are needed. The first is a test bench which generates and controls stimuli, monitors and stores responses and checks the simulation results. The second is a file, known as a standard delay format (SDF) file, which contains timing information of all cells in the design and can be produced by PAR tools. During simulation, a value change dump (VCD) file may also be produced in order to provide valuable information about value changes on selected variables in the design. The purpose of VCD files is to provide information for debug tools in order to help remove the function bugs and timing violations of the design.

In this chapter, we deal with these issues in more detail and give a comprehensive example based on FPGA design flow to illustrate how to enter, synthesis, implementation and configure the underlying FPGA device of a design. Along with the design flow, static timing analyses are also given and explained. In addition, design verification through dynamic timing simulations, incorporating the delays of logic elements and interconnect, is introduced.

Digital System Designs and Practices   Ming-Bo Lin
© 2008 John Wiley & Sons (Asia) Pte Ltd

13.1 FUNCTIONAL VERIFICATION

A design is the process that transforms a set of specifications into an implementation. Verification is the reverse process of design; that is to say, it begins with an implementation and confirms whether the implementation meets the specifications or not. The goal of verification is to ensure a design 100 % correct in both of its functionality and timing. On average, design teams usually spend about 50–70 % of their time to verify their designs. As a consequence, verification is a very important process that determines whether a design is successful or not.

Functional verification can be accomplished by simulation or formal proof (called formal verification). In the simulated-based functional verification, the design is placed under a test bench, input stimuli are applied to the design and the outputs from the design are compared with the reference outputs as well. The formal verification proves formally that a protocol, an assertion, a property or a design rule holds for all possible cases in the design. In other words, a simulation-based functional verification is based on the analysis of simulation results and code coverage whereas formal verification is built on the basis of mathematical algorithms.

13.1.1 Design Models

Before addressing the simulation-based verification, we need to explore the models of the *design under test* (DUT), also called *design under verification* (DUV). From the viewpoint of verification, a design under test can be considered as one of the following models based on whether the details of its internal functionality can be known from outside or not:

1. *Black box model*. In this model, only the external interfaces (namely, the input and output behavior of the design) are known. The internal signals and constructs are unknown (namely, black). Most simulation-based verifications begin with this model.

2. *White box model*. In this model, both the external interfaces and internal structures are known. Most formal verification environments use this model.

3. *Gray box model*. This model is a combination of both black box and white box. In this model, some of the internal signals in addition to the external interfaces are known. Most simulation-based verification environments use this model.

The major advantage of the black box model is that because the function of a design is independent of implementation, any structure changes inside the design (namely, DUT) would have little impact on the verification code. The disadvantage is that it lacks the control and observation points of the internal structure.

In contrast, the white box model has the capability of direct measurement of the DUV. Thus, it is more flexible and powerful to verify the design. It can flag a bug direct in the source instead of indirectly capturing the symptoms in a black box environment. The disadvantage of the white box model is that the verification code is dependent on the internal structures of DUV. If any signal name of internal constructs changes, the checker component must change as well.

Assertion-Based Verification   Assertion checking is a direct application of the white box model. The main purpose of assertion checking is to improve *observability*. Observability is the ease that incorrect behavior of a design can be identified. In general, a systematic application of assertions may catch about 30 % design bugs on large industrial projects.

Assertions are statements about a designer's intended behavior. They are usually used in a source description to describe the intended behavior of a group of statements. Some good examples are an FSM state should always use one-hot encoding, and the full and empty flags of an FIFO should never be asserted at the same time.

Assertions can be classified into the following types:

1. *Static assertion.* A static assertion is an atomic and simple check for the absence of an event. These events do not relate to any other events.

2. *Temporal assertion.* Several events occur in sequence and many events have to occur before the final asserted event can be checked. That is to say, they have a timing relationship.

13.1.2 Simulation-Based Verification

Simulation-based verification is the most widely used approach for verifying whether a design is functionally correct or not. The generic verification flow based on simulation is shown in Figure 13.1. Before a good design specification is obtained, the system architect needs to survey many possible ways to achieve the same purpose through simulating the architecture models of the design.

When the design specification is ready, a functional test plan is created. The functional test plan must at least include the following features: functions to be verified, tools required to verify it, completion criteria and resources and schedule details. The functions to be verified are derived by extracting the functionality from architecture specification, prioritizing functionality and creating the test cases. The functional test plan is the framework of functional verification of the design.

According to the framework, test signals can be applied to the DUT and the results can be stored and analyzed. If the result checking and code coverage analysis meet the

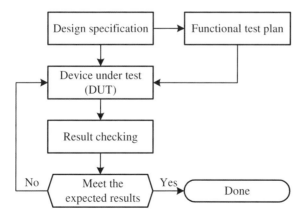

FIGURE 13.1 A simulation-based functional verification flow

expected results, then we have done it; otherwise, it has to modify the DUT until it may meet the expected results.

13.1.2.1 Hierarchy of Functional Verification   In general, the hierarchy of functional verification of an ASIC chip can be divided into many levels as follows:

1. *Designer level* (or block-level). In this level, verification is usually done by the designer using Verilog HDL or VHDL for both design and verification.

2. *Unit level.* A unit usually contains multiple designer blocks and has a more stable interface of it. Hence, in this level, verification is usually done by the design team using randomized stimuli and autonomous checking.

3. *Core level.* A core usually contains multiple units and has a complete functional specification and a stable interface. It is a special re-usable unit. In this level, verification employs a well-defined process coupled with a well-documented specification, including functions and interfaces, and possibly coverage items.

4. *Chip level.* A chip usually contains multiple cores and has very well-defined interface boundaries. In this level, verification is to ensure that all units are properly connected and the design adheres to the interface protocols of all units.

Functional verification usually combines both directed and random tests to achieve the maximum percentage of correctness. Directed tests are used to test a specific behavior of the design whereas random tests are used to simulate corner cases which may be missed by the designers.

The set of verification tests of a design includes the following:

1. *Compliance tests* are used to verify that the design complies with the specification.

2. *Corner case tests* try to find the corner cases that are most likely to break the design.

3. *Random tests* are essentially a complement to compliance and corner case tests.

4. *Real code tests* are used to uncover the errors that may have arisen from misunderstanding the specification by hardware designers.

5. *Regression tests* are tests generated for previous versions of a design. Using regression tests helps ensure that old bugs do not reappear, as well as helps uncover new bugs.

6. *Property check* is used to check certain properties of a design, such as a particular FIFO should never be read when it is empty.

13.1.2.2 Tools for Writing Test Benches   The essential component of simulation-based verification is the test bench. A test bench can be written by using Verilog HDL or other verification-assisted tools. As mentioned previously, Verilog HDL is not only a modeling language but also a verification language. Therefore, in a design based on Verilog HDL, the test bench is usually written in Verilog HDL.

However, with the advent of the system-on-a-chip (SoC) era, a design may exceed million gates. The test bench built on Verilog HDL becomes less effective due to the following reasons. Because of the decreasing controllability and observability (will be defined in the last chapter) of the design, it becomes harder and more time consuming to write a test bench and more difficult to verify the correct behavior of the design. In addition, the maintenance of the test benches becomes difficult because their size increase dramatically along with the increasing complexity of chip function and size.

Due to the above difficulties, many verification-assisted tools have been introduced recently. These are often called *high-level verification languages* (HVLs) and at least include Open-Vera, *e* language and systemC. These HVLs are programming languages that combine the object oriented approach of C++ with the parallelism and timing constructs in HDLs. They help in the automatic generation of test stimuli and provide an integrated environment for functional verification, including input drivers, output drivers, coverage and so on.

OpenVera is an HVL developed by Synopsys. It is an object-oriented programming language that supports complex data types, classes and inheritance. It has a built-in data type that matches the four-value signal type $\{0, 1, x, z\}$ used in Verilog HDL. The syntax of the OpenVera language is similar to C++ or Java. OpenVera uses the concept of port, interface and binding to connect to the HDL model. The inputs coming from the model are declared as ports; the attributes such as width of these inputs are specified as an interface; by using hierarchical names, interface and port signals are binded to the HDL signals.

e language is an HVL developed by EDA vendor Verisity (now a part of Synopsys) and is at the heart of Specman tool. Like OpenVera, *e* supports all common concepts of object-oriented programming languages such as abstract data type and inheritance. *e* uses `struct` to declare a class or a derived class. Similar to C++, *e* classes contain declarations of member objects such as data structures and functions.

SystemC is initiated by Open SystemC Initiative (OSCI). It is a large C++ library that supports high-level hardware design, modeling, simulation and verification. Although SystemC originally targeted mostly the simulation and specification of designs in C++ as its main goals, the synthesis of a sub-set of SystemC to RTL code is available today.

13.1.2.3 *An HDVL Language*   Recently, a unified hardware description and verification language (HDVL) standard known as *SystemVerilog* has been established as IEEE Std. 1800-2005. SystemVerilog is a super-set of Verilog HDL and provides a set of extensions to improve the capability of hardware description and enhance the capability of verification of Verilog HDL. Namely, it unifies the flow of design and verification. On the design hand, through the use of advanced design constructs provided by SystemVerilog, designers can model their designs in a more readable and concise way. On the verification hand, through coverage-driven, constrained-random test benches and assertion, designers may locate the bugs easily from SystemVerilog reports without tracing back through the output waveform. Details of SystemVerilog can be found in LRM [7].

13.1.3 Formal Verification

Formal verification uses mathematical techniques to prove an assertion or a property of the design without the need of technological considerations such as timing and physical effects. It proves a design property by exploring all possible ways to manipulate the

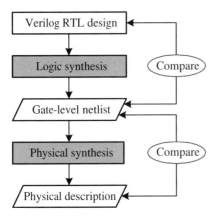

FIGURE 13.2 The operations of equivalence checking

design. The major advantage of formal verification is that it can prove the correctness of a design without doing any simulation.

At present, one of the most widely used applications of formal verification is equivalence checking, as shown in Figure 13.2. Equivalence checking is used to validate the RTL against RTL, gate-level netlist against the RTL code or the comparison between gate-level netlist to physical description.

The RTL to RTL verification is used to verify the new RTL code against the old function correct RTL code. This often occurs when new features need to be added to a function correct RTL code. The formal verification can then be used to verify whether the old function remains valid.

After a gate-level netlist and physical description of the RTL design are created by logic synthesis and physical synthesis (place and route) tools, it is necessary to check whether these implementations meet the functionality of the original RTL design. The straightforward approach is to re-run the test bench and apply all test stimuli used for RTL verification again with the gate-level netlist and the physical description.

Another method is by using formal verification to prove that their functionalities are equivalent. Equivalence checking ensures that the gate-level netlist or the physical description has the same functionality as the RTL description that was simulated. In other words, equivalence checking verifies that the logic synthesis tool accurately synthesizes the logic described in RTL code and the physical synthesis (place and route) tool properly transforms the gate-level netlists into physical layout. In order to work properly, equivalence checkers build a logical model of both the RTL and gate-level representations of the design and prove that they are functionally equivalent in a mathematical way.

Review Questions

Q13.1 Distinguish between design and verification.

Q13.2 Describe the meaning of formal verification.

Q13.3 Describe the meaning of the black box model.

Q13.4 Describe the meaning of the white box model.

Q13.5 Describe the meaning of the gray box model.

Q13.6 What are the two types of assertions?

Q13.7 Describe the functions of equivalence check.

Q13.8 What is the essential component of simulation-based verification?

Q13.9 Explain the reason why high-level verification languages have become increasingly popular recently.

Q13.10 Describe the meaning of regression tests.

13.2 SIMULATION

Up to now, simulation is still an efficient and dominated way for verifying both the functionality and timing of a design. Therefore, in this section, we deal with simulations of a design. We begin with a description of the generic types of simulations often encountered in designs, the simulation approaches in verifying digital systems and the architectures of simulators. Then, we introduce the principles of event-driven and cycle-based simulations.

13.2.1 Types of Simulations and Simulators

In general, the simulation of a digital design can be performed in a variety of levels, as described in the following:

1. *Behavioral simulation.* Behavioral simulation models large pieces of a system as black boxes with inputs and outputs accessible externally.

2. *Functional simulation.* Functional simulation ignores the timing of gates and assumes that each gate has a unit delay.

3. *Gate-level (logic) simulation.* This uses logic gates or logic cells as basic black boxes, where each of them may contain delay information.

4. *Switch-level simulation.* In this level of simulation, transistors are considered as switches; namely, they can only be turned on or off.

5. *Circuit-level (transistor-level) simulation.* In this level of simulation, transistors are considered as a set of nonlinear equations, which describe their nonlinear voltage and current characteristics.

For digital system designs, using simulation to verify whether a design can meet its specification is usually carried out at behavioral, functional or gate level. In terms of this sense, a design can be simulated in one of the following three ways:

1. *Software simulation.*

2. *Hardware acceleration.*

3. *Hardware emulation.*

Software simulation is typically used to run designs based on Verilog HDL or VHDL. It runs a software simulator on a personnel computer (PC), a workstation or

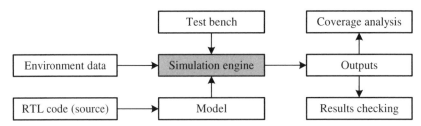

FIGURE 13.3 The simulation-based verification environment

a server. The simulation-based verification environment is shown in Figure 13.3. The heart of simulation-based verification is the simulation engine, which is software that runs on a workstation, a server or a PC. Before an RTL code (source) can be simulated, it is required to be transformed into models that can be recognized by the simulation engine. Software simulation is suitable for small designs and may be the most widely used approach for verifying a design. However, software simulation starts to consume large amounts of time and memory space and becomes a bottleneck in the verification process when designs start exceeding 100 k gates.

To speed up the simulation, an approach based on a mixed hardware and software system, named as hardware acceleration, is often used. In this method, the design is divided into two parts: synthesizable and not synthesizable. The synthesizable part is mapped onto a reconfigurable hardware system, which usually consists of CPLD or FPGA devices, as shown in Figure 13.4. The not synthesizable part runs on a software simulator, which may be a Verilog HDL simulator or an HVL simulator. The interaction between the simulator and hardware accelerator produces the final results, which can then be checked and analyzed. The advantages of hardware acceleration are as follows. First, it can speed up simulation by two to three orders of magnitude in comparison with software-based simulation. Second, it can run sequences of random transactions during functional verification.

The third widely used approach to verifying a design is known as hardware emulation. Hardware emulation is a dominated approach that is used to verify the design in a real-world environment with real system software running on the system. It is quite often that a design is verified by hardware emulation before it is transformed to an ASIC. At present, hardware emulations are often built on a reconfigurable CPLD or FPGA device. Sometimes, a system on a programmable chip (SoPC) is used. An SoPC is also a CPLD- or FPGA-based device. The general block diagram of hardware

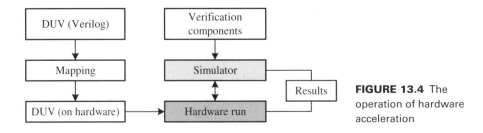

FIGURE 13.4 The operation of hardware acceleration

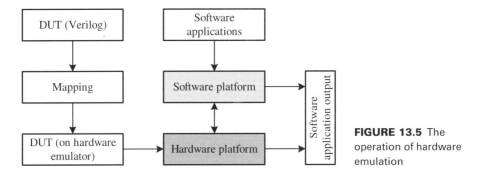

FIGURE 13.5 The operation of hardware emulation

emulation is shown in Figure 13.5. The operation of hardware emulation is as follows. The design to be simulated has to be mapped into the hardware platform. The software applications then run on the software platform which runs in turn on the hardware platform to produce the required results.

Hardware emulation can be used either as a prototype of the final product or as an emulation engine for verifying the design. When it is used as an emulation engine, the software application runs exactly as it would do on the real chip in a real circuit. As a result, software and hardware integration can start before the actual hardware is available. Hardware emulation can run at megahertz speeds or even faster, depending on the speed of the underlying hardware devices.

13.2.2 Architecture of HDL Simulators

The general architecture of HDL simulators is shown in Figure 13.6. An HDL simulator usually consists of a compiler and a simulation engine. The compiler is further composed of two parts: front end and back end. The front end consists of two phases: parsing and elaboration. In the parsing phase, the parser is responsible for checking the syntax of the source code and creates internal components to be used by the next phase. In

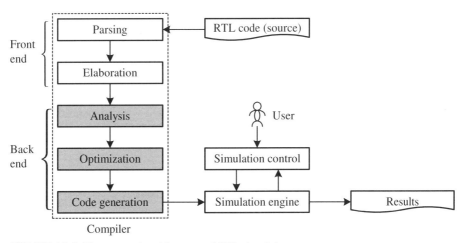

FIGURE 13.6 The general architecture of HDL simulators

the elaboration phase, the elaborator constructs a representation of the input circuit by connecting the internal components, unrolling loops, expanding generate-loops, setting up parameters passing for tasks and functions, and so on. The end results from an elaborator is a complete description of the input circuit, which can then input to the back end for generating the final code.

The back end determines the type of simulator and is mainly composed of three phases: analysis, optimization and code generation. The detailed operations of the analysis phase depend on the type of simulation engine used. For example, in a cycle-based simulator, it carries out the levelization and clock domain analysis.

The code generation phase generates the code to be used in a simulation engine. According to the types of simulation engines used, there are three classes of generated code: interpreted code, compiled code and native code. As a result, the simulators corresponding to each pair of these three codes and simulation engines are referred to as interpreted, compiled code and native code simulators, respectively. These three types of Verilog HDL simulators are listed as follows:

1. *Interpreted simulators*. These simulators read in the Verilog HDL design, create data structures in memory and run the simulation interpretively. For example, the Cadence Verilog-XL simulator.

2. *Compiled code simulators*. These simulators read in the Verilog HDL design and convert it to the equivalent C code. The C code is then compiled by a standard C compiler to get the binary executable. The binary is executed to run the simulation. For example, the Synopsys VCS simulator.

3. *Native code simulators*. These read in the Verilog HDL design and convert it directly to binary code for a specific machine platform. For example, the Cadence Verilog-NC simulator.

Verilog HDL simulators can also be categorized into two types based on the way that they are triggered. These types are as follows:

1. *Event-driven simulators*. These simulators process elements in the design only when signals at the inputs of these elements change.

2. *Cycle-based simulators*. These simulators process all elements in the design on a cycle-by-cycle basis, irrespective of changes in signals. They collapse combinational logic into equations. They are useful for synchronous designs where operations happen only at active clock edges. However, timing information between two clock edges is lost. In order to make up this flaw, most commercial cycle-based simulators are integrated with an event-driven simulator.

In the rest of this section, we will describe each of these two types of simulators in more detail.

13.2.3 Event-Driven Simulation

The basic concept of event-driven simulation is that the evaluation of gate or a block of code is performed only when an event occurs. As defined previously, an event is a change of value in a signal or a variable. In order to keep track of the events and to make sure they are processed in the correct order, the events are kept on a circular event

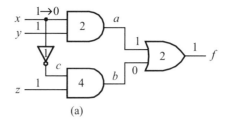

(a)

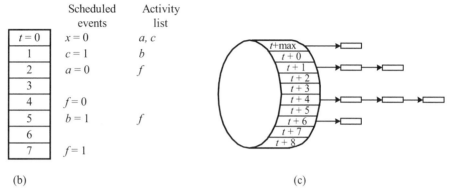

(b)   (c)

FIGURE 13.7 The operation of an event-driven simulator: (a) a circuit example; (b) scheduled events and the activity list; (c) timing wheel

queue, ordered by *simulation time.* The simulation time is maintained by the simulator to model the actual time that would be taken by the circuit being simulated. The general structure of an event-driven simulator is shown in Figure 13.7, which is based on a data structure known as a *timing wheel.* The timing wheel has time slots and each time slot points to a linear event queue that stores events occurring at that time.

The timing wheel is indeed a two-dimensional queue consisting of a circular time queue and many linear event queues. The circular time queue is used to schedule events occurring at different times and each linear event queue is used to schedule events occurring at the same time. The operation of putting an event on the queue is called *scheduling* an event. Events can occur at different times. Events that occur at the same time are processed in an arbitrary order.

Before describing the detailed operations of an event-driven simulator, the following two terms need to be defined first: *update event* and *evaluation event.* An update event means every change in value of a net or variable as well as the named event in the circuit being simulated. When the value of a net or variable changes, all processes (`always` blocks) dependent on the net, variable and named event must be evaluated. The evaluation of a process is called an evaluation event.

The detailed operation of an event-driven simulator is as follows:

while (there are events)
    if (there are no events at current simulation time) advance simulation time;
    for (each event at current simulation time)
    if (event is update) { // an update event

```
                    update the variables or nets;
                    schedule evaluation events;
            }
      else { // an evaluation event
                    evaluate the processes;
                    schedule update events;
      }
endwhile
```

The simulation progresses along time slots. At each time slot, the events in the event queue are processed one at a time until the queue is empty. If an event is an update event, then it updates the variables or nets and schedules evaluation events. If an event is an evaluation event, then it evaluates the processes and schedules update events. For example, the event that x changes its value from 1 to 0, as shown in Figure 13.7, is an update event. When it is executed, the evaluation events a and c are scheduled at simulation times 2 and 1, respectively, because their corresponding gates have their own propagation delays. These two evaluation events then trigger other update events. This process repeats until it reaches the end of the simulation time, set by the $finish or $stop system task.

In the above discussion, we assume that all events in the same time slot are processed in an arbitrary order. In practice, they may be processed in some predefined order according to the features of the events. In Verilog HDL, the events at the same simulation time are classified into the following five types according to the order of processing:

1. active;
2. inactive;
3. nonblocking assignment update;
4. monitor;
5. future events .

Active events are those ones that occur at the current simulation time and can be processed in any order. The processing of all active events is defined as a *simulation cycle*. Events that occur at the current simulation time, but will be processed after all active events have been processed, are called *inactive events*. An example of an inactive event is an explicit zero-delay assignment, such as "#0 c = b + 1; , ", which occurs at the current simulation time but is processed after all active events at the current simulation time are processed. As mentioned earlier, the nonblocking assignment events are processed in three steps: read variables, evaluate the right-hand-side expression and update the left-hand-side variable. The first two steps are active events and the last step is executed only after both active and inactive events have been processed. The monitor events are generated by system tasks such as $monitor or $strobe and are processed as the last events so as to capture the stable values of variables at the current simulation time. The operation of a Verilog HDL event-driven simulator [5] is as follows:

```
while (there are events)
      if (there are no active events) {
```

if (there are inactive events)
    activate all inactive events;
else if (there are nonblocking assignment update events)
    activate all nonblocking assignment update events;
else if (there are monitor events)
    activate all monitor events;
else // future events
    advance simulation time and activate all inactive events; }
for (each active event)
if (event is update)
    update the modified objects and schedule evaluation events;
else // an evaluation event
    evaluate the process and schedule update events;
endwhile

13.2.4 Cycle-Based Simulation

Although event-driven simulators are universal in the sense that they can be applied to a variety of applications, their simulation efficiency is quite low. A specialized technique to improve the simulation efficiency is cycle-based simulation. Cycle-based simulation is a technique that does not simulate detailed circuit timing, but instead computes the steady state response of a circuit at each clock cycle.

The speedup of cycle-based simulation may be over 100 compared to event-driven simulation. The reasons are the simplicity of algorithms used and the total optimization towards a synchronous hardware design style. As a result, the cycle-based simulation puts severe constraints on the DUT, which is described as follows. They ignore delay controls and hence the glitch behavior of signals between clock cycles cannot be detected, limit sequential constructs to an extent of only allowing synchronous designs and do not allow most test bench features.

The principles of cycle-based simulation can be illustrated by way of an example. Consider the logic circuit depicted in Figure 13.7(a) again. When a cycle-based simulator runs simulation on the circuit, it will evaluate a, c, b and f only at each clock cycle. In other words, the combinational logic is evaluated at each clock cycle boundary and a gate is evaluated only once. In the case of an event-driven simulator, a, c, b and f are evaluated not only at the clock cycle, but also when any of the events at the input of the gates occur.

In order to evaluate a stable value, the gates within a combinational logic circuit must be evaluated in the proper order. Considering the logic circuit shown in Figure 13.7(a) again, the proper evaluation order of nets a, b, c and f are as follows: first evaluate c and then a and b or b and a, and finally f. That is to say, it needs to arrange the gates into levels. This operation is known as *levelization*. One method to arrive at this is by using a topological sort, which transforms a directed acyclic graph (DAG) into a linearly ordered list. An extra feature of a topological sort is that it can also detect the cycle in a DAG. As a result, the feedback loop in a combinational logic circuit can be detected by a topological sort.

When multiple clocks are used in a circuit, not all gates are required to evaluate at each clock cycle. Only those gates associated with the same clock are evaluated at the clock transition. Consequently, we have to determine the part of the circuit which requires evaluation at each clock cycle. This process is called *clock domain analysis*. The set of gates belonging to the same clock are called the domain of the clock.

In summary, for cycle-based simulation, the back end of the simulator analyzes the clock domain, topologically sorts the gates into a linearly ordered list and generates codes accordingly for executing by the simulator.

Review Questions

Q13.11 Distinguish between design and verification.

Q13.12 Distinguish between behavioral simulation and functional simulation.

Q13.13 Describe the features of software simulation.

Q13.14 Describe the features of hardware acceleration.

Q13.15 Describe the general architecture of an HDL simulator.

Q13.16 What are the three types of HDL simulators?

Q13.17 What is an event-driven simulator?

Q13.18 What is a cycle-based simulator?

13.3 TEST BENCH DESIGN

As described in Section 1.4.1, a test bench comprises the following major components: generating and controlling stimuli, monitoring and storing responses and checking the results. In this section, we discuss how to generate stimuli and check results, how to generate clock signals and how to generate reset signals. In addition, verification coverage is introduced briefly.

13.3.1 Test Bench Design

The basic principles of test bench design are: the test bench should generate stimuli and check responses in terms of test cases, the test bench should employ reusable verification components whenever possible, rather than coding from scratch, and the response checking must be automatic (self-checking test bench). A test case is a set of statements used to exercise a set of stimuli for checking a specific function of the DUT. A test bench is usually composed of many test cases.

The two most common types of test benches are *deterministic* and *self-checking*. The deterministic test benches are often used to verify the basic functionality of the DUT at an early stage of the development cycle. The self-checking test benches are those that place the knowledge of the DUT function into the test bench environment so that they can automate the tedious result checking process. The self-checking test benches are used to verify the final functionality of the DUT.

Three types of automated response checking are often used in practice. These are *golden vectors*, *reference model* and *transaction based*, as shown in Figure 13.8.

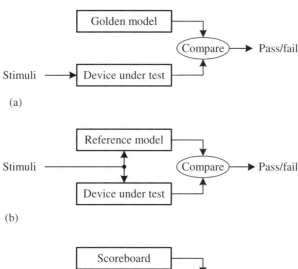

FIGURE 13.8 The operations of automated response checking: (a) golden vectors; (b) reference model; (c) transaction-based model

1. *Golden vectors.* In an environment of using golden vectors, some known output vectors are stored somewhere for comparing with the outputs from the DUT, as shown in Figure 13.8(a).

2. *Reference model.* In a reference model, a module or device called the *reference model* is used to accept the same stimuli as the DUT. Both outputs from the reference model and the DUT are then compared to determine whether the output of the DUT is correct, as shown in Figure 13.8(b).

3. *Transaction-based model.* In a transaction-based model, the DUT is supposed to have identifiable transactions in which commands and data are acted on and forwarded to appropriate output signals. A *scoreboard* is used to keep track commands and data driven on the inputs of the DUT. In this model, the test bench generates stimuli and checks responses in terms of transactions. The known transaction results are compared with the outputs from the DUT, as shown in Figure 13.8(c). The most common examples of network protocol devices, such as Ethernet, USB and IIC, that route and forward data packets of data, must use this kind of checking.

After introducing the response-checking strategies of test benches, we deal with how and when to generate stimuli. Stimuli can be generated in a deterministic manner or randomly and they can be generated on the fly during simulation or prior to simulation. As a result, two basic choices of stimulus generation are:

1. *Deterministic* versus *random* stimulus generation.

2. *Pregenerated test* case versus *on-the-fly test* case generation.

The result checking can be done during simulation or at the end of a test case, which are called *on-the-fly checking* and *end-of-test checking*, respectively.

1. *On-the-fly checking*. In this way, the result checking is done throughout the life of the test case. It is suitable for transaction-based operations. The advantage of on-the-fly checking is that it requires less memory for simulation because it need not store the entire simulation outputs and runs faster than the end-of-test checking.

2. *End-of-test checking*. In this way, the result checking is done at the end of the test. End-of-test checking is often used when the checking components require to check the state of the test bench after the test completes. It usually needs more memory for simulation and runs slower than the on-the-fly checking.

Two simple approaches are usually used to analyze the data value and the data protocol:

- *Waveform viewers* are used to view the dump files. It is used to debug a small size code.
- *Log files* contain traces of the simulation run.

The following examples separately illustrate three possible options of choosing test signals in the test sets: exhaustive test, random test and golden vectors.

EXAMPLE 13.1 *Test Bench Example 1: Exhaustive Test*

In this example, deterministic test signals are generated on-the-fly during simulation. Of course, generation of stimuli exhaustively can only be applied in a small design with small input space. In addition, the end-of-test checking is used in this example.

```
// test bench design example 1: exhaustive test.
'timescale 1 ns / 100 ps
module nbit_adder_for_tb;
parameter N = 4;
reg  [N-1:0] x, y;
reg  c_in;
wire [N-1:0] sum;
wire c_out;
// Unit Under Test port map
   nbit_adder_for UUT (
       .x(x), .y(y), .c_in(c_in), .sum(sum),
       .c_out(c_out));
reg [2*N-1:0] i;
initial
   for (i = 0; i <= 2**(2*N)-1; i = i + 1) begin
       x[N-1:0] = i[2*N-1:N]; y[N-1:0] = i[N-1:0];
       c_in =1'b0;
```

```
        #20;
   end
initial
   #1280 $finish;
initial
   $monitor($realtime,"ns %h %h %h %h", x, y, c_in,
    {c_out, sum});
endmodule
```
■

The following is an example of using random stimulus generation on the fly during simulation.

EXAMPLE 13.2 *Test Bench Example 2: Random Test*

In this example, the test signals are generated randomly on the fly during simulation. In addition, the on-the-fly checking is used in this example.

```
// Testbench design example 2: Random test.
'timescale 1 ns / 100 ps
module nbit_adder_for_tb1;
parameter N = 4;
reg  [N-1:0] x, y;
reg  c_in;
wire [N-1:0] sum;
wire c_out;
// Unit Under Test port map
   nbit_adder_for UUT (
      .x(x), .y(y), .c_in(c_in), .sum(sum),
      .c_out(c_out));
integer i;
reg [N:0] test_sum;
initial
   for (i = 0; i <= 2*N ; i = i + 1) begin
      x = $random % 2**N;
      y = $random % 2**N;
      c_in =1'b0;
      test_sum = x + y;
      #15;
      if (test_sum != {c_out, sum}) $display("Error
       iteration %h", i);
      #5;
   end
initial
   #200 $finish;
initial
```

```
    $monitor($realtime,"ns %h %h %h %h", x, y, c_in,
        {c_out, sum});
endmodule                                                    ■
```

The following example combines the use of golden vectors and on-the-fly checking.

EXAMPLE 13.3   *Test Bench Example 3: Golden Vectors*

In this example, the golden vectors are read in through using the $readmemh system task and stored in memory. The stimuli are applied to the DUT one by one and the outputs from the DUT are then compared with the golden vectors on the fly during simulation. If any error is detected, then a message is displayed on the standard ouput.

```
// Testbench design example 3: Using golden vectors.
'timescale 1 ns / 100 ps
module nbit_adder_for_tb2;
//Internal signals declarations:
parameter N = 4;
parameter M = 8;
reg  [N-1:0] x, y;
reg  c_in;
wire [N-1:0] sum;
wire c_out;
// Unit Under Test port map
   nbit_adder_for UUT (
    .x(x), .y(y), .c_in(c_in), .sum(sum), .c_out(c_out));
integer i;
reg [N-1:0] x_array [M-1:0];
reg [N-1:0] y_array [M-1:0];
reg [N:0] expected_sum_array [M-1:0];
initial begin  // reading verification vector files
   $readmemh("inputx.txt", x_array);
   $readmemh("inputy.txt", y_array);
   $readmemh("sum.txt", expected_sum_array);
end
initial
   for (i = 0; i <= M - 1 ; i = i + 1) begin
     x = x_array[i];
     y = y_array[i];
     c_in =1'b0;
     #15;
     if (expected_sum_array[i] != {c_out, sum})
        $display("Error iteration %h", i);
     #5;
   end
```

```
initial
   #200 $finish;
initial
   $monitor($realtime,"ns %h %h %h %h", x, y, c_in,
   {c_out, sum});
endmodule
```
∎

The contents of the files, `inputx.txt`, `inputy.txt` and `sum.txt`, are as follows:

inputx.txt	inputy.txt	sum.txt
4	1	05
9	3	0c
d	d	1a
5	2	07
1	d	0e
6	d	13
d	c	19
9	6	0f

13.3.2 Clock Signal Generation

The clock signal is the heart of any digital system. Any operation of digital systems is progressed by the trigger of a clock signal. Hence, the clock signal must be generated with care. For digital systems, many types of clock signals are often required. The following are some of them:

1. A general clock signal.
2. Aligned derived clock signals.
3. Clock multipliers.
4. Asynchronous clock signals.
5. Retriggerable monostable signal.

A General Clock Signal   There are many ways to generate a (general) clock signal. The simplest one is by using an `initial` statement to set the initial value of `clk` and an `always` statement to toggle the clock signal.

```
initial clk <=1'b0;
always #10 clk <= ~clk;
```

Another commonly used scheme is to assign 1 and 0 to the clock signal explicitly:

```
reg  clk;
always begin
   #5 clk <= 1'b0;
```

```
      #5 clk <= 1'b1;
end
```

The statement `forever` is also often used to generate a clock signal, such as the following example:

```
initial begin
   clk <= 1'b0;
   forever #10 clk <= ~clk;
end
```

In this example, the function of the `forever` statement is exactly like that of an `always` statement.

In most practical applications, it is better to define the period as a parameter. In these cases, we need to pay some attention to the timescale, which may affect the timing of the edge. Usually, two errors may be raised when a division is used to derive the duty cycle of the clock signal from a defined parameter. These two errors are *truncation error* and *rounding error*. The truncation error occurs when an integer division is used, such as the following example.

```
`timescale 1 ns / 1 ns
reg  clk;
parameter clk_period = 25;
// derive the clock signal
always begin
   #(clk_period/2) clk <= 1'b0;
   #(clk_period/2) clk <= 1'b1;
end
```

Due to the effect of time precision and truncation error of integer division, the duty cycle of the clock signal is 12 time units.

The rounding error occurs when a real division is used, such as the following example.

```
`timescale 1 ns / 1 ns
reg  clk;
parameter clk_period = 25;
// derive the clock signal
always begin
   #(clk_period/2.0) clk <= 1'b0;
   #(clk_period/2.0) clk <= 1'b1;
end
```

Due to the effect of time precision and rounding error of real division, the duty cycle of the clock signal is 13 time units.

In order to avoid truncation or rounding error, it needs to use proper time precision in the `timescale` compiler directive. An example to illustrate this is as follows.

```
`timescale 1 ns / 100 ps
reg  clk;
parameter clk_period = 25;

always begin
    #(clk_period/2) clk <= 1'b0;
    #(clk_period/2) clk <= 1'b1;
end
```

Although the time precision is set to 100 ps, there still remains a truncation error of integer division. Hence, the duty cycle of the clock signal is 12 time units.

Aligned Derived Clock Signals   Because the purpose of a test bench is to mimic the actual environment where the DUT would work, it often needs to generate two aligned clock signals that the hardware actually do. An improper approach may arise in a delta delay between the derived clock signals. For example:

```
always begin
    if (clk == 1'b1) clk2 <= ~clk2;
end
```

where the clock `clk2` toggles whenever clock signal `clk` is 1. As a result, there is some delay between `clk` and `clk2`—this delay is called a *delta delay*.

To avoid the delta delay, it needs to mimic the operations of a hardware device used to generate the aligned derived clock signals, as depicted in Figure 13.9. From this figure, we see that two clock signals `clk1` and `clk2` are derived from a common clock signal `clk`. Based on this idea, a proper approach for generating a derived clock signals is as follows.

```
// both clk1 and clk2 are derived from clk.
always begin
   clk1 <= clk;
   if (clk == 1'b1) clk2 <= ~clk2;
end
```

Explain the reason why there is no delta delay when two clock signals are derived in this way.

FIGURE 13.9 The proper derived clock scheme

Clock Multipliers   The clock multipliers in actual circuits are implemented using internal or external phase-locked loops (PLLs). A clock signal can be derived from another via a frequency divider or a frequency multiplier. The `repeat` statement can be used to derive a clock signal from another. Generating clock multiples by division does not need to know the frequency of the reference clock. The following example generates a clock signal `clk1` with one-fourth of the frequency of the clock signal `clk4`.

```
initial begin
   clk1 <= 1'b0;
   clk4 <= 1'b0;
   forever begin
      repeat(4) #(period_clk4/2) clk4 <= ~clk4;
      clk1 <= ~clk1;
   end
end
```

To generate a clock frequency by n times the reference clock, the `repeat` statement can be used to repeatedly generate $2n$ transitions. For example:

```
always @(posedge clk)
   repeat(2N) clkN <= #(period_clk/2N) ~clkN;
```

Another approach using a `forever` statement is as follows.

```
forever clkN <= #(period_clk/2N) ~clkN;
```

Asynchronous (Unrelated) Clock Signals   The word "asynchronous" means random. There is no way to accurately model unrelated clock signals. Unrelated clock signals should be modeled as a separate `initial` or `always` block.

```
initial begin
   clk100 <= 1'b0;
   #2;
   forever #5 clk100 <= ~clk100;
end
```

```
initial begin
   clk33 <= 1'b0;
   #5;
   forever #15 clk33 <= ~clk33;
end
```

In order to emphasize the random effect, a jitter may be added to one of the clocks to simulate the nondeterministic effect of phase. The jitter may be generated by using the random number generator system function `$random`.

Retriggerable Monostable Signal The following example is a behavioral description of a retriggerable monostable. The monostable outputs a high pulse with a width of 100 time units when it is triggered. The named event `retrig` restarts the monostable time period. If `retrig` continues to occur within 100 time units, then `qout` will remain at 1.

```
always begin: monostable
   #100 qout = 0;
end
always @retrig begin
   disable monostable;
   qout = 1;
end
```

Modify the above retriggerable monostable into a non-retriggerable one.

13.3.3 Reset Signal Generation

The hardware reset signal is the first signal to be applied to the DUT after the clock signal. Therefore, it is important to create the reset signal in the test bench. In particular, the following issues are needed to pay more attention to: *race condition*, *maintainability* and *reusability*.

Race Condition Race conditions are often created unintentionally between synchronized signals. To avoid the race problems, it is desirable to use nonblocking rather than blocking assignments. A good example to illustrate the race problem is as follows. In this example, a clock signal `clk` with a period of 10 time units and 50 % duty cycle is generated by an `always` block using blocking assignments. The reset signal `reset` is also generated by using blocking assignments with an `initial` block. The `reset` signal intends to last for 40 time units, from simulation time 20 to 60. Depending on the execution order of the `always` and `initial` blocks, the clock and reset signals, `clk` and `reset`, a race problem exists between them.

```
always begin
   #5 clk = 1'b0;
   #5 clk = 1'b1;
end
initial begin // has race condition.
   reset = 1'b0;
   #20 reset = 1'b1;
   #40 reset = 1'b0;
end
```

As discussed in Section 4.2.3, the race problem of the clock and reset signals can be avoided by using nonblocking assignments, as shown in the following example:

```
always begin
   #5 clk <= 1'b0;
   #5 clk <= 1'b1;
end
initial begin // no race condition.
   reset <= 1'b0;
   #20 reset <= 1'b1;
   #40 reset <= 1'b0;
end
```

Maintainability It is often desirable to change the clock period during the developing stage of a design. When this is the case, the reset signal generated by using absolute values of simulation time might not properly reset the device under test. As a result, a functional error may arise, especially, at the startup time of the simulation. This is the case of introducing a functional error due to lack of maintainability of the reset signal. In order to avoid this problem, the reset signal must be written in a way such that it is a function of the clock signal rather than an absolute value of the simulation time. For example, the reset signal must last for two clock periods instead of 40 time units.

The following example shows how to write a reset signal with maintainability.

EXAMPLE 13.4 *A Reset Signal with Maintainability*

In this example, the reset signal is set to 1 for two clock cycles, starting at the negative edge of clk. Consequently, it is independent of the period of the clock signal clk. Of course, due to the asynchronous feature of the reset signal, it is often desirable to add some phase jitter to the reset signal reset. This can be done by using a random number generator system function $random to generate a random number as the value of delay control. The details are left to the reader as an exercise.

```
always begin
   #(clk_period/2) clk <= 1'b0;
   #(clk_period/2) clk <= 1'b1;
end
// reset signal is always activated two clock cycles
initial begin
   reset <= 1'b0;
   wait (clk !== 1'bx); // wait until clock is active
   // set reset to 1 for two clock cycles
   // starting at falling edge of clk
   repeat (3) @(negedge clk) reset <= 1'b1;
   reset <= 1'b0;
end
```

Reusability   When verifying a design, the reset signal may need to be applied many times during a simulation. In order to re-use the reset process, it is convenient to encapsulate the generation of the reset signal with a task. The following example manifests this idea.

EXAMPLE 13.5 *A Reset Signal Generated by Using a Task*

This example is simply to rewrite the `initial` block used to generate the reset signal in the preceding example as a task. Consequently, in any place within a test bench, the task `hardware_reset` may be called to generate a reset signal.

```
always begin
   #(clk_period/2) clk <= 1'b0;
   #(clk_period/2) clk <= 1'b1;
end
// encapsulating the generation of a synchronous reset in
// a task.
task hardware_reset;
begin // reset signal is always activated two clock
      // cycles.
   reset = 1'b0;
   wait (clk !== 1'bx);
    // set reset to 1 for two clock cycles starting at
    // falling edge of clk.
   repeat (3) @(negedge clk) reset <= 1'b1;
   reset <= 1'b0;
end
endtask
```

Review Questions

Q13.19 What are the major components of a test bench?

Q13.20 What is a self-checking test bench?

Q13.21 What is the meaning of a golden vector?

Q13.22 What are the two types of result checking?

Q13.23 Describe how to generate two asynchronous clock signals.

Q13.24 Describe how to generate a clock signal, `clk5`, with five times the frequency of the original clock signal, `clk`.

Q13.25 Explain why the race problem may occur between reset and clock signals.

13.3.4 Verification Coverage

Verification coverage is a measurement of state space that a simulation-based verification has touched. The percentage of a coverage only means what fraction of coverage points is included in the simulation. It does not mean the percentage of functional correctness in the design. As a matter of fact, the quality of verification coverage strongly depends on how well the test bench is. Verification coverage includes two major types: *structural coverage* and *functional coverage*. Structural coverage denotes the representation of the design to be covered whereas functional coverage means the semantics of the design implementation to be covered.

13.3.4.1 Structural Coverage Structural (code) coverage deals with the structure of the Verilog HDL code and indicates which key parts of that structure have been covered. Structural coverage mainly includes the following: *statement coverage*, *branch or conditional coverage*, *toggle coverage*, *trigger coverage*, *expression coverage*, *path coverage* and *finite-state machine coverage*.

Statement Coverage Statement coverage gives the statistics about executable statements that are executed in a simulation. In other words, it gives the count of how many times statements were executed. The rationale behind the importance of statement coverage is based on the assumption that an unexercised code potentially bears bugs. It should be noted that statement coverage is a useful but not a complete metric.

Branch or Conditional Coverage Conditional coverage verifies that all branch sub-conditions have triggered the conditional branch and gives a count of how many times each condition occurred.

Toggle Coverage Toggle coverage measures how many times the signals have changed their logic values in a simulation.

Trigger Coverage Trigger coverage counts how many times an `always` block that was activated by each signal in the sensitivity list changes its value during simulation.

Expression Coverage Expression coverage measures various ways of an expression that paths through the code are executed during simulation.

Path Coverage Path coverage measures all possible ways that can execute a sequence of statements in a simulation. It is a refinement of branch coverage.

Finite-state Machine Coverage FSM coverage usually contains two types of coverage: state coverage and transition (arc) coverage. The state coverage measures which of the states have been discovered and transition coverage measures which of the state transitions have been visited.

Table 13.1 shows an example of code coverage analysis by typical simulation tools. Here, due to the simplicity of the design, the coverage of all three statements is 100 %; namely, all statements in the design are touched by the verification test.

In summary, an analysis of code coverage is only to let you know if you are not done. A 100 % code coverage is by no means an indication that the job is over or the design can work perfectly correctly.

TABLE 13.1 The results of code coverage analysis of the Booth algorithm

Module	Stmt count	Stmt hits	Stmt miss	Stmt %
Three-step Booth	35	35	0	100
Controller	21	21	0	100
Datapath	5	5	0	100

13.3.4.2 Functional Coverage Functional coverage makes sure that all possible legal values of input stimuli are exercised in all possible combinations at all possible times. Functional coverage lets you know whether you have done the job.

Item Coverage Item coverage records the individual scalar values, such as packet length, instruction opcode, interrupt level and so on.

Cross Coverage Cross coverage measures the presence or occurrence of combinations of values.

Transition Coverage Transition coverage measures the presence or occurrence of sequences of values.

Coding Styles

1. All response checking should be done automatically, rather than have the designer view waveforms and determine whether they are correct.

2. The time unit set in timescale must be matched with the actual propagation delay of gate-level circuitry.

3. The reset signal must be set properly. Especially, the time interval of the reset signal must be large enough. Otherwise, the initial operation of the gate-level simulation may not work properly.

Review Questions

Q13.26 What does 100 % functional coverage mean?

Q13.27 What does structural coverage mean?

Q13.28 What does functional coverage mean?

Q13.29 Explain the meaning of trigger coverage.

Q13.30 Explain the meaning of toggle coverage.

Q13.31 Explain the meaning of transition coverage.

13.4 DYNAMIC TIMING ANALYSIS

The simulation-based verification is still the most widely used approach to verifying both functionality and timing of a design. In the previous sections, we have dealt with functional verification through using simulation and formal proof. In this section, we address the timing verification used to verify whether the timing of a design can meet its timing specification or not.

13.4.1 Basic Concepts of Timing Analysis

Roughly speaking, the goal of timing analysis is to estimate when the output of a given circuit becomes stable. As shown in Figure 13.10, the output of the combinational logic, i.e. the data input D of the second flip-flop, needs to be stable at the time $t = T$ for the correct functionality. However, how can we make sure of this? At least two approaches can be used for this purpose: *dynamic timing analysis* and *static timing analysis* (STA). The dynamic timing analysis is achieved by carrying out timing analysis through simulating the design in a cycle-by-cycle manner whereas the static timing analysis is by performing timing analysis on the basis of signal paths.

In practical applications, the timing analysis has the following purposes. The first is timing verification, which verifies whether a design meets a given timing constraint, such as cycle-time constraint. The second is timing optimization, which needs to identify the critical part of a design for further optimization, such as *critical paths*. A critical path is a timing critical signal path that limits the performance of the entire design.

To verify the timing relationship between various components of the design, a standard delay format (SDF) containing timing information of all cells in the design is used to provide timing information for simulating the gate-level netlist or performing timing analysis of signal paths without doing simulation. Recall that the timing analysis with simulation is called dynamic timing analysis and without simulation is called static timing analysis.

An illustration of dynamic timing simulation using delay back-annotation is shown in Figure 13.11. Both pre-layout (after placement is completed) and post-layout (after routing is completed) delays are extracted from their corresponding outputs, as depicted in this figure. These delays are stored in separate SDF files and annotated backward to the gate-level netlist so as to take the actual delay into account. The process by which timing values from the SDF file update specify path delays, specparam values,

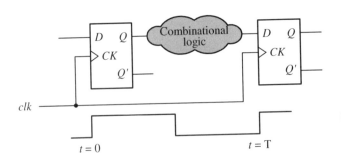

FIGURE 13.10 The meaning and requirement of timing analysis

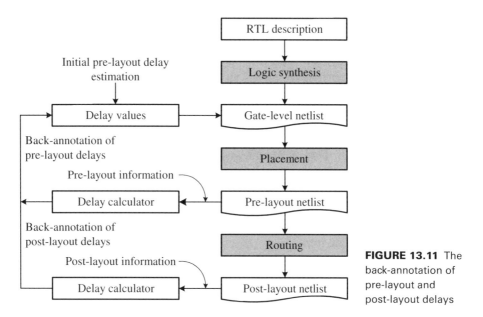

FIGURE 13.11 The back-annotation of pre-layout and post-layout delays

timing constraint values and interconnect delays is known as *delay back-annotation*. Detailed examples of how to use SDF files in dynamic timing simulations can be found in Section 13.7.2.

13.4.2 SDF and SDF Generation

The standard delay format (SDF) file contains timing information of all cells in the design. The timing information includes values for specify path delays, specparam values, timing check constraints and interconnect delays. The timing values in SDF files usually come from ASIC delay calculation tools that take connectivity, technology and layout geometry information into account.

The basic timing information of an SDF file comprises the following:

1. *IOPATH delay*. The IOPATH delay specifies the cell delay, which is computed based on the transition of the input signal and the output wire loading.

2. *INTERCONNECT delay*. The INTERCONNECT delay is a point-to-point, path-based delay, including the *RC* delay between the driving gate and the driven gate.

3. *Timing checks*. Timing checks contain values that determine the required setup time and hold time of each sequential cell. The values are based on the characterized values in the technology library.

The SDF file is used to provide timing information for simulating the gate-level netlist (namely, dynamic timing analysis). The SDF file can be generated in two phases:

- *Pre-layout SDF file*. The pre-layout SDF file is generated by using wire-load models but does not include clock trees in general. As a result, the delay values contained in the pre-layout SDF file only include the logic-cell delays and the estimated interconnect delays. For example, `*_map.sdf` (contains gate delay only) in ISE design flow.

- *Post-layout SDF file.* The post-layout SDF file contains delay values that are based on the actual layout, including logic-cell delays, interconnect delays and clock tree delays. For example, `*_timesim.sdf` (contains both gate and interconnect delays) in ISE design flow.

13.4.3 Delay Back-Annotation

As described above, during delay back-annotation, timing values are read from an SDF file into a specified region of the design to update specify path delays, specparam values, timing constraint values and interconnect delays. This operation can be accomplished by using the `$sdf_annotate` system task. The `$sdf_annotate` system task has the general form:

```
$sdf_annotate("sdf_file" [, [module_instance][,
  ["config_file"] [, ["log_file" ][, ["mtm_spec"][,
  [ "scale_factors"] [, ["scale_type"]]]]]]]);
```

where `sdf_file` is the SDF file generated from delay calculators; `module_instance` is the name of the design module. The rest arguments are optional.

The `config_file` is the configuration file used to provide detailed control over annotation. The `log_file` is the log file used to record results from each individual annotation of timing data during SDF annotation. The `mtm_spec` specifies which of the min/typ/max triples is used to annotate. It can be any of MAXIMUM, MINIMUM, TOOL_CONTROL (default) and TYPICAL, and overrides any MTM_SPEC in the configuration file. The `scale_factors` specifies the scale factors to be used and the default values are 1.0:1.0:1.0. The `scale_type` specifies how the scale factors should be applied to the min/typ/max triples. It can be any of FROM_MAXIMUM, FROM_MINIMUM, FROM_MTM (default) and FROM_TYPICAL. The `scale_factors` and `scale_type` arguments override the SCALE_FACTORS and SCALE_TYPE in the configuration file, respectively.

For example, to annotate the delay information of pre-layout back to the gate-level netlist, the following statement is included in the test bench:

```
$sdf_annotate ("four_bit_adder_map.sdf",
  four_bit_adder);
```

To annotate the delay information of post-layout, the following statement is included in the test bench:

```
$sdf_annotate ("four_bit_adder_timesim.sdf",
  four_bit_adder);
```

The above files `*.sdf` are generated automatically by the place and route (PAR) tools.

The detailed pre-layout and post-layout timing simulations will be discussed in more detail in Section 13.7.2.

13.4.4 SDF Format

SDF is a file format used to carry timing information to the simulator. The SDF is based on the IEEE Standard 1497 specification, which describes an ASCII file format that contains propagation and interconnect delays, as well as timing constraints.

An SDF file usually contains two parts: a header and cell descriptions. The following example gives an outline of a typical SDF file.

EXAMPLE 13.6 *A Typical SDF Format*

An SDF file contains a header and one or many cell descriptions. The header begins with `DELAYFILE` and usually ends with `TIMESCALE`. Except `SDFVERSION`, all other fields are optional. The `SDFVERSION` specifies the version of the SDF format used. The next two fields, `DESIGN` and `DATE`, describe the name of the design and the date of the file created, respectively. The `VENDOR`, `PROGRAM` and `VERSION` fields describe the originator, the name and the version of program used to generate the file, respectively. The `DIVIDER` field is used to separate the elements of the hierarchical path to each cell. It has two optional values, ".". and "/". It defaults to "." if the separator is omitted. The `VOLTAGE`, `PROCESS` and `TEMPERATURE` separately specify the voltage, process and temperature conditions. The `TIMESCALE` specifies the units for all values used in the SDF file. The default value is 1 ns.

The rest of an SDF file is a list of cells. The cells may be listed in any order. A cell description includes a cell type, instance name and delay specifications. It also includes a timing specification if the cell is a sequential one.

```
(DELAYFILE
  (SDFVERSION "3.0")
  (DESIGN "CRC_generator_detector")
  (DATE "[Fri Dec 23 19:34:37 2005] ")
  (VENDOR "Xilinx")
  (PROGRAM "Xilinx SDF Writer")
  (VERSION "G.28")
  (DIVIDER /)
  (VOLTAGE 1.425:1.425:1.425)
  (PROCESS   )
  (TEMPERATURE 85:85:85)
  (TIMESCALE 1 ps)
  (CELL (CELLTYPE "X_TRI")
   ...
  (CELL (CELLTYPE "X_FF")
   ...
```

The following example is an instance of a combinational logic cell.

EXAMPLE 13.7 *A Typical Combinational Cell Description*

A cell description begins with CELL and is followed by CELLTYPE and INSTANCE. The rest of a cell description is timing specifications. The CELLTYPE is the name of the component model appearing in the Verilog HDL netlist. The INSTANCE indentifies the particular instance of the cell, including the hierarchical cell path. The DELAY part specifies PATHPULSE and PORT, as well as IOPATH delays. The IOPATH specifies the delay of an input–output path. It is followed by the names of the input and output ports and an ordered list of delay specifications. In this example, two delays are used and each delay uses the min:typ:max delay specifier.

```
(CELL (CELLTYPE "X_TRI")
  (INSTANCE qout_obuf\[2\])
    (DELAY
      (PATHPULSE (605:605:605))
      (ABSOLUTE
        (PORT I (190:190:190)(190:190:190))
        (PORT CTL (190:190:190)(190:190:190))
        (IOPATH I O (3185:3185:3185)(3185:3185:3185))
        (IOPATH CTL O (3185:3185:3185)(3185:3185:3185))
      )
    )
  )
```
∎

In general, DELAY has four types [6]: PATHPULSE, PATHPULSEPERCENT, ABSOLUTE and INCREMENT. The keywords PATHPULSE and PATH-PULSEPERCENT specify how pulses will propagate across paths in this cell. The keywords ABSOLUTE and INCREMENT mean that the delay data in the SDF file will replace and add to the existing delay values in the design during annotation, respectively.

The following example is an instance of a sequential logic cell.

EXAMPLE 13.8 *A Typical Sequential Cell Description*

In a typical sequential cell description, a timing specification is included to specify the timing constraints. It begins with the keyword TIMINGCHECK. The timing specification is usually used to check the following timing constraints: setup and hold time, clock period, recovery time and pulse width. Details can be found in Section 7.3.3.

```
(CELL (CELLTYPE "X_FF")
  (INSTANCE crc_generator\[0\]\.qout\[2\])
    (DELAY
      (ABSOLUTE
```

```
          (IOPATH CLK O (449:449:449) (449:449:449))
          (IOPATH SET O (1019:1019:1019) (1019:1019:1019))
          (IOPATH RST O (1019:1019:1019) (1019:1019:1019))
        )
      )
      (TIMINGCHECK
        (SETUPHOLD (posedge I) (posedge CLK) (182:182:182)
          (33:33:33))
        (SETUPHOLD (negedge I) (posedge CLK) (182:182:182)
          (33:33:33))
        (SETUPHOLD (posedge CE) (posedge CLK) (95:95:95)
          (28:28:28))
        (SETUPHOLD (negedge CE) (posedge CLK) (95:95:95)
          (28:28:28))
        (PERIOD (posedge CLK) (1210:1210:1210))
        (RECOVERY (negedge SET) (posedge CLK)
          (341:341:341))
        (RECOVERY (negedge RST) (posedge CLK)
          (341:341:341))
        (WIDTH (posedge SET) (1210:1210:1210))
        (WIDTH (posedge RST) (1210:1210:1210))
      )
    )
```

Review Questions

Q13.32 What can dynamic timing simulation do?

Q13.33 Explain the meaning of delay back-annotation.

Q13.34 Distinguish the differences between DTA and STA.

Q13.35 Define the term 'critical path'.

Q13.36 What does an SDF file contain?

13.5 STATIC TIMING ANALYSIS

As we mentioned previously, timing verification is often performed through dynamic timing simulation. However, it has the following drawbacks. First, it poses a bottleneck for large complex designs because the simulation requires a lot of time to execute. Second, it relies on the quality and coverage of the test bench used for verification. Third, it is very difficult to figure a critical path out through the results from dynamic timing simulation. Hence, in this section, we introduce an alternative approach to verifying the timing specification of a logic without doing simulation. This approach is called *static timing analysis* (STA).

13.5.1 Fundamentals of Static Timing Analysis

Static timing analysis (STA) is an alternative approach to dynamic timing analysis. It is widely used to determine if a circuit meets timing constraints without having to simulate the design in the manner of cycle by cycle. In other words, static timing analysis analyzes a design in a static way. It computes the delay for each path of the design. As a result, the critical path can be easily pointed out. Compared to DTA, STA is independent of the input values.

To perform STA, the following two basic assumptions are made:

1. No combinational feedback loops are allowed.

2. All register feedback paths are broken by the clock boundary.

Based on these assumptions, the delay of each path can then be easily calculated. All path delays are checked to see whether timing constraints have been met or not. However, it is necessary to note that comprehensive sets of test benches are still needed to verify the functionality of the design because STA is used to verify timing specification only but it does not verify the functionality of the design. As discussed previously, a formal verification technique, equivalence checking, may also be used as an alternative way to verify the functionality of the gate-level netlist against the source RTL of the design.

In STA, designs are broken into sets of signal paths where each path has a start point and an endpoint, as shown in Figure 13.12. The start point may be either an input port or the clock input of a register, and the end point may be either an output port or the data input of a register.

There are four types of paths that can be combined from the start and end points. These are:

1. *Entry path* (input-to-D path)—starts at an input port (pad) and ends at the data input of a register such as a flip-flop or latch.

2. *Stage path* (register-to-register path or clock-to-D path)—starts at the clock input to a register and ends at the data input of another register.

3. *Exit path* (clock-to-output path)—starts at the clock input to a register and ends at an output port.

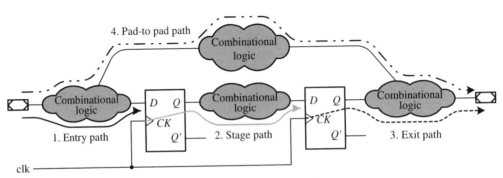

FIGURE 13.12 Four analysis types of static timing analysis

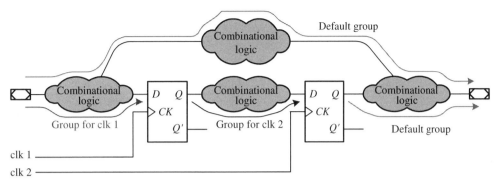

FIGURE 13.13 The clock domain and groups of clock domain

4. *Pad-to-pad path* (port-to-port path)—starts at an input port (pad) and ends at an output port (pad).

The above paths are defined over a single clock domain. When a circuit has multiple independent clock domains, the paths are grouped according to the clocks controlling their endpoints, as shown in Figure 13.13. Two new groups of paths are defined: *path group* and *default path group*. Each clock will be associated with a set of paths called a path group and the default path group comprises all paths not associated with a clock.

13.5.2 Timing Specifications

Several timing related constraints are defined when using STA to verify the timing specification of a design. These most commonly used constraints can be divided into two classes: port (or pad)-related constraints and clock-related constraints. The port (or pad)-related constraints are *input delay* (*offset-in*) *output delay* (*offset-out*), and *input–output* (*pad to pad*); the clock-related constraints contain *clock period* (cycle time) *setup time* and *hold time*.

13.5.2.1 *Port-Related Constraints*   The input delay (offset-in) constraint applies to the path from an input port (pad) to the data input of a register. It specifies the arrival time of the input signal relative to the active edge of the clock.

The output delay (offset-out) constraint applies to the path from the clock input of a register to an output port (pad). It specifies the latest time that a signal from the output of a register may reach to the output port (pad). The definitions of input and output delays are illustrated in Figure 13.14. More details of these two delays can be found in Section 12.2.2.

The input–output (pad-to-pad) constraint applies to the path from an input port (pad) to an output port (pad) without passing through any register.

13.5.2.2 *Clock-Related Constraints*   Before describing clock-related constraints in detail, it is required to explore the factors that cause the uncertainty of a clock signal. As depicted in Figure 13.15, several factors can affect the timing of a clock

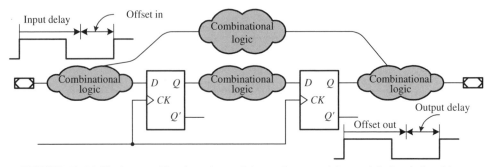

FIGURE 13.14 Timing specifications: input delay and output delay and their relationship to offset-in and offset-out

signal. These are *clock jitter*, *clock-to-Q delay*, *input capacitance*, *slew rate*, *interconnect capacitance*, *self-loading capacitance*, *clock skew* and *temperature*. We explain each factor in more detail as follows:

- *Clock jitter* is the temporal variation of clock signal at a given point on chip or system. In other words, the clock period may shrink or expand on a cycle-by-cycle basis.

- *Clock-to-Q* delay (t_q) is the maximum propagation delay from the clock edge (sampling edge) to the new value of the Q output.

- *Input capacitance* comprises gate capacitance of both nMOS and pMOS transistors.

- *Slew rate* is a metric of the current sinking and sourcing capability of the driving gate. The larger current, the faster slew rate.

- *Interconnect capacitances* consist of all parasitic capacitances of interconnect between any two nodes.

- *Self-loading capacitances* comprise the output capacitances of both nMOS and pMOS transistors.

- *Clock skew* is the misalignment of clock edges in a synchronous system or circuit. Clock skew reduces the timing margin between the data and the clock at the destination register.

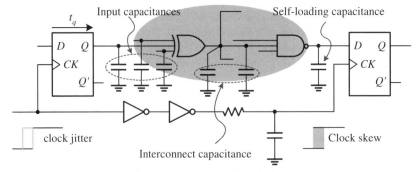

FIGURE 13.15 Factors affecting the timing of clock signals

- *Temperature* affects the currents of both nMOS and pMOS transistors in an inversely proportional manner.

Clock-related constraints include *clock period*, *setup time* and *hold time*. In the following, we describe each of these in more detail.

Clock period constraint applies to the paths between registers and specifies the maximum period of the clock of a synchronous circuit.

As defined before, the (minimum) setup time is the amount of time that data must be stable before the active clock transition is coming. The maximum data path is used to determine whether a setup constraint is met or not.

The (minimum) hold time is the amount of time that data must remain stable after the active clock transition. The minimum data path is used to determine whether the hold time is met or not.

13.5.2.3 Timing Analysis After timing specification is given, the static timing analysis can be proceeded to measure the critical path. A *critical path* is the path of longest propagation delay in a design. More formally, a critical path can be defined as follows. A critical path is a combinational logic path that has a negative or smallest slack time, where slack time is defined as:

$$\text{slack time} = \text{required time} - \text{arrival time}$$

$$= \text{requirement} - \text{datapath (in ISE)}$$

Critical paths limit the system performance. Critical paths not only tell us the system cycle time but also point out which part of the combinational logic should be changed to improve the system performance.

13.5.3 Timing Exceptions

Recall that timing analysis tools usually treat all paths in the design as single-cycle by default and perform STA accordingly. However, in most designs, there may be paths that exhibit timing exceptions. Two most common timing exceptions are *false paths* and *multiple-cycle paths*.

A false path is identified as a timing path that does not actually propagate a signal, as shown in Figure 13.16. The path indicated is never activated because both multiplexers cannot select their inputs 0 to their output end at the same time. However, the path has a delay of 200 ns, which is the longest one among the four possible paths. Therefore, a false path is the path that static timing analysis tools identify as failing timing, but the designer knows that it is not actually failing because it is never the path

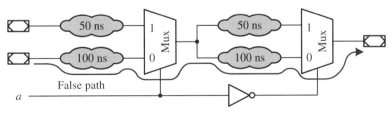

FIGURE 13.16 An example of false paths

being to propagate the desired signals. In short, a false path is a timing path that does not propagate a signal.

Another timing exception which often causes static timing analysis to be troubled is multiple-cycle paths. Simply speaking, in a multiple-cycle path, data may take more than one clock cycle to reach their destination. To manifest this idea, let us consider the following example.

EXAMPLE 13.9 *A Multiple-cycle Path Example*

In this example, variable qout_a is updated every clock cycle, variable qout_b is updated every two clock cycles and variable qout_c is updated every three clock cycles. The synthesized result is depicted in Figure 13.17.

```verilog
// a multiple cycle example
module multiple_cycle_example(clk, data_a, data_b, data_c,
                                qout_a, qout_b, qout_c);
parameter n = 8;
input   clk;
input   [n-1:0] data_a, data_b, data_c;
output reg [n-1:0] qout_a, qout_b, qout_c;
// trivial multiple-cycle operations.
always @(posedge clk) begin
    qout_a <= data_a * 5;
```

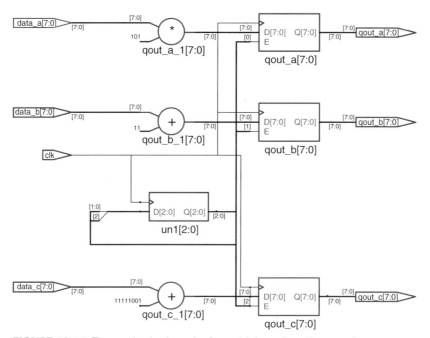

FIGURE 13.17 The synthesized result of a multiple-cycle path example

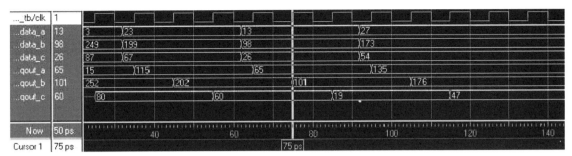

FIGURE 13.18 The simulation results of the multiple-cycle path example

```
    @(posedge clk) qout_b <= data_b + 3;
    @(posedge clk) qout_c <= data_c - 7;
end
endmodule
```

Because STA tools usually treat all paths in the design as a single-cycle by default and perform the STA accordingly, it is required to tell them which paths are multiple-cycle paths in order to avoid getting results falsely, namely, obtaining a timing violation. For example, as shown in Figure 13.17, both paths from inputs data_b, and data_c to their destinations qout_b and qout_c, respectively, will be treated as a single-cycle as the path from input data_a to its destination qout_c when no timing exception is set. However, as can be seen from Figures 13.18 and 13.19, the path from input data_b to qout_b is a two-cycle path and the path from input data_c to qout_c is a three-cycle path. They have more loose timing constraints than the path from input data_a to qout_a.

Coding Styles

1. It should avoid multiple-cycle paths and other timing exceptions in the design.

2. If timing exceptions must be used in a design, it should be sure to use start and end points which are guaranteed to exist and be valid at the chip level.

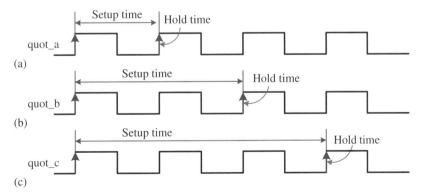

FIGURE 13.19 The example of a multiple-cycle path timing exception: (a) single-cycle timing relationship; (b) two-cycle timing relationship; (c) three-cycle timing relationship

Review Questions

Q13.37   What are the limitations of dynamic timing analysis?

Q13.38   What are the four types of paths of STA?

Q13.39   What do the port-related constraints include?

Q13.40   What do the clock-related constraints include?

Q13.41   Explain the meanings of false paths.

Q13.42   Explain the meanings of multiple-cycle paths.

13.6 VALUE CHANGE DUMP (VCD) FILES

A value change dump (VCD) file is a text file that contains information about value changes on selected variables in a design. The major purpose of VCD files is to provide information for other post-processing tools such as debug tools. Two types of VCD files are:

1. *Four-state VCD file*. The selected variable changes in {0, 1, x, z} without any strength information.

2. *Extended VCD file*. The selected variable changes in all states and strength information.

13.6.1 Four-State VCD Files

Verilog HDL provides the following system tasks to dump the required value changes in a design.

Specifying the Dump File   The $dumpfile system task specifies the name of the VCD file. It has the form:

```
$dumpfile (file_name);
```

Specifying the Variables to be Dumped   The $dumpvars system task is used to select which variables to dump into the specified file. It can be used with or without arguments and has the forms:

```
$dumpvars;
$dumpvars (levels[, modules_or_variables]);
```

When no arguments are specified, the $dumpvars system task dumps by default all variables to the VCD file. When arguments are specified, the $dumpvars system task dumps variables in a specified module and in all modules a specified number of levels below. The levels argument specifies how many levels of the hierarchy below each specified module instance to dump to the VCD file. A value 0 of the levels argument means all variables in the specified module and in all module instances below the specified module. The rest of the arguments can specify entire modules or individual variables within a module. The following example addresses some uses of the $dumpvars system task:

```
// all levels below adder_nbit_tb
$dumpvars (0, adder_nbit_tb);
// one level below module adder_nbit_tb
$dumpvars (1, adder_nbit_tb);
$dumpvars(0, UUT.sum, UUT.c_in); // all levels below
```

Stopping and Resuming the Dump   The dump operations can be paused and resumed later by using the system tasks: $dumpoff and $dumpon. Namely, they provide a mechanism to control the simulation period. The $dumpoff system task stops the dump whereas the $dumpon system task resumes the dump. They have the forms:

```
$dumpoff;
$dumpon;
```

When the $dumpoff system task is executed, a checkpoint is made by dumping an x value to every selected variable. When the $dumpon system task is executed later, each variable is dumped with its value at that time. No value changes are dumped to the specified file between the $dumpoff and $dumpon system tasks.

Checkpoint Generation   The $dumpall system task creates a checkpoint, by dumping the values of all selected variables at that time regardless of whether the port values have changed. It has the form:

```
$dumpall;
```

Setting the Maximum Size of the Dump File   The maximum size of the specified VCD file can be set by using the $dumplimit system task. It has the form:

```
$dumplimit(file_size);
```

The unit of file_size is bytes. The dumping stops and a comment is inserted in the file when the size of the VCD file reaches this number of bytes.

Empty File Buffer   The $dumpflush system task empties the VCD file buffer of the operating system to the VCD file so that no value changes are lost. It has the form:

```
$dumpflush;
```

The following example describes how the above system tasks can be used to create a VCD file in practice.

EXAMPLE 13.10 *A VCD File Example*

In this example, we suppose the module adder_nbit is instantiated as the unit under test. A VCD file named "adder_nbit.vcd" is opened by the $dumpfile system task. The file size is limited to 4k bytes by the $dumplimit system task. The selected variables to be dumped are those input and output ports of the module adder_nbit. The value changes observed are the first and the last 20 time units of simulation time. The durations between them are turned off and on by the $dumpoff and $dumpon system tasks, respectively.

```
`timescale 1 ns / 1 ns
// an example illustrates the VCD file
module adder_nbit_tb;
//Internal signals declarations:
reg [3:0] x;
reg [3:0] y;
reg c_in;
wire [3:0] sum;
wire c_out;
// Unit Under Test port map
   adder_nbit UUT (
           .sum(sum), .c_out(c_out), .x(x), .y(y),
           .c_in(c_in));

reg [7:0] i;
initial begin
   $dumpfile("adder_nbit.vcd");
   $dumplimit(4096);
   $dumpvars(0, UUT.sum, UUT.c_out, UUT.x, UUT.y,
    UUT.c_in);
end
initial
   for (i = 0; i <= 255; i = i + 1) begin
       x[3:0] = i[7:4]; y[3:0] = i[3:0]; c_in =1'b0;
   #5;
   end
initial begin
   #20    $dumpoff;
   #1000 $dumpon;
end
initial #1040 $dumpflush;
initial
   #1280 $stop;
initial
   $monitor($realtime,"ps %h %h %h %h", x, y, c_in,
    {c_out, sum});
endmodule
```
■

13.6.2 VCD File Format

The VCD file consists of three parts: *header information*, *node information* and *value changes*. The header information gives the date, the version number of the simulator and the timescale used. The node information contains definitions of the scope and type of the variables being dumped. The part of value changes comprises their values of those variables with values being changed at each simulation time. The contents of the VCD file obtained from the preceding example are shown in Figure 13.20.

```
$date                                   0(              $end
        Wed Jul 12 15:57:11 2006        0'              #1020
$end                                    0&              $dumpon
$version                                0-              1$
        ModelSim Version 6.0            0,              1#
$end                                    0+              1"
$timescale                              0*              0!
        1ns                             0.              1%
$end                                    $end            0)
$scope module adder_nbit_tb $end        #5              0(
$scope module UUT $end                  1-              1'
$var wire 1 ! sum [3] $end              1$              1&
$var wire 1 " sum [2] $end              #10             1-
$var wire 1 # sum [1] $end              0-              1,
$var wire 1 $ sum [0] $end              1,              0+
$var wire 1 % c_out $end                0$              1*
$var wire 1 & x [3] $end                1#              0.
$var wire 1 ' x [2] $end                #15             $end
$var wire 1 ( x [1] $end                1-              0-
$var wire 1 ) x [0] $end                1$              0,
$var wire 1 * y [3] $end                #20             1+
$var wire 1 + y [2] $end                $dumpoff        0$
$var wire 1 , y [1] $end                x$              0#
$var wire 1 - y [0] $end                x#              0"
$var wire 1 . c_in $end                 x"              1!
$upscope $end                           x!              #1025
$upscope $end                           x%              1-
$enddefinitions $end                    x)              1$
#0                                      x(              #1030
$dumpvars                               x'              0-
0$                                      x&              1,
0#                                      x-              0$
0"                                      x,              1#
0!                                      x+              #1035
0%                                      x*              1-
0)                                      x.              1$
(Continued next column)                 (Continued next column)
```

FIGURE 13.20 A VCD file example

Because VCD files are usually processed by debug tools but not analyzed manually, we do not discuss it furthermore. The detailed file format can be referred to the LRM [5].

13.6.3 Extended VCD Files

The system tasks used to create and dump extended VCD files are roughly the same as four-state VCD files with the following exceptions:

1. Each system task must also provide a specified file name except the `$dumpportsflush` system task, which means all files when no argument is specified. This also means that many extended VCD files may coexist at the same time during simulation.
2. The variables that can be dumped are only those ports of the specified modules.
3. The values dumped not only include all states $\{0, 1, x, z\}$ but also contain strength information.

Specifying the Dump File and Ports   The `$dumpports` system task specifies the name of the extended VCD file and the ports to be dumped. It has the form:

```
$dumpports (module_names, file_name);
```

where `module_names` are one or more module identifiers, separated by a comma.

Stopping and Resuming the Dump   The dump operations can be paused and resumed later by using the system tasks: `$dumpportsoff` and `$dumpportson`. That is to say, they provide a mechanism to control the simulation period. The `$dumpportsoff` system task stops the dump whereas the `$dumpportson` system task resumes the dump. They have the forms:

```
$dumpportsoff(file_name);
$dumpportson(file_name);
```

When the `$dumpportsoff` system task is executed, a checkpoint is made by dumping an x value to every port. When the `$dumpportson` system task is later executed, each port is dumped with its value at that time. No value changes are dumped to the specified file between the `$dumpportsoff` and `$dumpportson` system tasks.

Checkpoint Generation   The `$dumpportsall` system task creates a checkpoint, by dumping the values of all ports at that time regardless of whether the port values have changed. It has the form:

```
$dumpportsall(file_name);
```

Setting the Maximum Size of the Dump File   The maximum size of the specified extended VCD file can be set by using the `$dumpportslimit` system task. It has the form:

```
$dumpportslimit(file_size, file_name);
```

The unit of file_size is bytes. The dumping stops and a comment is inserted in the file when the size of the extended VCD file reaches this number of bytes.

Empty File Buffer   The $dumpportsflush system task empties the extended VCD file buffer of the operating system to the extended VCD file so that no value changes are lost. It has the form:

```
$dumpportsflush(file_name);
```

When no argument is specified, the extended VCD buffers for all files opened by the $dumpports system task are flushed.

The following example describes how the above system tasks can be used to create an extended VCD file in practice.

EXAMPLE 13.11   *An Extended VCD File Example*

In this example, we suppose the module adder_nbit is instantiated as the unit under test. An extended VCD file named "adder_nbit_evcd.vcd" is opened by the $dumpports system task, which also specifies the ports of the module to be dumped. The file size is limited to 4k bytes by the $dumpportslimit system task. The value changes observed are the first and the last 20 time units of the simulation time. The duration between them are turned off and on by the $dumpportsoff and $dumpportson system tasks, respectively.

```
'timescale 1 ns / 1 ns
// an example illustrates the extended VCD file
module adder_nbit_evcd_tb;
// internal signals declarations:
reg [3:0] x;
reg [3:0] y;
reg c_in;
wire [3:0] sum;
wire c_out;
// Unit Under Test port map
   adder_nbit UUT (
       .sum(sum), .c_out(c_out), .x(x), .y(y),
       .c_in(c_in));

reg [7:0] i;
initial begin
   $dumpports(adder_nbit, "adder_nbit_evcd.vcd");
   $dumpportslimit(4096, "adder_nbit_evcd.vcd");
end
initial
   for (i = 0; i <= 255; i = i + 1) begin
       x[3:0] = i[7:4]; y[3:0] = i[3:0]; c_in =1'b0;
       #5 ;
   end
initial begin
```

```
    #20   $dumpportsoff("adder_nbit_evcd.vcd");
    #1000 $dumpportson("adder_nbit_evcd.vcd");
end
initial #1040 $dumpportsflush("adder_nbit_evcd.vcd");
initial
    #1280 $stop;
initial
    $monitor($realtime,"ps %h %h %h %h", x, y, c_in,
        {c_out, sum});
endmodule                                                    ■
```

It is instructive to run this program and observe the differences between the extended VCD file and VCD file.

Review Questions

Q13.43 Describe the function of a value change dump (VCD) file.

Q13.44 What are the contents of a VCD file?

Q13.45 What are differences between four-state and extended VCD files?

Q13.46 Describe the operations of the $dumpfile system task.

Q13.47 Describe the operations of the $dumpports system task.

13.7 A CASE STUDY: FPGA-BASED DESIGN AND VERIFICATION FLOW

In this section, we describe the FPGA-based design flow and verification through simulation. First, we summarize the basic operations of Xilinx's Integrated Software Environment (ISE), which includes design entry, synthesis, implementation and configure FPGA. Along with the design flow, static timing analyses are also given and explained. Next, we consider how to verify the design through dynamic timing simulations, incorporating the delays of logic elements and interconnect. Finally, we give a general RTL-based verification flow.

13.7.1 ISE Design Flow

When using an FPGA device to implement a design, we follow a design flow that optimizes the logic to fit into the logic elements, places logic elements in the FPGA fabrics, routes the wires to connect the logic elements and generates a programming file. The Xilinx's ISE design flow, as shown in Figure 13.21, can be divided into four major phases:

1. *Design entry.* Two design entry methods, including HDL and schematic drawings, can be used.

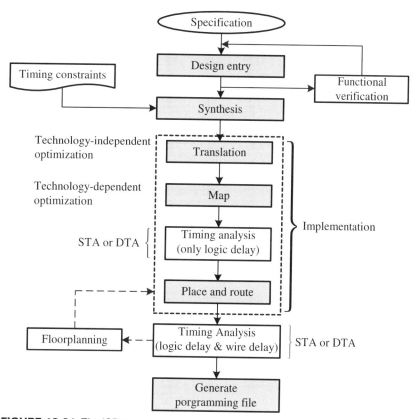

FIGURE 13.21 The ISE design flow

2. *Synthesis to create a gate netlist*. This phase translates the design in HDL and/or schematic files into an industry standard format known as an *electronic design interchange format* (EDIF) file (*.edf).

3. *Implementation*. It includes three main steps: translation, map and place and route (PAR).

4. *Configure FPGA*. It downloads a BIT file into the specified FPGA device to set the fabrics to function as desired.

In the following we describe each of these steps in more detail.

Design Entry The most commonly used design entry approach is based on HDL, including Verilog HDL and VHDL. After a design is entered into the design flow, the functional verification of the design is usually followed through simulation or formal proof, as described previously.

Synthesis The function of the synthesis phase is to extract logic from the HDL model and translate the design in HDL and/or schematic files into an EDIF file (*.edf). The synthesis phase can be divided into three main steps: parsing, synthesis and optimization. In the parsing step, possible syntactic errors are identified. The function of the synthesis step is to identify macro, extract finite-state machine and identify resources that can be shared. The operations of the optimization step include macro implementation, timing optimization, LUT mapping and register replication.

The macro identification (or inference) consists of two major steps: macro recognization and macro implementation. In the macro recognization, the Xilinx Synthesis Technology (XST) tries to recognize as many macros as possible. In the macro implementation, XST makes technology dependent choices, which are based on macro types and sizes according to the criteria which improve the design performance and decrease the area.

Currently, most synthesis tools, including XST and SynplifyPro, are able to recognize finite-state machines from the source description regardless of the modeling style used. In addition, these synthesis tools are able to optimize the finite-state machines on the basis of state assignment and flip-flop type. The most commonly used state assignment codes include one-hot code for speed optimization, binary code for area optimization, Gray code for hazards and glitches minimization and other user-defined codes such as Johnson code.

The output from synthesis is an EDIF file, a portion of which is shown as follows:

```
...
(cell OBUF (cellType GENERIC)
   (view PRIM (viewType NETLIST)
     (interface
       (port O (direction OUTPUT))
       (port I (direction INPUT))
     )
   )
)
(cell LUT3 (cellType GENERIC)
   (view PRIM (viewType NETLIST)
     (interface
       (port I0 (direction INPUT))
       (port I1 (direction INPUT))
       (port I2 (direction INPUT))
       (port O (direction OUTPUT))
     )
   )
)
```

Implementation   The implementation phase includes three main steps: translation, map and place and route (PAR). The translation step prepares for the map step including managing the design hierarchy. Namely, it merges multiple design files into a single gate-level netlist. The function of the map step is to optimize and fit the logic into the logic elements. In other words, it groups logical symbols from the gate-level netlist into physical components, including CLBs and IOBs. The PAR step places synthesized logic components into FPGA fabrics of the underlying FPGA device, connect them and extract timing data into reports. Placement chooses which fabrics will hold each of the logic elements while routing chooses which wire segments will be used to make necessary connections between the logic elements.

The map step finishes the logic synthesis operations. A resource used report is obtained after the map step finishes. The following gives the resource used of the example of a 4×4 `booth_array_multiplier`.

```
Design Summary
--------------
Number of errors:      0
Number of warnings:    0
Logic Utilization:
  Number of 4 input LUTs:     31 out of 3,072 1%
Logic Distribution:
  Number of occupied Slices: 17 out of 1,536 1%
  Number of Slices containing only related logic:
   17 out of 17 100%
  Number of Slices containing unrelated logic:
   0 out of 17   0%
*See NOTES below for an explanation of the effects
   of unrelated logic
Total Number 4 input LUTs:    31 out of 3,072 1%

  Number of bonded IOBs:      16 out of   200 8%

Total equivalent gate count for design:  186
Additional JTAG gate count for IOBs:  768
Peak Memory Usage:   137 MB
```

From this report, we can see the feasibility of design. We need to rethink the design if the LUTs used are more than we expected.

In addition, the map step also outputs a pre-layout static timing analysis about the design. The post-map STA of the 4×4 `booth_array_multiplier` is shown as follows.

```
Data Sheet report:
-----------------
All values displayed in nanoseconds (ns)

Pad to Pad
--------------+---------------+---------+
Source Pad    |Destination Pad| Delay   |
--------------+---------------+---------+
x[0]          |product[0]     |   4.943|
              ...
x[0]          |product[6]     |   7.625|
x[0]          |product[7]     |   7.625|
```

```
x[1]                |product[1]     |    4.943|
                      . . .
x[1]                |product[6]     |    7.625|
x[1]                |product[7]     |    7.625|
x[2]                |product[2]     |    4.943|
                      . . .
x[2]                |product[6]     |    6.731|
x[2]                |product[7]     |    6.731|
x[3]                |product[3]     |    4.943|
                      . . .
x[3]                |product[6]     |    5.837|
x[3]                |product[7]     |    5.837|
                      . . .
y[0]                |product[6]     |    7.625|
y[0]                |product[7]     |    7.625|
y[1]                |product[1]     |    4.943|
                      . . .
y[1]                |product[6]     |    7.625|
y[1]                |product[7]     |    7.625|
y[2]                |product[2]     |    5.837|
                      . . .
y[2]                |product[6]     |    7.625|
y[2]                |product[7]     |    7.625|
y[3]                |product[3]     |    6.284|
                      . . .
y[3]                |product[6]     |    7.178|
y[3]                |product[7]     |    7.178|
```

From this report, we can see one critical path is from `x[1]` to `product[6]`, which needs 7.625 ns. Due to the fact that in the map step the design is only mapped into logic elements without placing into a real FPGA device, the delay value obtained is only the cumulative delay of logic elements.

In the PAR step, logic elements are placed onto the device and connected according to the required function. After PAR, the implementation tools extract timing data and give a post-layout static timing analysis about the design. For example, the following is the post-PAR STA of the 4×4 `booth_array_multiplier`.

```
Data Sheet report:
----------------
All values displayed in nanoseconds (ns)

Pad to Pad
--------------+--------------+---------+
Source Pad    |Destination Pad|  Delay  |
--------------+--------------+---------+
x[0]                |product[0]     |    7.620|
```

	. . .		
x[0]	|product[6]		12.866|
x[0]	|product[7]		12.385|
x[1]	|product[1]		7.181|
	. . .		
x[1]	|product[6]		13.102|
x[1]	|product[7]		12.621|
x[2]	|product[2]		7.563|
	. . .		
x[2]	|product[6]		11.470|
x[2]	|product[7]		10.989|
x[3]	|product[3]		6.848|
	. . .		
x[3]	|product[6]		9.851|
x[3]	|product[7]		9.370|
	. . .		
y[0]	|product[6]		12.643|
y[0]	|product[7]		12.162|
y[1]	|product[1]		7.330|
	. . .		
y[1]	|product[6]		12.551|
y[1]	|product[7]		12.070|
y[2]	|product[2]		8.670|
	. . .		
y[2]	|product[6]		12.298|
y[2]	|product[7]		11.817|
y[3]	|product[3]		9.459|
	. . .		
y[3]	|product[6]		12.251|
y[3]	|product[7]		11.770|

From this report, we can see the critical path is from x[1] to product[6], which needs about 13.102 ns. This delay is the actual delay of the path, which includes both logic elements and interconnect (wire) delays. The interconnect delays along with each path can be easily estimated by subtracting the delay obtained from the post-map output from that obtained from the post-PAR output. For example, the interconnect delay of the path from x[1] to product[6] is equal to $13.102 - 7.625 = 5.477$ ns.

Configure FPGA   This generates a programming file (.bit), called a BIT file, which can be used to configure the specified FPGA device so that it would work as expected. The BIT file can be downloaded directly to the FPGA, or converted into a PROM (flash memory) file which stores the programming information.

Design Constraints   Even though HDLs, Verilog HDL and VHDL, capture the function of the logic, some mechanism is still needed to specify non-functional requirements: timing, area and power consumption. The means used to convey non-functional requirements to the synthesis tools and implementation tools is a file known as *design constraints*. Details of the design constraints can be found in Section 12.2.2.

For FPGA devices, design constraints include timing, area and I/O pin assignments. The timing constrains improve design performance by placing logic elements closer together so that shorter routing wires may be used. Timing constraints cover pad-to-pad constraint and period constraint, as well as offset-in and offset-out constraints. The pad-to-pad constraint is essential for completely constraining a design. The period constraint improves the path delays between synchronous elements. The offset-in constraint improves path delays from input ports to synchronous elements and the offset-out improves path delays from synchronous elements to output ports.

The area constraint is achieved by carefully placing logic blocks on the underlying FPGA device. Placement can also affect the system performance (timing) and I/O pin assignments of the final system.

I/O pin assignments not only assign IOBs to I/O ports of the design but also allow one to specify the direction (input, output or bidirectional), slew rate, drive strength and I/O standard (voltage level). I/O pin assignments are essential to the design of a printed-circuit board (PCB) on which the FPGA device will be used.

13.7.2 Dynamic Timing Simulations

As we have described previously, dynamic simulation is still an effective and widely used approach for verifying a design. In the following, we introduce how to perform a variety of dynamic simulations along with the design flow. As shown in Figure 13.22, there are four types of simulations along with the design flow. Except the behavioral simulation used to verify the functional correctness of design entry at the beginning of the design flow, the remaining three types of simulations are at gate-level. These are based on the outputs of various steps of the implementation phase of ISE. As a result,

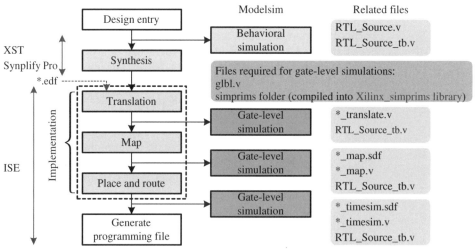

FIGURE 13.22 The ISE-based simulation flow

different but functionally equivalent source files are used in different steps. However, the same test bench is used in all steps, from behavioral down to post-PAR.

Gate-level Timing Simulation   There are three types of timing simulations at gate-level. Before these simulations can be done, there are one file and one folder that must be prepared: *glbl.v* and *simprims folder*. The glbl.v file is usually compiled into the work library and the simprims folder is compiled into a library, say Xilinx_simprims.

Post-translation Simulation   The translation step prepares the design for mapping. It expands the design hierarchy and reads design constraints. The translation step generates a post-translation simulation model, which is built from primitives that more closely relate to the structure of the logic to be used in the FPGA. A snap of the result from the translation phase is as follows.

```
X_AND2 \bdatapath/acc/qout_6_cry_6  (
   .I0(product_c[14]),
   .I1(ps_d_i[3]),
   .O(\bdatapath/acc/qout_6_cry_6_0_25 )
);
X_XOR2 \bdatapath/acc/qout_6_s_6  (
   .I0(\bdatapath/acc/qout_6_cry_5_26 ),
   .I1(\bdatapath/acc/qout_6_axb_6_30 ),
   .O(\bdatapath/acc/qout_6 [6])
);
defparam \bdatapath/acc/N_97_i.INIT = 16'hF600;
X_LUT4 \bdatapath/acc/N_97_i  (
   .ADR0(mp_qb10[0]),
   .ADR1(product_c[0]),
   .ADR2(ps[0]),
   .ADR3(ps[1]),
   .O(\bdatapath/acc/N_97_i_28 )
);
```

Of course, this is different from the original source description. However, they are functional equivalent. At this point, it is worth noting that if you find a discrepancy between your source description and the post-translation model, it may be because some constructs in your source description do not exactly translate in the way that you thought they should do.

In order to perform post-translation simulation (gate-level simulation), the following files are required. In the rest of this section, we use `booth_array_multiplier` as the running example.

- **Post-synthesis source file**: `booth_array_multiplier.vm` (SynplifyPro) or `booth_array_multiplier_translate.v` (Xinlix ISE).
- **Library files**:
  1. *glbl.v* ($Xilinx/verilog/src/);

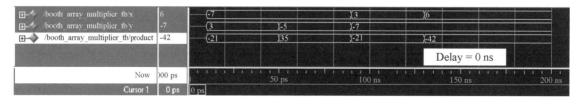

FIGURE 13.23 The results of post-translation simulation

> **2.** *Xilinx_simprims* (simprims folder ($Xilinx/verilog/src/simprims) compiled into the Xilinx_simprims library).

- **Test bench file**: `booth_array_multiplier_TB.v`.

The simulation results of the 4×4 `booth_array_multiplier` are shown in Figure 13.23. The translation step does not take delay of logic elements into account; the post-translation simulation is only used for verifying whether the function is still equivalent to that of the design entry.

> **Post-map Simulation** The second type of gate-level simulations is post-map simulation, which incorporates into the delays of logic elements but not the delays of interconnect. In order to perform this simulation, two files are required in addition to the library files: a post-map simulation model (*_map.v) and a standard delay format (*_map.sdf). These two files are generated automatically by the implementation tool at the map step. Like the post-translation simulation model, this synthesized model is functionally equivalent to our original design but one more step closely matches the structure of the logic elements of the underlying FPGA device. The SDF file backannotates the delays of logic elements into the simulation model.

- **Post_map Source file**: `booth_array_multiplier_map.v`.
- **Library files**:

  > **1.** *glbl.v* ($Xilinx/verilog/src/);
  >
  > **2.** *Xilinx_simprims* (simprims folder ($Xilinx/verilog/src/simprims) compiled into the Xilinx_simprims library).

- **SDF file**: `booth_array_multiplier_map.sdf`.
- **Test bench file**: `booth_array_multiplier_TB.v`.

The simulation results of the 4×4 `booth_array_multiplier` are shown in Figure 13.24. Although the map step does take delay of logic elements into account,

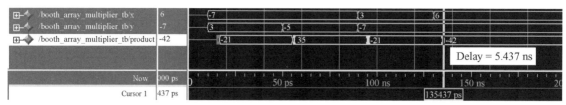

FIGURE 13.24 The results of post-map simulation

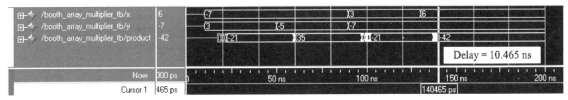

FIGURE 13.25 The results of post-PAR simulation

it is still very difficult to find the longest delay of logic elements because a lot of data need to be searched manually. Consequently, the post-map simulation is only used for verifying whether the function is still equivalent to that of the design entry.

Post-PAR Simulation   The third and final type of gate-level simulations is post-PAR simulation, which incorporates into both the delays of logic elements and interconnect. In order to perform this simulation, two files are required in addition to the library files: a post-PAR simulation model (*_timesim.v) and a standard delay format (*_timesim.sdf). These two files are generated automatically by the implementation tool at the PAR step. Like the post-map simulation model, this synthesized model is functionally equivalent to our original design but most closely matches the structure of the logic elements of the underlying FPGA device. The SDF file back-annotates the delays of the logic elements and interconnect into the simulation model.

- **PAR Source file**: `booth_array_multiplier_timesim.v`.
- **Library files**:

    1. *glbl.v* ($Xilinx/verilog/src/);

    2. *Xilinx_simprims* (simprims folder ($Xilinx/verilog/src/simprims) compiled into the Xilinx_simprims library).

- **SDF file**: `booth_array_multiplier_timesim.sdf`.
- **Test bench file**: `booth_array_multiplier_TB.v`.

The simulation results of the 4×4 `booth_array_multiplier` are shown in Figure 13.25. Although the PAR step does take both delays of logic elements and interconnect into account, like post-map simulation it is still very difficult to find the longest delay of logic elements and its associated interconnect, namely, critical path, because a lot of data need to be searched manually. Consequently, the post-PAR simulation is only used for verifying whether the function is equivalent to that of the design entry. As for the critical path, it usually resorts to the static timing analysis.

After PAR, we can also examine the power dissipation of the design. The power report from the implementation tool of ISE is shown as follows.

```
Power summary:                                      I(mA)    P(mW)
-------------------------------------------------------------
Total estimated power consumption:                            345
                                  ---
                     Vccint 1.50V:            8       12
                     Vccaux 3.30V:          100      330
```

```
                    Vcco33 3.30V:          1          3
                            ---
                       Inputs:          0          0
                        Logic:          0          0
                      Outputs:
                        Vcco33          0          0
                      Signals:          0          0
                            ---
          Quiescent Vccint  1.50V:          8         12
          Quiescent Vccaux  3.30V:        100        330
          Quiescent Vcco33  3.30V:          1          3

Thermal summary:
------------------------------------------------------------
    Estimated junction temperature:                    32C
              Ambient temp:   25C
                 Case temp:   31C
             Theta J-A range:  20 -  23C/W

Decoupling Network Summary:      Cap Range (uF)        #
------------------------------------------------------------
Capacitor Recommendations:
Total for      Vccint :                               16
                              470.0  - 1000.0 :    1
                              0.470  -  2.200 :    1
                              0.0470 - 0.2200 :    3
                              0.0100 - 0.0470 :    5
                              0.0010 - 0.0047 :    6
                                     ---
Total for      Vccaux :                                8
                              470.0  - 1000.0 :    1
                              0.0470 - 0.2200 :    1
                              0.0100 - 0.0470 :    2
                              0.0010 - 0.0047 :    4
                                     ---
Total for      Vcco33 :                                8
                              470.0  - 1000.0 :    1
                              0.0470 - 0.2200 :    1
                              0.0100 - 0.0470 :    2
                              0.0010 - 0.0047 :    4

Analysis completed: Sun Oct 29 08:26:00 2006
------------------------------------------------------------
```

There are three groups of entries. The first group covers the estimated current and power dissipation of the design. The second group provides the estimated temperature at which the device will operate. This gives an indication whether an additional cooling is required. The third group tell us how much decoupling capacitance is required to help filter out power supply noise. The decoupling capacitors are placed between the power supply terminals.

13.7.3 An RTL-Based Verification Flow

As we have discussed in the previous sections, before an RTL-based design is said to be completed, there are many verifications which must be done. The general RTL-based verification flow includes three major steps: simulation, static timing analysis and prototyping. We list each in detail as follows:

1. Simulation.
   (a) Functional (behavioral) simulation.
   (b) Code coverage analysis.
   (c) Assertion (property) checking.
   (d) Gate-level simulation.
   (e) Dynamic timing simulation (gate-level simulation + SDF back annotation).
2. Static timing analysis.
   (a) Critical paths.
   (b) Timing violations.
3. Prototyping.
   (a) FPGA prototyping.
   (b) Cell-based prototyping.

Review Questions

Q13.48 What is the basic design flow of Xilinx's ISE?

Q13.49 Describe the function of the synthesis step of the synthesis phase.

Q13.50 Describe the function of the optimization step of the synthesis phase.

Q13.51 What are the three major steps of the implementation phase?

Q13.52 Describe the operations of place and route.

Q13.53 What are the differences between the post-map and post-PAR SDF files?

Q13.54 Describe the differences between the post-map and post-PAR simulations in terms of timing.

SUMMARY

Verification is a necessary process that makes sure a design can meet its specifications. Due to the inherent features of the timing relationship between various components of the hardware module in question, the verification process needs to assure that the design works not only correctly in function but also in timing. As a consequence, the verification of a design can be divided into two types: functional verification and timing verification. Functional verification only considers whether the logic function of the design meets the specifications and timing verification considers whether the design meets the timing constraints.

Functional verification can be done either by simulation or by formal proof. In simulated-based functional verification, the design is placed under a test bench, input stimuli are applied to the design and the outputs from the design are compared with the reference outputs as well. Formal verification proves formally that a protocol, an assertion, a property or a design rule holds for all possible cases in the design.

Simulation-based verification is based on the analysis of simulation results and code coverage. The quality of this type of verification relies considerably on the test bench, which generates and controls stimuli, monitors and stores responses, and checks the simulation results. The two most common types of test benches are deterministic and self-checking. The deterministic test benches are often used to verify the basic functionality of the device under test (DUT) in an early stage of the development cycle. The self-checking test benches are those that place the knowledge of the DUT function into the test bench environment so that they can automate the tedious result checking process. The self-checking test benches are used to verify the final functionality of the DUT.

Formal verification can be further classified into equivalence checking and property verification. Equivalence checking determines whether two implementations are equivalent and property verification determines whether a property is in a design or not. The major drawback of formal verification is that it can only be applied to a design of limited size.

Timing verification is often carried out through simulation, known as dynamic timing simulation, which can be used to verify both function and timing of a design. However, dynamic timing simulation has the following shortcomings. First, it poses a bottleneck for large complex designs because the simulation requires a lot of time to execute. Second, it relies on the quality and coverage of the test bench used for verification. Third, it is very difficult to figure a critical path out through using the result from dynamic timing simulation. Hence, an alternative way, called static timing analysis (STA), is widely used to determine if a circuit meets timing constraints without having to simulate the design in the manner of cycle-by-cycle. In other words, static timing analysis analyzes a design in a static way and computes the delay for each path of the design. Therefore, the critical paths can be easily pointed out.

In order to use dynamic timing simulation to verify the timing constraints of a design, besides test benches, a file, known as a standard delay format (SDF) file, is also needed. The SDF file contains timing information of all cells in the design and can be produced by PAR tools. In addition, when using simulation to verify the function and timing of a design, a value change dump (VCD) file may also be produced during simulation in order to provide valuable information about value changes on selected variables in the design for debug tools to help remove the function bugs and timing violations of the design.

A comprehensive example based on FPGA design flow is also given to illustrate how to enter, synthesis, implementation and configure the underlying FPGA device of a design. Along with the design flow, static timing analyses are also given and explained. In addition, design verification through dynamic timing simulations, incorporating the delays of logic elements and interconnect, is introduced.

REFERENCES

1. J. Bergeron, *Writing Testbenches: Functional Verification of HDL Models,* 2nd Edn, Kluwer Academic Publishers, Boston, MA, USA, 2003.
2. J. Bhasker, *A Verilog HDL Primer,* 3rd Edn, Star Galaxy Publishing, 2005.
3. H. Bhatnagar, *Advanced ASCI Chip Synthesis: Using Synopses Design Compiler and Prime Time,* Kluwer Academic Publishers, Boston, MA, USA, 1999.
4. M.D. Ciletti, *Modeling, Synthesis and Rapid Prototyping with the Verilog HDL,* Prentice-Hall, Upper Saddle River, NJ, USA, 1999.
5. IEEE 1364-2001 Standard, *IEEE Standard Verilog Hardware Description Language,* 2001.
6. IEEE 1497-2001 Standard, *IEEE Standard for Standard Delay Format (SDF) for the Electronic Design Process,* 2001.
7. IEEE 1800-2005 Standard, *IEEE Standard for SystemVerilog—Unified Hardware Design, Specification and Verification Language,* 2005.
8. M. Keating and P. Bricaud, *Reuse Methodology Manual: For System-on-a-Chip Designs,* Kluwer Academic Publishers, Boston, MA, USA, 2002.
9. W.K. Lam, *Hardware Design Verification: Simulation and Formal Method-Based Approaches,* Prentice-Hall, Upper Saddle River, NJ, USA, 2005.
10. R. Munden, *ASIC and FPGA Verification: A Guide to Component Modeling,* Morgan Kaufmann, 2005.
11. S. Palnitkar, *Verilog HDL: A Guide to Digital Design and Synthesis,* 2nd Edn, SunSoft Press, 2003.
12. B. Wile, J.C. Goss and W. Roesner, *Comprehensive Functional Verification: The Complete Industry Cycle,* Morgan Kaufmann, 2005.
13. W. Wolf, *FPGA-Based System Design,* Prentice-Hall, Upper Saddle River, NJ, USA, 2004.

PROBLEMS

13.1 Replace the parameter definitions in the n-bit CRC generator example in Section 9.5.2 with the following two values:

```
parameter N = 16;
parameter [N:0]tap = 17'b11000000000000101;
```

(a) Describe how to verify whether the CRC-16 circuit works correctly.

(b) Write a test bench to generate the stimuli, monitor the responses and check the results at the end of test.

13.2 Considering the Problem 13.1 again, write a self-checking test bench to read the stimuli from a file, monitor the responses and check the results on the fly.

13.3 Considering the Problem 13.1 again, write a self-checking test bench to randomly generate the stimuli, monitor the responses and check the results on the fly.

13.4 Instantiate the Booth array multiplier discussed in Section 11.3.4 with the following statement:

```
booth_array_multiplier #(8, 8) booth_array
                        (.x(x), .y(y),
                         .product(product));
```

    (a) Describe how to verify whether the `booth_array_multiplier` module works correctly.

    (b) Write a test bench to generate the stimuli, monitor the responses and check the results at the end of the test.

13.5 Considering the Problem 13.4 again, write a self-checking test bench to read the stimuli from files, monitor the responses and check the results on the fly.

13.6 Considering the Problem 13.4 again, write a self-checking test bench to randomly generate the stimuli, monitor the responses and check the results on the fly.

13.7 Consider the Problem 13.4 again, by properly setting the options of the simulator to activate the code coverage analysis of the `booth_array_multiplier` module .

13.8 In order to mimic the jitter effects of two asynchronous clock sources in practical applications, a random number is often used to generate a random phase of the clock signals. Modify the following codes so that they have the random phases added to the clock signals. Assume that the random phase of each clock is in the range of 5 % of the clock period.

```
initial begin
   clk100 <= 1'b0;
   forever #5 clk100 <= ~clk100;
end
// the other clock source
initial begin
   clk33 <= 1'b0;
   forever #15 clk33 <= ~clk33;
end
```

13.9 Consider the tutorial example discussed in Section 1.4.3. Modify the test bench to dump out variables within a module of your choice using the `$dumpvars` system task.

    (a) Dump out variables under the scope of the current module.

    (b) Dump out variables at all levels under the scope of the current module.

13.10 Consider the parameterized multiplexer discussed in Section 8.3.1. Instantiate it with the size specified as follows. Then synthesize it and observe the LUTs used, as well as the critical path delay. Perform post-PAR simulation.

    (a) An 8-to-1 multiplexer.

(b) A 16-to-1 multiplexer.

(c) A 32-to-1 multiplexer.

13.11 Consider the parameterized magnitude comparator discussed in Section 8.5.1. Instantiate it with the size specified as follows. Then synthesize it and observe the LUTs used, as well as the critical path delay. Perform post-PAR simulation.

(a) An 8-bit magnitude comparator.

(b) A 16-bit magnitude comparator.

(c) A 32-bit magnitude comparator.

13.12 Consider the Booth algorithm described in Section 11.2.2. Instantiate it with the size specified as follows. Then synthesize it and observe the LUTs used, as well as the minimum clock period. Perform post-PAR simulation.

(a) An 8×8 multiplier.

(b) A 16×16 multiplier.

(a) A 32×32 multiplier.

13.13 Consider the Booth array multiplier discussed in Section 11.3.4 again. Instantiate it with the size specified as follows. Then synthesize it and observe the LUTs used, as well as the critical path delay. Perform post-PAR simulation.

(a) An 8×8 multiplier.

(b) A 16×16 multiplier.

(c) A 32×32 multiplier.

13.14 Within a test bench, write a program segment to generate two asynchronous clock signals. One generates a clock signal with a frequency of 25 MHz and the other 50 MHz.

13.15 Within a test bench, write a program segment to generate two synchronous clock signals. One generates a clock signal with a frequency of 20 MHz and the other 100 MHz.

13.16 Write a program segment to realize a retriggerable monostable. The `out` generates a pulse of 50 time units when it is triggered each time.

13.17 Write a program segment to realize a non-retriggerable monostable. The `out` generates a pulse of 20 time units when it is triggered each time.

ARITHMETIC MODULES

IN **THIS** chapter, we will examine many frequently used arithmetic modules, including addition, multiplication, division, ALU, shift and two digital-signal processing (DSP) filters as well. Along with these arithmetic operations and their algorithms, we also re-emphasize the concept that a hardware algorithm can usually be realized by using either a multiple-cycle or a single-cycle structure.

The bottleneck of a conventional n-bit ripple-carry adder is on the generation of carry needed in each stage. To overcome this, many schemes have been proposed. Among these, the carry-look-ahead (CLA) adder and two parallel-prefix adders, the Kogge–Stone adder and the Brent–Kung adder, will be explored.

The shift-and-add (subtract) technique is applied to construct both multiplication and division algorithms, which may process unsigned and signed inputs. In this chapter, we are concerned with an unsigned array multiplier and the modified Baugh–Wooley signed array multiplier, as well as two basic division algorithms: restoring and nonrestoring division algorithms.

An arithmetic and logic unit (ALU) is often the major component for the datapath of many applications, in particular, for central processing unit (CPU) design. An ALU comprises arithmetic operations and logical operations. In some applications, the shift operations, including both logical and arithmetic shifts, are also included as an important part of its ALU functions. A multiplexer-based single-cycle structure, known as a barrel shifter, is considered in detail. Finally, two basic DSP techniques, finite-impulse response (FIR) and infinite-impulse response (IIR) filters, are introduced.

14.1 ADDITION AND SUBTRACTION

The importance of addition is manifested not only on itself but also on constructing other related operations such as multiplication and division. In modern digital systems, subtraction is usually done by two's complement arithmetics, which is also based on

Digital System Designs and Practices   Ming-Bo Lin
© 2008 John Wiley & Sons (Asia) Pte Ltd

the addition. Hence, we usually use addition to mean both addition and subtraction. In this section, we are concerned with carry-look-ahead (CLA) adders and parallel-prefix adders.

14.1.1 Carry-Look-Ahead Adder

As we know, the performance bottleneck of an n-bit ripple-carry adder is on the generation of carriers needed in all stages. In order to explore the problem of carry generation, let us consider the full adder shown in Figure 14.1, which is the full adder implemented in terms of the two half adders that we have introduced in Chapter 1.

If we define two new functions, *carry generate* (g_i) and *carry propagate* (p_i), in terms of inputs x_i and y_i as follows:

$$g_i = x_i \cdot y_i$$

$$p_i = x_i \oplus y_i$$

then the sum and carry-out can be represented as a function of both carry generate (g_i) and carry propagate (p_i) and are as follows:

$$s_i = p_i \oplus c_i$$

$$c_{i+1} = g_i + p_i \cdot c_i$$

As a result, the carry input of the $(i + 1)$th-stage full adder can be generated by using the recursive equation of the c_{i+1}. For example, the first four carry signals are:

$$c_1 = g_0 + p_0 \cdot c_0$$
$$c_2 = g_1 + p_1 \cdot c_1$$
$$\quad = g_1 + p_1(g_0 + p_0 \cdot c_0) = g_1 + p_1 \cdot g_0 + p_1 \cdot p_0 \cdot c_0$$
$$c_3 = g_2 + p_2 \cdot c_2$$
$$\quad = g_2 + p_2(g_1 + p_1 \cdot g_0 + p_1 \cdot p_0 \cdot c_0)$$
$$\quad = g_2 + p_2 \cdot g_1 + p_2 \cdot p_1 \cdot g_0 + p_2 \cdot p_1 \cdot p_0 \cdot c_0$$
$$c_4 = g_3 + p_3 \cdot c_3$$
$$\quad = g_3 + p_3 \cdot (g_2 + p_2 \cdot g_1 + p_2 \cdot p_1 \cdot g_0 + p_2 \cdot p_1 \cdot p_0 \cdot c_0)$$
$$\quad = g_3 + p_3 \cdot g_2 + p_3 \cdot p_2 \cdot g_1 + p_3 \cdot p_2 \cdot p_1 \cdot g_0 + p_3 \cdot p_2 \cdot p_1 \cdot p_0 \cdot c_0$$

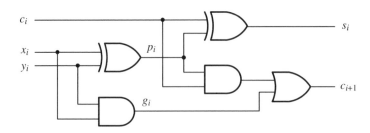

FIGURE 14.1 The logic diagram of the ith full adder in an n-bit adder

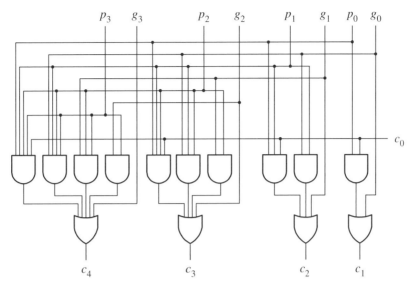

FIGURE 14.2 The logic diagram of a 4-bit carry-look-ahead generator

The resulting carry signals are only functions of both inputs x and y, and the carry-in (c_0). The two-level logic circuit used to realize the above carry-look-ahead (CLA) generator is shown in Figure 14.2. By using this carry-look-ahead generator, the four-bit adder may be represented as a function of the carry propagate signals p_i and carry signals c_i. The resulting circuit is shown in Figure 14.3.

The following two examples describe the above carry-look-ahead adder by employing continuous assignments and generate-loop statements, respectively.

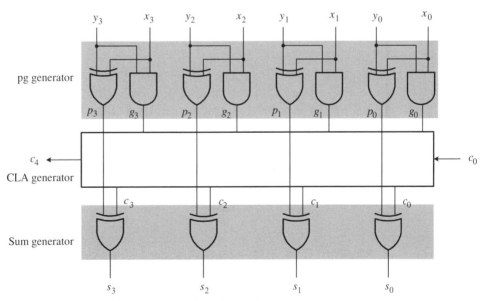

FIGURE 14.3 The logic diagram of a 4-bit CLA adder

EXAMPLE 14.1 *A 4-bit CLA Adder Using* `assign` *Statements*

In this example, we use continuous assignments `assign` to code the logic diagrams shown in Figures 14.2 and 14.3 into a Verilog HDL module.

```
// a 4-bit CLA adder using assign statements.
module cla_adder_4bits(x, y, cin, sum, cout);
// inputs and outputs
input  [3:0] x, y;
input  cin;
output [3:0] sum;
output cout;
// internal wires
wire  p0,g0, p1,g1, p2,g2, p3,g3;
wire  c4, c3, c2, c1;

// compute the p for each stage
assign p0 = x[0] ^ y[0],  p1 = x[1] ^ y[1],
       p2 = x[2] ^ y[2],  p3 = x[3] ^ y[3];
// compute the g for each stage
assign g0 = x[0] & y[0], g1 = x[1] & y[1],
       g2 = x[2] & y[2], g3 = x[3] & y[3];
// compute the carry for each stage
assign c1 = g0 | (p0 & cin),
       c2 = g1 | (p1 & g0) | (p1 & p0 & cin),
       c3 = g2 | (p2 & g1) | (p2 & p1 & g0) |
            (p2 & p1 & p0 & cin),
       c4 = g3 | (p3 & g2) | (p3 & p2 & g1) |
            (p3 & p2 &  p1 & g0) |
            (p3 & p2 & p1 & p0 & cin);
// compute Sum
assign sum[0] = p0 ^ cin, sum[1] = p1 ^ c1,
       sum[2] = p2 ^ c2,  sum[3] = p3 ^ c3;
// assign carry output
assign cout = c4;
endmodule
```

■

One disadvantage of using continuous assignments to model the CLA adder is that it lacks the parameterized capability. The most common way to design a parameterized module is through using generate-loop statements, as shown in the following example.

EXAMPLE 14.2 *A 4-bit CLA Adder Using Generate-Loop Statements*

In this example, we rewrite the operations of each continuous assignment in the preceding example with a generate-loop statement. Through properly setting the parameter n, an arbitrary width CLA adder can be obtained.

```
// an N-bit CLA adder using generate loops
module cla_adder_generate(x, y, cin, sum, cout);
// inputs and outputs
parameter N = 4; // define the default size
input   [N-1:0] x, y;
input   cin;
output [N-1:0] sum;
output cout;
// internal wires
wire    [N-1:0] p, g;
wire    [N:0] c;
// assign input carry
assign c[0] = cin;
genvar i;
// compute carry generate and carry propagate functions
generate for (i = 0; i < N; i = i + 1) begin: pq_cla
   assign p[i] = x[i] ^ y[i];
   assign g[i] = x[i] & y[i];
end endgenerate
// compute carry for each stage
generate for (i = 1; i < N+1; i = i + 1) begin: carry_cla
   assign c[i] = g[i-1] | (p[i-1] & c[i-1]);
end   endgenerate
// compute sum
generate for (i = 0; i < N; i = i + 1) begin: sum_cla
   assign sum[i] = p[i] ^ c[i];
end endgenerate
// assign final carry
assign cout = c[N];
endmodule
```

14.1.2 Parallel-Prefix Adder

Although in theory the CLA generator can be implemented in two-level logic, in practice it cannot work as fast as we expect. The reason is that in a practical gate circuit, the propagation delay of the gate is not independent of the number of its inputs (namely, fanin). As a result, it is preferred to use bounded-fanin rather than unbounded-fanin gates when designing wide-width adders. In this section, we consider another method that is also frequently used to design wide-width adders. This method is known as the *parallel-prefix approach*, which is described in detail in the rest of this section.

The prefix computation is defined as follows. Consider a sequence of n elements $\{x_{n-1}, \cdots, x_1, x_0\}$ with a binary associative operator, denoted by $\odot$. The *prefix sums* of this sequence are the n partial sums defined by:

$$s_{[i,0]} = x_i \odot \cdots \odot x_1 \odot x_0 \tag{14.1}$$

where $0 \leq i < n$.

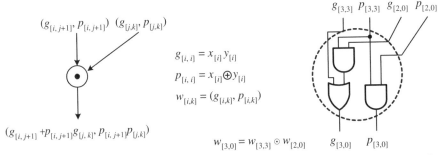

$$(g_{[i,j+1]}, p_{[i,j+1]}) \; (g_{[j,k]}, p_{[j,k]})$$

$$g_{[i,i]} = x_{[i]} y_{[i]}$$

$$p_{[i,i]} = x_{[i]} \oplus y_{[i]}$$

$$w_{[i,k]} = (g_{[i,k]}, p_{[i,k]})$$

$$(g_{[i,j+1]} + p_{[i,j+1]} g_{[j,k]}, \, p_{[i,j+1]} p_{[j,k]})$$

$$w_{[3,0]} = w_{[3,3]} \odot w_{[2,0]} \qquad g_{[3,0]} \quad p_{[3,0]}$$

FIGURE 14.4 The logic diagram of the ith parallel-prefix cell used in n-bit parallel-prefix adders

Let $g_{[i,k]}$ and $p_{[i,k]}$ represent the carry generation and propagation from bits k to i, respectively. Both $g_{[i,k]}$ and $p_{[i,k]}$ can then be defined recursively as follows:

$$g_{[i,k]} = g_{[i,j+1]} + p_{[i,j+1]} \cdot g_{[j,k]}$$

$$p_{[i,k]} = p_{[i,j+1]} \cdot p_{[j,k]} \tag{14.2}$$

where $0 \le i < n$, $k \le j < i$, $0 \le k < n$, $g_{[i,i]} = x_{[i]} \cdot y_{[i]}$ and $p_{[i,i]} = x_{[i]} \oplus y_{[i]}$. Based on this definition, the carry of the ith-bit adder, $c_i = g_{i-1} + p_{i-1} \cdot c_{i-1}$, can be written as $g_{[i,0]} = g_{[i,j+1]} + p_{[i,j+1]} \cdot g_{[j,0]}$.

In order to parallelize the operations, let us group $g_{[i,j]}$ and $p_{[i,j]}$ as a group, denoted as $w_{[i,j]} = (g_{[i,j]}, p_{[i,j]})$. It is easy to show that the operator $\odot$ in the following equation:

$$w_{[i,k]} = w_{[i,j+1]} \odot w_{[j,k]}$$

$$= (g_{[i,j+1]}, p_{[i,j+1]}) \odot (g_{[j,k]}, p_{[j,k]})$$

$$= (g_{[i,j+1]} + p_{[i,j+1]} g_{[j,k]}, \, p_{[i,j+1]} p_{[j,k]}) \tag{14.3}$$

is a binary associative operator, where $k \le j < n$. The logic diagram used to implement the operator $\odot$ is shown in Figure 14.4.

EXAMPLE 14.3 *The CLA Generator as Prefix Sums*

Let $w_{[i,i]} = (g_{[i,i]}, p_{[i,i]})$, for all $0 \le i \le 3$. By the definition of the binary operator $\odot$, the $w_{[i,0]}$, where $0 \le i \le 3$, can be calculated in a variety of ways. Some of them are shown in the following:

$$w_{[1,0]} = w_{[1,1]} \odot w_{[0,0]} = (g_{[1,1]}, p_{[1,1]}) \odot (g_{[0,0]}, p_{[0,0]})$$

$$= (g_{[1,1]} + p_{[1,1]} g_{[0,0]}, \, p_{[1,1]} p_{[0,0]})$$

$$w_{[3,2]} = w_{[3,3]} \odot w_{[2,2]} = (g_{[3,3]}, p_{[3,3]}) \odot (g_{[2,2]}, p_{[2,2]})$$

$$= (g_{[3,3]} + p_{[3,3]} g_{[2,2]}, \, p_{[3,3]} p_{[2,2]})$$

$$w_{[2,0]} = w_{[2,2]} \odot w_{[1,0]}$$

$$= (g_{[2,2]}, p_{[2,2]}) \odot (g_{[1,1]} + p_{[1,1]} g_{[0,0]}, \, p_{[1,1]} p_{[0,0]})$$

$$= (g_{[2,2]} + p_{[2,2]} g_{[1,1]} + p_{[2,2]} p_{[1,1]} g_{[0,0]}, \, p_{[2,2]} p_{[1,1]} p_{[0,0]})$$

$$w_{[3,0]} = w_{[3,2]} \odot w_{[1,0]}$$

$$= (g_{[3,3]} + p_{[3,3]}g_{[2,2]}, p_{[3,3]}p_{[2,2]}) \odot (g_{[1,1]} + p_{[1,1]}g_{[0,0]}, p_{[1,1]}p_{[0,0]})$$

$$= (g_{[3,3]} + p_{[3,3]}g_{[2,2]} + p_{[3,3]}p_{[2,2]}g_{[1,1]} + p_{[3,3]}p_{[2,2]}p_{[1,1]}g_{[0,0]}, p_{[3,3]}p_{[2,2]}p_{[1,1]}$$

$$p_{[0,0]})$$

As we have mentioned above, $g_{[i,0]}$ corresponds to the carry of the ith-bit adder c_i. The reader may compare the above result to the carries derived from the recursive equation c_{i+1}. It is worth noting that here $g_{[0,0]}$ corresponds to the carry input c_{in} and the indices used are always 1 greater than those used in the recursive equation. That is, $g[i, i] = g_{i-1}$ and $p[i, i] = p_{i-1}$, where $0 < i < n$. ∎

There are many possible parallel-prefix networks on which fast adders are based. In the following, we introduce the two most common networks: the *Kogge–Stone network* and the *Brent–Kung network*.

14.1.2.1 Kogge–Stone Adder
The Kogge–Stone parallel-prefix network is shown in Figure 14.5. An n-input Kogge–Stone parallel-prefix network has a propagation delay of $log_2 n$ levels and a cost of $n log_2 n - n + 1$ cells. For instance, there are 4-level propagation delays and 49 cells in a 16-input Kogge–Stone parallel-prefix network, which can be easily justified from Figure 14.5.

In the following, we give an example showing how the Kogge–Stone parallel-prefix network is used to construct an n-bit fast adder.

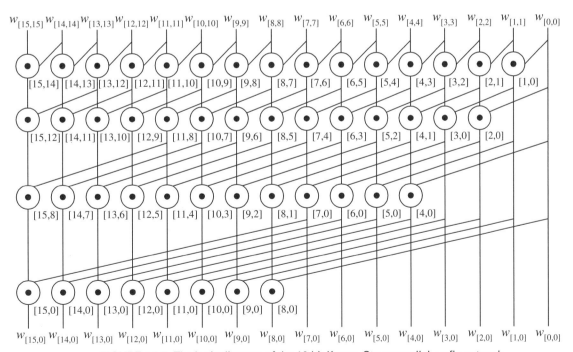

FIGURE 14.5 The logic diagram of the 16-bit Kogge–Stone parallel-prefix network

EXAMPLE 14.4 *A Fast Adder Based on the Kogge–Stone Parallel-Prefix Network*

Based on the observation from Figure 14.5, it is easy to know that there are $m = log_2 n$ levels for an n-input network. Level 0 is the input and level m is the output. Level j starts from the cell with an index of 2^{j-1} and each cell within it combines with the 2^{j-1}th previous cell's output. After completing the last-stage computing, the sum can simply be derived by using the following two equations:

$$sum[0] = p[0][0] \oplus cin,$$

$$sum[i] = g[i-1][m] \oplus p[i][0], \quad \text{for } 0 < i < n$$

```
// an N-bit Kogge-Stone adder using generate loops
module Kogge_Stone_adder(x, y, cin, sum, cout);
// inputs and outputs
parameter  N = 16; // define the default size
parameter  M = 4;  // M = log2 N
input   [N-1:0] x, y;
input   cin;
output [N-1:0] sum;
output cout;
// internal wires
wire    p[N-1:0][M:0];
wire    g[N-1:0][M:0];
// assign input carry
genvar i, j;
// compute carry generate and carry propagate functions
generate for (i = 0; i < N; i = i + 1) begin:
  pg_KoggeStone
    assign p[i][0] = x[i] ^ y[i];
    assign g[i][0] = x[i] & y[i];
end endgenerate
// compute carry for each stage using parallel prefix
generate for (j = 1; j <= M; j = j + 1) begin:
  carry_prefix
    for (i = 0 ; i < N; i = i + 1) begin: parrallel_prefix
      if (i < 2**(j-1)) begin
        assign p[i][j] = p[i][j-1];
        assign g[i][j] = g[i][j-1];
      end
      else begin
        assign p[i][j] = p[i][j-1] & p[i-2**(j-1)][j-1];
        assign g[i][j] = g[i][j-1] |
                         (p[i][j-1] & g[i-2**(j-1)][j-1]);
      end
    end
end  endgenerate
// compute sum and the final-stage carry
```

```
generate for (i = 1; i < N; i = i + 1) begin:
  sum_KoggeStone
    assign sum[i] = g[i-1][M] ^ p[i][0];
end endgenerate
assign sum[0] = p[0][0] ^ cin;
assign cout = g[N-1][M];
endmodule
```

14.1.2.2 Brent–Kung Adder

The Brent–Kung parallel-prefix network is shown in Figure 14.6. An n-input Brent–Kung parallel-prefix network has a propagation delay of $2log_2n - 2$ levels and a cost of $2n - 2 - log_2n$ cells. For instance, there are 6-level propagation delays and 26 cells in a 16-input Brent–Kung parallel-prefix network, which can be easily justified from Figure 14.6.

In the following, we give an example showing how the Brent–Kung parallel-prefix network is used to construct an n-bit fast adder.

EXAMPLE 14.5 *A Fast Adder Based on the Brent–Kung Parallel-Prefix Network*

Although from Figure 14.6 there are only $2m - 2 = 2log_2n - 2$ levels for an n-input network, we use $2m - 1$ levels, where the m and $m - 1$ levels are used for forward- and

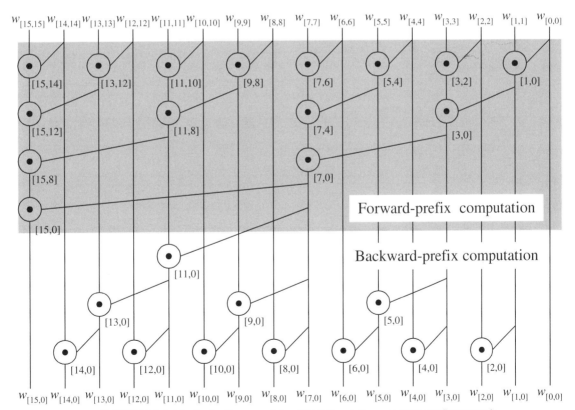

FIGURE 14.6 The logic diagram of the 16-bit Brent–Kung parallel-prefix network

backward-prefix computations, respectively. Level 0 is the input and level m is the output. In forward-prefix computation, the cells with index $2^j - 1$ at level j are combined with their 2^{j-1}th previous cell's output; otherwise, they keep the values from their previous level. In the backward-prefix computation, the cells with indices that have $m - j$ lower bits equal to $2^{m-j-1} - 1$ at level j are combined with their $i - 2^{m-j-1}$th previous cell if the value of $i - 2^{m-j-1}$ is greater than 0; otherwise, they keep the values from their previous level.

```verilog
// an N-bit Brent-Kung adder using generate loops
module Brent_Kung_adder(x, y, cin, sum, cout);
// inputs and outputs
parameter  N = 16; // define the default size
parameter  M = 4;  // M = log2 N
input   [N-1:0] x, y;
input   cin;
output [N-1:0] sum;
output cout;
// internal wires
wire    p[N-1:0][2*M-1:0];
wire    g[N-1:0][2*M-1:0];
// assign input carry
genvar i, j;
// compute carry generate and carry propagate functions
generate for (i = 0; i < N; i = i + 1) begin:
  pg_BrentKung
    assign p[i][0] = x[i] ^ y[i];
    assign g[i][0] = x[i] & y[i];
end endgenerate
// compute carry for each stage using parallel prefix
generate for (j = 1; j <= M; j = j + 1) begin:
  forward_prefix
    for (i = 0; i < N; i = i + 1) begin: prefix_a
      if ((i & {j{1'b1}}) != (2**j - 1)) begin
        assign p[i][j] = p[i][j-1]; // copy all p and g
                                    // values
        assign g[i][j] = g[i][j-1];
      end
      else begin
        assign p[i][j] = p[i][j-1] & p[i-2**(j-1)][j-1];
        assign g[i][j] = g[i][j-1] |
                         (p[i][j-1] & g[i-2**(j-1)][j-1]);
      end
    end
end endgenerate
generate for (j = 1; j < M; j = j + 1) begin:
  backward_prefix
```

```
    for (i = N - 1; i >= 0; i = i - 1) begin: prefix_b
        if (((i&{(M-j){1'b1}})==(2**(M-j-1)-1))&&
            (i-2** (M-j-1))>0) begin
            assign p[i][M+j] = p[i][M+j-1] &
                        p[i-2**(M-j-1)] [M+j-1];
            assign g[i][M+j] = g[i][M+j-1] |
                        (p[i][M+j-1] & g[i-2**(M-j-1)][M+j-1]);
        end
        else begin
            assign p[i][M+j]=p[i][M+j-1]; // copy all p and
                                          // g values
            assign g[i][M+j]=g[i][M+j-1];
        end
    end
end endgenerate
// compute sum and the final-stage carry
generate for (i = 1; i < N; i = i + 1) begin:
  sum_BrentKung
    assign sum[i] = g[i-1][2*M-1] ^ p[i][0];
end endgenerate
assign sum[0] = p[0][0] ^ cin;
assign cout = g[N-1][2*M-1];
endmodule
```

Review Questions

Q14.1 Explain the meanings of carry generate and carry propagate.

Q14.2 What are the propagation delay and cost of an n-bit CLA adder?

Q14.3 What are the propagation delay and cost of an n-bit Kogge–Stone adder?

Q14.4 What are the propagation delay and cost of an n-bit Brent–Kung adder?

14.2 MULTIPLICATION

The basic operations of the multiplication algorithms considered in this chapter are based on the shift-and-add (subtract) technique and may process unsigned and signed inputs. A multiplication algorithm may usually be realized by using either a multiple-cycle or a single-cycle structure. The single-cycle structure evolves into array structures such as an unsigned array multiplier and the modified Baugh–Wooley signed array multiplier. These two array multipliers will be discussed in detail in this section.

14.2.1 Unsigned Multiplication

Basically, the operation of multiplication in hardware is exactly like the 'pencil-and-paper' method that we use daily. This method is often called the *shift-and-add approach*.

In this section, we first present the approach and then derive an array multiplier based on it. In addition, two variations, including an (n/k)-cycle structure and a pipelined structure, of the basic array multiplier are addressed in detail.

14.2.1.1 *Shift-and-Add Multiplication*

The basic operation of shift-and-add multiplication is based on the following observations. A multiple-bit multiplicand times a 1-bit multiplier only has the following two rules:

1. the partial product is the same as the multiplicand if the multiplier is 1;

2. otherwise, the partial product is 0.

This case can be easily extended into the situation where the multiplier is also a multiple-bit number. If this is the case, the above two rules are applied to each individual bit of the multiplier and then sum up all partial products. The result is the final product. The detailed operations are summarized as the following algorithm.

Algorithm: Shift-and-added multiplication

Input: An m-bit multiplicand and an n-bit multiplier.
Output: The $(m + n)$-bit product.
Begin

    1. Load multiplicand and multiplier into registers M and Q, respectively; clear register A and set loop count CNT equal to n.

    2. repeat

        2.1 if $(Q[0] == 1)$ **then** $A = A + M$;

        2.2 Right shift register pair $A : Q$ one bit;

        2.3 $CNT = CNT - 1$;

    until $(CNT == 0)$;

End

According to the algorithm, it is easy to derive a multiple-cycle hardware structure, i.e. a sequential implementation, as shown in Figure 14.7. The detailed ASM chart and Verilog HDL module are left to the reader as exercises. Another consequence of the above algorithm immediately follows. If the multiplier is in the form of a bit-serial

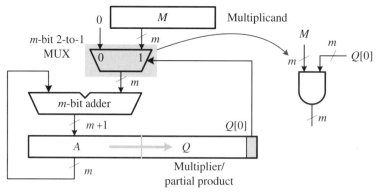

FIGURE 14.7 The sequential implementation of the shift-and-add multiplication

$$
\begin{array}{cccccccc}
 & & & x_3 & x_2 & x_1 & x_0 & = X \text{ (multiplicand)} \\
 & & \times & y_3 & y_2 & y_1 & y_0 & = Y \text{ (multiplier)} \\
\hline
 & & & x_3y_0 & x_2y_0 & x_1y_0 & x_0y_0 & \\
 & & x_3y_1 & x_2y_1 & x_1y_1 & x_0y_1 & & \\
 & x_3y_2 & x_2y_2 & x_1y_2 & x_0y_2 & & & \\
+ & x_3y_3 & x_2y_3 & x_1y_3 & x_0y_3 & & & \\
\hline
P_6 & P_5 & P_4 & P_3 & P_2 & P_1 & P_0 & \\
\end{array}
$$

} Partial product

Product

FIGURE 14.8 An example illustrating the idea of an array multiplier

input, then the above algorithm can be easily adapted to perform such a bit-serial multiplication. The details of this circuit are also left to the reader as an exercise.

14.2.1.2 Array Multipliers

As we have introduced, a multiple-cycle structure can also be implemented by using an iterative logic structure. In order to illustrate this idea, let us consider the 4×4 example shown in Figure 14.8, which can be taken as the case that we expand the shift-and-add multiplication in the temporal dimension and arrange each partial product to the proper position without doing any summation. Each partial product is simply generated by ANDing the appropriate bit of the multiplier with the multiplicand. For an $m \times n$ multiplication, this can be done with $m \times n$ two-input AND gates.

Once all partial products are generated, the next step is to sum up them together in order to obtain the final product. Depending on how to sum up the partial products, a lot of algorithms have been reported in the last decades. In the rest of this section, we introduce two basic approaches, which are based on the straightforward method and use full adders but with a slightly different concept. The first uses ripple-carry adders (RCAs, also called carry-ripple adders) and the second uses carry-save adders (CSAs). Both RCAs and CLA adders are often called carry-propagate adders (CPAs) because the carry-out from the lower-order bit is added to its higher-order bit in these adders.

An example of a 4×4 array multiplier using ripple-carry adders is shown in Figure 14.9. In this figure, each cell corresponds to an item in Figure 14.8, and consists of an AND gate that forms a partial product and a full adder to add the partial product into the running sum. The arrangement of the cells are according to Figure 14.8 in a one-by-one manner. The following example shows how to code this structure in Verilog HDL using generate-loop statements.

EXAMPLE 14.6 *An $m \times n$ Unsigned Array Multiplier Using Generate-Loop Statements*

Since the multiplication structure is a two-dimensional array, we need a nested generate-loop structure. In order to record internal temporary net values, two two-dimensional arrays `sum` and `carry` are declared for this purpose. Due to its simplicity, each cell is described by a continuous assignment. The boundary cells are required to pay some attention. These boundary cells include the first row, and each LSB cell as well as each MSB cell of the rest of the rows. After carefully taking these cells into consideration, the rest cells are easily described.

```verilog
// an M-by-N unsigned array multiplier using generate
// loops
module unsigned_array_multiplier_generate(x, y, product);
// inputs and outputs
parameter M = 4;
parameter N = 4;
input  [M-1:0] x;
input  [N-1:0] y;
output [M+N-1:0] product;
// internal wires
wire sum   [M-1:0][N-1:0]; // declare internal nets
wire carry [M-1:0][N-1:0];
genvar i, j;
generate for (i = 0; i < N; i = i + 1) begin:
  unsigned_multiplier
    if (i == 0)  // describe the first row
      for (j = 0; j < M; j = j + 1) begin: first_row
        assign sum[j][i] = x[j] & y[i], carry[j][i]=0;
      end
    else         // describe the rest rows
      for (j = 0; j < M; j = j + 1) begin: rest_rows
        if (j == 0)   // LSB of each row
          assign {carry[j][i],sum[j][i]}=sum[j+1][i-1]+
            (x[j]&y[i]);
        else if (j == M - 1) // MSB of each row
          assign {carry[j][i],sum[j][i]}=
                (x[j]&y[i]) + carry[M-1][i-1] +
                carry[j-1][i];
        else   // the rest bits of each row
          assign {carry[j][i],sum[j][i]}=
                (x[j]&y[i])+sum[j+1][i-1]+
                carry[j-1] [i];
      end
end endgenerate
// generate product bits
generate for (i = 0; i < N ; i = i + 1) begin: product_
  lower_part
  assign product[i] = sum[0][i];
end endgenerate
generate for (i = 1; i < M ; i = i + 1) begin: product_
  upper_part
  assign product[N-1+i] = sum[i][N-1];
end endgenerate
  assign product[M+N-1] = carry[M-1][N-1];
endmodule
```

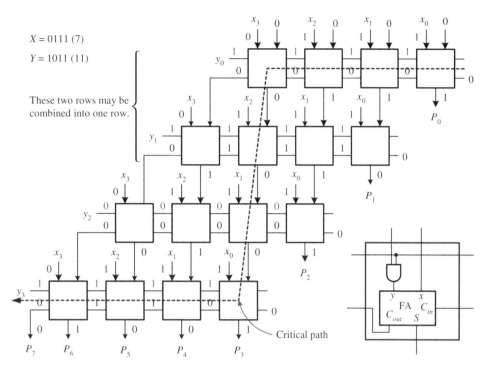

$X = 0111$ (7)

$Y = 1011$ (11)

These two rows may be combined into one row.

FIGURE 14.9 The logic diagram of a 4×4 unsigned array multiplier

One disadvantage of the basic array multiplier is that it has a longer propagation delay. It is easy to see the critical path depicted by the dashed line of the basic array multiplier shown in Figure 14.9. The propagation delay of the critical path for an $m \times n$ array multiplier is $[2(m - 1) + n]t_{FA}$ when the first two rows are not combined and is $[2(m - 1) + (n - 1)]t_{FA}$ when the first two rows are combined, where t_{FA} is the propagation delay of a full adder. The hardware cost of an $m \times n$ array multiplier is as follows. It needs $m \times n$ AND gates and $m \times n$ full adders when the first two rows are not combined and $m \times (n - 1)$ full adders when the first two rows are combined.

Another widely used array multiplier uses carry-save adders (CSAs) to replace the RCAs used in the basic array multiplier. The resulting structure is known as a *CSA array multiplier*. A CSA is simply an n-bit full adder placed in parallel. It is similar to an RCA but leaves the carry-out of each cell to the output rather than input to the next higher significant bit cell. As a result, an n-bit CSA accepts three n-bit inputs and generates two n-bit outputs: carry and sum. In order to combine these two outputs into a result, a CPA has to be applied to the output carry and sum. Figure 14.10 shows a 4×4 CSA unsigned array multiplier. Like the basic array multiplier, it is easy to design a parameterized $m \times n$ CSA array multiplier. However, this is left to the reader as an exercise.

From the CSA array multiplier shown in Figure 14.10, it is easy to figure out the critical path, which is marked by the dashed line shown in this figure. In general, the critical path of an $m \times n$ CSA array multiplier is composed of n or $(n - 1)$ CSA cells, depending on whether the first two rows are combined or not, and an m-bit CPA, namely, $(m + n)$ or $(m + n - 1)$ full adders. That is, the critical path has a propagation delay of

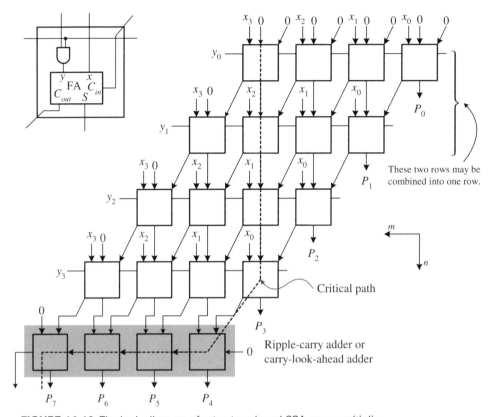

FIGURE 14.10 The logic diagram of a 4×4 unsigned CSA array multiplier

$(m + n)t_{FA}$, when the first two rows are not combined and of $[m + (n - 1)]t_{FA}$, when the first two rows are combined. Compared to the basic array multiplier, the $m \times n$ CSA array multiplier has less propagation delay by a factor of $m - 2$ at the cost of m full adders or an m-bit CLA adder.

Review Questions

Q14.5 Explain the operations of the sequential implementation of the shift-and-add multiplication shown in Figure 14.7.

Q14.6 What are the features of RCAs and CLA adders?

Q14.7 Why are both RCAs and CLA adders called CPAs?

Q14.8 Define an n-bit CSA.

Q14.9 What are the differences between RCAs and CSAs?

14.2.1.3 Variations of Array Multipliers Now that we have considered both multiple-cycle (sequential) and single-cycle (array) structures of the multiplication algorithm based on the shift-and-add approach, in what follows we first explore a compromise approach known as an (n/k)-cycle structure and then describe how to pipeline the single-cycle structure by using the unsigned array multiplier as an example.

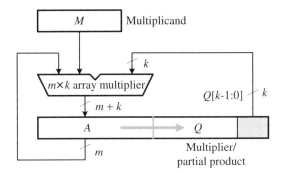

FIGURE 14.11 An *(n/k)*-cycle structure for an *m* × *n* unsigned multiplier

(n/k)-Cycle Multiplication Structure The general (n/k)-cycle structure of $m \times n$ multiplication is shown in Figure 14.11. Compared to the circuit shown in Figure 14.7, an $m \times k$ array multiplier is used instead of the trivial $m \times 1$ array multiplier, which is actually an m-bit adder with the multiplicand input ANDed by the proper bit of multiplier. The operation of Figure 14.11 is similar to that of Figure 14.7 except that now it processes k bits rather than a single bit each time. Hence, the accumulator (A) concatenated with the multiplier register (Q) are shifted right k-bit positions after each time the partial product is loaded into the accumulator. The multiplication is finished after (n/k) steps, with each step needing two clock cycles, one for computation of the $m \times k$ partial product and loading into the accumulator and the other for the k-bit shift-right operation. It is worth noting that the (n/k)-cycle structure is reduced to the multiple structure depicted in Figure 14.7 when k is set to 1 and reduced to the array structure when k is set to n. This denotes a tradeoff between the cost of hardware and performance.

The hardware required for this (n/k)-cycle structure can be calculated as follows. In addition to three registers, an m-bit multiplicand register, an $(m + k)$-bit accumulator and an n-bit multiplier/partial product register, there is an $m \times k$ array multiplier which requires $m \times k$ full adders and AND gates. The total running time of this structure is $2(n/k)t_{clk}$, without considering various housekeeping operations such as clearing the accumulator and loading the multiplicand as well as the multiplier into their individual register, where t_{clk} is determined by the propagation delay of the $m \times k$ array multiplier, as well as the t_q and t_{setup} of the registers used. Hence, it needs $2(n/k)\{[2(m - 1) + k]t_{FA} + t_q + t_{setup}\}$. The constant factor 2 counts both the computation of the $m \times k$ partial product and the shift-right k-bit positions of the multiplier register, as well as the accumulator, as can be seen from Figure 14.11.

Pipelined Multiplication Structure Another variation of the array multiplier structure is the pipelined multiplication structure. The general pipeline structure for an $m \times n$ unsigned multiplier is shown in Figure 14.12. Basically, this pipelined structure can be considered either as the expansion of the (n/k)-cycle multiplier structure in temporal dimension or as the partition of an $m \times n$ unsigned array multiplier into (n/k) stages of $m \times k$ unsigned array multipliers with pipeline registers being inserted after each stage.

As in the (n/k)-cycle structure, the value of k in the pipelined structure may affect both the pipeline clock period, T_i, and the pipeline latency. Hence, its value should be

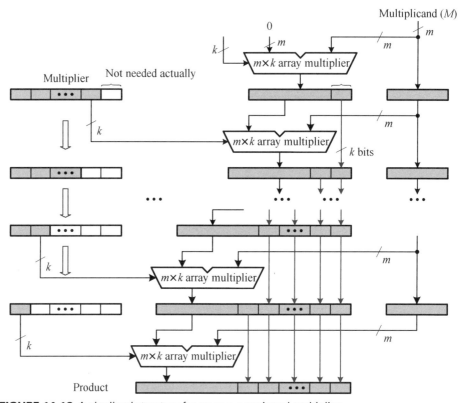

FIGURE 14.12 A pipelined structure for an $m \times n$ unsigned multiplier

chosen very carefully according to the actual requirements. The pipeline clock period is determined by the propagation delay of the $m \times k$ unsigned array multiplier and the t_q and t_{setup} of the registers used. That is, T_i is equal to $[2(m - 1) + k]t_{FA} + t_q + t_{setup}$. The hardware required in the pipelined multiplier structure is calculated as follows. In addition to the $m \times n$ full adders and AND gates. Three types of register are required. These are the $(n/k - 1)$ m-bit multiplicand register, the $0.5(n/k - 1)$ n-bit multiplier register and the $(n/k)(m + k) + 0.5(n/k - 1) \times n$ bits of the partial product register. The total register required is then equal to $(n/k)(m + k) + (n/k - 1)(m + n)$ bits.

Review Questions

Q14.10 Design a shift register with the capability of shifting its content right k-bit positions in one clock cycle, where k is a constant.

Q14.11 Explain the operation of the (n/k)-cycle structure of the $m \times n$ unsigned multiplier shown in Figure 14.11.

Q14.12 Explain the operation of the $m \times n$ pipelined unsigned array multiplier depicted in Figure 14.12.

Q14.13 Show that the total registers required in the pipelined structure of an $m \times n$ unsigned multiplier is $(n/k)(m + k) + (n/k - 1)(m + n)$ bits.

14.2.2 Signed Multiplication

Like unsigned multiplication, signed multiplication is also widely used in digital systems. There are many algorithms that can be used to implement signed two's complement multiplication. In this book, we only consider the two most common approaches: the *Booth algorithm* and the *modified Baugh–Wooley algorithm*. The Booth algorithm has been dealt with in greater detail in Sections 11.2.2 and 11.3.4. Hence, in this subsection, we only focus on the modified Baugh–Wooley algorithm.

In order to illustrate the operation of the modified Baugh-Wooley algorithm, let the multiplicand and multiplier be $X = (x_{m-1}, \cdots, x_1, x_0)$ and $Y = (y_{n-1}, \cdots, y_1, y_0)$, respectively. Recall that the most significant bit of a number has a negative weight in the two's complement representation. Based on this property, the two numbers X and Y can be represented as follows:

$$X = -x_{m-1}2^{m-1} + \sum_{i=0}^{m-2} x_i 2^i,$$

$$Y = -y_{n-1}2^{n-1} + \sum_{j=0}^{n-2} y_j 2^j,$$

and the product of X and Y can then be represented as:

$$\begin{aligned} P &= XY \\ &= \left(-x_{m-1}2^{m-1} + \sum_{i=0}^{m-2} x_i 2^i\right)\left(-y_{n-1}2^{n-1} + \sum_{j=0}^{n-2} y_j 2^j\right) \\ &= \sum_{i=0}^{m-2}\sum_{j=0}^{n-2} x_i y_j 2^{i+j} + x_{m-1}y_{n-1}2^{m+n-2} - \\ &\quad \left(\sum_{i=0}^{m-2} x_i y_{n-1} 2^{i+n-1} + \sum_{j=0}^{n-2} x_{m-1} y_j 2^{j+m-1}\right) \end{aligned} \tag{14.4}$$

A numerical example of a 4×4 modified Baugh–Wooley algorithm is shown in Figure 14.13. The array implementation of it is depicted in Figure 14.14. It is easy to see that two kinds of cell are required in the array. Some cells use NAND gates

					x_3	x_2	x_1	x_0	$= X$ (multiplicand)
				$\times$	y_3	y_2	y_1	y_0	$= Y$ (multiplier)
				1	$\overline{y_3x_0}$	x_2y_0	x_1y_0	x_0y_0	Partial product
				$\overline{y_3x_1}$	x_2y_1	x_1y_1	x_0y_1		Partial product
			$\overline{y_3x_2}$	x_2y_2	x_1y_2	x_0y_2			Partial product
$+$	1	y_3x_3	$\overline{y_2x_3}$	$\overline{y_1x_3}$	$\overline{y_0x_3}$				Partial product
	P_7	P_6	P_5	P_4	P_3	P_2	P_1	P_0	Product

FIGURE 14.13 The idea of a modified Baugh–Wooley array multiplier

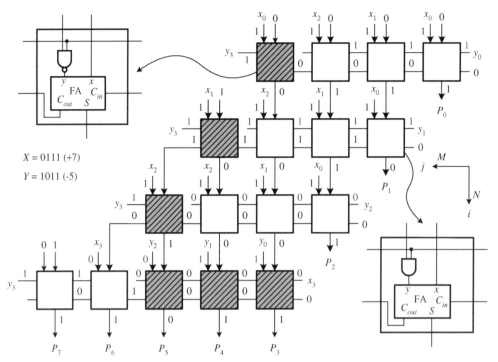

FIGURE 14.14 The block diagram of a 4 × 4 modified Baugh–Wooley array multiplier

instead of AND gates to generate partial products. By comparing both Figures 14.14 and 14.9, we can see that the unsigned and modified Baugh–Wooley arrays are so similar that a single array can be used for both purposes if XOR gates are used to conditionally complement some terms depending on the mode.

In the following, we only give an example to show how to design a parameterized two's complement array multiplier based on the modified Baugh–Wooley algorithm. The problem of using a single array for performing both unsigned and signed multiplication is left to the reader as an exercise.

EXAMPLE 14.7 *An m × n Two's Complement Multiplier Based on the Modified Baugh–Wooley Algorithm*

Like the unsigned array multiplier, the modified Baugh–Wooley array multiplier also needs $m \times n$ cells. As shown in Figure 14.14, the structure of the modified Baugh-Wooley array multiplier is much similar to that of the unsigned array multiplier except that it requires two types of cell. The shaded cells use NAND gate rather than AND gate to produce partial products. By properly coding these cells as special cases, a parameterized module results.

```
// an M-by-N two's complement array multiplier
module modified_Baugh_Wooley(x, y, product);
// inputs and outputs
parameter M = 4;
```

```
parameter N = 4;
input   [M-1:0] x;
input   [N-1:0] y;
output [M+N-1:0] product;
// internal wires
wire sum    [M-1:0][N-1:0]; // declare internal nets
wire carry [M-1:0][N-1:0];
genvar i, j;
generate for (i = 0; i < N; i = i + 1) begin:
  signed_multiplier
   if (i == 0)  // describe the first row
      for (j = 0; j < M; j = j + 1) begin: first_row
         if (j == M - 1)
            assign sum[j][i] = !(x[i] & y[N-1]),
                   carry[j][i] = 1;
         else
            assign sum[j][i] = x[j] & y[i],
                   carry[j][i] =   0; end
   else if (i == N - 1) // describe the last rows
      for (j = 0; j < M; j = j + 1) begin: last_row
         if (j == M - 1)
            assign {carry[j][i],sum[j][i]} =
                   (x[M-1] &y[N-1]) +
                   carry[j][i-1] + carry[j-1][i];
         else if (j == 0)
            assign {carry[j][i],sum[j][i]}=
                   !(x[M-1]&y[j])+sum[j+1][i-1];
         else
           assign {carry[j][i],sum[j][i]} = !(x[M-1]& y[j])
                   +sum[j+1][i-1] + carry[j-1][i]; end
   else                   // describe the rest rows
      for (j = 0; j < M; j = j + 1) begin: rest_rows
         if (j == 0) // LSB of each row
            assign {carry[j][i],sum[j][i]} =
                   (x[j]&y[i])+sum[j+1][i-1];
         else if (j == M - 1) // MSB of each row
            assign {carry[j][i],sum[j][i]} =
                   !(x[i] &y[N-1])+
                   carry[M-1][i-1]+ carry[j-1][i];
         else // the rest bits of each row
            assign {carry[j][i],sum[j][i]}=
                   (x[j]& y[i])+sum[j+1][i-1]+
                   carry[j-1][i];   end
end endgenerate
// generate product bits
```

```
generate for (i = 0; i < N ; i = i + 1)
begin: product_lower_part
   assign product[i] = sum[0][i];
end endgenerate
generate for (i = 1; i < M ; i = i + 1)
begin: product_upper_part
   assign product[N-1+i] = sum[i][N-1];
end endgenerate
   assign product[M+N-1] = carry[M-1][N-1] + 1'b1;
endmodule                                              ∎
```

Unlike both the unsigned array multiplier and the Booth array multiplier, which can accommodate the situation that $m \neq n$, the above discussed modified Baugh–Wooley array multiplier can only work properly when $m = n$. However, the modified Baugh–Wooley algorithm can work well for any m and n if we modify the array shown in Figure 14.14 a little bit. Please examine the example given in Figure 14.13 very carefully and exploit how the inverted partial products are generated, especially, in the case of $m \neq n$. The detailed operations when $m \neq n$, along with its Verilog HDL module, are left as exercises for the reader.

Review Questions

Q14.14 Compare the unsigned array multiplier with the modified Baugh–Wooley array multiplier in terms of the cost and propagation delay.

Q14.15 How many inverted partial product terms are needed in an $m \times m$ modified Baugh–Wooley array multiplier?

Q14.16 How many AND gates are required in an $m \times n$ unsigned array multiplier?

Q14.17 How many full adders are needed in an $m \times n$ unsigned array multiplier?

14.3 DIVISION

The essential operation of multiplication is a sequence of additions. In contrast, the essential operation of division is a sequence of subtractions. Based on this idea, we introduce in this section two basic division algorithms known as the *restoring division algorithm* and *nonrestoring division algorithm*, respectively. For simplicity, we only consider the case of unsigned input numbers. Like multiplication algorithms, a division algorithm may also be realized by using either a multiple-cycle or a single-cycle structure.

14.3.1 Restoring Division Algorithm

It is general to compare the magnitude of the dividend and the divisor when performing division. The quotient bit is set to 1 and the divisor is subtracted from the dividend if the dividend is greater than or equal to the divisor; otherwise, the quotient bit is set to 0 and the next bit is proceeded. In digital systems, the comparison is usually carried out directly by subtraction in order to simplify the circuit design.

Based on the above idea, the basic techniques of division circuit designs can be classified into two types: restoring and nonrestoring methods. In the restoring division method, the quotient bit is set to 1 whenever the result after the divisor is subtracted from the dividend is greater than or equal to 0; otherwise, the quotient bit is set to 0 and the divisor is added back to the dividend so as to restore the original dividend before proceeding the next bit.

The above operations can be summarized as the following algorithm:

Algorithm: Unsigned restoring division

Input: An n-bit dividend and an m-bit divisor.
Output: The quotient and remainder.
Begin
    1. Load divisor and dividend into registers M and D, respectively;
        clear partial-remainder register R and
        set loop count CNT equal to $n - 1$.
    2. Left shift register pair $R : D$ one bit.
    3. repeat
        3.1 Compute $R = R - M$;
        3.2 if $(R < 0)$ **begin** $D[0] = 0$; $R = R + M$; **end**
           else $D[0] = 1$;
        3.3 left shift $R : D$ one bit;
        3.4 $CNT = CNT - 1$;
           until $(CNT == 0)$
End

14.3.2 Nonrestoring Division Algorithm

In the restoring division method, the divisor (M) needs to be added back to the dividend (D) so as to restore the dividend to its value before the subtraction is performed whenever $D - M < 0$. As a result, it requires $3/2n$ m-bit additions and subtractions on average to complete an n-bit division.

In the nonrestoring division method, the divisor is not added back to the dividend when $D - M < 0$. Instead, the result $D - M$ is shifted left one bit (corresponding to shifting M right one bit) and then adding to the divisor M. The result is the same as that the divisor (M) is added back to the dividend (D) and then performing the subtraction. To see this, let $X = D - M$ and assume that $X < 0$. In the restoring division method, the divisor is added back to the X and then the result is shifted left one bit and subtracted by M, namely, $2(X + M) - M = 2X + M$, which is equivalent to shifting the X left one bit and then adding to the divisor M. Consequently, both approaches yield the same result. However, the nonrestoring division method needs only n m-bit additions and subtractions. The detailed operations of the nonrestoring division method are summarized as the following algorithm.

Algorithm: Unsigned nonrestoring division

Input: An n-bit dividend and an m-bit divisor.
Output: The quotient and remainder.
Begin
    1. Load divisor and dividend into registers M and D, respectively;
        clear partial-remainder register R and
        set loop count CNT equal to $n - 1$.
    2. Left shift register pair $R : D$ one bit.
    3. Compute $R = R - M$;
    4. repeat
           4.1 if $(R < 0)$ **begin**
                $D[0] = 0$; left shift $R : D$ one bit; $R = R + M$; **end**
           else begin
                $D[0] = 1$; left shift $R : D$ one bit; $R = R - M$; **end**
           4.2 $CNT = CNT - 1$;
        until $(CNT == 0)$
    5. if $(R < 0)$ **begin** $D[0] = 0$; $R = R + M$; **end else** $D[0] = 1$;
End

A numerical example to illustrate the operation of the nonrestoring division algorithm is shown in Figure 14.15.

Like other arithmetic operations dealt with in this chapter, the nonrestoring division algorithm can also be implemented by using either a multiple-cycle or a single-cycle

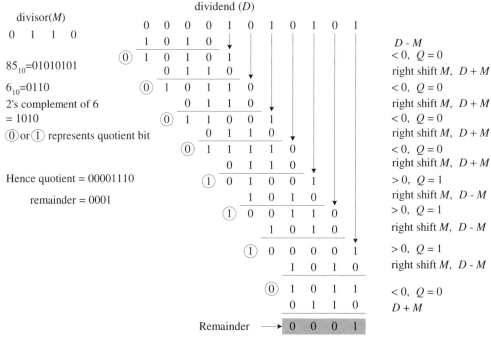

FIGURE 14.15 A numerical example of an unsigned nonrestoring division

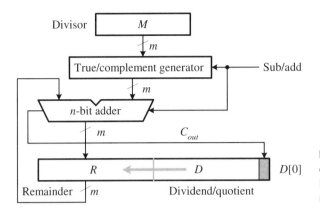

FIGURE 14.16 The logic diagram of a sequential implementation of unsigned nonrestoring division

structure. The basic multiple-cycle structure for the nonrestoring division algorithm is depicted in Figure 14.16. It consists of an m-bit two's complement adder, two m-bit registers, M for storing the divisor and R for the partial remainder, and one n-bit register D for the dividend and quotient register. The partial remainder R and register D also form a $(m + n)$-bit left-shift register. The quotient bit at each round is inserted into $D[0]$. The division operation is completed after n cycles. The detailed ASM chart and its related Verilog HDL module are easily derived from the nonrestoring division algorithm and Figure 14.16. Consequently, they are left to the reader as exercises.

14.3.3 Nonrestoring Array Divider

As mentioned previously, a single-cycle structure can be easily derived by expanding its multiple-cycle structure in temporal dimension. A 4-divided-by-4 example of an unsigned array divider based on this idea is shown in Figure 14.17. The basic limitations of this array divider are as follows. First, it does not check the divide-by-0 case. Second, the MSB of the divisor must be set to 0. Third, the length of the dividend must not be less than that of the divisor.

The following example shows how to describe an unsigned array nonrestoring divider based on the algorithm described above using generate-loop statements.

EXAMPLE 14.8 *An Example of an Array Nonrestoring Divider*

As shown in Figure 14.17, only one boundary condition and the remainder correction require to be specially considered. In the first row, one operand is the divisor and the other operand has all bits set to 0s except the LSB, which is the MSB of the dividend. The carry input must be set to 1 so that subtraction is performed in the first row. Starting from the second row until to the last row, one operand is the divisor and the other operand is the sum from its preceding row except the LSB, which is an appropriate dividend bit. The carry input is the carry output of the MSB of its preceding row. The remainder correction is simply to add 0 or the divisor to the remainder depending on whether the sign bit of the remainder is 0 or 1.

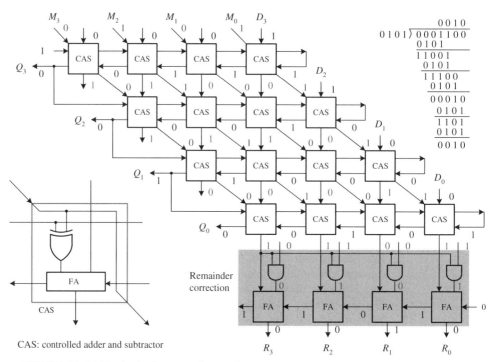

FIGURE 14.17 The logic diagram of an unsigned array nonrestoring divider

```
// an M-by-N unsigned array nonrestoring divider
module unsigned_array_divider(x, y, quotient, remainder);
// inputs and outputs
parameter M = 4;   // default divisor width
parameter N = 4;   // default dividend width
input   [M-1:0] x; // divisor
input   [N-1:0] y; // dividend
output  [N-1:0] quotient;
output  [M-1:0] remainder;
wire    [M-1:0] rem_carry;
// internal wires
wire sum    [M-1:0][N-1:0]; // declare internal nets
wire carry [M-1:0][N-1:0];
genvar i, j;
generate for (i = N - 1; i >= 0; i = i - 1) begin:
  unsigned_divider
    if (i == N - 1)   // describe the first row
      for (j = 0; j < M; j = j + 1) begin: first_row
        if (j == 0)
          assign {carry[j][i], sum[j][i]} =
                  y[i] + !x[j] + 1;
    else
```

```
                    assign {carry[j][i], sum[j][i]} =
                            !x[j] + carry[j-1][i];
        end
    else                    // describe the rest rows
        for (j = 0; j < M; j = j + 1) begin: rest_rows
            if (j == 0) // LSB of each row
                assign {carry[j][i],sum[j][i]} =
                        y[i] + (x[j]^carry[M-1][i+1]) +
                        carry[M-1][i+1];
            else          // the rest bits of each row
                assign {carry[j][i],sum[j][i]} =
                        sum[j-1][i+1] + (x[j]^carry[M-1][i+1])
                        +carry[j-1][i];
        end
end endgenerate
// generate quotient
generate for (i = 0; i < N ; i = i + 1)
begin: product_quotient
    assign quotient[i] = carry[M-1][i];
end endgenerate
// generate and adjust the final remainder
generate for (j = 0; j < M ; j = j + 1)
begin: remainder_adjust
    if (j == 0)
        assign {rem_carry[j],remainder[j]} =
                sum[j][0] + (sum[M-1][0]&x[j]);
    else
        assign {rem_carry[j],remainder[j]} =
                sum[j][0] + (sum[M-1][0]&x[j]) +
                rem_carry[j-1];
end endgenerate
endmodule
```

We only consider the case of unsigned input numbers on which restoring and nonrestoring division algorithms are applied. The nonrestoring division algorithm is also often applied to signed input numbers with only a minor modification needed. The interested reader can refer to Parhami [5] for details.

Review Questions

Q14.18   Describe the basic operation of a restoring division algorithm.

Q14.19   Explain the rationale behind the nonrestoring division algorithm.

Q14.20   Describe the basic operation of a nonrestoring division algorithm.

Q14.21   In what case does the remainder need to be corrected in the nonrestoring division algorithm?

14.4 ARITHMETIC AND LOGIC UNIT

An arithmetic and logic unit (ALU) is often the major component for the datapath of many applications, especially for central processing unit (CPU) design. An ALU contains two portions: arithmetic unit and logical unit. The arithmetic unit sometimes also includes multiplication, even division, in addition to the two basic operations, addition and subtraction; the logical unit simply comprises the three basic Boolean operators, AND, OR and NOT. In some more general applications, the shift operation is often also an important function of their ALU. Therefore, we first consider the shift operation in what follows.

14.4.1 Shift Operations

The shift operation is to shift an input number a specified number of bit positions with zeros or a sign bit filled in the vacancies. Depending on whether the vacancy is filled with a 0 or the sign bit, the shift operation can be cast into *logical shift* and *arithmetic shift.*

Logical shift   In logical shift operation, the vacancy bits are filled with 0s. Depending on the shift direction, it can be divided into the following two types:

- *Logical left shift.* The input is shifted left a specified number of bit positions and all vacancy bits are filled with 0s.
- *Logical right shift.* The input is shifted right a specified number of bit positions and all vacancy bits are filled with 0s.

Arithmetic shift   The basic features of an arithmetic shift are that its shifted result is the input number divided by 2 when the operation is right shift and is the input number multiplied by 2 when the operation is left shift. In order to maintain these features, the vacancy bits of the shifted result are filled with 0s when the operation is left shift and filled with sign bits when the operation is right shift. More formally, arithmetic shift can be divided into the following two types:

- *Arithmetic left shift.* The input is shifted left a specified number of bit positions and all vacancy bits are filled with 0s. Indeed, this is exactly the same as logical left shift.
- *Arithmetic right shift.* The input is shifted right a specified number of bit positions and all vacancy bits are filled with the sign bit.

The device used to perform the shift operation described above is known as a *shifter.* In general, an n-bit shifter is a device that can shift its input number at most n bits. Like other arithmetic operations, an n-bit shifter can also be implemented with either a multiple-cycle or a single-cycle structure.

The basic multiple-cycle structure is a universal shift register, such as the one described in Section 9.3.2. It loads the number to be shifted and then performs the desired number of shift operations at the cost of equal number of cycles. Detailed operations of the universal shift register can be found in the related section again.

Like other arithmetic operations discussed in this book, the multiple-cycle structure is essentially a sequential logic. On the contrary, the single-cycle structure is simply a combinational logic. In the following, we introduce a combinational logic for implementing a shifter with an arbitrary number of shifts. This device is known as a *barrel shifter*.

A barrel shifter is a device that consists of an input $I = \langle I_{n-1}, I_{n-2}, \cdots, I_1, I_0 \rangle$, an output $O = \langle O_{n-1}, O_{n-2}, \cdots, O_1, O_0 \rangle$ and the number of shifts $S = \langle S_{m-1}, S_{m-2}, \cdots, S_1, S_0 \rangle$. In logical left shift, the relationship between the output and the input can be described as follows:

$$O_i = \begin{cases} 0 & \text{for all } i < s \\ I_{i-s} & \text{for all } i \geq s \end{cases} \tag{14.5}$$

where $i = 0, 1, \cdots, n - 1$ and $s = 0, 1, \cdots, m - 1$.

An example of an 8-bit barrel logical left shifter implemented by three-stage multiplexer columns, with each stage containing eight 2-to-1 multiplexers, is shown in Figure 14.18. Each stage of the multiplexers is controlled by $s[i]$ and shifts the input a

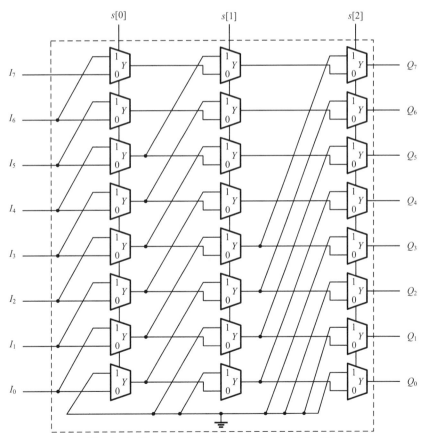

FIGURE 14.18 The logic diagram of an 8-bit barrel logical left shifter

number of $s[i] \times 2^i$ bit positions. In other words, each stage shifts its input the number of 0 or 2^i bit positions depending on whether the value of $s[i]$ is 0 or 1.

In the following, we show how to describe a barrel logical left shifter using generate-loop statements.

EXAMPLE 14.9   *A Parameterized Barrel Logical Left Shifter*

In a barrel shifter, each stage shifts its input the number of 0 or 2^i bit positions depending on whether the value of $s[i]$ is 0 or 1. The first stage accepts the data input and the last stage outputs the shifted result to data output.

```
// a barrel logical left shifter
module barrel_shifter_left_logical(din, amount, dout);
// inputs and outputs
parameter N = 8;  // default data width
parameter S = 3;  // S = log2 N
input  [N-1:0] din;    // data input
output [N-1:0] dout;   // data output
input  [S-1:0] amount; // shift amount
// internal wires
wire   [N-1:0] temp_data [S-1:0];
genvar i, j;
generate for (i = 0; i < S ; i = i + 1)
begin: barrel_shifter
 if (i == 0)    // the first column
   for (j = 0; j < N; j = j + 1) begin: first_column
     if ( j < 2**i)
        assign temp_data[i][j] = (amount[i]) ? 1'b0 : din[j];
      else
        assign temp_data[i][j] =
              (amount[i]) ? din[j-2**i]: din[j];
    end
 else
   for (j = 0; j < N; j = j + 1) begin: rest_columns
     if ( j < 2**i)
        assign temp_data[i][j] =
              (amount[i]) ? 1'b0 : temp_data[i-1][j];
      else
        assign temp_data[i][j] = (amount[i]) ?
              temp_data[i-1][j-2**i] : temp_data[i-1][j];
    end
end endgenerate
// generate data output
assign dout = temp_data[S-1];
endmodule
```

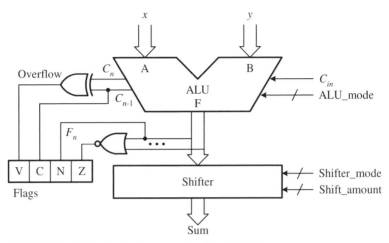

FIGURE 14.19 The logic diagram of a simple ALU

14.4.2 ALUs

An ALU comprises two major function groups: arithmetic and logic. Sometimes, a shifter is also incorporated into an ALU in order to facilitate both logical and arithmetic shift operations. In this section, we address two simple ALUs. One is constructed in a straightforward manner and the other uses the logic function of a full-adder to implement the required logical operations.

A Simple ALU Example A simple ALU example is shown in Figure 14.19, which is composed of a flag register, an ALU and a shifter. The shifter is a barrel shifter which can perform logical and arithmetic shifts. Both shifts can be in both directions: left and right. These options are selected by the Shifter_mode and the shift amount is specified by the Shift_amount. The ALU can carry out both arithmetic and logical operations selected by the ALU_mode.

The flag register includes *negative* (N), *zero* (Z), *overflow* (V) and *carry* (C). It indicates the status of the ALU after an arithmetic or logical operation is carried out. In some applications, if the status of the shifted result needs to be indicated, both negative (N) and zero (Z) circuits can be applied to the output of the shifter instead of the out of the ALU.

The flags N, Z, V and C are defined as follows:

- The *negative* (N) flag is set to 1 when the result of the ALU is negative; namely, the MSB bit is 1. Otherwise, it is cleared.
- The *zero* (Z) flag is set to 1 when the result of the ALU is zero; namely, all bits are zero. Otherwise, it is cleared.
- The *overflow* (V) flag is set when the sign bit of the result is changed; otherwise, it is cleared. In other words, the overflow occurs whenever the number of carry input to the sign bit does not balance the number of carry output from the sign bit.

- The *carry* (*C*) flag is set to 1 when the carry-out from the MSB of the result of the ALU is 1. Otherwise, it is cleared.

The logical operations can be easily implemented by using the three basic gates: AND, OR and NOT. Due to its intuitive simplicity, we will not further consider their details here.

A Full-Adder-Based ALU   Another ALU example relies on the function of a full-adder to implement three basic logical operations. To demonstrate this, recall that the sum and carry-out switching functions of a full adder can be represented as follows:

$$S = (x \oplus y) \oplus C_{in}$$

$$C_{out} = x \cdot y + (x \oplus y) \cdot C_{in} \tag{14.6}$$

where *S* implies both XOR and XNOR operations and C_{out} implies OR and AND operations. Consequently, all three logical operations can be obtained through appropriately controlling the value of input carry C_{in}:

- If $C_{in} = 0$;

$$S = (x \oplus y) \qquad \text{(XOR)}$$

$$C_{out} = x \cdot y \qquad \text{(AND)} \tag{14.7}$$

- If $C_{in} = 1$;

$$S = (x \odot y) \qquad \text{(XNOR)}$$

$$C_{out} = x + y \qquad \text{(OR)} \tag{14.8}$$

As a result, by properly controlling the value of C_{in} and routing *S* and C_{out} to the output, the desired logical operations can be easily obtained. A complete logic diagram of a 4-bit full-adder-based ALU is depicted in Figure 14.20(a). Figures 14.20(b) and (c) give its function table and logic symbol, respectively. The reader is encouraged to verify the function table shown in Figure 14.20(b) by checking the details of the logic circuit depicted in Figure 14.20(a).

The following example uses a generate-loop statement to describe the full-adder-based ALU.

EXAMPLE 14.10   *An n-bit Full-Adder-Based ALU Using a Generate-Loop Statement*

Each bit of the ALU is composed of a 3-to-1 multiplexer, an XOR gate, a 2-to-1 multiplexer and a full adder. These components are all combinational logic. Thus, they are easily described by using continuous assignments. The resulting module is as follows.

```
// an N-bit ALU using generate loops
module ALU_generate(x, y, s1, s0, m1, m0, cout, sum);
// inputs and outputs
parameter N = 4; //define the default size
```

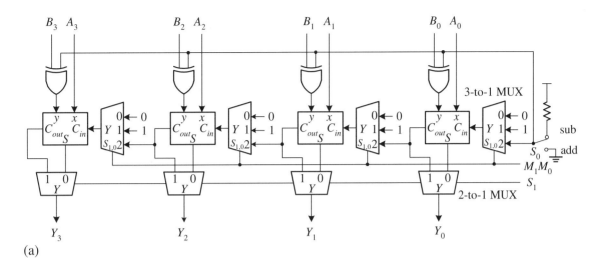

(a)

(b)

M_1	M_0	S_1	S_0	Function
0	0	0	0	$Y_i = A_i \oplus B_i$ (XOR)
0	0	1	0	$Y_i = A_i B_i$ (AND)
0	1	0	0	$Y_i = A_i \odot B_i$ (XNOR)
0	1	1	0	$Y_i = A_i + B_i$ (OR)
1	ϕ	0	0	$Y = A + B$ (addition)
1	ϕ	0	1	$Y = A - B$ (substrction)

(c)

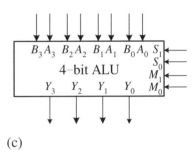

FIGURE 14.20 The logic block diagram (a) along with its function table (b) and logic symbol (c) of a full-adder-based ALU

```
input  [N-1:0] x, y;
input  s1, s0, m1, m0;  // mode selection
output cout;
output [N-1:0] sum;
// internal wires
wire   [N-1:0] cin, c, s;
// using a generate block to generate the entire ALU
genvar i;
generate for (i = 0; i < N; i = i + 1) begin: ALU
   if (i == 0) // LSB
      assign cin[i] = (~m1 & ~m0 & 1'b0) |
                      (~m1 & m0 & 1'b1) |
                      ( m1 & ~m0 & s0);
   else        // the rest bits
      assign cin[i] = (~m1 & ~m0 & 1'b0) |
                      (~m1 & m0 & 1'b1) |
                      ( m1 & ~m0 & c[i-1]);
```

```
        assign {c[i], s[i]} = x[i] + (y[i] ^ s0) + cin[i];
        assign sum[i] =  (~s1 & s[i]) | (s1 & c[i]);
end endgenerate
assign cout = c[N-1];  // assign final carry
endmodule
```

The function of practical ALUs are often much more complicated than what we have dealt with in this section. However, their operations are quite similar to the ones introduced in this section.

Review Questions

Q14.22   Describe the functions of a typical ALU.

Q14.23   Describe the means by which logical operations can be performed in the full-adder-based ALU.

Q14.24   What is the operation of barrel shifters?

Q14.25   What is the function of the flag register associated with an ALU?

Q14.26   How would you detect whether the overflow occurs in an ALU operation?

Q14.27   Show that $C_{out} = x \cdot y + (x \oplus y) = x + y$.

14.5 DIGITAL-SIGNAL PROCESSING MODULES

With the mature development of digital-signal processing (DSP) techniques and algorithms over the last decades, DSP has become an ubiquitous technique in designing modern digital systems, including image processing, instrumentation/control, speech/audio, military, telecommunications, biomedical applications, and so on. Hence, we introduce in this section the most basic DSP techniques which are widely used in many practical systems. These basic techniques include finite-impulse response (FIR) and infinite-impulse response (IIR) filters.

14.5.1 Finite-Impulse Response Filters

In an FIR filter, the output is a weighted sum of present and past samples of its input. The output of an FIR can generally be described by the following equation:

$$y(n) = \sum_{k=0}^{N-1} h(k)x(n - k) \tag{14.9}$$

where $h(k)$ represent the tap (weighting) coefficients and $x(n - k)$ are input samples. Tap coefficients determine the features of an FIR. The details of how to determine the tap coefficients $h(k)$ from a given specification are beyond the scope of this present book. However, the interested reader can refer to related DSP textbooks, such as Oppenheim and Schafer [4] and Schilling and Harris [7].

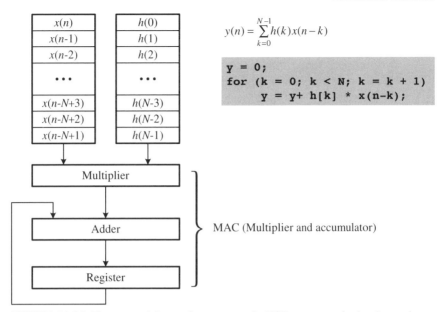

FIGURE 14.21 The general datapath structure of a DSP processor for implementing an *N*th-order FIR filter

The general datapath structure for implementing FIR filters is shown in Figure 14.21. It comprises two memories, tap coefficient memory and input sample memory, a multiplier, an adder and a register as well. It is easy to derive an ASM chart of the controller for this datapath and use Verilog HDL to model it. However, these are left to the reader as exercises.

The single-cycle implementation of the above general datapath structure of an FIR filter is shown in Figure 14.22. Here, three basic components, register, multiplier and adder, are repeated in use at each stage. Due to its shape, the structure is often called a *transversal filter*. The multiplier and adder associated with each tap is often combined into one basic unit known as a *multiplier and accumulator* (MAC), which is usually constructed from a special hardware design technique in order to speed up its operation.

In what follows, we give a simple example of a constant-coefficient FIR filter to illustrate how to describe an FIR filter in practice.

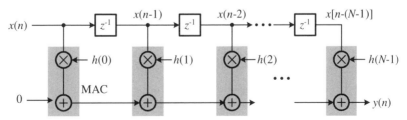

FIGURE 14.22 The general structure of an *N*th-order FIR transversal filter

EXAMPLE 14.11 *An Nth-Order W-bit FIR Filter with Constant Coefficients*

In this example, we assume that tap coefficients are constant, which is often the case in practice. Hence, they are defined by using a `localparam` statement. The evaluation of the FIR equation is then performed by a simple continuous assignment. Besides, it is necessary to sample the new inputs at each clock and shift the old samples one step forward. These two operations are completed by using an `always` block. Note that the input to register `x[1]` is `data_in`.

```
// an Nth order W-bit generic FIR filter with constant
// coefficients
module FIR (clock, reset, data_in, data_out);
parameter N = 4;  // define order
parameter W = 8;  // define word size
input clock, reset;
input  [W-1:0] data_in;
output [N-1:0] data_out;
reg    [W-1:0] x[N:1]; // samples
// define coefficients for the FIR filter
localparam h0 = 8'd51, h1 = 8'd23, h2 = 8'd58,
           h3 = 8'd26, h4 = 8'd34;
integer i;
assign
   data_out = h0 * data_in   + h1 * x[1] +
              h2 * x[2] + h3 * x[3] + h4 * x[4];
// update input samples
always @(posedge clock or posedge reset)
   if (reset) for (i = 1; i <= N; i = i + 1)
     x[i] <= 0;
   else begin
     x[1] <= data_in;
     for (i = 2; i <= N; i = i + 1) x[i] <= x[i-1];
   end
endmodule
```
∎

It is worth noting that the for-loop in the `else` part of the above module cannot be replaced with a single assignment: `x[N:2] <= x[N-1:1]`. Why? Try to explain this.

14.5.2 Infinite-Impulse Response Filters

Another widely used digital filter structure is the IIR filter. In IIR filters, the output at a given time is a function of their inputs and their previously computed outputs. The output of an IIR filter can be represented as a weighted sum of inputs according to the

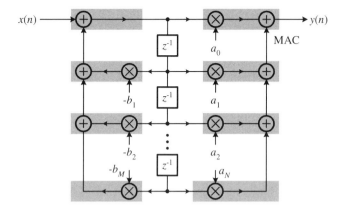

FIGURE 14.23 The general structure of the direct form realization of an IIR filter

recursive difference equation shown in the following:

$$y(n) = \sum_{k=0}^{N-1} a(k)x(n-k) - \sum_{k=0}^{M} b(k)y(n-k) \qquad (14.10)$$

For simplicity, we assume that $N = M$. A direct form realization of an IIR filter is shown in Figure 14.23. Like the traversal FIR filter, an Nth-order IIR filter in direct form is also composed of three basic components: register, multiplier and adder. The multiplier and adder associated with each tap is also often combined into a single unit MAC. Consequently, regardless of whether the structure is for FIR or IIR, only two basic components, register and MAC, are required.

In the following, we give a simple example of a constant-coefficient IIR filter to illustrate how to describe an IIR filter in practice.

EXAMPLE 14.12 *An Nth-Order W-bit IIR Filter with Constant Coefficients*

Like the preceding example, the tap coefficients are assumed to be constant, which is often the case in practice. Hence, they are defined by using two separate `localparam` statements. The evaluation of the IIR equation is then performed by two simple continuous assignments and uses a third continuous assignment to combine them together. Besides, it is necessary to sample the new inputs at each clock and shift the old samples one step forward. These two operations are completed by using an `always` block. Note that in a direct form IIR filter, the input to register `x[1]` is `data_in - b_out` rather than `data_in` only.

```
// an Nth order generic IIR filter with constant
// coefficients
module IIR (clock, reset, data_in, data_out);
parameter N = 4;   // define order
parameter W = 8;   // define word size
```

```
input clock, reset;
input   [W-1:0] data_in;
output  [W-1:0] data_out;
wire    [W-1:0] a_out, b_out;
reg     [W-1:0] x[N:1]; // samples
// define coefficients for the FIR filter
localparam a0 = 8'd51, a1 = 8'd78, a2 = 8'd12, a3 = 8'd84,
           a4 = 8'd25;
localparam b1 = 8'd23, b2 = 8'd58, b3 = 8'd26, b4 = 8'd45;
integer i;
// compute the data_out from a_out and b_out
assign a_out = a0 * data_in    + a1 * x[1] +
               a2 * x[2] + a3 * x[3] + a4 * x[4];
assign b_out = b1 * x[1] + b2 * x[2] +
               b3 * x[3] + b4 * x[4];
assign data_out = a_out - b_out;
// update input samples
always @(posedge clock or posedge reset)
   if (reset) for (i = 1; i <= N; i = i + 1)
      x[i] <= 0;
   else begin
      x[1] <= data_in - b_out;
      for (i = 2; i <= N; i = i + 1) x[i] <= x[i-1];
   end
endmodule                                                    ■
```

Review Questions

Q14.28 What are the basic arithmetic operations of an FIR filter?

Q14.29 What are the basic arithmetic operations of an IIR filter?

Q14.30 How many registers are required in an Nth-order FIR traversal filter?

Q14.31 How many multipliers are required in an Nth-order FIR traversal filter?

Q14.32 How many registers are required in an Nth-order direct form IIR filter?

Q14.33 How many multipliers are required in an Nth-order direct form IIR filter?

SUMMARY

In this chapter, we have examined many frequently used arithmetic modules, including addition, multiplication, division, ALU, shift and two digital-signal processing (DSP) filters. Along with these arithmetic operations and their algorithms, we also emphasized the concept again that a hardware algorithm can often be realized by using a multiple-cycle or a single-cycle structure.

The bottleneck of a conventional n-bit ripple-carry adder is on the generation of carry needed in each stage. To overcome this, many schemes have been proposed.

Among these, the carry-look-ahead (CLA) carry generator and its associated adder are first considered. Then two parallel-prefix-based carry generators and their associated adders, the Kogge–Stone adder and the Brent–Kung adder, are discussed.

The multiplication algorithm considered in this chapter is a shift-and-add-based approach, which is essentially the 'pencil-and-paper' algorithm we use daily. The single-cycle structure of this algorithm evolves into an array structure, which includes an unsigned array multiplier and a modified Baugh–Wooley signed array multiplier. These two multipliers have a similar structure and can be combined into a single one with a mode selection to choose whether the operation is unsigned or signed. Moreover, an (n/k)-cycle structure and a pipelined structure of the $m \times n$ unsigned multiplication are explored in great detail.

The essential operation of multiplication is a sequence of additions. In contrast, the essential operation of division is a sequence of subtractions. Based on this idea, two basic division algorithms, known as a restoring division algorithm and a nonrestoring division algorithm are introduced. In reality, these two algorithms are evolved from the 'pencil-and-paper' algorithm we use daily.

An arithmetic and logic unit (ALU) is often the major component for the datapath of many applications, especially for central processing unit (CPU) design. An ALU contains two portions: arithmetic and logical units. The arithmetic unit sometimes also includes multiplication, even division, in addition to the two basic operations: addition and subtraction; the logical unit simply comprises the three basic Boolean operators, AND, OR and NOT. In some more general applications, the shift operation is also included as an important part of their ALU functions.

The shift operation is to shift an input number a specified number of bit positions with zeros or sign bits filled in the vacancies. Depending on whether the vacancy is filled with a 0 or the sign bit, the shift operation can be cast into logical shift and arithmetic shift. Multiple-cycle and single-cycle structures can be employed to implement these shift operations. The former is inherently sequential and based on a universal shift register while the latter is inherently combinational and based on a multiplexer-based network known as a barrel shifter.

With the mature development of digital-signal processing (DSP) techniques and algorithms over the last decades, DSP has become an ubiquitous technique in designing modern digital systems, ranging from image processing, instrumentation/control, speech/audio, military, telecommunications, to biomedical applications. For this, the two most basic DSP techniques which are frequently used in many practical systems are introduced. These two basic techniques are finite-impulse response (FIR) and infinite-impulse response (IIR) filters.

REFERENCES

1. M.D. Ciletti, *Advanced Digital Design with the Verilog HDL,* Prentice-Hall, Upper Saddle River, NJ, USA, 2003.
2. K. Hwang, *Computer Arithmetic Principles, Architecture and Design,* John Wiley & Sons, Inc., New York, NY, USA, 1979.
3. E.C. Ifeachor and B. W. Jervis, *Digital Signal Processing: A Practical Approach,* Addison-Wesley, Reading, MA, USA, 1993.

4. A.V. Oppenheim and R. W. Schafer, *Discrete-Time Signal Processing*, 2nd Edn, Prentice-Hall, Upper Saddle River, NJ, USA, 1999.

5. B. Parhami, *Computer Arithmetic: Algorithms and Hardware Designs*, Oxford University Press, New York, NY, USA, 2000.

6. B. Parhami, *Computer Architecture: From Microprocessors to Supercomputers*, Oxford University Press, New York, NY, USA, 2005.

7. R. J. Schilling and S. L. Harris, *Fundamentals of Digital Signal Processing Using Matlab*, Thomson Publishers, 2005.

PROBLEMS

14.1 Consider all three fast adders discussed in Section 14.1. Compare the performances of these three fast adders in terms of the number of LUTs and propagation delays with different word widths, ranging from 4 to 64 spaced with 2^k, where k is 3, 4 or 5.

14.2 Consider the logic diagram of the 16-bit hybrid Brent–Kung/Kogge–Stone carry network shown in Figure 14.24.

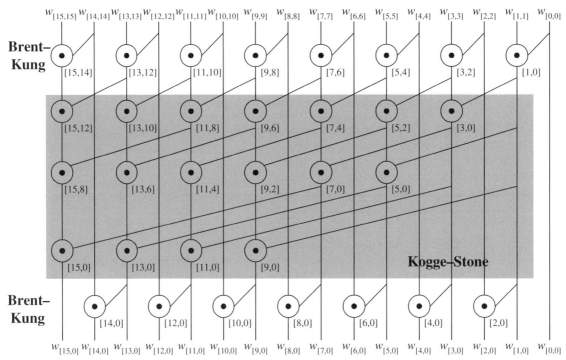

FIGURE 14.24 The logic diagram of a 16-bit hybrid Brent–Kung/Kogge–Stone carry network

(a) Write a parameterized Verilog HDL module to describe an *n*-bit adder based on the hybrid carry network.

(b) Write a test bench to verify the functionality of the parameterized module.

14.3 Consider the problem of a Gray code to binary code conversion. As we have described in Section 6.2.1, the rule of converting a Gray code into its corresponding binary code is that if the number of 1s of the input Gray code, counting from

MSB to the current position, i, is odd, then the binary bit is 1; otherwise, the binary bit is 0.

  (a) Show that the Gray code to binary code conversion problem can be cast into a prefix-sum problem.

  (b) Draw a parallel-prefix network to show how to compute each binary bit from the Gray code in parallel.

  (c) Write a parameterized module using the parallel-prefix approach to describe the conversion operations.

  (d) Write a test bench to verify the functionality of the parameterized module.

14.4 Consider the shift-and-add multiplier described in Section 14.2.1.

  (a) Derive an ASM chart from the algorithm given in Section 14.2.1.

  (b) Write a parameterized Verilog HDL module to describe the ASM chart obtained.

  (c) Write a test bench to verify the functionality of the module.

14.5 Consider the 4-bit bit-serial multiplier shown in Figure 14.25.

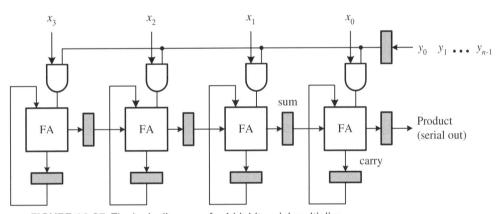

FIGURE 14.25 The logic diagram of a 4-bit bit-serial multiplier

  (a) Derive an ASM chart from observation of the figure.

  (b) Write a parameterized Verilog HDL module to describe the ASM chart obtained.

  (c) Write a test bench to verify the functionality of the module.

14.6 Write a parameterized module to describe the unsigned CSA array multiplier shown in Figure 14.10. You also need to write a test bench to verify the functionality of the module.

14.7 Consider the (n/k)-cycle structure for an $m \times n$ multiplier described in Figure 14.11.

  (a) Derive an ASM chart by referring to the algorithm given in Section 14.2.1.

  (b) Write a parameterized Verilog HDL module to describe the ASM chart obtained. The parameters used are m, k and n.

  (c) Write a test bench to verify the functionality of the module.

14.8 Write a parameterized module to describe the pipelined structure of the $m \times n$ unsigned multiplier shown in Figure 14.12. The parameters used are m, k and n. You also need to write a test bench to verify the functionality of the module.

14.9 Consider an $m \times n$ unsigned CSA array multiplier. Suppose that an m-bit RCA is used in the last stage to sum up the sum and carry in order to yield the final product.

(a) Partition the array into l stages with the same or almost the same propagation delay. Represent the l as a function of m and n.

(b) Insert pipeline registers and draw the logic diagram of the resulting pipelined structure.

(c) Write a parameterized Verilog HDL module to describe the above pipelined structure.

(d) Write a test bench to verify the functionality of the module.

14.10 Consider the (n/k)-cycle structure for the $m \times n$ Booth multiplier described in Section 11.2.2.

(a) Derive an ASM chart by referring to the Booth array structure given in Section 11.3.4.

(b) Write a parameterized Verilog HDL module to describe the ASM chart obtained. The parameters used are m, k and n.

(c) Write a test bench to verify the functionality of the module.

14.11 Refer to Figure 14.12, pipeline the structure of the $m \times n$ Booth array multiplier described in Section 11.3.4 and write a parameterized module to describe the resulting pipelined structure. The parameters used are m, k and n. You also need to write a test bench to verify the functionality of the module.

14.12 Write a parameterized $m \times n$ array multiplier module that can perform both unsigned and signed multiplication controlled by a `mode` input and only use a single array. Write a test bench to verify the functionality of the module.

14.13 Consider the modified Baugh–Wooley module discussed in Section 14.2.2.

(a) Give a numerical example to exploit the operation of a modified Baugh–Wooley algorithm with the case of $m \neq n$.

(b) Modify the modified Baugh–Wooley module so that it can accommodate the situation of $m \neq n$.

(c) Write a test bench to verify the functionality of the module.

14.14 Consider the unsigned nonrestoring division algorithm described in Section 14.3.2.

(a) Derive an ASM chart from the algorithm given in Section 14.3.2.

(b) Write a parameterized Verilog HDL module to describe the above ASM chart.

(c) Write a test bench to verify the functionality of the module.

14.15 Consider the (n/k)-cycle structure for the nonrestoring division algorithm discussed in Section 14.3.2.

(a) Derive an ASM chart by referring to the nonrestoring array divider described in Figure 14.17.

(b) Write a parameterized Verilog HDL module to describe the ASM chart obtained. The parameters used are m, k and n.

(c) Write a test bench to verify the functionality of the module.

14.16 Refer to Figure 14.12, pipeline the structure of the nonrestoring array divider described in Figure 14.17 and write a parameterized module to describe the resulting pipelined structure. The parameters used are m, k and n. You also need to write a test bench to verify the functionality of the module.

14.17 In a barrel logical right shifter, the relationship between the output and the input can be described as follows:

$$O_i = \begin{cases} 0 & \text{for all } i > (n-1) - s \\ I_{i+s} & \text{otherwise} \end{cases} \qquad (14.11)$$

where $i = 0, 1, \cdots, n-1$, and $s = 0, 1, \cdots, m-1$.

Design a barrel shifter that can perform logical right shift and write a test bench to verify its functionality.

14.18 In a barrel arithmetic right shifter, the relationship between the output and the input can be described as follows:

$$O_i = \begin{cases} I_{n-1} & \text{for all } i > (n-1) - s \\ I_{i+s} & \text{otherwise} \end{cases} \qquad (14.12)$$

where $i = 0, 1, \cdots, n-1$, and $s = 0, 1, \cdots, m-1$.

Design a barrel shifter that can perform arithmetic right shift and write a test bench to verify its functionality.

14.19 Design a barrel shifter that can perform both logical left and right shifts. Write a test bench to verify its functionality.

14.20 Design a barrel shifter that can perform both arithmetic and logical shifts, including both left and right directions. Write a test bench to verify its functionality.

14.21 Suppose that a datapath composed of a register file, an ALU and a shifter is required. The ALU has the arithmetic functions +, -, * and /, as well as the three basic logical operations, AND, OR and NOT. The register file has eight 8-bit registers. The shifter can perform arithmetic and logical shift operations, including left and right shift operations. Model this datapath in behavioral style and write a test bench to verify the functionality of the datapath module.

14.22 Consider the general datapath structure for the FIR filters shown in Figure 14.21.

(a) Using the datapath and controller design paradigm, derive an ASM chart of the controller.

(b) Write Verilog HDL modules to describe the controller and the datapath, respectively.

(c) Write a test bench to verify the functionality of the FIR filter module.

14.23 Observe the direct form of the IIR filter shown in Figure 14.23.

(a) Design a general datapath structure (modify the one depicted in Figure 14.21) suitable for implementing IIR filters.

(b) Using the datapath and controller design paradigm, derive an ASM chart of the controller.

(c) Write Verilog HDL modules to describe the controller and the datapath, respectively.

(d) Write a test bench to verify the functionality of the IIR filter module.

14.24 In floating-point multiplication, the exponents of the two operands are added and their significands are multiplied:

$$(\pm 2^{e1} s1) \times (\pm 2^{e2} s2) = \pm 2^{e1+e2}(s1 \times s2) \tag{14.13}$$

Suppose that both inputs and the result are represented as the single-precision format of the IEEE-754 standard, which is shown in Figure 14.26. Design a floating-point multiplication circuit and describe it in Verilog HDL. Write a test bench to verify the functionality of the resulting circuit.

```
31 30          23 22                          0
┌─┬───────────┬──────────────────────┐
│S│Exponent (e)│     Mantissa (m)     │   value = (−1)^S × 2^{e−127} × 1.m
└─┴───────────┴──────────────────────┘
```

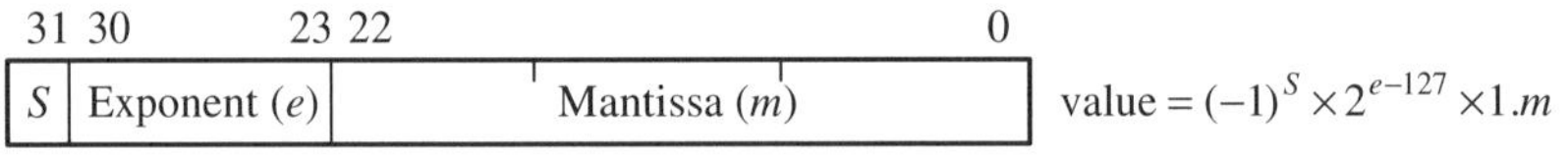

FIGURE 14.26 The single-precision format of the IEEE-754 floating-point standard

14.25 In floating-point addition, the exponents of the two operands are equalized and their significands are added:

$$(\pm 2^{e1} s1) + (\pm 2^{e1}(s2/2^{e1-e2})) = \pm 2^{e1}(s1 \pm s2/2^{e1-e2}) \tag{14.14}$$

Assume that $e1 > e2$. Suppose that both inputs and the result are represented as the single-precision format of the IEEE-754 standard, which is shown in Figure 14.26. Design a floating-point addition circuit and describe it in Verilog HDL. Write a test bench to verify the functionality of the resulting circuit.

14.26 In floating-point division, the exponents of the two operands are subtracted and their significands are divided:

$$(\pm 2^{e1} s1)/(\pm 2^{e2} s2) = \pm 2^{e1-e2}(s1/s2) \tag{14.15}$$

Suppose that both inputs and the result are represented as the single-precision format of the IEEE-754 standard, which is shown in Figure 14.26. Design a floating-point division circuit and describe it in Verilog HDL. Write a test bench to verify the functionality of the resulting circuit.

DESIGN EXAMPLES

WE BEGIN this chapter with an introduction of the bus structure and its related issues such as bus arbitration and data transfer modes. The bus structure can be either a tristate bus or a multiplexer-based bus. In order to schedule the use of a shared bus, bus arbitration logic is required for any bus system. Daisy-chain and radial arbitration are the two most widely used bus schedule schemes. The essential of bus arbitration is the rule to choose a device from the many ones requested with the use of a bus at the same time. Both fixed priority and round-robin priority are usually used for this purpose. Data transfer on a bus can be proceeded in a synchronous or an asynchronous manner and in a bundle of bits (namely, parallel) or a single bit at a time (namely, serial).

Next, we consider a real-world example that illustrates the design of a small μC system, which is the most complex design example in the book. This system includes a general-purpose input and output (GPIO), timers and a universal asynchronous receiver and transmitter (UART) being connected by a system bus composed of an address bus and a data bus, as well as a control bus. The GPIO is an 8-bit parallel input/output port that can be used as a single 8-bit port or as eight 1-bit ports. A timer is a device that can be used to count events. The basic operations of timers and how to design such timers are discussed in greater detail. A UART is a device that supports the serial communication between two devices in an asynchronous manner. We also give an example to illuminate how to design such a device. The 16-bit CPU is the heart of the system, which provides 27 instructions and 7 addressing modes. A detailed description of it is given in this chapter.

15.1 BUS

A bus is a set of wires used to transport information between two or more devices in a digital system. In practice, multiple devices are usually required to connect onto a bus

Digital System Designs and Practices   Ming-Bo Lin
© 2008 John Wiley & Sons (Asia) Pte Ltd

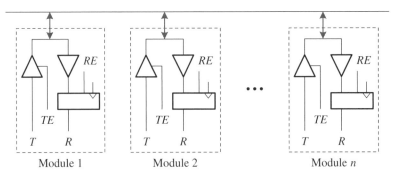

FIGURE 15.1 The block diagram of a typical tristate bus structure

in order to transfer messages through the bus to their destinations. However, this may cause a problem called *multiple drivers*, which is often solved by using a technique known as *multiplexing*. Through the multiplexing technique, only one device can use the bus for sending its message to the specified destination. In fact, the multiplexing technique consists of two related issues: *bus structure* and *bus arbitration*. The bus structure is the physical organization of the bus and the bus arbitration is a mechanism used to schedule the use of the bus. In other words, it only allows one device to send a message at a time. In this section, we deal with these two issues in more detail.

15.1.1   Bus Structures

In practice, a bus can be realized by using either tristate buffers or multiplexers. The bus is called a *tristate bus* when using tristate buffers and a *multiplexer-based bus* when using multiplexers. The tristate bus is often called *bus* for short. In the following, we deal with these two structures in greater detail.

15.1.1.1   Tristate Bus   A typical tristate bus structure used in digital systems is illustrated in Figure 15.1, where n modules are connected to the bus. Each module is connected to the bus through a bidirectional interface that enables it to drive a signal T to the bus when the transmit enable control signal TE is asserted and to sample a signal off the bus onto an internal signal R when the receive enable control signal RE is asserted. For instance, when module 1 wants to send a message a to module 2, module 1 asserts its transmit enable TE to drive its transmit signal T onto the bus. During the same cycle, module 2 asserts its receive enable control signal RE to sample the signal off the bus onto its internal received signal R.

In the following, we give an example to explain how to model a tristate bus.

EXAMPLE 15.1   *An Example of a Tristate Bus*

Suppose that an n-bit tristate buffer is connected onto a bus. When the enable control is asserted, the data is placed on the bus. Otherwise, the output of the buffer is in a high-impedance state.

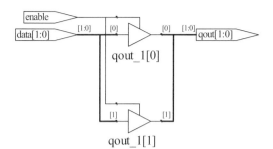

FIGURE 15.2 The block diagram of a tristate bus

```
// a tristate bus example.
module tristate_bus(data, enable, qout);
parameter N = 2;    // define bus width
input   enable;
input   [N-1:0] data;
output [N-1:0] qout;
wire   [N-1:0] qout;
// the body of tristate bus
assign qout = enable ? data : {N{1'bz}};
endmodule
```

The synthesized result is depicted in Figure 15.2. You may try it on your system. It is worth noting that each module has one copy of the above tristate_bus module. All tristate buffers which drive the same bus must be turned on exclusively; otherwise, the data on the bus will be in an erroneous state due to the confliction of multiple driven sources.

Using a proper combination of tristate buffers, a bidirectional bus system can be easily constructed. For example, a bidirectional bus system can be formed by connecting two tristate buffers in a back-to-back manner. The resulting module is given as follows.

EXAMPLE 15.2 *An Example of a Bidirectional Bus*

A bidirectional bus is indeed a connection of two tristate buffers in such a way that the input of one buffer is connected to the output of the other and the enable control of each buffer is used to control the data transfer direction. In this example, we do not connect together the enable controls of the two buffers and separately denote these two controls as send and receive.

```
// bidirectional bus example
module bidirectional_bus(data_to_bus,send,receive,data_
                         from_bus,qout);
parameter N = 2;                 // define bus width
input   send, receive;
input   [N-1:0] data_to_bus;
output [N-1:0] data_from_bus;
inout   [N-1:0] qout;            // bidirectional bus
wire   [N-1:0] qout, data_from_bus;
```

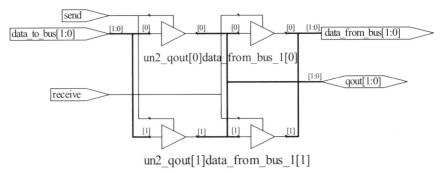

FIGURE 15.3 The block diagram of a bidirectional bus

```
// the body of bidirectional bus
assign data_from_bus = receive ? qout : {N{1'bz}};
assign qout = send ? data_to_bus : {N{1'bz}};
endmodule
```

The synthesized result is depicted in Figure 15.3. You are encouraged to try it on your own system. It is worth noting that each module has one copy of the above `bidirectional_bus` module.

The objective of both bus and multiplexer is the same. Namely, they select one data source from multiple ones and then route the data to its destination. When using a bus, each module only needs a bidirectional interface and an appropriate bus arbiter to schedule the sequence of bus activities. However, in some applications using a bus structure may not be a good choice, especially when the capacitive loading of the driver within the bidirectional interface of the active module is large. For example, in the typical tristate bus structure shown in Figure 15.1, each transmit buffer needs to drive an amount of $n \times (C_{bout} + C_{bin})$ capacitive load, where C_{bout} and C_{bin} are the output capacitance of the tristate output buffer and the input capacitance of the input buffer, respectively. This amount of capacitive load may be intolerant in some applications.

15.1.1.2 Multiplexer-Based Bus An alternative approach to avoiding the large amount of capacitive load is to use a multiplexer, as illustrated in Figure 15.4.

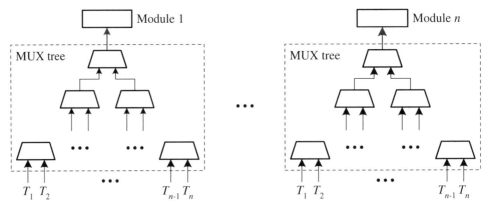

FIGURE 15.4 The typical multiplexer-based bus structure

The output signals T_i of n modules are routed to their destination through a multiplexer tree. It can be shown that the propagation delay of a multiplexer tree is much less than the bus structure when the modules attached to it are large enough. Consequently, for many practical systems such as an ARM processor, a multiplexer tree is often used instead of a tristate bus to obtain a better performance.

Coding Styles

1. It should only allow one output driver on a tristate bus to be activated at a time.

15.1.2 Bus Arbitration

Due to the fact that a bus is a shared and an exclusively used resource among modules, at any time only one module may send messages onto the bus. In order to work properly, a bus needs to be sequenced, namely, to be scheduled for use when multiple transmitters onto it wish to use it. Generally, buses may be internally sequenced or externally sequenced. When a bus is sequenced externally, a center controller is needed to manage the *TE* and *RE* control signals of all modules appearing on the bus. When a bus is sequenced internally, each module generates its own *TE* and *RE* control signals according to a bus protocol. Both strategies are often used in digital systems. For instance, the internal bus of a microprocessor is usually an externally sequenced bus whereas the bus between the processor and its peripherals, such as memory, a *direct memory access* (DMA) controller or a *universal serial bus* (USB) controller, is often an internally sequenced bus.

From the above discussion, there must exist some mechanisms to arbitrate the use of the bus when multiple transmitters initiate data transfer. The operation that chooses one transmitter from multiple ones attempting to transmits data on the bus is called a *bus arbitration*. The device used to perform the function of bus arbitration is known as a *bus arbiter*.

Currently, the two most widely used bus arbitration schemes are *daisy-chain arbitration* and *radial arbitration*, as shown in Figure 15.5. When using daisy-chain

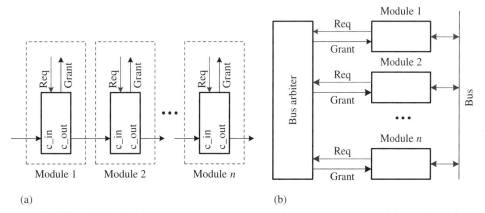

FIGURE 15.5 Block diagrams of the two bus arbitration schemes: (a) daisy-chain arbitration; (b) redial arbitration

arbitration, as shown in Figure 15.5(a), each module has an arbitration logic called *daisy-chain logic*, consisting of two inputs and two outputs. The inputs are *carry-in* and *request*; the outputs are *carry-out* and *grant*. The carry-in indicates whether the preceding stage has granted the bus or not. When it is high, the preceding stage has not granted the bus. Otherwise, the bus has been granted by its preceding stage. When both carry-in and request of the module are high, the grant signal is set high to indicate that the bus is granted by the module. At the same time, the carry-out is set low to inhibit the succeeding stage to grant the bus.

One widely used implementation of the daisy-chain logic is shown in Figure 15.6(a), which is an iterative logic and consists of two basic gates. The logic expressions of carry-out and grant are as follows.

$$g_i = r_i c_i$$
$$c_{i+1} = r'_i c_i \tag{15.1}$$

An example of daisy-chain logic using this iterative logic cell is shown in Figure 15.6(b), which is a 4-request daisy-chain bus arbiter. The request r_0 has the highest priority while request r_3 has the lowest priority. When a device with request r_i requests the use of the bus, it will inhibit the request of the bus of its succeeding devices with request r_{i+1} by clearing the carry-out c_{i+1}. For instance, when device 0 requests the bus, it inhibits the requests of other devices by setting the carry-out c_1 to 0.

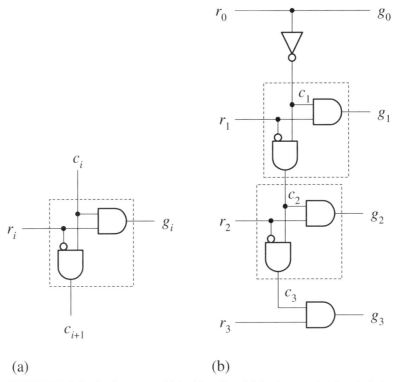

(a)   (b)

FIGURE 15.6 Logic diagrams of (a) a bit cell and (b) a 4-request daisy-chain bus arbiter

The daisy-chain arbitration has the following two features. First, the priority is fixed with the highest priority associated with the first module. Second, the bus arbitration time is determined by the modules cascaded in the daisy-chain. This can be easily seen from Figure 15.5(a). The second feature may limit the system performance when a large number of modules are connected onto the bus.

The second bus arbitration scheme is radial arbitration, which is also known as *independent-requested line arbitration*, that uses separate request and grant lines for each module. As shown in Figure 15.5(b), each module has a set of request and grant lines. The request lines of all modules sharing the bus are connected to a bus arbiter through which, at most, one grant line is activated. The structure of a bus arbiter with fixed-priority is simply composed of a priority encoder and a decoder. The priority encoder determines which module is granted for using the bus and the decoder generates the grant signal for the module. This fixed-priority logic is quite easily implemented and therefore is left to the reader as an exercise.

An approach besides the fixed-priority used to construct the bus arbiter is to use *variable priority*. Figure 15.7 shows such an example, where a little modification of the iterative logic cell shown in Figure 15.6(a) is made by adding an OR gate along with a priority input p_i. Each priority input p_i determines the priority of the device associated with it.

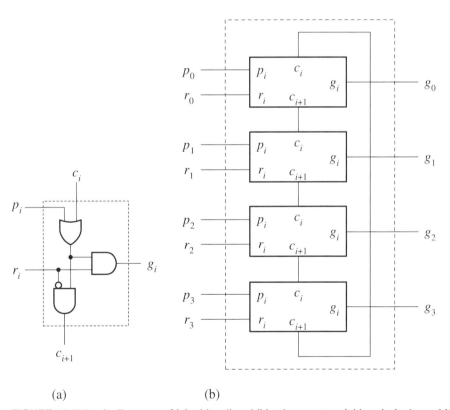

(a)   (b)

FIGURE 15.7 Logic diagrams of (a) a bit cell and (b) a 4-request variable-priority bus arbiter

The means to generating the priority p_i determines the priority of the devices. For example, the priority of the devices is to shift one position if a shift register is used to generate the priority p_i. The priority is random if it is generated by using a random number. The random number can be easily generated by using the PR-sequence generators described in Section 9.5.1.

Although the above two possible variable-priority arbiters can determine the priority dynamically, they are still not good enough for practical use because they cannot provide strong fairness. Consider the case that two adjacent devices with requests r_i and r_{i+1} repeatedly request the bus. Request r_{i+1} wins the arbitration only when p_{i+1} is true and request r_i wins the arbitration for the other possible priority inputs. The request r_i wins the arbitration $n - 1$ times more often than request r_{i+1}.

To overcome the above unfairness, in most digital systems a priority scheme known as *round-robin priority* is often used. In this scheme, the device just being served is made the lowest priority whereas the device succeeding it is made the highest priority. Based on this rule, an iterative logic cell can be constructed, as shown in Figure 15.8. The priority p_i is set to 1 if its preceding device has been granted its request on the use of the bus. The logic function of the next-p_i is as follows.

$$\text{next-}p_i = anyg' p_i + g_{(i-1) \bmod n} \tag{15.2}$$

An example of a 4-request round-robin priority bus arbiter is shown in Figure 15.8. If no grant was issued, the *anyg* signal is 0, which causes the priority to remain unchanged. If a grant was issued at the current cycle, one of g_i must be 1. Hence, *anyg* will be 1, which causes p_{i+1} to be 1 at the next cycle. As a result, the request next to the one receiving the grant has the highest priority and the request being served has the lowest priority.

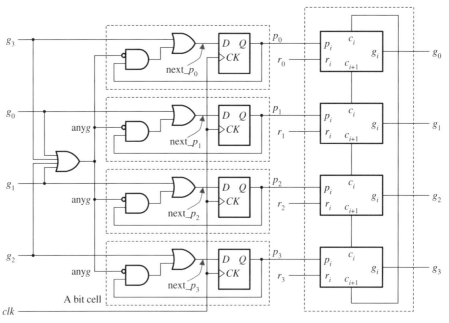

FIGURE 15.8 The logic diagram of a 4-request round-robin bus arbiter

Review Questions

Q15.1 Explain what a bus is.

Q15.2 What are the two basic bus structures?

Q15.3 Compare the differences between tristate bus and multiplexer-based bus structures.

Q15.4 What are the two most widely used bus arbitration schemes?

15.2 DATA TRANSFER

When the data transfer between two devices is synchronized by a common clock, it is called a *synchronous mode*; otherwise, it is called an *asynchronous mode.* The actual data transfer can be proceeded in parallel or in serial. Parallel data transfer means a set of data bits is transferred at the same time while serial data transfer means the data is transferred in a bit-by-bit manner. In this section, we will deal with all of these kinds of data transfers.

15.2.1 Synchronous Data Transfer

As we have discussed, a bus is a set of wires and its goal is to transfer messages from one source to one or more destinations. In general, a bus operates in units of cycles, messages and transactions. A message is a logical unit of information transferred between a source and destination(s). A transaction consists of a sequence of messages that are strongly related. It is initiated by one message and consists of a chain of related messages generated in response to the initiating message.

As mentioned before, a bus must be scheduled before it can transfer messages. Once a bus is sequenced, the transfer of messages onto it can then be proceeded in a synchronous or an asynchronous way. Despite which way is used, there must be a signal to indicate when the data signal is valid or stable. In synchronous data transfer, the data transfer is controlled directly by the clock signal, whereas in asynchronous data transfer the data transfer may be controlled by using either a *strobe* or a *handshaking* scheme.

15.2.1.1 *Parallel Data Transfer*   In synchronous data transfer mode, each transfer is in synchronism with the clock signal, as shown in Figure 15.9. In other words, the receiver samples and latches the data in the specified edge of the clock signal such as at the positive edge. In a synchronous bus system, the device generating the address and command is called a *bus master* and the device receiving the address and command from the bus is called a *bus slave.*

Synchronous bus transfers can be further divided into two types: single-clock bus cycle and multiple-clock bus cycle. The single-clock bus cycle, as shown in Figure 15.9(a), needs only one clock cycle to complete a data transfer. At the positive edge of the clock signal, the bus master sends the address and command, and latches the data read on the previous cycle; at the negative edge of the clock signal, the bus slave sends the data.

The multiple-clock bus cycle is shown in Figure 15.9(b), which requires multiple clock cycles to complete a data transfer. An example of a two-clock bus cycle is shown

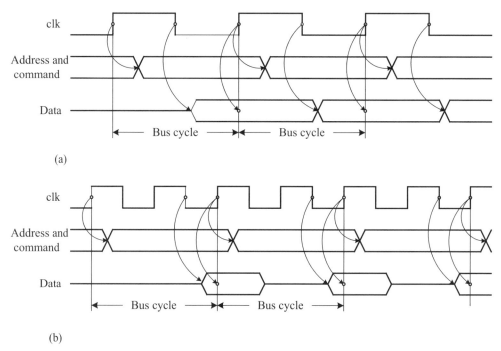

FIGURE 15.9 Typical timing diagrams of (a) single-clock and (b) multiple-clock synchronous bus cycles

in Figure 15.9(b). At the positive edge of the first clock signal, the bus master sends address and command, and latches the data read on the previous cycle; at the negative edge of the second clock signal, the bus slave sends the data. The number of clock cycles required is usually 2 or 3. Sometimes, many more clock cycles are required in a multiple-clock bus cycle. The actual clock cycles needed is determined by the operating speeds of the devices onto the bus.

15.2.1.2 *Serial Data Transfer*   Like parallel data transfer, serial data transfer can also be proceeded in either an asynchronous or a synchronous way. In synchronous serial data transfer, the clock signal is being sent along with the data implicitly or explicitly. Sending the clock signal with data explicitly as a separate signal is an intuitive solution for synchronous serial data transfer and is often known as an *explicitly clocking scheme*. A simple example is shown in Figure 15.10(a), where the negative edges of the clock signal are used to capture the data. The major drawback of an explicitly clocking scheme is that at high signaling speeds, where the wire delay becomes significant, the different delays for both data and clock wires may cause incorrect data to be sampled at the destination. Hence, the explicitly clocking scheme is usually used in low-speed serial data transfer, especially for those cases below several Mbps.

To overcome the above difficulty, a widely used approach is to encode the clock signals into the data and the clock signal is then extracted at the receiver before sampling the data. Such an approach is known as an *implicitly clocking scheme*. The non-return-to-zero (NRZ) is such a widely used encoding scheme. An instance of NRZ coding is shown in Figure 15.10(b). The implicitly clocking scheme is an approach

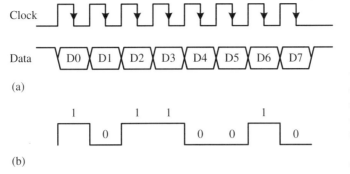

FIGURE 15.10 The data format of synchronous serial data transfer: (a) serial data transfer with explicitly clocking; (b) serial data transfer with implicity clocking (ZRN code)

that may be used in high-speed serial data transfer, especially for those applications of a modern high-speed I/O bus.

Review Questions

Q15.5 Describe the operations of synchronous bus transfer.

Q15.6 Describe the meaning of an explicitly clocking scheme.

Q15.7 What is the major drawback of an explicitly clocking scheme?

Q15.8 Describe the two kinds of synchronous serial data transfers.

15.2.2 Asynchronous Data Transfer

In the asynchronous data transfer mode, each data transfer occurs at random; that is to say, it cannot be predicted in advance. The data transfer may be controlled by using either a strobe or handshaking scheme. It is worth noting that both strobe and handshaking are used extensively on numerous occasions that require the transfer of data between two asynchronous (namely, independent) devices. In the following, we consider these two schemes in more detail.

15.2.2.1 Strobe In the strobe control scheme, only one control signal known as a *strobe* is needed. When there are data to be transferred, the strobe signal is enabled by either the source device or destination device, depending on the actual application.

Source Initiated Transfer The detailed operations of source-initiated data transfer are depicted in Figure 15.11(a). The source device places the data onto the data bus and then asserts the strobe control signal to notify the destination device that the data is available. The destination device then samples and stores the data onto its internal register at the negative edge of the strobe control signal. The data transfer cycle is completed. The data transfer from the CPU to a memory location is such an example, where the write control signal is the strobe signal.

Destination Initiated Transfer The detailed operations of destination-initiated data transfer are depicted in Figure 15.11(b). The destination device asserts the strobe control signal to request data from the source device. Once it has received this strobe signal, the source device places the data onto the data bus for a duration long enough for the destination device to read it. Usually, the destination device samples and stores the data onto its internal register at the negative edge of the strobe control signal. The strobe

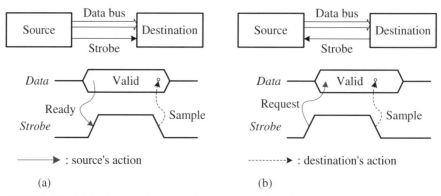

FIGURE 15.11 The timing diagrams of strobe-controlled data transfers: (a) source-initiated transfer; (b) destination-initiated transfer

signal is deasserted by the destination device and the data transfer cycle is completed. The data transfer from a memory location to the CPU is such an example, where the read control signal is the strobe signal.

In summary, the strobe scheme presumes that the requested device is always ready for transferring data once it receives the request. Unfortunately, this is not always the case. In so many applications, the requested device is not ready when it is requested to send data. Therefore, another control signal must be used for the requested device to indicate whether it is ready or not. The resulting scheme is known as handshaking. It is indeed a two-way control scheme.

15.2.2.2 Handshaking

Conceptually, handshaking is a technique that provides a two-way control scheme for asynchronous data transfer. In this kind of data transfer, each transfer is sequenced by the edges of two control signals: request *req* (or valid) and acknowledge *ack*, as shown in Figure 15.12. In the handshaking transfer, four events are proceeded in a cyclic order. These events are *ready* (*request*), *data valid*, *data acceptance* and *acknowledge* in that order.

Source-Initiated Transfer An asynchronous data transfer can be initiated by either a source device or a destination device. The one shown in Figure 15.12(a) is initiated by a source device. The sequence of events is as follows:

1. *Ready*. The destination device deasserts the acknowledge signal and is ready to accept the next data.
2. *Data valid*. The source device places the data onto the data bus and asserts the valid signal to notify the destination device that the data on the data bus is valid.
3. *Data acceptance*. The destination device accepts (latches) the data from the data bus and asserts the acknowledge signal.
4. *Acknowledge*. The source device invalidates data on the data bus and deasserts the valid signal.

Due to the inherent back and forth operations of the scenario, the asynchronous data transfer described above is known as *handshaking protocol* or *handshaking transfer*.

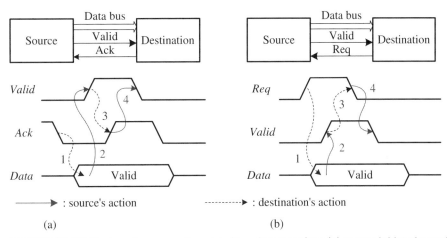

FIGURE 15.12 Timing diagrams of handshaking data transfers: (a) source-initiated transfer; (b) destination-initiated transfer

Destination-Initiated Transfer   The destination-initiated data transfer is shown in Figure 15.12(b). The sequence of events is as follows:

1. *Request*. The destination device asserts the request signal to request data from the source device.

2. *Data valid*. The source device places the data on the data bus and asserts the valid signal to notify the destination device that the data are valid now.

3. *Data acceptance*. The destination device accepts (latches) the data from the data bus and deasserts the request signal.

4. *Acknowledge*. The source device invalidates data on the data bus and deasserts the valid signal to notify the destination device that it has removed the data from the data bus.

Two-Phase Handshaking   The above handshaking is known as a *four-phase handshaking* scheme because it needs four events to complete an active cycle. In fact, the handshaking can be accomplished by only using two events. This kind of handshaking scheme is known as a *two-phase handshaking*.

In the two-phase handshaking, only two phases of operations can be distinguished. These operations are the active cycle of the source device and the active cycle of the destination device. As shown in Figure 15.13, once the source device has recognized the acknowledge signal from the destination device, it places valid data onto the data bus and then signals the destination device by generating a transition on the request signal. The destination device generates a transition on the acknowledge signal to recognize this event and latch into the data on the data bus. Consequently, the request event terminates the active cycle of the source device and the acknowledge event terminates the active cycle of the destination device.

The important features of two-phase handshaking data transfer are of being simple and fast. However, it needs the detection of transitions, which may not be applicable to some applications, such as modern CMOS technologies that typically

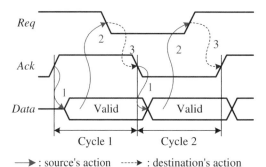

FIGURE 15.13 Timing diagrams of two-phase handshaking data transfers

tend to be sensitive to levels or transitions on only one direction. As a result, the (four-phase) handshaking data transfer depicted in Figure 15.12 is more widely used.

A major problem of handshaking data transfer, regardless of whether a four-phase or two-phase handshaking scheme is used, is that the data transfer will not be completed if one device is faulty. Such an error can be detected by using a timeout mechanism, which will give an alarm if the destination device does not respond within a specified time interval. In practice, the timeout mechanism can be realized by using a timer that is restarted each time the source device begins a data transfer. A variety of timers will be introduced in detail later in this chapter.

15.2.2.3 *Asynchronous Serial Transfer*   In asynchronous serial data transfer, the clock signal is not sent with the data. Instead, the receiver generates its local clock which is then used to capture the data being received. The standard data format for asynchronous serial data transfer is shown in Figure 15.14. The data are transmitted one byte at a time. Each byte is packed as a frame consisting of a start bit, data bits, an optional parity bit and one or more stop bits. The start bit is set to 0 whereas the stop bit is set to 1 so that there is at least one transition for each byte. When there are no data to be sent, the transmitter continuously sends 1s in order to maintain a continuous communication channel. The receiver monitors the channel continuously until the start bit is detected. Then, it starts to receive a new data frame.

Due to the fact that in asynchronous serial data transfer, the data being received (called RxD) and the sampling pulse (known as RxC) are operated in their own timing, the sampling point might be placed in any position of the bit time of the receiving data. In order to sample the data being received as possible at or nearby the center of the bit time, the sampling clock frequency at the receiver is often set to n times the transmission clock frequency, TxC, where n may be 1, 4, 16 or 64. Based on this idea, the receiver samples the start bit after $n/2$ counts of the RxC cycles once the

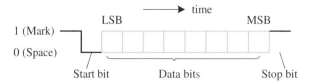

FIGURE 15.14 The data format of asynchronous serial data transfer

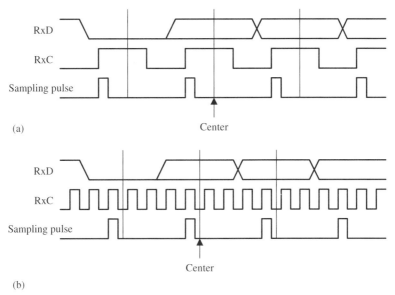

FIGURE 15.15 The reception of asynchronous serial data transfer at two different sampling frequencies: (a) $R \times C = T \times C$; (b) $R \times C = T \times C^* 4$

negative edge of the RxD is detected. Then, it samples each bit thereafter every n RxC cycles.

The need that a sampling clock frequency is often higher than that of the transmission clock frequency can be illustrated by Figure 15.15. Figure 15.15(a) shows the case where the sampling clock frequency is set to the transmission clock frequency. In this case, only one sampling pulse is used for each bit time. At the worst case, the sampling pulse may have occurred at the edge of the bit time so that the resulting sample may be easily erroneous. One way to overcome this drawback is to increase the sampling clock frequency of RxC. For instance, in Figure 15.15(b), there are four RxC pulses within each bit time. The maximum deviation of the sampling pulse is only 25 % of the bit time. This situation is much better when a higher sampling clock frequency is used.

Review Questions

Q15.9 Describe the operations of asynchronous data transfer.

Q15.10 What are the four events associated with handshaking data transfer?

Q15.11 Describe the operations of two-phase handshaking data transfer.

Q15.12 Describe the operations of four-phase handshaking data transfer.

Q15.13 Distinguish between synchronous and asynchronous serial data transfers.

15.3 GENERAL-PURPOSE INPUT AND OUTPUT

In any general μC system, both input and output ports are commonly used in most applications. Hence, a device that can be programmed into either input, output or even bidirectional is often a necessary module for such a system. This device is known as a *general-purpose input and output* (GPIO) module.

15.3.1 Basic Principles

As we have mentioned previously, a port is a physical place where a signal can be passed through among modules. A port is called a *parallel port* when it can process a bundle of signals as a single unit at the same time. A port is called a *serial port* when it can only process signals in a serial way, that is, one bit at a time. A port that can only allow the signal to be inputed from the outside world of the system is called an *input port*; a port that can only allow the signal to flow out of the system is called an *output port*. A port that can be used as an input or an output port, but not both at the same time, is known as a *bidirectional port*.

A general-purpose parallel port is a device or module that can be used as an input or output as required. It generally has eight bidirectional input/output (I/O) pins and two 8-bit registers: *data direction register* (DDR) and *port register* (PORT). The DDR controls the data direction of the bidirectional I/O pins. In other words, it determines which I/O pins are inputs and which are outputs. The PORT register holds the data written to the port. It may or may not be read back, depending on how the GPIO is designed.

When using a general-purpose device, two common user's viewpoints of the device need to be taken into account: the hardware interface, also known as the *hardware model*, and the software interface, also known as the *programming model*. For example, consider the parallel port module shown in Figure 15.16. The hardware model of the parallel port is shown in Figure 15.16(a). It has two groups of interface

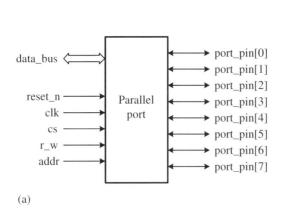

FIGURE 15.16 The block diagram of a parallel port: (a) hardware model; (b) programming model; (c) function table

signals. One group is the interface signals to the system such as the CPU and the other is the port pins for use with I/O devices. The interface with the system includes an 8-bit *data bus*, an *address signal* (*addr*) for selecting the PORT register or DDR, a *chip select* (*cs*) to enable the chip, a *read and write* (*r_w*) control, a *clock input* (*clk*) and a *reset* (*reset_n*) control signal. The function table of a typical parallel port is shown in Figure 15.16(c). The programming model of a device is the user's view of the device while using it. It consists of two registers, the data direction register (DDR) and the port register (PORT), as shown in Figure 15.16(b).

In order to design the GPIO port described above, the following issues are often taken into consideration:

- *Readback capability of PORT register.* The port register usually holds the output data written into port pins. It may or may not allow the data to be read back depending on how the GPIO port is designed.

- *Group or individual bit control.* Several options can be made to control the programmability of the direction of port pins. One extreme is to allow the direction of the entire port pins to be controlled by a single-bit DDR; the other extreme is to use one DDR-bit for each port pin, thus allowing the direction of each port pin to be set individually. Of course, the compromise is to group port pins into several disjoint sets and then allow each set to be programmed into input or output as a unit. The cardinality of sets is usually set to 2 or 4.

- *Selection of the value of the DDR.* Although the value of DDR used to control the direction of a GPIO pin is arbitrary, one way is to use 0 for output and 1 for input. The other way uses the reverse values. The former is usually found in devices from Intel and the latter from Motorola.

- *Handshaking control.* The handshaking control may also be provided along with the GPIO port. These control signals can be provided by two specific I/O pins, which may reside in another GPIO port, or use two dedicated I/O pins, in addition to the GPIO pins.

- *Readback capability of the DDR.* The data written into the DDR may or may not allow the data to be read back depending on how the GPIO port is designed.

- *Input latch.* The input may be latched when the pin is read.

- *Input/Output pull-up.* For a bidirectional I/O port, it usually needs to pull-up the input/outputs pin of the GPIO.

- *Drive capability.* What is the amount of current it can source and sink? The I/O pins are usually used to drive an external LED, or other devices such as relay. When the drive capability is insufficient, some extra circuits are needed to make up the limitation and hence increases the cost of the application systems.

One example of a GPIO port is shown in Figure 15.17(a), which is the port-1 structure of the MCS-51 system. It only contains a port register without the DDR. When writing 1 to port register, the nMOS transistor is turned off and as a result the port_pin[i] is pulled-up to high (V_{DD}). When writing 0 to port register, the nMOS transistor is turned on and therefore the port_pin[i] is pulled-down to ground. The port

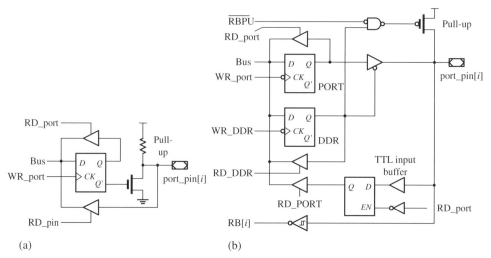

FIGURE 15.17 Logic diagrams of the ith-bits of two GPIO examples: (a) MCS-51 Port 1; (b) PIC 18Fxx2 Port RB2:RB0

register and the port_pin[i] can be separately read by enabling the RD_port and RD_pin control signals, respectively.

Another example is shown in Figure 15.17(b), which is found in the pin structure of port RB2:RB0 in the PIC 18Fxx2 system. From this figure, it is apparent that both port register and DDR are used and both registers can not only be written but also be read back under the control of appropriate control signals. In addition, the output also has an active pull-up transistor controlled by $\overline{RBPU}$ when the port is configured as an input. Moreover, both TTL input and Schmitt trigger buffers are used for driving two different internal signals from the same external pin. The input is also latched when reading the port pin.

15.3.2 A Design Example

In this section, we consider a simple GPIO port which is widely used in most microcontroller systems, ranging from 8 bits to 32 bits. In this GPIO, we assume that the direction of each port pin is separately controlled by an individual DDR bit, as shown in Figure 15.18. The DDR is cleared when the reset_n is activated so as to set the port_pin[i] as input for the reason of failure safe. The port_pin[i] is set as output when DDR is 1 and as input when DDR is 0. The port register does not allow the data to be read back whereas the DDR does. The shaded multiplexer should be removed and the output of the active-low tristate buffer is connected directly to the data_bus[i] if the DDR does not allow the data to be read back.

The following example explores how an 8-bit GPIO port can be designed by using the above mentioned principles.

EXAMPLE 15.3 *An Example of General-purpose Input and Output*

In this example, an 8-bit GPIO port is designed. The hardware model and programming model are shown in Figure 15.16(a). The module first generates four internal control

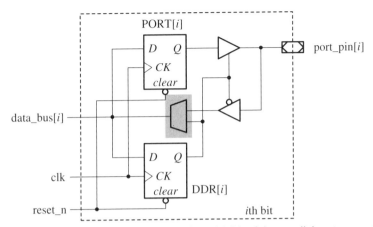

FIGURE 15.18 The logic diagram of the *i*th-bit of the parallel port example

signals. Then, it uses a generate-loop statement to produce an interface logic circuit and employs an `always` block to clear or update the port register and *DDR*.

```
// an example of parallel port
module parallel_port(reset_n, clk, r_w, cs, addr0,
                     data_bus, port_pin);
input reset_n, clk, r_w, cs, addr0;
inout  [7:0] data_bus;
output [7:0] port_pin;
// internal control signals
reg    [7:0] DDR, PORT; // directional and output
                        // registers
wire   read_DDR, load_DDR, read_PORT, load_PORT;
// generate internal control signals
assign read_DDR  = ((cs == 1) && (addr0 == 1)
                    && (r_w == 1)),
       load_DDR  = ((cs == 1) && (addr0 == 1)
                    && (r_w == 0)),
       read_PORT = ((cs == 1) && (addr0 == 0)
                    && (r_w == 1)),
       load_PORT = ((cs == 1) && (addr0 == 0)
                    && (r_w == 0));
// generate interface logic circuit
genvar i;
generate for (i = 0; i < 8; i = i + 1) begin: interface
   assign port_pin[i] = DDR[i]?PORT[i]:8'bz ;
   assign data_bus[i] = read_DDR?DDR[i]:(read_PORT?
                        port_pin[i]:8'bz);
   // if not allow to read DDR, use the following
   // statement
```

```
   // assign data_bus[i] = read_PORT ? port_pin[i] : 8'bz;
end endgenerate
// clear or update port register and DDR
always @(posedge clk or negedge reset_n)
   if (!reset_n) begin
      DDR <= 8'b0; PORT <= 8'b0; end
   else begin
      if (load_DDR)  DDR <= data_bus;
      if (load_PORT) PORT <= data_bus;
   end
endmodule
```
∎

Review Questions

Q15.14 Describe the hardware model of a typical GPIO module.

Q15.15 What is the programming model of a typical GPIO module?

Q15.16 Describe the operations of the port structure shown in Figure 15.17(a).

Q15.17 Describe the operations of the port structure shown in Figure 15.17(b).

Q15.18 Describe the operations of the port structure shown in Figure 15.18.

15.4 TIMERS

The timers are essential modules in any μC system because they provide at least the following important applications: time-delay creation, event counting, time measurement, period measurement, pulse-width measurement, time-of-day tracking, waveform generation and periodic interrupt generation as well. In this section, we deal with the basic operations of timers and the issues of how to design such timers.

15.4.1 Basic Timer Operations

The essential operations of both timer and counter are exactly the same. A counter is called a *timer* if it is operated at a known clock of fixed frequency. In practice, the timers used in most μC systems are counters with programmable operation modes. The basic operation modes of a timer are as follows: *terminal count* (binary/BCD event counter), *rate generation, (digital) monostable* (or called one-shot) and *square-wave generation*.

A typical timer like the 82C54, is a 16-bit presettable counter and has three terminals: *clk*, *gate* and *out*. In addition, the timer has a data bus and a set of appropriate control signals, which enable access to the internal registers of the timer, as shown in Figure 15.19. The timer consists of two 16-bit registers: *latch* and *timer*. The latch register stores the initial value to be loaded into the timer for counting down. It may be reloaded into the timer as many times as needed or only once depending on the operating mode selected. The timer performs the actual counting operations. The

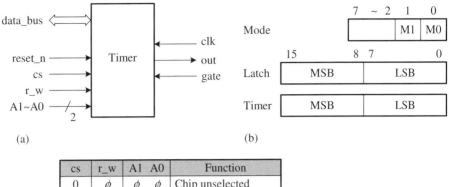

FIGURE 15.19 The block diagram of a typical timer: (a) hardware model; (b) programming model; (c) function table

initial value, called *count*, is written into the latch register and then written into the timer register in the next clock cycle automatically. In the following, we illustrate the detailed operations of each basic timer operating mode in order.

Terminal Count   The terminal-count mode is used to count a specific number of clock cycles. After the write operation is complete, the count is loaded into the timer on the next clock pulse presented in clock input and the count is decremented by 1 for each clock pulse that follows. When the count reaches 0, the terminal count, a 0-to-1 transition occurs at the out terminal. The gate input is an enable control. The timer is enabled when the gate is 1; otherwise, it is disabled. Figure 15.20(a) displays a waveform to illustrate how the timer operates.

The logic circuit used to implement the above operations is shown in Figure 15.20(b). The timer_load control signal is generated by an FSM described as the ASM chart shown in Figure 15.20(c).

In the following example, we describe how to model the timer being operated in terminal-count mode.

EXAMPLE 15.4   *A Timer Example to Illustrate the Terminal-count Mode*

The module consists of five `always` blocks. The first `always` block describes the latch register, which is loaded from the data bus controlled by `wr`. This `wr` signal also activates the `timer_load` control signal for one clock cycle. The second `always` block describes the timer register, which is a binary down counter with parallel load controlled by `timer_load` and enabled by `timer_enable`. The third and fourth `always` blocks generate the `out` and `timer_stop` signals, respectively. The fifth `always` block produces a one-cycle `timer_load` signal triggered by a `wr` input signal.

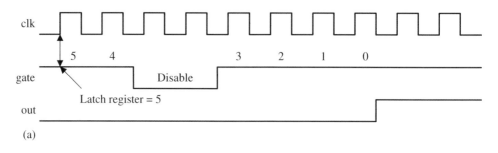

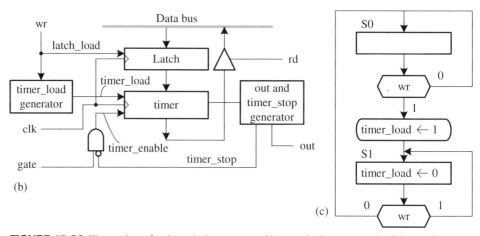

FIGURE 15.20 Illustration of a timer being operated in terminal-count mode: (a) waveform example of terminal-count mode; (b) block diagram of terminal-count mode; (c) generate one-cycle timer_load pulse

```verilog
// an example of timer with terminal-count mode
module timer_tc(clk, reset_n, gate, wr, rd, data_bus,
                out);
parameter N = 16; // default size of timer
input clk, reset_n, gate, wr, rd;
inout [N-1:0] data_bus;
output reg out;
// internal used reg variables
reg  [N-1:0] timer, latch;
wire timer_enable, clk_out, out_reset, timer_stop_reset;
reg  timer_load, timer_stop;
// the state used to generate a one-cycle timer_load
// signal
reg  state;
localparam S0 = 1'b0, S1 = 1'b1;
// describe the latch register
always @(posedge clk or negedge reset_n)
   if (!reset_n) latch <= 0;
   else if (wr) latch <= data_bus;
```

```
// describe the timer register
assign data_bus = rd ? timer : {N{1'bz}};
always @(posedge clk or negedge reset_n)
   if (!reset_n) timer <= 0;
   else if (timer_load) timer <= latch;
        else if (timer_enable) timer <= timer - 1;
assign clk_out = ~clk;
// generate out and timer_stop signals
assign out_reset = !reset_n || wr;
always @(posedge clk_out or posedge out_reset)
   if (out_reset) out <= 1'b0;
   else if (~|timer) out <= 1'b1;
// produce the timer_stop control signal to stop the timer
// whenever the terminal count is reached.
assign timer_stop_reset = !reset_n || timer_load;
always @(posedge clk or posedge timer_stop_reset or
         posedge out)
   if (timer_stop_reset) timer_stop <= 1'b0;
   else if (out) timer_stop <= 1'b1;
        else timer_stop <= 1'b0;
assign timer_enable = gate & ~timer_stop;
// generate one-cycle timer_load control signal
always @(posedge clk or negedge reset_n)
   if (!reset_n) begin timer_load <= 1'b0;
               state <= S0; end
   else case (state)
      S0: if (wr) begin state <= S1;
              timer_load <= 1'b1; end
          else state <= S0;
      S1: begin timer_load <= 1'b0;
          if (wr) state <= S1; else state <= S0; end
   endcase
endmodule                                                ■
```

Rate Generator   In this mode, the out terminal outputs one clock pulse for every N clock pulses, as shown in Figure 15.21. In other words, it works exactly as a modulo N counter. The gate input should be fixed at the logic 1 to enable the timer.

This mode is implemented by the timer register from the latch register being reloaded whenever the terminal count is reached. This is easily done by a minor modification from the timer with terminal-count mode. Therefore, the implementation details of this mode are left as an exercise to the reader.

Retriggerable Monostable (One-shot)   In this mode, the out terminal outputs a high-level pulse for a duration equal to the number of clock pulses, which is equal to the number (N) preset to the timer. The gate input is used as the trigger. When the gate

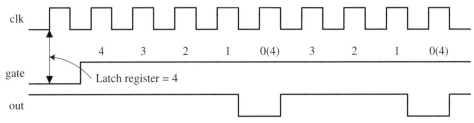

FIGURE 15.21 Illustration of the timer being operated in rate-generation mode

(trigger) input is switched from logic 0 to 1, the out terminal switches from 0 to 1 on the next pulse and remains in that logic value until timeout. This mode is generally retriggerable. An illustration of this operating mode is shown in Figure 15.22(a).

Figure 15.22(b) shows the block diagram of the timer operated in one-shot mode. The duration of the output pulse is stored in the latch register. This duration is loaded into the timer each time the trigger input gate is switched from 0 to 1 by the timer_load control signal. At the same time, the out terminal is set to high, which in turn enables the timer to count down. Once the timer counts down to 0, the *JK* flip-flop is cleared, which in turn resets the out terminal to 0 and disables the timer. The reader is encouraged to compare this monostable circuit with the one described in Section 9.6.2.

The following example shows how to model the timer being operated in one-shot mode.

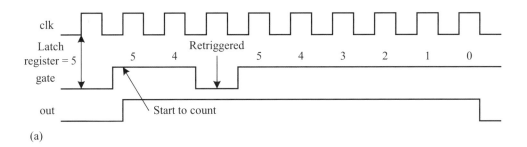

(a)

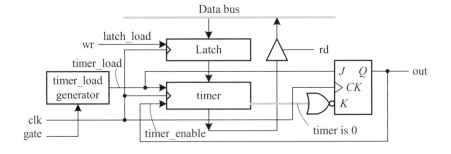

(b)

FIGURE 15.22 Illustration of the timer being operated in retriggerable monostable (one-shot) mode: (a) a waveform example of one-shot mode; (b) block diagram of one-shot mode

EXAMPLE 15.5 *A Timer Example to Illustrate the Monostable Mode*

As described above, the timer is reloaded from the latch register each time the trigger input gate is switched from 0 to 1 by the timer_load control signal, which is generated by the last two always blocks. The first two always blocks describe the latch and timer register, respectively. The third always block describes the output *JK* flip-flop, which is set to 1 when the timer is loaded a new value and reset to 0 when the timer reaches 0.

```
// an example of timer with retriggerable one-shot mode
module timer_one_shot(clk, reset_n, gate, wr, rd,
                      data_bus, out);
parameter N = 16; // default size of timer
input clk, reset_n, gate, wr, rd;
inout [N-1:0] data_bus;
output reg out;
// internal used reg variables
reg  [N-1:0] timer, latch;
wire timer_enable, out_d, timer_load_reset;
reg  timer_load, timer_load_clear;
// the state used to generate a one-cycle timer_load
// signal
reg  state;
localparam S0 = 1'b0, S1 = 1'b1;
// describe the latch register
always @(posedge clk or negedge reset_n)
   if (!reset_n) latch <= 0;
   else if (wr) latch <= data_bus;
// describe the timer register
assign data_bus = rd ? timer : {N{1'bz}};
always @(posedge clk or negedge reset_n)
   if (!reset_n) timer <= 0;
   else if (timer_load) timer <= latch;
       else if (timer_enable) timer <= timer - 1;
// generate output pulse
// describe the JK flip-flop and its related circuits.
assign out_d = (timer_load & ~out) | (|timer & out);
always @(posedge clk or negedge reset_n)
   if (!reset_n) out <= 1'b0;
   else out <= out_d;
assign timer_enable = out;
// generate one-cycle timer_load_clear control signal
always @(posedge clk or negedge reset_n)
   if (!reset_n) begin timer_load_clear <= 1'b0;
                       state <= S0; end
```

```
    else case (state)
      S0: if (gate) begin state <= S1;
                           timer_load_clear <= 1'b1; end
           else state <= S0;
      S1: begin timer_load_clear <= 1'b0;
           if (gate) state <= S1; else state <= S0; end
    endcase
// generate one-cycle timer_load control signal
assign timer_load_reset = !reset_n || timer_load_clear;
always @(posedge gate or posedge timer_load_reset)
    if  (timer_load_reset) timer_load <= 1'b0;
    else timer_load <= 1'b1;
endmodule
```

Square-Wave Generator   The out terminal outputs a square wave with 50 % duty cycle whenever the timer is loaded with an even number. If an odd number (N) is loaded into the timer instead of an even number, the out terminal will be low for $N/2$ clock pulses and high for $(N + 1)/2$ clock pulses. An illustration is given in Figure 15.23(a).

The block diagram of one possible implementation of the square-wave mode is depicted in Figure 15.23(b). The basic idea behind the operation of this mode is on the value being loaded into the timer each time the timer reaches 1. According to the above

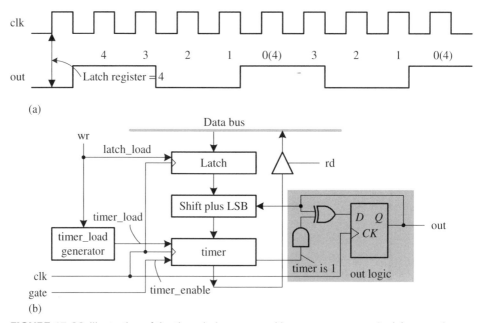

FIGURE 15.23 Illustration of the timer being operated in square-wave mode: (a) a waveform example of square-wave form; (b) block diagram of square-wave mode

description, the output is low for $N/2$ cycles and high for $N/2$ or $(N + 1)/2$ cycles, depending whether the N is even or odd. This is equivalent to saying that the value in the latch register is shifted right one position and then loaded into the timer when the output is low or added with the LSB of the latch register before it is loaded into the timer when the output is high. This is done by the block named as shift plus LSB in the figure.

The following example shows how to model the timer being operated in square-wave mode.

EXAMPLE 15.6 *A Timer Example to Illustrate the Square-Wave Mode*

The output out logic consists of a T-type flip-flop triggered by the terminal count with 1 of the timer. This output pulse is also used to determine whether the value loaded into the timer is $N/2$ or $N/2 + LSB$. The out logic is accomplished by the third always block and the function of shift and plus LSB block is combined into the second always block directly.

```verilog
// an example of timer with square-wave mode
module timer_square_wave(clk, reset_n, gate, wr, rd,
                          data_bus, out);
parameter N = 16; // default size of timer
input clk, reset_n, gate, wr, rd;
inout [N-1:0] data_bus;
output reg out;
// internal used reg variables
reg  [N-1:0] timer, latch;
wire timer_enable, out_d;
reg  timer_load;
// the state used to generate a one-cycle timer_load
// signal
reg  state;
localparam S0 = 1'b0, S1 = 1'b1;
// describe the latch register
always @(posedge clk or negedge reset_n)
   if (!reset_n) latch <= 0;
   else if (wr) latch <= data_bus;
// describe the timer register
assign data_bus = rd ? timer : {N{1'bz}};
always @(posedge clk or negedge reset_n)
   if (!reset_n) timer <= 0;
   else if (timer_load || out_d) begin
        if (out == 0) timer <= (latch >> 1);
        else timer <= (latch >> 1) + latch[0]; end
        else if (timer_enable) timer <= timer - 1;
assign out_d = (timer == 1);
// generate output pulse --- square wave
```

```
always @(posedge clk or negedge reset_n)
   if (!reset_n) out <= 0;
   else out <= out_d ^ out;
assign timer_enable = gate;
// generate one-cycle timer_load control signal
always @(posedge clk or negedge reset_n)
   if (!reset_n) begin timer_load <= 1'b0;
                       state <= S0; end
   else case (state)
      S0: if (wr) begin state <= S1;
                        timer_load <= 1'b1; end
          else state <= S0;
      S1: begin timer_load <= 1'b0;
          if (wr) state <= S1; else state <= S0; end
   endcase
endmodule
```
■

15.4.2 Advanced Timer Operations

In addition to the four basic operation modes described previously, modern timers are often facilitated with the following three additional modes: *input capture mode*, *output compare mode* and *pulse-width modulation* mode. In the rest of this section, we discuss these modes in detail.

Input-Capture Mode   In the input-capture mode, the timer latches the content of the timer into a latch whenever the predefined event occurs. This event is often represented as the positive or negative edge of an external signal. By capturing the timer value, many measurements can then be made. The following measurements are most common: period, pulse width, duty cycle, timing reference and event arrival time.

The input-capture mode can be easily implemented. A possible one is shown in Figure 15.24, where a free-running timer and a capture register coupled with a few logic gates are used. The n-bit timer is a modulo 2^n binary up counter, which is cleared when the reset signal is asserted or each time the write signal wr is activated. Once the write signal wr returns to its inactive state, the counter starts to count up. The content of the timer is captured into the capture register whenever the positive edge of the trigger

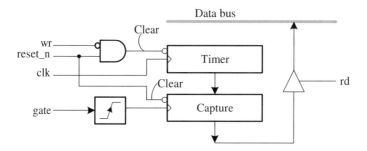

FIGURE 15.24 Block diagram of the timer being operated in input-capture mode

control signal gate is coming. This captured value can then be read by asserting the read signal *rd*.

The following example explains how to describe a timer being operated in the input-capture mode.

EXAMPLE 15.7 *A Timer Being Operated in Input-Capture Mode*

This example only contains two `always` blocks. The first one describes the operations of the timer register and the second one describes the operations of the capture register.

```
// an example of timer with capture mode
module timer_capture(clk, reset_n, gate, wr, rd,
                     data_bus);
parameter N = 16; // default size of timer
input clk, reset_n, gate, wr, rd;
inout [N-1:0] data_bus;
// internal used reg variables
reg  [N-1:0] timer, capture;
wire timer_reset;
assign data_bus = rd ? capture : {N{1'bz}};
// describe the timer register
assign timer_reset = !reset_n || wr;
always @(posedge clk or posedge timer_reset)
   if (timer_reset) timer <= 0;
   else if (!wr) timer <= timer + 1;
// capture the timer into capture register
always @(posedge gate or negedge reset_n)
   if (!reset_n) capture <= 0;
   else capture <= timer;
endmodule
```
■

Output-Compare Mode   In the output-compare mode, the timer compares the content of the timer against a value stored in the latch register, as illustrated in Figure 15.25. The out terminal is set to 1 for a one-clock cycle if both values are

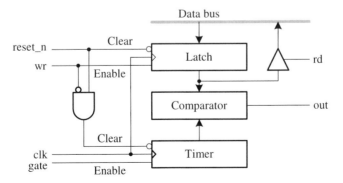

FIGURE 15.25 Block diagram of the timer being operated in output-compare mode

equal; otherwise, it is cleared to 0. The applications of the output-compare mode at least include the following: to generate a time delay, to trigger an action at some future time and to generate a digital waveform.

A possible implementation of this mode is shown in Figure 15.25, which is composed of a latch register, a timer and a comparator. The latch stores the value to be compared against the content of the timer. The timer, an n-bit binary-up counter, starts from 0 enabled by the control input gate. The timer is reset to 0 each time the reset signal is asserted or a new value is written into the latch register. The comparator compares both the contents of the latch register and timer. It generates a one-cycle high-level pulse output when they are matched; otherwise, it remains at low-level.

The following example illustrates how to describe the timer being operated in output-compare mode.

EXAMPLE 15.8   *A Timer Being Operated in Output-Compare Mode*

This example contains two `always` blocks to separately describe the latch register and the timer. The compare operation is performed by a continuous assignment.

```
// an example of timer with compare mode
module timer_compare(clk, reset_n, gate, wr, rd, data_bus,
                     out);
parameter N = 16; // default size of timer
input clk, reset_n, gate, wr, rd;
inout [N-1:0] data_bus;
output out;
// internal used reg variables
reg  [N-1:0] timer, latch;
wire timer_reset;
assign data_bus = rd ? latch : {N{1'bz}};
// describe the latch which stores the compare value
always @(posedge clk or negedge reset_n)
   if (!reset_n) latch <= 0;
   else if (wr) latch <= data_bus;
// describe the timer --- reset by reset_n and wr
// and enabled by gate input
assign timer_reset = !reset_n || wr;
always @(posedge clk or posedge timer_reset)
   if (timer_reset) timer <= 0;
   else if (gate) timer <= timer + 1;
// compare the timer against the value stored in latch
assign out = (timer == latch);
endmodule
```

Pulse-Width Modulation (PWM) Mode   In the PWM mode, the timer generates the output signal with the specified duty-cycle and frequency. The PWM mode is

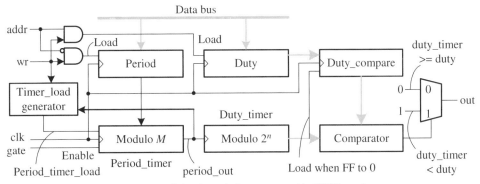

FIGURE 15.26 Block diagram of the timer being operated in PWM mode

employed in a wide variety of applications, ranging from measurement and communications to power control and conversion.

In essence, PWM is a way of digitally encoding analog signal levels through the use of high-resolution counters. When implemented with timers, two timers are required. One is for controlling the period and the other for controlling the *duty cycle*. The duty-cycle of a signal is defined as follows:

$$\text{duty-cycle} = \frac{\text{duration of high output}}{\text{period}} \times 100\,\% \qquad (15.3)$$

One simple possible implementation is illustrated in Figure 15.26. Two timers are separately used to define the period and the duty cycle. The timer used to control the period is a module M down counter, named as a period_timer, which reloads the count value stored in the period register whenever the period_timer counts down to 0, in which a period_out is generated, or a new value is written into the period register.

The timer used to control the duty cycle is a module 2^n binary-up counter, which is denoted as a duty_timer. The content of the duty_timer is compared to a duty value stored in the duty_compare register. The out terminal is set to high if the content of the duty_timer is less than the duty value and set to low otherwise. In order to avoid the output glitch, the duty register is used to accept a duty value from the data bus. The duty value is then loaded into the duty_compare register each time the output of the duty_timer changes from *FF* to 00.

The following example describes how to model the timer being operated in PWM mode depicted in Figure 15.26.

EXAMPLE 15.9 *A Timer Being Operated in PWM Mode*

The first two `always` blocks describe the period and duty registers, respectively. The next `always` block models the `duty_compare` register, which is loaded when the `duty_timer` changes from *FF* to 00. The next two `always` blocks describe the period_timer and duty_timer, respectively. The last always block generates a single-cycle period_timer_load control signal for loading the period_timer. The output signal is generated by a continuous assignment.

```verilog
// an example of timer being operated in pwm mode
module timer_pwm(clk, reset_n, gate, wr, addr, data_bus,
                 out);
parameter N = 16; // default size of timer
input clk, reset_n, gate, wr, addr;
inout [N-1:0] data_bus;
output out;
reg period_timer_load, state;
wire period_out, duty_out;
localparam S0 = 1'b0, S1 = 1'b1;
// internal used reg variables
reg  [N-1:0] period_timer, period, duty_timer, duty,
             duty_compare;
// describe the period register which stores the period
always @(posedge clk or negedge reset_n)
   if (!reset_n) period <= 0;
   else if (wr && (addr == 0)) period <= data_bus;
// describe the duty register which stores the duty
always @(posedge clk or negedge reset_n)
   if (!reset_n) duty <= 0;
   else if (wr && (addr == 1)) duty <= data_bus;
// describe the duty-compare register
always @(posedge clk or negedge reset_n)
   if (!reset_n) duty_compare <= 0;
   else if (duty_out) duty_compare <= duty;
// describe the period_timer, a presettable down counter
always @(posedge clk or negedge reset_n)
   if (!reset_n) period_timer <= 0;
   else if (period_timer_load || period_out)
                period_timer <= period;
      else if (gate) period_timer <= period_timer - 1;
assign period_out = ~|period_timer;
// describe the duty_timer, a modulo 2^N up counter
always @(posedge clk or negedge reset_n)
   if (!reset_n) duty_timer <= 0;
   else if (period_out) duty_timer <= duty_timer + 1;
assign duty_out = &duty_timer;
// generate one-cycle period_timer_load control signal
always @(posedge clk or negedge reset_n)
   if (!reset_n) begin period_timer_load <= 1'b0;
                       state <= S0; end
```

```
      else case (state)
        S0: if (wr && (!addr)) begin state <= S1;
                period_timer_load <= 1'b1; end
            else state <= S0;
        S1: begin period_timer_load <= 1'b0;
            if (wr && (!addr)) state <= S1;
            else state <= S0; end
      endcase
// generate the output signal
assign out = (duty_timer < duty_compare) ? 1'b1 : 1'b0;
endmodule                                                    ■
```

Review Questions

Q15.19 Explain the function of the timer_load generator in Figure 15.20.

Q15.20 Explain the rationale behind the timer being operated in one-shot mode.

Q15.21 Explain the rationale behind the timer being operated in square-wave mode.

Q15.22 Define the duty-cycle.

Q15.23 Modify the module `timer_capture` so that it can capture the timer value at the negative edge of the external trigger signal.

15.5 UNIVERSAL ASYNCHRONOUS RECEIVER AND TRANSMITTER

Most μC systems have one or more serial data ports used to communicate with serial devices such as μC emulators. The serial data port can be asynchronous or synchronous. In this section, we are concerned with the design of an asynchronous serial port.

15.5.1 UART

Currently, the most widely used device for asynchronous serial data transmission and receiving is known as the *universal asynchronous receiver and transmitter* (UART). Its hardware and software models are shown in Figure 15.27. The hardware model includes the CPU interface and the I/O interface, as depicted in Figure 15.27(a). The CPU interface consists of *data_bus*, *reset_n*, *chip select* (*cs*), *read and write control* (*r_w*), *clock* (*clk*) and *register select* (*rs*). The I/O interface includes two data lines and two clock signals. The transmitter data line *TxD* is employed to send the data out of the device and the receiver data line *RxD* is used to receive data from outside of the device. Both data lines, *TxD* and *RxD*, are controlled by two local clock signals: *TxC* and *RxC*, respectively. The frequency of the clock *RxC* is often much higher than *TxC*, which is usually set to 4, 16 or 64 times.

The software model consists of four registers, *receiver data register* (*RDR*), *transmitter data register* (*TDR*), *status register* (*SR*) and *control register* (*CR*), as

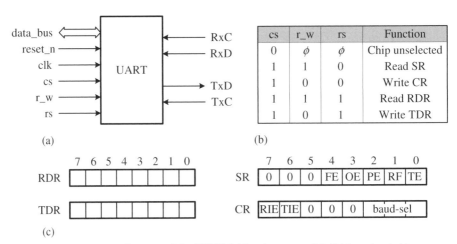

(a)

(b)

cs	r_w	rs	Function
0	ϕ	ϕ	Chip unselected
1	1	0	Read SR
1	0	0	Write CR
1	1	1	Read RDR
1	0	1	Write TDR

(c)

FIGURE 15.27 Logic diagram of the UART: (a) hardware model; (b) function table; (c) programming model

shown in Figure 15.27(c). The RDR and SR can only be read out while TDR and CR can only be written into. As a consequence, only one register select (rs) is required. Both registers SR and CR are selected when rs is 0 and both RDR and TDR are selected when rs is 1. In addition, a chip select (cs) is also used to enable the chip; more precisely, it only controls the read and write of the four registers but does not affect the operations of TxD and RxD. The function table is shown in Figure 15.27(b). The details of each bit of both SR and CR will be described in their related sub-sections.

In order to design a useful UART device, the following design issues must be taken into consideration:

- *Baud rate.* The baud rate is the number of signaling events per unit time. If only one bit is sent per signaling interval, the baud rate is equal to bits per second (bps). For example, in a four-level voltage signaling scheme, the bit rate is twice the baud rate. In this present book, we only consider the case of binary transmission, for which only one bit is sent per signaling interval. As a result, the baud rate is used interchangeably with the bit rate. It ranges from 300 to 19 200, or even more.

- *Sampling clock frequency.* The sampling clock frequency determines the error tolerance in transmitter and receiver clock mismatch. The most common sampling clock frequency is $RxC = n \times TxC$, where n is often set to 1, 4, 16 or 64.

- *Stop bits.* The stop bits are mandatory and used coupled with the start bit to define a character (byte) frame, as shown in Figure 15.14. The number of stop bits may be 1, 1.5 or 2 bits. The most often used is one bit.

- *Parity check.* The parity check may be either even or odd. A parity is called even parity if the number of 1s of the information and the parity bit is even. Otherwise, the parity is called odd parity. Parity check is optional. In some UART devices, this bit can also be programmed as the ninth data bit.

In the rest of this section, we deal with how to design such a serial data transmission and receiving device in detail. The typical device is composed of three major parts:

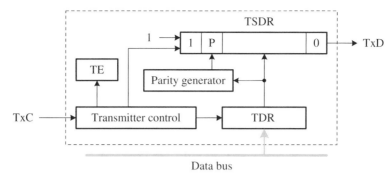

FIGURE 15.28 Logic diagram of the transmitter of a typical UART

transmitter, *receiver* and *baud-rate generator*. For simplicity, we will use a character frame that consists of one start bit, eight data bits, one even parity bit and one stop bit. In addition, the sampling clock frequency is assumed to be 8 times the transmission clock frequency.

15.5.2 Transmitter

The essential component of the transmitter is a shift register that shifts right its content out into TxD continuously with the LSB first and the MSB bit filled with 1. Figure 15.28 shows the basic components of a typical transmitter. The transmitter is composed of a *TDR*, a *transmitter shift data register* (*TSDR*), a *TDR empty flag* (*TE*), a *parity generator* and a *transmitter control circuit*. The *TDR* accepts the data to be transmitted from the data bus. Once the *TDR* is filled, the content of the *TDR*, along with its even parity bit, is loaded into the *TSDR* at the next positive edge of the transmitter clock (*TxC*). The content of the *TSDR* is then continuously shifted out to TxD by the TxC clock.

The detailed operations of the transmitter are illustrated by the ASM chart shown in Figure 15.29. Figure 15.29(a) shows the operations of the CPU, which must detect the flag *TE* before it can write data into the *TDR*. Once the data is in the *TDR*, it is then transferred into the *TSDR*. Then, the *TE* flag is cleared and the transmitter starts to send the content of the *TSDR* out through TxD in the way of a bit-by-bit manner. After the entire frame is finished, the transmitter goes back to the idle state and is ready for sending further bytes. In the idle state, if it has no data to be sent, it continuously sends 1s.

The following example realizes the ASM chart shown in Figure 15.29(b). Here, the standard paradigm introduced in Section 11.2.2 is employed.

EXAMPLE 15.10 *An Example of the Transmitter of a UART*

Basically, the module exactly follows the ASM chart depicted in Figure 15.29(b), except the addition of the first two `always` blocks that are employed to load the TDR and update the TDR empty flag (*TE*), respectively.

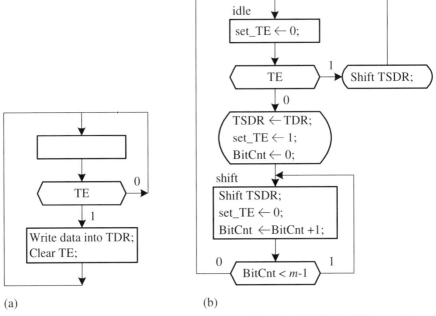

FIGURE 15.29 The ASM chart of the transmitter of a typical UART: (a) CPU operations; (b) TSDR operations

```
// an example of UART transmitter
module UART_transmitter(clk, reset_n, load_TDR, data_bus,
                        TE, TxC, TxD);
parameter M = 10; //the frame size excluding the stop bit
input clk, reset_n, TxC, load_TDR;
input [7:0] data_bus;
output TxD, TE;
// internal used registers
reg [7:0] TDR;
reg [M-1:0] TSDR;
reg [3:0] BitCnt;
reg ps, ns, set_TE, TE;
localparam idle = 1'b0, shift = 1'b1;
// load TDR when load_TDR is activated
always @(posedge clk  or negedge reset_n)
   if (!reset_n) TDR <= 8'b0;
   else if (load_TDR) TDR <= data_bus;
// update TDR empty flag (TE)
always @(posedge clk or negedge reset_n)
   if (!reset_n) TE <= 1'b1;
   else TE <= (set_TE && !TE) || (!load_TDR && TE);
// load TSDR from TDR and perform data transmission
// step 1: initialize and update state registers
always @(posedge TxC or negedge reset_n)
```

```
   if (!reset_n) ps <= idle;
   else ps <= ns;
// step 2: compute next state
always @(*)
   case (ps)
      idle: if (TE == 1) ns = idle;
            else ns = shift;
      shift: if (BitCnt < M - 1) ns = shift;
            else ns = idle;
   endcase
// step 3: execute RTL operations
always @(posedge TxC or negedge reset_n)
   if (!reset_n) begin TSDR <= {M{1'b1}};
      BitCnt <= 0; set_TE <= 1'b0; end
   else case (ps)
      idle: begin set_TE <= 1'b0;
            if (TE == 1) TSDR <= {1'b1, TSDR[M-1:1]};
            else begin
               TSDR <= {^TDR, TDR, 1'b0};
               set_TE <= 1'b1;
               BitCnt <= 0; end  end
      shift: begin
               TSDR <= {1'b1, TSDR[M-1:1]};
               set_TE <= 1'b0;
               BitCnt <= BitCnt + 1; end
   endcase
assign TxD = TSDR[0] & (ps == shift) | (ps == idle);
endmodule
```

15.5.3 Receiver

The essential component of the receiver is also a shift register that samples and shifts right the content of RxD into a shift register with the LSB first whenever the negative edge of RxD is detected. Figure 15.30 shows the basic components of a typical receiver. The receiver is composed of an RDR, a *receiver shift data register* ($RSDR$), an RDR

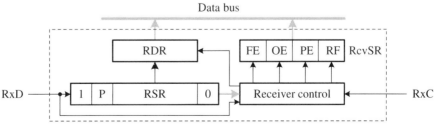

FIGURE 15.30 Logic diagram of the receiver of a typical UART

full flag (*RF*), along with another three flags, and a *receiver control circuit*. Once the entire frame is received, the data part is extracted from the *RSDR* and loaded into the *RDR* to be ready for reading.

There are four flags associated with the receiver. These are *frame error* (*FE*), *overrun error* (*OE*), *parity error* (*PE*) and *RDR full* (*RF*). The flag *FE* indicates the fact that the frame is in error; that is to say, the stop bit is 0 rather than 1. The flag *OE* indicates the fact that a new byte is received before the previous one is read. The flag *PE* indicates that the parity of the byte received is incorrect. The flag *RF* is set to 1 whenever a new byte is extracted from the *RSDR* and loaded into the *RDR*.

The detailed operations of the receiver are illustrated by the ASM chart shown in Figure 15.31. Figure 15.31(a) shows the operations of the CPU, which must detect the flag *RF* before it can read data from the *RDR*. Each time the data is read, the read control signal also clears all four flags: *FE*, *OE*, *PE* and *RF*. There are two counters required in the receiver module. One is employed to count the number of sampling

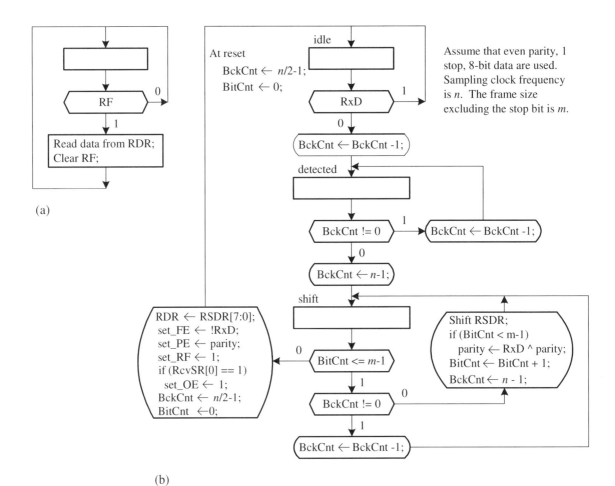

FIGURE 15.31 The ASM chart of the receiver of a typical UART: (a) CPU operations (b) RSDR operations

clocks during each bit time, called the *bit-clock counter* (*BckCnt*), and the other is used to count the number of bits already received, called the *bit-number counter* (*BitCnt*).

After reset, the *BckCnt* and *BitCnt* counters are set to $n/2$ and 0, respectively, where n is the sampling clock frequency. The receiver starts from the *idle* state and enters into the *detected* state once it has detected a negative edge transition of *RxD*. This event also triggers the *BckCnt* to decrease 1. The receiver will stay at this state until the counter *BckCnt* reaches 0. Then, the counter *BckCnt* is reset to $n - 1$ and the receiver enters into the *shift* state to start the reception of data. Once in the *shift* state, the receiver samples the *RxD* every n sampling clocks. This process is repeated until all data bits and the parity bit are received. When this situation is reached, the data is extracted from the register *RSDR* and loaded into the *RDR* register. At the same time, all four flags are updated. The flag *RF* is set to 1 to indicate that the data is ready in the *RDR* for reading. The other flags reflect the actual situations of the received data, that is, the frame error, overrun error and parity error.

The following example realizes the ASM chart shown in Figure 15.31(b). Here, we also use the standard paradigm introduced in Section 11.2.2.

EXAMPLE 15.11   *An Example of the Receiver of a UART*

Basically, this example exactly follows the ASM chart shown in Figure 15.31(b) except for the first two `always` blocks, which are used to update the `RcvSR` register and generate a single-clock-cycle `set_RF_1clk` from `set_RF`, respectively. The `set_RF_1clk` is employed to update the status register `RcvSR`. This is required because the duration of `set_RF` is one *RxC* clock cycle which is much larger than the system clock cycle. If we use `set_RF` directly as the load enable signal for the status register `RcvSR`, then the signal `read_RDR` will not clear the `RcvSR` correctly. Why? Explain this.

```
// an example of UART receiver -- RxC is 8 times TxC
// 1 stop, 8-bit data, and even parity are used
module UART_receiver(clk, reset_n, read_RDR, RDR, RcvSR,
                     RxC, RxD);
parameter M = 10; // default frame size excluding the
                  // stop bit
parameter N = 8;  // default sampling clock frequency
localparam K = 3; // K = log2 N
localparam L = 4; // L = log2 M
input  clk, reset_n, read_RDR, RxC, RxD;
output reg [3:0] RcvSR;
output reg [7:0] RDR; // receiver data register
// internal used registers
reg [M-1:0] RSDR;   // receiver shift data register
reg [L-1:0] BitCnt; // number of bits received
reg [K-1:0] BckCnt; // number of TxC clocks elapsed
reg [1:0] ps, ns;
wire RcvSR_reset;
reg parity, set_FE, set_OE, set_PE, set_RF, set_RF_1clk;
```

```
localparam idle = 2'b00, detected = 2'b01, shift = 2'b10;
// update status register (RcvSR)
assign RcvSR_reset = !reset_n || read_RDR;
always @(posedge clk  or posedge RcvSR_reset)
   if (RcvSR_reset) RcvSR <= 4'b0000;
   else if (set_RF_1clk) RcvSR <= {set_FE, set_OE, set_PE,
                                   set_RF};
// generate set_RF_1clk signal from set_RF
reg state;
localparam S0 = 1'b0, S1 = 1'b1;
always @(posedge clk or negedge reset_n)
   if (!reset_n) begin set_RF_1clk <= 1'b0;
                       state <= S0; end
   else case (state)
      S0: if (set_RF) begin state <= S1;
             set_RF_1clk <= 1'b1; end
          else state <= S0;
      S1: begin set_RF_1clk <= 1'b0;
          if (set_RF) state <= S1; else state <= S0; end
   endcase
// receive data and load RSDR into RDR
// step 1: initialize and update state registers
always @(posedge RxC or negedge reset_n)
   if (!reset_n) ps <= idle;
   else ps <= ns;
// step 2: compute next state
always @(*)
   case (ps)
      idle: if (RxD == 1) ns = idle;
            else ns = detected;
      detected: if (BckCnt != 0) ns = detected;
            else ns = shift;
      shift: if (BitCnt <= M-1)  ns = shift;
            else ns = idle;
      default: ns = idle;
   endcase
// step 3: execute RTL operations
always @(posedge RxC or negedge reset_n)
   if (!reset_n) begin RSDR <= {M{1'b0}};
     RDR <= 8'b00000000;
     BckCnt <= N/2-1; BitCnt <= 0; set_FE <= 1'b0;
     set_OE <= 1'b0; set_PE <= 1'b0;
   set_RF <= 1'b0; parity <= 0; end
   else case (ps)
```

```
      idle: begin set_FE <= 1'b0; set_OE <= 1'b0;
               set_PE <= 1'b0;
               set_RF <= 1'b0; parity <= 1'b0;
               if (RxD == 0) BckCnt <= BckCnt - 1; end
      detected: if (BckCnt != 0) BckCnt <= BckCnt - 1;
               else BckCnt <= N - 1;
      shift: begin
               if (BitCnt <= M - 1) begin
                  if (BckCnt != 0)
                     BckCnt <= BckCnt - 1;
                  else begin
                     RSDR <= {RxD, RSDR[M-1:1]};
                     if (BitCnt < M-1)
                        parity <= RxD ^ parity;
                     BitCnt <= BitCnt + 1;
                     BckCnt <= N - 1;   end end
               else begin RDR <= RSDR[7:0];
                  set_FE <= !RxD; set_PE <= parity;
                  BckCnt <= N/2 - 1; set_RF <= 1'b1;
                  if (RcvSR[0] == 1) set_OE <= 1'b1;
                  BitCnt <= 0; end
            end
   endcase
endmodule
```

15.5.4 Baud-Rate Generator

The baud-rate generator plays an important role in the operations of any UART module because it provides both clock sources, *TxC* and *RxC*, for transmitter and receiver modules, respectively.

The two most widely used approaches for designing baud-rate generators are *multiplexer-based* and *timer-based*, as shown in Figure 15.32. Of course, the essential components of both approaches are the same, that is, counters. Figure 15.32(a) shows the structure of a multiplexer-based baud-rate generator, which consists of a modulo M prescalar counter, a baud-rate counter and a multiplexer. The prescalar counter is utilized to scale down the system clock frequency so that its output is n times the maximum baud-rate allowed in the UART module, where n is the sampling clock frequency. The purpose of the multiplexer is to scale this maximum baud-rate down to many lower possible baud-rates which can then be selected by the baud-selection *baud_sel* set by the control register (*CR*).

Another widely used approach nowadays is by using a timer, with a specific mode known as a timer-based baud-rate generator, as shown in Figure 15.32(b). The timer-based baud-rate generator is composed of a reloadable timer, called a *baud-rate timer*, and a modulo n counter. The baud-rate timer is used to scale the system clock frequency directly down to the required baud-rate times the scale factor (n) of the sampling clock frequency (*RxC*) with respect to the transmission clock frequency (*TxC*).

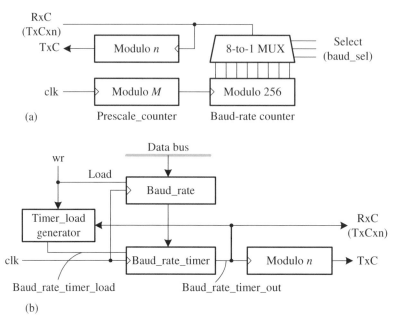

FIGURE 15.32 Logic diagram of a timer-based baud-rate generator: (a) multiplexer-based baud-rate generator; (b) timer-based baud-rate generator

An implementation of the multiplexer-based baud-rate generator is illustrated in the following example. Implementation of a timer-based baud-rate generator is left as an exercise for the reader.

EXAMPLE 15.12 *A Multiplexer-Based Baud-Rate Generator*

This implementation is quite straightforward. Each `always` block corresponds to the functional block shown in Figure 15.32(a). Therefore, we will not further explain it here. Note that in this example, we assume that the sampling clock frequency is 8 times the transmission clock frequency; namely, *n* is set to 8.

```
// an example of multiplexer-based baud-rate generator
module UART_baud_rate(clk, reset_n, baud_sel, TxC, TxCx8);
parameter BD_STEPS = 8; // default number of baud-rate
                        // steps
input clk, reset_n;
input [2:0] baud_sel;
output TxC, TxCx8;
reg TxCx8;
integer i;
// internal used reg variables
wire prescale_out, TxC, TxCx8_clk;
reg  [2:0] prescale_counter;
reg  [BD_STEPS-1:0] baud_rate_counter;
```

```
reg  [2:0] mod8_counter;
// describe the prescale_counter, a presettable down
// counter
always @(posedge clk or negedge reset_n)
   if (!reset_n) prescale_counter <= 0;
   else prescale_counter <= prescale_counter + 1;
assign prescale_out = &prescale_counter;
// describe the a modulo-256 up counter
always @(posedge clk or negedge reset_n)
   if (!reset_n) baud_rate_counter <= 0;
   else if (prescale_out)
            baud_rate_counter <= baud_rate_counter + 1;
// describe the 8-to-1 multiplexers for selecting the
// required baud rate
always @(baud_sel or baud_rate_counter) begin
   TxCx8 = 0;
   for (i = 0; i < BD_STEPS; i = i + 1)
      if (baud_sel == i) TxCx8 = baud_rate_counter[i];
   end
assign TxCx8_clk = TxCx8;
// describe the modulo8_counter
always @(posedge TxCx8_clk or negedge reset_n)
   if (!reset_n) mod8_counter <= 0;
   else mod8_counter <= mod8_counter + 1;
assign TxC = mod8_counter[2];
endmodule                                              ■
```

15.5.5 UART Top Module

After considering the detailed operations of all three separate modules, the top module of the UART can then be constructed. As we have mentioned previously, there are four registers within the underlying UART. Except that the TDR and RDR are associated with the transmitter and receiver, respectively, the status register SR is shared by both transmitter and receiver, and the control register CR is used to control all three modules. As a result, it is more convenient to take care of both the SR and CR registers at the top module. In addition, an interrupt logic is often associated with a useful UART module to make the CPU aware of events, such as TDR empty or RDR full, occurring in the UART. This logic is also easier to be built at the top module.

The following gives an example to show how to put together all three separate modules and construct the related logic circuits described above.

EXAMPLE 15.13 *An Example of the Top Module of a UART*

This top module, in addition to embracing the three separate modules discussed before, processes the interface logic, defines the control register and builds interrupt generation logic.

```
// a UART consists of transmitter, receiver, and baud-
// rate generator
module UART_top(clk, reset_n, r_w, cs, rs, data_bus, RxD,
                TxD, IRQ);
input clk, reset_n, r_w, cs, rs, RxD;
inout [7:0] data_bus;
output TxD, IRQ;
// internal used registers
wire [3:0] RcvSR;
wire RxC, TE, TxC;
wire read_UART, write_UART, load_TDR, read_SR, load_CR;
wire [7:0] RDR;
reg  [7:0] CR;
reg  IRQ;

// instantiate transmitter, receiver, and baud-rate
// generator modules
UART_transmitter my_transmitter (.clk(clk), .reset_n
  (reset_n), .load_TDR(load_TDR), .data_bus(data_bus),
  .TE(TE), .TxC(TxC), .TxD(TxD));
UART_receiver my_receiver (.clk(clk), .reset_n(reset_n),
  .read_RDR(read_RDR), .RDR(RDR), .RcvSR (RcvSR),
  .RxC(RxC), .RxD(RxD));
UART_baud_rate my_baud_rate(.clk(clk), .reset_n(reset_n),
          .baud_sel(CR[2:0]), .TxC(TxC), .TxCx8(RxC));

// generate interface control signals
assign read_UART  = (cs == 1) && (r_w == 1),
       write_UART = (cs == 1) && (r_w == 0);
assign read_RDR = read_UART && (rs == 1'b1),
       load_TDR = write_UART && (rs == 1'b1),
       read_SR  = read_UART && (rs == 1'b0),
       load_CR  = write_UART && (rs == 1'b0);

// read RDR or SR
assign data_bus = (read_RDR) ? RDR :
                  ((read_SR) ? {3'b000, RcvSR, TE}: 8'bz);
// define the control register
always @(posedge clk or negedge reset_n)
   if (!reset_n) CR <= 8'b0000_0000; // control register
   else if (load_CR) CR <= data_bus;

// irq generation logic
always @(posedge clk or negedge reset_n)
   if (!reset_n) IRQ <= 1'b0;
```

```
      else if (((CR[7] == 1) && (RcvSR[0] == 1 ||
            RcvSR[2] == 1)) ||
            (CR[6] == 1) && (TE == 1)) IRQ <= 1'b1;
        else IRQ <= 1'b0;
endmodule
```

Review Questions

Q15.24 Describe the operations of the transmitter of a typical UART.

Q15.25 Describe the operations of the receiver of a typical UART.

Q15.26 Distinguish between bit rate and baud rate.

Q15.27 What are the two basic structures of baud-rate generators?

15.6 A SIMPLE CPU DESIGN

In this section, an instruction set architecture (ISA) from a simple commercial 16-bit CPU embedded in a TI MSP430 system [8] is used as an example to illuminate how to design a hardware for realizing an ISA. In general, there are many implementation options, such as execution cycle and hardware resource, which can be made for a specified ISA. Different considerations of implementation options may often result in different structures of the final hardware.

15.6.1 Fundamentals of CPU

A basic μC system consists of a center processing unit (CPU), which is composed of a datapath and a controller in turn, memory and a set of peripherals interconnected together with a system bus, as shown in Figure 15.33. The system bus includes a data bus, an address bus and a control bus.

The CPU is a programmable digital module through which many different functions can be implemented by using appropriate sub-sets of instructions provided by the

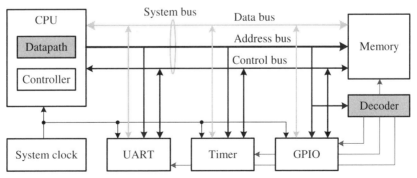

FIGURE 15.33 Block diagram of a typical small microcomputer system

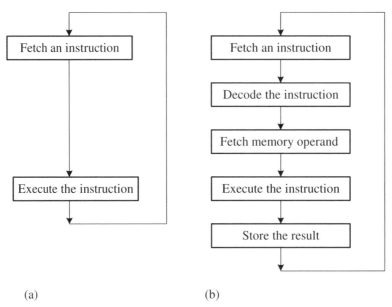

FIGURE 15.34 The basic operations of a typical CPU: (a) CPU basic operations; (b) more detailed CPU operations

ISA of the CPU. For instance, multiplication can be carried out with a shift-and-add algorithm, as we have described in the previous chapter by using a sub-set of instructions from the ISA.

The basic operations of a typical CPU are shown in Figure 15.34(a), in which two steps, *fetch instruction* and *execute instruction*, are repeated forever. After the system is reset, the CPU starts to fetch an instruction from the memory and then executes the instruction. Then, the CPU fetches another instruction and repeats the above steps forever. More detailed operations of a CPU are depicted in Figure 15.34(b). Here, three steps are added into the basic flow shown in Figure 15.34(a): *decode the instruction, fetch memory operand and store the result*. After the CPU fetches an instruction, it decodes the instruction and determines whether a memory operand is required or not. If it is required, the CPU fetches the memory operand before executing the instruction. After an instruction is executed, it may be required to store the result back to memory.

Now that we have realized the basic operations of a typical CPU, we are in a position to learn how to describe a CPU or how to use a CPU. When describing or using a CPU, the following issues are the most important to be taken into account: *the programming model, instruction formats, addressing modes* and *instruction set*.

Programming Model   The programming model is a set of registers that can be accessed by programmers through using instruction set from the CPU. The programming model of the 16-bit CPU considered is shown in Figure 15.35, which contains sixteen 16-bit registers. Among the registers, the first four registers, $R0$, $R1$, $R2$ and $R3$, have dedicated functions, while the other twelve registers, $R4$ to $R15$, are working registers for general use, including data registers, address pointers or index values. Each of these sixteen registers can work as either a 16-bit or an 8-bit (lower byte) register.

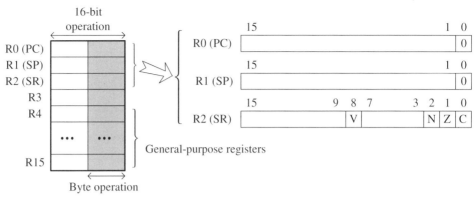

FIGURE 15.35 The programming model of the CPU under consideration

The programming counter (*PC*) (*R*0) always points to the next instruction to be executed. Each instruction uses an even number of bytes and the *PC* increments accordingly. In other words, the LSB of the *PC* is always set to 0.

The stack pointer (*SP*) (*R*1) is employed by the CPU to store the return addresses of sub-routine calls and interrupts. In the underlying CPU, the *SP* is supposed to use a predecrement, postincrement scheme. In addition, register *SP* (*R*1), like other general-purpose registers, can be used with all instructions and addressing modes.

The status register (*SR*) (*R*2) is employed by the CPU to store the ALU flags *N*, *Z*, *V* and *C*, among others. It can only be used as a source or destination register addressed by the register mode. The other addressing modes associated with it are utilized to generate constants, as shown in Table 15.1.

Register *R*3, along with source-operand addressing modes, are employed to generate four useful constants, 0, +1, +2 and −1, as shown in Table 15.1. Consequently, register *R*3 cannot be used as a source register. It can only be used as a destination register. For software programming, it is common practice to leave the first four registers as special registers and only use the rest of the twelve registers for general use.

Instruction Formats   Any instruction is composed of two major parts: *opcode* and *operand*. The opcode field defines the operations of the instruction and the operand field specifies the operands to be operated by the instruction. There are three different

TABLE 15.1 Constant generation

Reg.	As	Constant	Comments
R2	00	—	Register mode
R2	01	(0)	Absolute address mode
R2	10	0x0004	+4, bit processing
R2	11	0x0008	+8, bit processing
R3	00	0x0000	0, word processing
R3	01	0x0001	+1, bit processing
R3	10	0x0010	+2, bit processing
R3	11	0xFFFF	−1, word processing

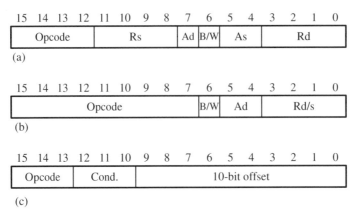

FIGURE 15.36 The instruction formats of the CPU under consideration: (a) double-operand instruction format; (b) single-operand instruction format; (c) jump instruction format

instruction formats as shown in Figure 15.36. The first is for double-operand instructions, which include data transfer, arithmetic operations and logical operations. The second is for single-operand instructions, which include rotate and arithmetic right shift into carry, push onto stack, swap bytes, sub-routine call, return from interrupt and sign extension instructions. The third is for jump instructions, which include a 10-bit offset. The jump instructions include both conditional and unconditional types.

Addressing Modes   Addressing modes are the ways that operands are fetched. The addressing modes include the following major types: *register*, *indexed*, *register indirect* and *immediate*. The variations of indexed mode include *symbolic mode* and *absolute mode*. The *autoincrement mode* is a variation of register indirect. Table 15.2 shows the addressing modes of the CPU under consideration.

Register addressing (also known as *register direct*) is probably the simplest method of referring to a data operand. In this addressing mode, the register contents are the operands.

The indexed addressing mode uses a special index register. In this mode, the address of the operand is formed by adding the content of the index register with a

TABLE 15.2 Addressing modes

As/Ad	Addressing mode	Syntax	Comments
00/0	Register mode	Rn	Register contents are operand
01/1	Indexed mode	X(Rn)	(Rn+X) points to the operand
	Symbolic mode	ADDR	(PC+X) points to the operand
	Absolute mode	& ADDR	X(SR) points to the operand
10/−	Register indirect mode	@ Rn	Rn points to the operand
	Autoincrement mode	@ Rn+	Rn points to the operand and increments 1 or 2
11/−	Immediate mode	#N	PC+

base address. There are three variations of this addressing mode depending on which registers are used.

- *Indexed mode.* In this mode, the address of the operand is at the memory location of $X + Rn$, where X is the next word and Rn is any register of the register file.
- *Symbolic mode.* This is also called the *PC-relative* addressing mode. In this mode, the PC ($R0$) is used as the index register. The address of the operand is at the memory location of $X + PC$, where X is the next word.
- *Absolute mode.* In this mode, the SR ($R2$) is used as the index register. The address of the operand is at the memory location of $X + 0$, where X is the next word.

Register indirect addressing is a way of referring to an operand in memory using a register. In this mode, the address of the operand in memory is the content of the specified register.

Register indirect autoincrement addressing (also known as *autoincrement*) operates much like the register indirect addressing mode except that the specified register is incremented afterward by 1 for .B instructions and 2 otherwise. One variation of this addressing mode is the immediate addressing, which is the case when *PC* is used as the address register.

Instruction Set There are three major sets of instructions of the underlying CPU. These sets are double-operand, single-operand and jump instruction sets.

The double-operand instruction set comprises twelve instructions, as shown in Table 15.3. They are employed to carry out data transfer, arithmetic and logical operations. For arithmetic operations, only addition and subtraction are provided. Seven addressing modes may be applied to obtain the source operand and four addressing modes can be used to specify the destination operand. All four types of operations, register to register, register to memory, memory to register and memory to memory, are provided equally well.

TABLE 15.3 The double-operand instruction set

Mnemonic	Operation	N	Z	V	C	Op code
MOV(.B) src, dst	dst ← src	—	—	—	—	0x4xxx
ADD(.B) src, dst	dst ← src + dst	*	*	*	*	0x5xxx
ADDC(.B) src, dst	dst ← src + dst + C	*	*	*	*	0x6xxx
SUB(.B) src, dst	dst ← .not.src + dst + 1	*	*	*	*	0x8xxx
SUBC(.B) src, dst	dst ← .not.src + dst + C	*	*	*	*	0x7xxx
CMP(.B) src, dst	dst ← −src	*	*	*	*	0x9xxx
DADD(.B) src, dst	dst ← src + dst + C (decimal)	*	*	*	*	0xAxxx
BIT (.B) src, dst	dst .and. src	*	*	0	*	0xBxxx
BIC(.B) src, dst	dst ← .not.src .and. dst	—	—	—	—	0xCxxx
BIS(.B) src, dst	dst ← src .or. dst	—	—	—	—	0xDxxx
XOR(.B) src, dst	dst ← src .xor. dst	*	*	*	*	0xExxx
AND(.B) src, dst	dst ← src .and. dst	*	*	0	*	0xFxxx

TABLE 15.4 The single-operand instruction set

Mnemonic	Operation	N	Z	V	C	Op code
RRA (.B) dst	Arithmetic shift right, C ← LSB	*	*	0	*	0x110x
RRC(.B) dst	Rotate right through carry	*	*	*	*	0x100x
PUSH(.B) src	SP ← Sp ← -2, @ SP src	–	–	–	–	0x120x
SWAPB	Swap bytes	–	–	–	–	0x108x
CALL dst	SP ← SP -2, @ SP ← PC + 2, PC ← dst	–	–	–	–	0x128X
RETI	SR ← TOS, SP ← SP + 2, PC ← TOS, SP ← SP +2,	*	*	*	*	0x130x
SXT dst	dst[15:8] ← dst[7]	*	*	0	*	0x118x

The single-operand instruction set consists of seven instructions, as shown in Table 15.4. They are employed to carry out right rotate with carry (RRC) one bit position, arithmetic right shift into carry (RRA) one bit position, push an operand onto stack (PUSH), swap higher and lower bytes in a word (SWAP), call a sub-routine (CALL), return from interrupt (RETI) and extend the sign bit of lower byte into high byte (SXT). All instructions except the RETI instruction are allowed to use all addressing modes.

The jump instruction set includes eight instructions, as listed in Table 15.5. These instructions are utilized to test the status of a single ALU flag (C, Z, N) or a combination of two flags (N and V). In addition, an unconditional jump is also provided. When the status of the flag being tested is valid, the value of PC is added by a 10-bit offset. The unit of the 10-bit offset is in word; that is, the signed 10-bit offset is doubled before it is added to the PC. As a result, the possible jump range is from -511 to $+512$ words relative to the current PC value.

Review Questions

Q15.28 Describe the basic operations of a typical CPU.

Q15.29 Define addressing mode.

Q15.30 When using a CPU, what are the important issues which need to be considered?

TABLE 15.5 The jump instruction set

Mnemonic	Operation	N	Z	V	C	Op code
JNE/JNZ label	Jump to label if zero bit is reset	–	–	–	–	0x20xx
JEQ/JZ label	Jump to label if zero bit is set	–	–	–	–	0x24xx
JC label	Jump to label if carry bit is set	–	–	–	–	0x2Cxx
JNC label	Jump to label if carry bit is reset	–	–	–	–	0x28xx
JN label	Jump to label if negative bit is set	–	–	–	–	0x30xx
JGE label	Jump to label if (N .xor. V) = 0	–	–	–	–	0x34xx
JL label	Jump to label if (N .xor. V) = 1	–	–	–	–	0x38xx
JMP label	Jump to label unconditionally	–	–	–	–	0x3Cxx

Q15.31 What is the meaning of the programming mode of a CPU?

Q15.32 Explain the meaning of the PC-relative addressing mode.

Q15.33 Can the immediate addressing mode be considered as an autoincrement addressing mode?

15.6.2 Datapath Design

The major components of the datapath of any CPU are a register file and an ALU coupled with some scratch registers and a few multiplexers used to route data to the right places within the datapath. For a specified ISA, the datapath is usually not unique, which will be self-explanatory in the later part of this section. A possible datapath of the underlying ISA is shown in Figure 15.37. In the following, we will detail how to derive this datapath from the underlying ISA.

The datapath is the shared resource for the instruction set. Its function is to facilitate the required operations of each instruction in the ISA so that each instruction can be efficiently carried out. These operations are centered around some registers and an ALU. Consequently, the most straightforward way to obtain a datapath from an ISA is to consider first its core parts, the register file and ALU, and then explore the associated routing paths through the use of multiplexers. The register file and ALU generally play the most important roles in the datapath of any CPU.

Register File   From the double-operand instruction set, it can be easily seen that the register file needs two read ports and one write port so as to support the execution of register-to-register operation in one cycle. However, from the programming model depicted in Figure 15.35 and the constant generator shown in Table 15.1, both registers $R2$ and $R3$ need to be considered specially. Register $R2$ contains the four ALU flags, N,

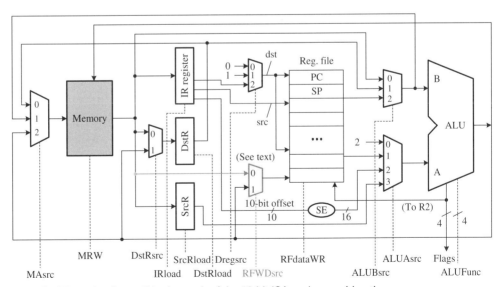

FIGURE 15.37 A possible datapath of the 16-bit ISA under consideration

Z, V and C, which need to be updated each time when the ALU performs an operation. Hence, register $R2$ needs to have the capability of update in addition to the normal write operation through the write port. Furthermore, both registers $R2$ and $R3$ are used to generate constants specified by the combinations of the source addressing modes in the double-operand instruction set. That is, we need to decode the source addressing modes and provide the proper constant outputs when a double-operand instruction being executed specifies the registers $R2$ or $R3$ as its source register.

From the above discussion, we realize that there are many options for implementing the register file. The most intuitive method is to employ flip-flops as building cells. Thus, each 16-bit register can be accessed independently. In this case, the problem of register $R2$ is easily solved. However, two 16-bit 16-to-1 multiplexers are required for providing the two read ports. This may consume much area in cell-based or many LUTs in FPGA-based design.

Another commonly used approach is to declare a multiple-port register file directly, like the one shown in Section 9.2.2, and then use an independent register for $R2$ to solve the update problem associated with it. This is a much better way since it consumes less area or fewer LUTs. Of course, there may exist many other possible implementations that may be used to realize the register file.

ALU　The second core part of the datapath of a CPU is the ALU module. The functions of an ALU used in a datapath can be obtained from observation of the desired operations from the ISA in question. From the instruction sets shown in Tables 15.3 to 15.5, the required ALU operations are listed as Table 15.6.

The simplest way to implement the ALU functions listed in this table is to utilize a case statement to describe the ALU functions in behavioral style. Another way is to modify the ALU described in Section 14.4.2 to fit the required functions of Table 15.6. Due to the simplicity of these two implementations, the details of them are left to the reader as exercises.

TABLE 15.6　The ALU functions required for the ISA in question

Instruction	ALU function	Mode selection				B/W	Mnemonic
		m3	m2	m1	m0		
MOV(.B)	F←A	0	0	0	0	—	pass A
MOV(.B)	F←B	0	0	0	1	—	pass B
ADD(.B)	A+B	0	0	1	0	0/1	add(b/w)
ADDC(.B)	A+B+C	0	0	1	1	0/1	addc(b/w)
SUB(.B), CMP(.B)	A+not B+1	0	1	0	0	0/1	sub(b/w)
SUBC(.B)	A+not B+not C	0	1	0	1	0/1	subc(b/w)
DADD(.B)	A+6H, A+60H conditional	0	1	1	0	0/1	dadd(b/w)
AND(.B), BIT(.B)	A and B	0	1	1	1	0/1	and(b/w)
BIC(.B)	not A and B	1	0	0	0	0/1	bic(b/w)
BIS(.B)	A or B	1	0	0	1	0/1	or(b/w)
XOR(.B)	A xor B	1	0	1	0	0/1	xor(b/w)
RRA(.B)	Arithmetic shift right into C	1	0	1	1	0/1	asrc(b/w)
RRC(.B)	Rotate right through C	1	1	0	0	0/1	rotatec(b/w)
SWAP	Swap byte	1	1	0	1	0	swap

The rest of the datapath besides the register file and the ALU are multiplexers and some scratch registers, as shown in Figure 15.37. These multiplexers and scratch registers are employed to route the data to an appropriate place for carrying out the required operations needed by the instructions. Two scratch registers, *SrcR* and *DstR*, are used to facilitate the operations associated with memory operands. For the source memory operand, it is fetched and stored in SrcR before the instruction is executed. For the destination memory operand, only the memory address of the operand is stored in DstR and the actual operand is deferred to be fetched until the execution steps of the instruction begin. The memory address of the destination operand is formed by adding together a word from memory and the content of a specified register. More details of the datapath will be explained in the 'control unit' sub-section below. The multiplexer with a light line is used only for speeding up the execution of the memory-to-register instructions. Here, we omit it for simplicity.

Review Questions

Q15.34 What is the major function of the datapath in a CPU?

Q15.35 Why do we say that the register file and ALU are the two core parts of a CPU?

Q15.36 Point out the possible path for implementing a memory-to-memory operation from the datapath shown in Figure 15.37.

Q15.37 Point out the possible path for implementing a memory-to-register operation from the datapath shown in Figure 15.37.

Q15.38 Point out the possible path for implementing a register-to-memory operation from the datapath shown in Figure 15.37.

15.6.3 Control Unit Design

The goal of a control unit is to generate the control signals required for executing instructions by the datapath. The control unit accepts the data from instruction register (*IR*) and the ALU flags (*N*, *Z*, *V*, and *C*) and generates the control signals as indicated in Figure 15.37 accordingly.

The most straightforward way to design the control unit is to use the decoder-based approach, as shown in Figure 15.38, where a opcode decoder is used after the

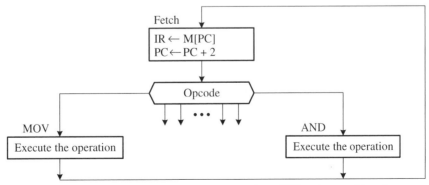

FIGURE 15.38 A possible control unit structure of the CPU under consideration

instruction fetch phase. Each instruction is then executed accordingly. Even though this approach is intuitively simple, the states required are rather large. For the running example, the total number of states will exceed seventy. The resulting state machine is not only hard to handle but also costs too much hardware.

To overcome the above difficulty, a widely used approach is to group the instruction in accordance with addressing modes and the operations of instructions, as shown in Figure 15.39. Here, the instructions are first divided into three different

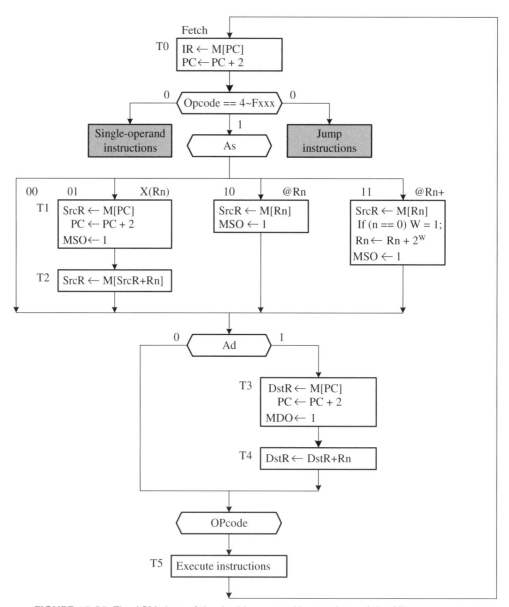

FIGURE 15.39 The ASM chart of the double-operand instructions of the CPU under consideration

sets: double-operand, single-operand and jump instructions. The instructions in the same set have a very similar addressing mode. Hence, they can be processed in a uniform way. As shown in Figure 15.39, all double-operand instructions have the same addressing modes. Therefore, we first handle the addressing mode in order to prepare the operands for the execution steps of the instruction. The total number of states in this approach is the fetch plus the maximum number of execution steps, which is less than ten.

As mentioned before, each instruction before it can be executed has to be fetched from memory first. The fetch step consists of the following two operations:

$$IR \longleftarrow M[PC]$$

$$PC \longleftarrow PC + 2$$

which corresponds to the following control signals sent to the controller: Dregsrc0, ALUBsrc2, MAsrc1, IRload, ALUAsrc0, ALUFunc0 and RFdataWR.

Double-Operand Instructions   The ASM chart of the double-operand instruction set is shown in Figures 15.39 and 15.40. Due to the fact that two operands are used in this instruction set, we first consider the source addressing mode and then deal with the destination addressing mode. As shown in Figure 15.39, for those instructions which use a source operand from memory, we first fetch the operand from the specified memory location and store it in the scratch register $SrcR$ to be ready for execution. In order to indicate this situation, a flag known as a memory source operand (MSO), is used.

The processing of the destination addressing mode is similar to that of the source addressing mode except that now we do not fetch the memory operand into the scratch register $DstR$. Instead, we store the address of the destination operand in $DstR$ and set the memory destination operand flag MDO to indicate this situation.

The actual execution of the double-operand instruction set is shown in Figures 15.40. Each instruction has four cases controlled by combinations of the flags, MSO and MDO. For instance, consider the following instruction:

```
ADD #5, &1000  ; mem[1000] <- mem[1000] + 5
```

which means an immediate data of 5 is added into the memory operand at location 1000. The As field of this instruction is 11 and the Ad field is 1. Thus, according to the ASM chart shown in Figure 15.39, at state $T1$ the scratch register $SrcR$ stores 5 and at state $T3$ the $DstR$ stores 1000. At state $T5$, data 5 are added into the memory location at 1000, as shown in Figure 15.40.

Single-Operand Instructions   The operations of the single-operand instructions are described in the ASM chart shown in Figure 15.41. In these instructions, only source addressing modes are allowed. The operations of these addressing modes are exactly the same as those in the double-operand instructions depicted in Figure 15.39. The detailed operations of each instruction in this set are described in the ASM chart shown in Figure 15.42.

MOV (4xxx)　　　　　　　　　T5

```
case ({MSO,MDO})
00: Rd ← Rs
01: M[DstR]←Rs
10: Rd ← SrcR
11: M[DstR]← SrcR
```

DADD (Axxx)　　　　　　　　　T5

```
case ({MSO,MDO})
00: Rd ← Rd + Rs + C
01: M[DstR]←M[DstR] + Rs + C
10: Rd ← Rd + SrcR + C
11: M[DstR]← M[DstR] + SrcR + C
```

ADD (5xxx)　　　　　　　　　T5

```
case ({MSO,MDO})
00: Rd ← Rd + Rs
01: M[DstR]←M[DstR] + Rs
10: Rd ← Rd + SrcR
11: M[DstR]← M[DstR] + SrcR
```

BIT (Bxxx)　　　　　　　　　T5

```
case ({MSO,MDO})
00: Rd .and. Rs
01: M[DstR] .and. Rs
10: Rd .and. SrcR
11: M[DstR] .and. SrcR
```

ADDC (6xxx)　　　　　　　　　T5

```
case ({MSO,MDO})
00: Rd ← Rd + Rs + C
01: M[DstR]← M[DstR] + Rs + C
10: Rd ← Rd + SrcR + C
11: M[DstR]← M[DstR] + SrcR + C
```

BIC (Cxxx)　　　　　　　　　T5

```
case ({MSO,MDO})
00: Rd ← Rd .and. .not.Rs
01: M[DstR]← M[DstR] .and. .not.Rs
10: Rd ← Rd .and. .not.SrcR
11: M[DstR]← M[DstR] .and. .not.SrcR
```

SUBC (7xxx)　　　　　　　　　T5

```
case ({MSO,MDO})
00: Rd ← Rd + .not. Rs + C
01: M[DstR]← M[DstR]+ .not. Rs + C
10: Rd ← Rd + .not. SrcR + C
11: M[DstR]← M[DstR] + .not. SrcR + C
```

BIS (Dxxx)　　　　　　　　　T5

```
case ({MSO,MDO})
00: Rd ← Rd .or. Rs
01: M[DstR]← M[DstR] .or. Rs
10: Rd ← Rd .or. SrcR
11: M[DstR]← M[DstR] .or. SrcR
```

SUB (8xxx)　　　　　　　　　T5

```
case ({MSO,MDO})
00: Rd ← Rd + .not. Rs + 1
01: M[DstR]← M[DstR]+ .not. Rs + 1
10: Rd ← Rd + .not. SrcR + 1
11: M[DstR]← M[DstR] + .not. SrcR + 1
```

XOR (Exxx)　　　　　　　　　T5

```
case ({MSO,MDO})
00: Rd ← Rd .xor. Rs
01: M[DstR]← M[DstR] .xor. Rs
10: Rd ← Rd .xor. SrcR
11: M[DstR]← M[DstR] .xor. SrcR
```

CMP (9xxx)　　　　　　　　　T5

```
case ({MSO,MDO})
00: Rd + .not. Rs + 1
01: M[DstR]+ .not. Rs + 1
10: Rd + .not. SrcR + 1
11: M[DstR] + .not. SrcR + 1
```

AND (Fxxx)　　　　　　　　　T5

```
case ({MSO,MDO})
00: Rd ← Rd .and. Rs
01: M[DstR]← M[DstR] .and. Rs
10: Rd ← Rd .and. SrcR
11: M[DstR]← M[DstR] .and. SrcR
```

FIGURE 15.40 The detailed operations of the double-operand instructions of the CPU under consideration

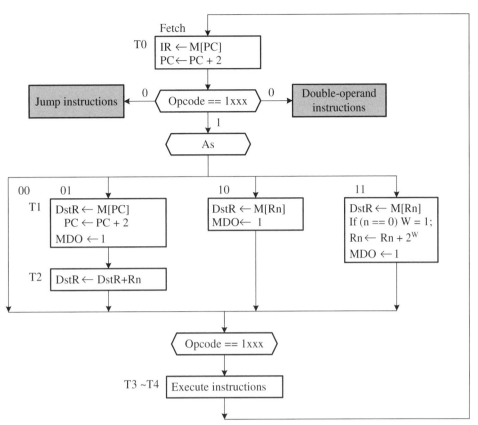

FIGURE 15.41 The ASM chart of the single-operand instructions of the CPU under consideration

Jump Instructions   The operations of the jump instructions are described in the ASM chart shown in Figure 15.43. Instead of processing addressing modes, these instructions take the conditional flags into account. The resulting operations are quite simple—they only need one state—as shown in Figure 15.43.

Review Questions

Q15.39   Show how the instruction `MOV R4, &addr` executes its operations on the datapath shown in Figure 15.37.

Q15.40   Show how the instruction `ADD 3(R4), 8(R5)` executes its operations on the datapath shown in Figure 15.37.

Q15.41   Show how the instruction `RRA &addr` executes its operations on the datapath shown in Figure 15.37.

Q15.42   Show how the instruction `SWAP R8` executes its operations on the datapath shown in Figure 15.37.

Q15.43   Show how the instruction `JGE 1200` executes its operations on the datapath shown in Figure 15.37.

		T3	T4
RRA (0001_0001_00)	MDO =0	Rd ← ARS(Rd)	
	MDO =1	SrcR ← M[DstR]	M[DstR] ← ARS(SrcR)
RRC (0001_0000_00)	MDO =0	Rd ← Rotate(Rd)	
	MDO =1	SrcR ← M[DstR]	M[DstR] ← Rotate(SrcR)
PUSH (0001_0010_00)	MDO =0	SP ← SP -2	M[SP] ← Rs
	MDO =1	SP ← SP -2 SrcR ← M[DstR]	M[SP] ← SrcR
SWAP (0001_0000_10)	MDO =0	Rd ← Swap(Rd)	
	MDO =1	SrcR ← M[DstR]	M[DstR] ← Swap(SrcR)
CALL (0001_0010_10)	MDO =0	SP, DstR ← SP - 2	M[DstR] ← PC + 2 PC ← Rd
	MDO =1	SP, DstR ← SP - 2 SrcR ← M[DstR]	M[DstR] ← PC + 2 PC ← SrcR
RETI (0001_0011_00)		SR ← M[SP] SP ← SP + 2	PC ← M[SP] SP ← SP + 2
SXT (0001_0001_10)	MDO =0	Rd ← SE(Rd)	
	MDO =1	SrcR ← M[DstR]	M[DstR] ← SE(SrcR)

FIGURE 15.42 The detailed ASM chart of single-operand instructions

SUMMARY

The bus structure can be either a tristate bus or a multiplexer-based bus. Due to the fact that a bus is a shared and exclusively used resource, only one transmitter can place data onto the bus at any time. As a result, a bus arbitration logic is required for any bus system to schedule its use. Daisy-chain and radial arbitration are the two most widely used bus schedule schemes. The former embeds a daisy-chain logic into each module through which a priority can be determined in sequence. Each device in the chain has a fixed priority. In contrast, the latter uses separate request and grant lines for each module. The priority of each request line may be fixed or variable. The most widely used variable priority scheme is known as round-robin priority, in which the device just being served is made the lowest priority whereas the device succeeding it is made the highest priority.

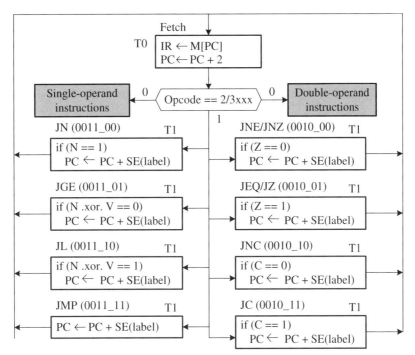

FIGURE 15.43 The ASM chart of the jump instructions of the CPU under consideration

Data transfer on a bus can be proceeded in a synchronous or an asynchronous manner and in a bundle of bits (namely, parallel) or a single bit at a time (namely, serial) as well. When the data transfer between two devices is synchronized by a common clock, it is a synchronous mode; otherwise, it is an asynchronous mode. When the data is transferred in a bundle of bits at the same time, it is a parallel data transfer; when the data is transferred in a bit-by-bit manner, it is a serial data transfer.

In this chapter, we also considered a real-world example that illuminates the design of a small μC system. This system includes a GPIO, timers, a UART and a system bus, including address bus and data bus, as well as control bus. The GPIO is an 8-bit parallel input/output port that can be used as a single 8-bit port or as eight 1-bit ports. The basic operations of timers and how to design such timers are considered in detail. UART is a device that supports the serial communication between two devices in an asynchronous manner. An example is given to illustrate how to design such a device.

The heart of the system is a 16-bit CPU, which provides 27 instructions and 7 addressing modes. In this chapter, we have dealt with the important issues of learning about and designing a CPU. These include the programming model, instruction formats, addressing modes and instruction set. In addition, we use the 16-bit ISA as an example to show how to design its datapath and control unit. For the datapath, we use a register file and ALU as the core components and then instruct how to derive the entire datapath from the operations of the ISA. For the control unit, we have shown how to design an efficient control unit by using ASM charts.

REFERENCES

1. J. Bhasker, *A Verilog HDL Primer,* 3rd Edn, Star Galaxy Publishing, 2005.
2. M. D. Ciletti, *Advanced Digital Design with the Verilog HDL,* Prentice-Hall, Upper Saddle River, NJ, USA 2003.
3. W. J. Dally and B. Towles, *Principles and Practices of Interconnection Networks,* Morgan Kaufmann, San Francisco, CA, USA, 2004.
4. J. P. Hayes, *Computer Architecture and Organization,* 3rd Edn, McGraw-Hill, New York, NY, USA, 1997.
5. S. Palnitkar, *Verilog HDL: A Guide to Digital Design and Synthesis,* 2nd Edn, SunSoft Press, 2003.
6. B. Parhami, *Computer Architecture: From Microprocessors to Supercomputers,* Oxford University Press, New York, NY, USA, 2005.
7. C. H. Roth, Jr, *Digital Systems Design Using VHDL,* PWS, Boston, MA, USA, 1998.
8. Texas Instruments, *MSP430x4xx Family User's Guide,* 2003.

PROBLEMS

15.1 Suppose that a bus system uses a radial arbitration scheme and allows at most eight devices attached onto it at the same time. Use an 8-to-3 priority encoder and a 3-to-8 decoder to realize the bus arbitration logic.

15.2 Consider the handshaking data transfer scheme shown in Figure 15.12. Design a module to facilitate this kind of data transfer. In addition, write a test bench to verify its functionality.

15.3 Consider the case of synchronous serial data transfer with an explicitly clocking scheme, as shown in Figure 15.10(a). Design a module to facilitate this kind of data transfer. In addition, write a test bench to verify its functionality.

15.4 Modify the GPIO introduced in Section 15.3.2 by adding two extra control signals so that it can facilitate the capability of handshaking. In addition, write a test bench to verify its functionality.

15.5 Combine the four basic timer operations into one single timer module with the mode selection as shown in Figure 15.19. Write a test bench to verify the functionality of the resulting timer.

15.6 Design a timer module that performs the rate-generation function described in Figure 15.21.

   (a) Draw the logic circuit of the timer in RTL style.

   (b) Model it using Verilog HDL.

   (c) Write a test bench to verify its functionality.

15.7 Suppose that a timer has two inputs, clock and gate, and one output out, exactly the same as the one introduced in Figure 15.19. However, this timer has only two operation modes. In mode 0, the timer is a 16-bit binary-up counter; in mode 1, it is an 8-bit reloadable binary-up counter. Design this timer and write a test bench to verify its functionality.

15.8 Design a real-time interrupt module that generates a periodic interrupt to remind the CPU to carry out some routine tasks, assuming that the interval of interrupt may be programmable. Write a test bench to verify its functionality.

15.9 Consider the PWM timer shown in Figure 15.26 which operates at a frequency of 12 MHz.

   (a) What is the frequency range that the output signal `out` can be generated?

   (b) What is the duty-cycle that the output signal `out` can be generated?

15.10 Design a UART that has the following features. There are 9 data bits without parity check; that is, the parity bit is used as the ninth data bit. The sampling clock frequency is 16 times the transmission clock frequency and the baud rate is fixed to 1/64 of the system clock frequency. Write a test bench to verify its functionality.

15.11 Design a UART that has the following features. There are 8 data bits with an even parity check. The sampling clock frequency is n times the transmission clock frequency, where n is the scale factor and can be set to 1, 4, 16 or 64. The baud rate can be set to any commonly used value. Write a test bench to verify its functionality.

15.12 The priority interrupt controller (PIC) is a device or a module that is widely used in μC systems. Design a PIC module which at least has the following features. The inputs are eight interrupt request inputs from I/O devices, each of which can be level-sensitive or edge-triggered. The outputs are an interrupt request (INT) signal to the CPU and an 8-bit data bus. The inputs are prioritized and numbered from 0 to 7 with an increasing priority order. Each input can be disabled if it is not allowed to generate an interrupt request signal. The one-byte interrupt vector is output to data bus when an acknowledge is received from the CPU.

15.13 Use D-type flip-flops as basic cells to construct the three-port register file required in the datapath shown in Figure 15.37.

   (a) Draw the logic circuit of the resulting datapath in RTL style.

   (b) Model it using Verilog HDL.

   (c) Write a test bench to verify its functionality.

15.14 The register file may also be designed with memory instead of registers. Based on this idea, design the three-port register file required in the datapath shown in Figure 15.37. It is useful to take the register $R2$ as a special case and construct it with D-type flip-flops.

   (a) Draw the logic circuit of the resulting datapath in RTL style.

   (b) Model it using Verilog HDL.

   (c) Write a test bench to verify its functionality.

15.15 Consider the first ALU introduced in Section 14.4.2.

   (a) Modify it to meet the functions required in Table 15.6. Draw the logic circuit of the result.

   (b) Model it using Verilog HDL.

   (c) Write a test bench to verify its functionality.

15.16 Consider the second ALU introduced in Section 14.4.2.

   (a) Modify it to meet the functions required in Table 15.6. Draw the logic circuit of the result.

(b) Model it using Verilog HDL.

(c) Write a test bench to verify its functionality.

15.17 Consider the datapath shown in Figure 15.37. Write a Verilog HDL module to model it and write a test bench to verify its functionality.

15.18 Consider the case when the multiplexer with light line as shown in Figure 15.37 is used. Analyze Tables 15.3 to 15.5 and show what kinds of instructions can be speeded up when they are executed.

15.19 Suppose that the CPU only supports the double-operand instructions as described in Figures 15.39 and 15.40. Design the control unit and write a test bench to verify its functionality.

15.20 Suppose that the CPU only supports the single-operand instructions as described in Figures 15.41 and 15.42. Design the control unit and write a test bench to verify its functionality.

15.21 Suppose that the CPU only supports the jump instructions as described in Figure 15.43. Design the control unit and write a test bench to verify its functionality.

15.22 The following `always` statement intends to update a status register (SR). The SR is cleared if *reset_n* is asserted or it is read by the control signal *read_SR*. The SR is updated with the new value of *Flags* each time when the load control signal *set_Flag* is asserted. Suppose that the pulse width of the control signal *read_SR* is one system clock (*clk*) period when it is asserted and the pulse width of the control signal *set_Flag* is much wider than one system clock (*clk*) period.

```
assign SR_reset = !reset_n || read_SR;
always @(posedge clk  or posedge SR_reset)
   if (SR_reset) SR <= 4'b0000;
   else if (set_Flag) SR <= Flags;
```

(a) Explain the reason why the above `always` block cannot correctly clear the status register SR when the *read_SR* control signal is asserted.

(b) Show a solution to correct the problem.

DESIGN FOR TESTABILITY

TESTING IS an essential step in any PCB-based, cell-based or FPGA-based system, even in a simple digital logic circuit because the only way to ensure that a system or a circuit may function properly is through a careful testing process. The goal of testing is to find any existing faults in a system or a circuit.

In order to test any faults in a system or a circuit, some fault models must be first defined. The most common fault models found in CMOS technology include stuck-at faults, bridge faults and stuck-open faults. Based on a specific fault model, the system or circuit can then be tested by applying stimuli, observing the responses and analyzing the results. The input stimulus specific for use in a given system or circuit is often called a test vector, which is a combination of input values. A collection of test vectors is called a test set. Except in an exhaustive test, a test set is usually a sub-set of all possible combinations of input values. Four basic types of test vector generations for combinational logic circuits, including fault table, fault simulation, Boolean difference and path sensitization, are described concisely in this chapter. In addition, a simplified D algorithm is introduced and illustrated by an example.

Nowadays, the most effective way to test sequential circuits is by adding some extra circuits to the circuit under test in order to increase both controllability and observability. Such a design is known as a testable circuit design or as a design for testability. The widely used approaches include the ad hoc approach, scan-path method and built-in self-test (BIST). The scan-path method is also extended to system-level testing, such as SRAM, the core-based system and system-on-a-chip (SoC).

16.1 FAULT MODELS

Any test must be based on some kind of fault model. Even though there are many fault models that have been proposed over the last decades, the most widely used fault model in logic circuits is the stuck-at fault model. Moreover, two additional fault

Digital System Designs and Practices Ming-Bo Lin
© 2008 John Wiley & Sons (Asia) Pte Ltd

models, bridging fault and stuck-open fault, are also common in CMOS technology. In this section, we introduce some useful terms, along with their definitions.

16.1.1 Fault Models

A *fault* is the physical defect in a circuit. The physical defect may be caused by process defects, material defects, age defects or even package defects. When the fault manifests itself in the circuit, it is called a *failure*. When the fault manifests itself in the signals of a system, it is called an *error*. In other words, a defect means the unintended difference between the implemented hardware and its design. A fault is a representation of a defect at an abstract function level and an error is the wrong output signal produced by a defective circuit.

The most common fault model used today is still the *stuck-at fault* model, due to its simplicity and the fact that it can model many faults arising from physical defects, such as broken line, opened diode, shorted diode and short-circuit between the power supply and ground. The stuck-at fault model can be *stuck-at-0* or *stuck-at-1*, or both. 'Stuck-at' means that the net will adhere to a logic 0 or 1 permanently. A stuck-at-0 fault is modeled by assigning a fixed value of 0 to a signal line or net in the circuit. Similarly, a stuck-at-1 fault is modeled by assigning a fixed value of 1 to a signal line or net in the circuit. For instance, Figure 16.1(a) shows the stuck-at-0 fault. Because the logic gate under consideration is AND, its output is always 0 whenever one of its inputs is stuck-at-0, regardless of the value of the other input. Figure 16.1(b) shows the stuck-at-1 fault. Because the logic gate under consideration is OR, its output is always 1 whenever one of its inputs is stuck-at-1, regardless of the value of the other input.

In Figure 16.1(a), the output of the AND gate is always 0 whenever one or both inputs, a, b, or the output, f, is stuck-at-0. In Figure 16.1(b), the output of the OR gate is always 1 whenever one or both inputs, a, b, or the output, f, is stuck-at-1. These kinds of indistinguishable faults are called *equivalent faults*. More precisely, an equivalent fault means that two or more faults of a logic circuit transform the circuit in a way such that two or more faulty circuits have the same output function. The size of the test set can be reduced considerably if we make good use of equivalent faults. A process used to select one fault from each equivalence set is known as *fault collapse*. The metric of this is the *collapse ratio*, which is defined as follows:

$$\text{Collapse ratio} = \frac{|\text{set of collapsed faults}|}{|\text{set of faults}|}$$

where $|x|$ denotes the cardinality of the set x.

When a circuit has only one net occurring a stuck-at fault, it is called a *single fault*; when a circuit has many nets occurring stuck-at faults at the same time, this is known as *multiple faults*. For a circuit with n nets, it can have $3^n - 1$ possible stuck-at net

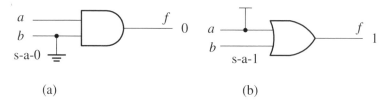

(a)   (b)

FIGURE 16.1 The stuck-at fault models: (a) stuck-at-0 fault; (b) stuck-at-1 fault

combinations but at most $2n$ single stuck-at faults. Of course, a single fault is a special case of multiple faults. Besides, when all faults in a circuit are either stuck-at-0 or stuck-at-1, but cannot be both, the fault phenomenon is named as a *unidirectional fault*.

In a CMOS circuit, there are two additional common fault models: *bridge fault* and *stuck-open fault*. A bridge fault is a shorting between any set of nets unintentionally, such as the example shown in Figure 16.2(a). Under the assumption of ignoring the effect of pMOS transistors, the output f will function as $[(x + z)(w + y)]'$ rather than its original function $(wx + yz)'$ when the bridge fault S_1 occurred and function as $(yz)'$ rather than its original function $(wx + yz)'$ when the bridge fault S_2 occurred. Consequently, bridge faults often change the logic function of the circuit.

In general, the effect of a bridge fault is determined completely by the underlying technique of the logic circuit. In CMOS technology, a bridge fault may evolve into a stuck-at fault or a stuck-open fault, depending on where the bridge fault occurs.

A stuck-open fault is a feature of CMOS circuits. It means that some net is broken during manufacturing or after a period of operation. A major difference between it and stuck-at fault is that the circuit is still a combinational logic when it is in a stuck-at fault but the circuit will be converted into a sequential circuit when it is in a stuck-open fault. To make this point more clear, let us consider Figure 16.2(b), which is a two-input NOR gate. When the nMOS transistor Q_{1n} is at stuck-open fault, the output f will change its function from $(x + y)'$ into a sequential logic function. (Why? Try to explain this.)

Sometimes a *stuck-closed* (*stuck-on*) fault is also considered in CMOS technology. A stuck-open fault is modeled as a switch being permanently in the open state. A stuck-closed (stuck-on) fault is modeled as a switch being permanently in the shorted state. The effect of a stuck-open fault is to produce a floating state at the output of the faulty logic circuit while the effect of a stuck-closed fault is to produce a conducting path from power supply to ground.

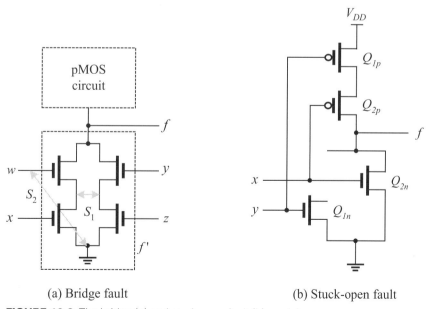

(a) Bridge fault   (b) Stuck-open fault

FIGURE 16.2 The bridge (a) and stuck-open fault (b) models

Review Questions

Q16.1   Define stuck-at-0 and stuck-at-1 faults.

Q16.2   Define bridge fault and stuck-open fault.

Q16.3   What is the meaning of a fault model?

Q16.4   How many single stuck-at faults can an m-input logic gate have?

16.1.2 Fault Detection

Before starting this sub-section, we first define three related terms: *test*, *fault detection* and *fault location*. A test is a response generated by a procedure that applies proper stimuli to a specified fault-free circuit. Fault detection is a procedure which determines whether the under-test circuit is a fault or not by applying a test. The fault location means locating the exact position, or nearby, of a fault. However, before a fault can be located, it is generally necessary to detect whether the fault has occurred. As a result, locating a fault often requires more resources and is more difficult than detecting a fault.

As we have mentioned in Section 13.1.1, the design model for the circuit under test (CUT) can be one of the black box model, white box model and gray box model. In the rest of this chapter, we will assume that the gray box model is used throughout. In other words, the CUT is a black box with a known logic function, even the logic circuit, but all stimuli must be applied at the (primary) inputs and all responses must be detected at the (primary) outputs. The basic model of the circuit under test is shown in Figure 16.3.

In order to test whether a fault has occurred, we need to apply an appropriate stimulus from the inputs to set the net with a fault to be detected to an opposite logic value and propagate the net value to the outputs so as to observe the value and determine whether the fault has occurred. More precisely, the capability of fault detection is based on the following two features of the circuit under test—the *controllability* of a particular node in a logic circuit is a measure of the ease of setting the node to a 1 or a 0 from the inputs, while the *observability* of a particular node in a logic circuit is the degree to which you can observe the node at the outputs.

In combinational logic, if at least one test can be found to determine whether a specified fault has occurred, the fault is called a *detectable fault* or a *testable fault*; otherwise, the fault is known as an *undetectable fault*. In other words, a detectable fault means that at least one input assignment can be found to make the outputs different between the fault-free and faulty circuits. An untestable fault means that we cannot find any input assignment to make the outputs different between the fault-free and faulty circuits, namely, a fault for which no test can be found. The metric of faults that can be

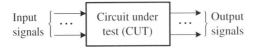

FIGURE 16.3 The basic model of the circuit under test

detected is known as *fault coverage*, which is defined as follows:

$$\text{Fault coverage} = \frac{\text{number of detected faults}}{\text{total number of faults}}$$

To achieve a world-class quality level, the fault coverage of a circuit has to be in excess of 98.5 %. In theory, the condition that all stuck-at faults of a combinational logic can be detected is that the combinational logic must be an irredundant (or irreducible) logic circuit; namely, its logic function must be an irredundant switching expression.

EXAMPLE 16.1 *Detectable and Undetectable Stuck-at Faults*

Figure 16.4 is a logic circuit with a redundant gate; that is, it is not an irredundant logic circuit. From the basic features of AND and OR gates, in order to make the output of an AND gate equal to the value set by a specified input, the rest of the inputs of the gate must be set to 1; to make the output of an OR gate equal to the value set by a specified input, the rest of the inputs of the gate must be set to 0. As a result, nets α and β can be set independently to 0 or 1. Both α_{s-a-0} and β_{s-a-0} are unobservable faults but α_{s-a-1} and β_{s-a-1} are observable faults (why?). Hence, α_{s-a-0} and β_{s-a-0} are undetectable faults but α_{s-a-1} and β_{s-a-1} are detectable faults.   ■

From the above example, we can draw a conclusion. A fault of a CUT that is detectable or undetectable depends on the following two parameters of the fault: controllability and observability. As mentioned before, when a combinational logic is an irredundant logic circuit, any physical defect occurring on any net of the circuit will cause the output value to change so as to differ from its normal value. Hence, it can be detected by a test.

Review Questions

Q16.5 Define test, fault detection and fault location.

Q16.6 Define detectable and undetectable faults.

Q16.7 Define controllability and observability.

Q16.8 How many single stuck-at faults does Figure 16.4 have, once the redundant AND gate is removed? Can they all be detectable?

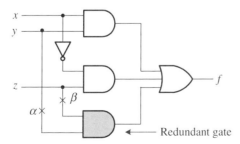

FIGURE 16.4 A redundant gate may cause some nets associated with it to be untestable

16.1.3 Test Vector

In the course of testing a circuit, it is necessary to find a simple set of input signals that can be applied to the input of the circuit. This set of input signals is called a *test set.* A combination of input signals used to test a specified fault is called the *test vector* for the fault. A test set of a circuit is a union of test vectors for the circuit. If a test set can test all testable (detectable) faults of a circuit, the test set is known as a *complete test set.*

In general, a truth table of a circuit is a complete test set of the circuit. However, there are 2^n possible combinations of an n-input combinational circuit. Consequently, it is impossible or ineffective for a large n in practice. Fortunately, we do not generally need to use the entire truth table as a complete test set. Through carefully examining the logic circuit, we usually can find a complete test set with a much smaller size than that of the truth table. The following example illustrates this idea.

EXAMPLE 16.2 *An Example of a Complete Test Set*

Figure 16.5 shows a two-input NAND gate. As mentioned above, the simplest way to test this circuit is to apply the four combinations of inputs a and b one by one and then compare the result of the output with its truth table to determine whether it is faulty or not.

When the combination of inputs a and b is 01, the output is 1 if it is fault-free but is 0 if net α is stuck-at 1 or γ is stuck-at 0. When the combination of inputs a and b is 10, the output is 1 if it is fault-free but is 0 if net β is stuck-at 1 or γ is stuck-at 0. When the combination of inputs a and b is 11, the output is 0 if it is fault-free but is 1 if nets α or β are stuck-at 0 or γ is stuck-at 1. Due to the fact that the above three combinations have completely tested all possible single stuck-at faults, they constitute a complete test set. That is, the complete test set of the two-input NAND gate is {01, 10, 11}. It saves a 25% test cost compared to an exhaustive test with the entire truth table. ■

In summary, the technique of using all possible combinations of input values of a circuit to test the circuit is known as an *exhaustive test*, which belongs to a kind of complete test. However, it can usually find a complete test set of smaller size than that of the truth table if we examine the circuit under test more carefully.

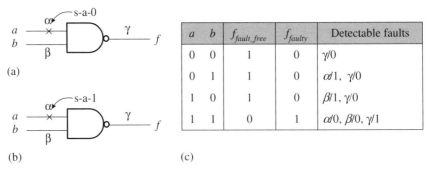

a	b	f_{fault_free}	f_{faulty}	Detectable faults
0	0	1	0	$\gamma/0$
0	1	1	0	$\alpha/1,\ \gamma/0$
1	0	1	0	$\beta/1,\ \gamma/0$
1	1	0	1	$\alpha/0,\ \beta/0,\ \gamma/1$

(a)

(b)

(c)

FIGURE 16.5 An example of a complete test

Review Questions

Q16.9 Define test vector, test set and complete test set.

Q16.10 Define exhaustive test.

Q16.11 Why is the size of a complete test set usually smaller than that of the truth table of the circuit?

16.1.4 Difficulty of Sequential Circuit Test

As mentioned previously, the behavior of a sequential circuit is not only determined by the current inputs but also by the previous outputs which are determined by the previous inputs in turn. This makes the test of the circuit much more difficult. In combinational logic, the purpose of the test is to verify whether the circuit behavior conforms with the truth table. In a sequential circuit, the purpose of the test is to verify whether the circuit behavior conforms with the state table. As a result, the fault detection of sequential circuits can be divided into two phases:

1. Transfer the state of the circuit under test to a known state.

2. Test all state transitions of the circuit under test.

The first phase is often accomplished by using a sequence known as a *homing sequence*, which is an input sequence that produces an output response that uniquely identifies the final state of a machine after the homing sequence has been applied. For a minimal state diagram, there exists at least one homing sequences. One variation of homing sequences is called a *preset homing sequence*, which is a homing sequence that does not employ the output response to determine subsequent inputs in the sequence.

EXAMPLE 16.3 *Homing Sequence Illustration*

Suppose that in the state table shown in Figure 16.6(a), the initial state is unknown. The state of the state machine is transferred to state B or D after a 0-value input x is applied. The state is transferred to state B if the 0-value input x is applied again and transferred to state A or C, depending on whether the output value is 0 or 1, if the 0-value input x is applied one more time. The other input sequences and state transitions are shown in Figure 16.6(b).

Figure 16.6(c) shows the output sequences and final state under the input sequences: 00, 01 and 11. The final state can be uniquely determined from the output sequences under the input sequences, 00 and 01. However, the final state may be A or C if the output sequence is 11 under the input sequence 11. Consequently, both input sequences 00 and 01 are homing sequences but 11 is not. ∎

In order to test all state transitions of a state machine, it is necessary to transfer from one state to another. The sequence used for this purpose is called a *transfer sequence*. In a formal definition, a transfer sequence for state S_i and S_j of a state machine is the shortest input sequence that will take the machine from state S_i to S_j.

(a)

PS	NS, z, x	
	0	1
A	B,0	A,1
B	B,0	A,0
C	D,0	C,1
D	B,0	C,1

(b)

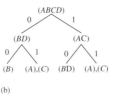

(c)

Initial state	Input sequence					
	00		01		11	
A	00	B	00	A	11	A
B	00	B	00	A	01	A
C	00	B	01	C	11	C
D	00	B	00	A	11	C

FIGURE 16.6 An example to illustrate the difficulty of testing a sequential circuit: (a) state table; (b) homing sequence; (c) output sequence and final state

EXAMPLE 16.4 *Transfer Sequence*

In Figure 16.6(a), to transfer the machine from state C to D, a 0-value input x has to be applied; to transfer the machine from state D to B, a 0-value input x has to be applied; to transfer the machine from state C to A, a sequence 001 may be applied. These sequences are all transfer sequences. ■

A state diagram (machine) is strongly connected if and only if there exists for each ordered pair of states (S_i, S_j) of the machine an input sequence that will transfer the machine from state S_i to S_j. For a strongly connected state diagram, a proper sequence can transfer the state diagram from one state to another. As a result, for any strongly connected minimal state diagram, the following procedure may be used to transfer it to a known state.

Procedure: Initial State Setting of a Sequential Circuit

1. Select an appropriate homing sequence.
2. Apply the homing sequence and observe the output sequence.
3. Determine the final state from the output sequence observed.
4. If the final state S_i is not the desired state S_j, apply a proper transfer sequence to transfer the circuit to the final state S_j.

From the above discussion, we can realize that it is quite difficult to test an arbitrary sequential circuit directly because the state diagram of the sequential circuit is not guaranteed to be strongly connected and hence it might be impossible to transfer from a known state to the state to be tested. Consequently, at present, the most widely used approach to overcoming this difficulty of testing is by adding some extra circuits so that it can directly set the state of the circuit to the desired one without resorting to homing and transfer sequences. We will describe this important technique in more detail later in this chapter.

Review Questions

Q16.12 Define homing sequence and transfer sequence.

Q16.13 What is a strongly connected state diagram?

Q16.14 Is the state diagram of Figure 16.6(a) strongly connected?

16.2 TEST VECTOR GENERATION

As described previously, in order to test a combinational logic circuit, it is necessary to derive a test set with a minimal number of test vectors so as to save the test cost and speed up the test process. Over several decades of development, a great deal of test vector generation methods have been reported. In this section, we only survey the four most basic approaches: *fault table*, *fault simulation*, *Boolean difference* and *path sensitization*. The interested reader can refer to Abramovici *et al.* [1], Bushnell and Agrawal [3] and Wang *et al.* [17] for further details. In industry and in the literature, the terms of test vector and *test pattern* are often used interchangeably.

16.2.1 Fault Table Approach

The fault table approach is a tabular method that uses the truth table of a combinational circuit to aid the generation of test vectors for the circuit. The essential idea behind this approach is on the observation that when a test vector can test a specified fault, it will generate an output value deviated from that of the fault-free output. Consequently, by comparing the fault-free and faulty output values, the test vectors can then be found from the truth table. More precisely, let $f(x_{n-1}, \ldots, x_1, x_0)$ be the fault-free output value under the input values $(x_{n-1}, \ldots, x_1, x_0)$ and $f_\alpha(x_{n-1}, \ldots, x_1, x_0)$ be the output value of the faulty circuit in which the net α is stuck-at fault. Then when:

$$f(x_{n-1}, \ldots, x_1, x_0) \oplus f_\alpha(x_{n-1}, \ldots, x_1, x_0) = 1$$

denotes the input combination $(x_{n-1}, \ldots, x_1, x_0)$ we can detect the stuck-at fault at the net α. That is, the input combination $(x_{n-1}, \ldots, x_1, x_0)$ is a test vector of the stuck-at fault of the net α. In the following, we use the circuit shown in Figure 16.7 as an example to illustrate how the fault table approach is used to derive the test vectors.

In the fault table approach, a fault table is derived by calculating the output values of each input combination under each possible stuck-at fault. The resulting table is known as a *fault truth table* or a *fault table* for short. That is, a fault table is a table that displays a set of faults and a set of test inputs. The fault table of the circuit under consideration is shown in Figure 16.8.

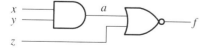

FIGURE 16.7 A circuit example illustrating the fault table approach

x	y	z	f	f_{x0}	f_{x1}	f_{y0}	f_{y1}	f_{z0}	f_{z1}	f_{a0}	f_{a1}	f_{f0}	f_{f1}
0	0	0	1	1	1	1	1	1	0	1	0	0	1
0	0	1	0	0	0	0	0	1	0	0	0	0	1
0	1	0	1	1	0	1	1	1	0	1	0	0	1
0	1	1	0	0	0	0	0	1	0	0	0	0	1
1	0	0	1	1	1	1	0	1	0	1	0	0	1
1	0	1	0	0	0	0	0	1	0	0	0	0	1
1	1	0	0	1	0	1	0	0	0	1	0	0	1
1	1	1	0	0	0	0	0	0	0	0	0	0	1

Truth table   Single stuck-at-fault truth table

FIGURE 16.8 The fault (truth) table of the example circuit

The next step of the fault table approach is to derive a minimal set of test vectors. For this, it is necessary to calculate the value of $f_\alpha \oplus f$, where α represents any possible stuck-at fault. When the value is 1, the corresponding input combination is a test vector for the stuck-at fault of net α. The resulting table is called a *fault detection table*. In other words, a fault detection table is a table that identifies which input combinations can detect some certain faults. It is formed by performing the XOR operation of each output of the stuck-at-fault column in the fault table with the output of a fault-free circuit. An example of a fault detection table is shown in Figure 16.9.

The final step is to reduce the fault detection table by employing the same technique used in reducing the prime implicant table found in the process of minimizing a switching function. The resulting table is often called a *reduced fault detection table*. That is, a reduced fault detection table is the one that combines equivalent faults and deletes the redundant rows in the fault detection table. An example of a reduced fault detection table is shown in Figure 16.10. Be definition, a test set is a minimal set of test vectors that covers all stuck-at faults, which can then be obtained from the reduced fault detection table. In the following, we give a complete example to illustrate how the fault table approach is used to find a test set.

EXAMPLE 16.5 *Fault Table Approach for Finding a Test Set*

In Figure 16.7, there are five nets. Thus, there are ten possible stuck-at faults. The truth table for each possible stuck-at fault is shown in Figure 16.8. By calculating $f_\alpha \oplus f$ for

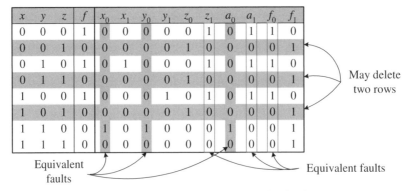

FIGURE 16.9 The fault detection table of the example circuit

x	y	z	f	$\tilde{g}$	$\tilde{x}_1$	$\tilde{y}_1$	$\tilde{z}_0$	$\tilde{h}$	$\tilde{f}_1$
0	0	0	1	0	0	0	0	1	0
0	0	1	0	0	0	0	①	0	1
0	1	0	1	0	①	0	0	1	0
1	0	0	1	0	0	①	0	1	0
1	1	0	0	①	0	0	0	0	1
1	1	1	0	0	0	0	0	0	1

$g = \{x_0, y_0, a_0\}$

$h = \{z_1, a_1, f_0\}$

FIGURE 16.10 The reduced fault detection table of the example circuit

each stuck-at fault, we obtain the fault detection table as depicted in Figure 16.9. The outputs of the stuck-at faults x_0, y_0, and a_0 are the same under all input combinations. Hence, they are equivalent faults. In addition, z_1, a_1 and f_0 are also equivalent faults. Let $g = \{x_0, y_0, a_0\}$ and $h = \{z_1, a_1, f_0\}$. Furthermore, because the output values of the input combinations, 001, 011 and 101, are the same, they can be combined into one row, say 001, without loss of any information when finding the test vectors. The resulting reduced fault detection table is shown in Figure 16.10.

Because each fault of g, x_1, y_1 and z_0, can be only detected by a specific input combination separately, these input combinations are essential and are the test vectors for the above faults. After selecting the essential test vectors, any faults that can be detected by these test vectors are also checked. Since all faults are checked after all essential test vectors are selected, the test vectors of the circuit are $\{(0, 0, 1), (0, 1, 0),$ $(1, 0, 0), (1, 1, 0)\}$, which is the test set. ∎

Review Questions

Q16.15 Describe the basic technique of the fault table approach.

Q16.16 Define fault table and fault detection table.

Q16.17 Define reduced fault detection table.

16.2.2 Fault Simulation

The major drawback of the fault table approach is its time and the space complexities are proportional to an exponential function of the input size. An alternative way to the fault table approach is by using a fault simulation technique much like the ways used to verify the functionality of a design. Fault simulation is to select a primary input combination and determine its fault detection capabilities by simulation.

The basic idea of using fault simulation to find test vectors for a specified fault is as follows. As shown in Figure 16.11, for each combination of inputs, both fault-free and faulty circuits are simulated in parallel and their outputs are observed. When an input

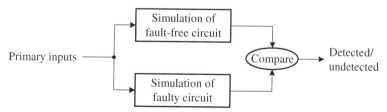

FIGURE 16.11 The basic model of the fault simulation approach

x	y	z	f	f_{x0}	f_{x1}	f_{y0}	f_{y1}	f_{z0}	f_{z1}	f_{a0}	f_{a1}	f_{f0}	f_{f1}	Faults detected
0	0	0	1	1	1	*	1	*	0→1	0→0→1			1	$z1, a1, f0$
0	0	1	0	0	0	*	0	*	*	0	*	*	1	$f1$
0	1	0	1	1	0	1	1	*	*	1	*	*	*	$x1$
0	1	1	0	0	*	0	0	*	*	0	*	*	*	
1	0	0	1	1	*	1	0	1	*	1	*	*	*	$y1$
1	0	1	0	0	*	0	*	1	0	0	*	*	*	$z0$
1	1	0	0	1	0	1	0	*	0	1	0	*	*	$x0, y0, a0$
1	1	1	0	*	0	*	0	*	0	*	0	*	*	

Truth table   * --- not simulated for this input pattern.

FIGURE 16.12 The column method of fault simulation

combination causes both circuits to produce different outputs, the input combination is the test vector and it then proceeds to simulate another fault; otherwise, another input combination is applied next. If both circuits produce the same output for all input combinations, the specified fault is undetectable; otherwise, the fault is detectable. The process is repeated until all faults are covered, at least an acceptable number of faults are covered or some predefined stopping point is reached.

There are two fundamental methods of generating a test pattern using the fault simulation approach: *the column method* and *the row method*. We describe each of these in the following in order.

In the column method of fault simulation, we select a fault to be simulated and apply the input combination in sequence until a test vector for the fault is found, as shown in Figure 16.12, and then continue to the next fault from the current input combination until the end and then return to the initial input combination. This process is repeated until all faults are considered.

The row method of fault simulation is proceeded in an alternative direction. Rather than fixing a fault, it fixes input combination and considers the faults to be tested one by one, as shown in Figure 16.13, until the end of the fault list and then applies the next input combination. This process is repeated until all faults are considered. During the course of simulation, a test vector for the fault is found when both circuits produce different output values.

x	y	z	f	f_{x0}	f_{x1}	f_{y0}	f_{y1}	f_{z0}	f_{z1}	f_{a0}	f_{a1}	f_{f0}	f_{f1}	Faults detected
0	0	0	1	1	1	1	1	1	0	1	0	0	1	$z1, a1, f0$
0	0	1	0	0	0	0	0	1	*	0	*	*	1	$z0, f1$
0	1	0	1	1	0	1	1	*	*	1	*	*	*	$x1$
0	1	1	0	0	*	0	0	*	*	0	*	*	*	
1	0	0	1	1	*	1	0	*	*	1	*	*	*	$y1$
1	0	1	0	0	*	0	*	*	*	0	*	*	*	
1	1	0	0	1	*	1	*	*	*	1	*	*	*	$x0, y0, a0$
1	1	1	0	*	*	*	*	*	*	*	*	*	*	

Truth table   * --- not simulated for this input pattern.

FIGURE 16.13 The row method of fault simulation

After the test vectors are found regardless of whether the column or row method is used, a table like the reduced fault detection table used in the fault table approach is then constructed to find the minimal test set. The test set found from the column method is $\{(0, 0, 0), (0, 0, 1),(0, 1, 0),(1, 0, 0), (1, 0, 1),(1, 1, 0)\}$ and from the row method is $\{(0, 0, 0), (0, 0, 1),(0, 1, 0),(1, 0, 0),(1, 1, 0)\}$, which are not the minimal sets. This is the general case when using simulation to find test vectors because both methods do not explore the detection capability of an input combination to its maximum extent. In either method, only one test vector is found for each fault. Although both column and row methods of simulation produce a nonminimal test set, they are faster and consume much less memory space than the fault table approach.

16.2.3 Boolean Difference

Although the fault table approach can provide a complete way to find a complete test set for any combinational logic circuit, the sizes of both the truth table and fault detection table are exponentially proportional to the number of inputs. That is, they are proportional to 2^n, where n is the number of inputs. As a result, it is impractical when n is large. In this section, we introduce an alternative approach based on an algebraic technique known as the *Boolean difference* method. Even though Boolean difference is not an effective way to compute test vectors for large circuits, it provides us with another taste of test pattern generation approaches.

Boolean difference is an algebraic technique for developing test patterns for combinational circuits. It is sometimes also called *Boolean partial derivative*. The Boolean difference of a switching function $f(x_{n-1}, \ldots, x_1, x_0)$ with respect to variable x_i is defined as:

$$\frac{df(X)}{dx_i} = f(x_{n-1}, x_{n-2}, \ldots, x_{i+1}, 0, x_{i-1}, \ldots, x_1, x_0) \oplus$$

$$f(x_{n-1}, x_{n-2}, \ldots, x_{i+1}, 1, x_{i-1}, \ldots, x_1, x_0)$$

$$= f_i(0) \oplus f_i(1) \tag{16.1}$$

The following example illustrates how to compute the Boolean difference of a switching function with respect to a given variable.

EXAMPLE 16.6 *A Boolean Difference Example*

Let $f(X) = x_1 x_2 + x_3$ and then the Boolean difference with respect to variable x_1 is

$$\frac{df(X)}{dx_1} = (1 \ldots x_2 + x_3) \oplus (0 \ldots x_2 + x_3)$$

$$= (x_2 + x_3) \oplus x_3$$

$$= (x_2 + x_3)' \ldots x_3 + (x_2 + x_3) \ldots x_3'$$

$$= x_2 x_3'$$

When $df(x)/dx_i = 1$, it means $f(x_{n-1}, \ldots, x_i, \ldots, x_0) \neq f(x_{n-1}, \ldots, x'_i, \ldots x_0)$; that is, $f(X)$ is dependent on the variable x_i. When $df(x)/dx_i = 0$, it means $f(x_{n-1}, \ldots, x_i, \ldots, x_0) = f(x_{n-1}, \ldots, x'_i, \ldots x_0)$; that is, $f(X)$ is independent of the variable x_i. As a result, in order to test the stuck-at fault at net x_i, it is necessary to find an input combination such that $df(x)/dx_i = 1$. That is to say, to test the stuck-at-0 fault at net x_i, it is required to compute:

$$x_i \frac{df(X)}{dx_i} = 1 \tag{16.2}$$

and to test the stuck-at-1 fault at net x_i, it is required to compute:

$$x'_i \frac{df(X)}{dx_i} = 1 \tag{16.3}$$

The following example illustrates how the above technique can be used to find the test vectors for both stuck-at-0 and stuck-at-1 faults of a specified net.

EXAMPLE 16.7 *The Boolean Difference Approach*

As shown in Figure 16.14, the logic function of the circuit is:

$$f(X) = (x_1 + x_2)x'_3 + x_3 x_4$$

To find the stuck-at-0 fault at net α, it needs to calculate:

$$x_3 \frac{df(X)}{dx_3} = 1 = x_3[f_3(0) \oplus f_3(1)]$$

$$= x'_1 x'_2 x_3 x_4 + x_1 x_3 x'_4 + x_2 x_3 x'_4$$

Therefore, the test vector set is $\{(0, 0, 1, 1), (1, \phi, 1, 0), (\phi, 1, 1, 0)\}$.
To find the stuck-at-1 fault at net α, it needs to calculate:

$$x'_3 \frac{df(X)}{dx_3} = 1 = x'_3[f_3(0) \oplus f_3(1)]$$

$$= x'_1 x'_2 x'_3 x_4 + x_1 x'_3 x'_4 + x_2 x'_3 x'_4$$

Therefore, the test vector set is $\{(0, 0, 0, 1), (1, \phi, 0, 0), (\phi, 1, 0, 0)\}$. ∎

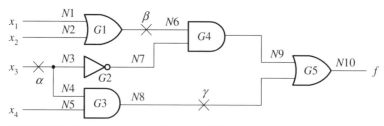

FIGURE 16.14 An example of Boolean difference

To find the test vectors of an internal net β of a circuit, it is necessary to represent the net β as a function of primary inputs first and then denote the output of the circuit as a function of the net β. Finally, the Boolean difference with respect to the net β is found and then the function of β is replaced with the primary inputs. The following example explains how this technique is employed.

EXAMPLE 16.8 *The Boolean Difference Approach*

In order to find the test vectors of net β in Figure 16.14, it is necessary to represent the net β as a function of the primary inputs, namely, $y = x_1 + x_2$, and then represent the f as a function of y:

$$f(X, y) = yx'_3 + x_3x_4$$

The test vectors for the stuck-at-0 fault at net β can be found by evaluating the following equation:

$$y\frac{df(X, y)}{dy} = 1 = y[x_3x_4 \oplus (x'_3 + x_3x_4)]$$

$$= yx'_3 = (x_1 + x_2)x'_3$$

$$= x_1x'_3 + x_2x'_3$$

Hence, the test vector set is $\{(1, \phi, 0, \phi), (\phi, 1, 0, \phi)\}$.

Similarly, the test vectors for the stuck-at-1 fault at net β can be found by evaluating the following equation:

$$y'\frac{df(X, y)}{dy} = 1 = y'[x_3x_4 \oplus (x'_3 + x_3x_4)]$$

$$= y'x'_3 = (x_1 + x_2)'x'_3$$

$$= x'_1x'_2x'_3$$

The test vector set is $\{(0, 0, 0, \phi)\}$. ∎

In summary, the Boolean difference approach can only find the test vectors for a stuck-at fault at a time. Hence, it is necessary to find the test vectors for each fault and then construct a fault detection table to find the minimal test set of the circuit under consideration.

Review Questions

Q16.18 Define the Boolean difference of a switching function.

Q16.19 Find the test vectors of the stuck-at-0 fault at net γ of Figure 16.14.

Q16.20 Find the test vectors of the stuck-at-1 fault at net γ of Figure 16.14.

16.2.4 Path Sensitization and *D*-Algorithm

As described previously, the essence of detecting a fault is to apply a proper set of input signals (test vector) to set the net of interest with an opposite value from its fault value, propagate the fault effect to the primary outputs and observe the output value to determine whether the fault at the net has occurred or not. The path sensitization approach is a direct application of this idea and has the following three steps [3, 15]:

1. *Fault sensitization.* This step sets the net to be tested (namely, the test point) to an opposite value from the fault value. Fault sensitization is also known as *fault excitation.*

2. *Fault propagation.* This step selects one or more paths, starting from the test point (i.e. the specified net) to the outputs, in order to propagate the fault effect to the output(s). Fault propagation is also known as *path sensitization.*

3. *Line justification.* This step sets the inputs to justify the internal assignments previously made to sensitize a fault or propagate its effect. Line justification is also known as *consistency operation.*

The following examples further explain how the path sensitization approach works.

EXAMPLE 16.9   *The Path Sensitization Approach*

As mentioned above, the stuck-at-0 fault at net β of Figure 16.15 is tested by first setting the logic value of net β to 1, which requires us to set inputs x_1 or x_2 to 1 in turn. Then, in order to propagate the logic value at net β to the output, a path from net β to the output should be established. This means that the input x_3 has to be set to 0, which also sets the net γ to 0. As a result, the value of the input x_4 is irrelevant. Therefore, the test vectors are the set: $\{(1, \phi, 0, \phi), (\phi, 1, 0, \phi)\}$, which is the same as that of the preceding example. ∎

A path is called a *sensitizable path* for a stuck-at-fault net if it can propagate the net value to the primary outputs with a consistency operation. Otherwise, the path is called an *unsensitizable path* for the stuck-at-fault net. The following example illuminates how the sensitizable and unsensitizable paths occurs in an actual logic circuit.

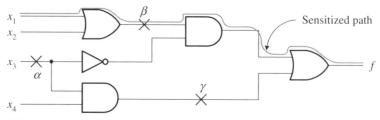

FIGURE 16.15 An example of path sensitization

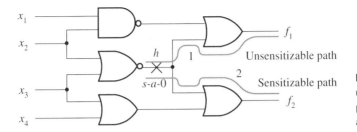

FIGURE 16.16 An unsensitizable path in the path sensitization approach

EXAMPLE 16.10   *The Hazard of the Path Sensitization Approach*

When the stuck-at-0 fault at net h shown in Figure 16.16 is to be tested, both inputs x_2 and x_3 have to be set to 0 in order to set the logic value at net h to 1. Two possible paths from net h could be set as the sensitization paths to the output. If path 1 is used, both inputs x_1 and x_2 need to be set to 1 in order to propagate the logic value at net h to the output. As a result, we have an inconsistent setting of input x_2. The path 1 is said to be an unsensitizable path for the stuck-at-0 fault at net h. If path 2 is used, both inputs x_3 and x_4 need to be set to 0 in order to propagate the logic value at net h to the output. This is a consistent operation. Therefore, the test vectors are the set: $\{(\phi, 0, 0, 0)\}$, which is independent of input x_1. The path 2 is a sensitizable path for the stuck-at-0 fault at net h. ∎

In the path sensitization approach, when only one path is selected, the result is known as a *one-dimensional* or *single-path sensitization* approach; when multiple paths are selected at the same time, the result is called a *multiple-dimensional* or *multiple-path sensitization* approach. The following example demonstrates the difficulty of single-path sensitization and how multiple-path sensitization can be used to overcome it.

EXAMPLE 16.11   *Multiple-Path Sensitization*

As shown in Figure 16.17(a), an inconsistent setting of the input values x_3 and x_3' is obtained when testing the stuck-at-0 fault at the input a. In this case, both inputs x_3 and x_3' must be set to 0. However, when all paths related to the input x_1 are sensitized at the same time, both inputs x_3 and x_3' are separately set to be 1 and 0, which is a consistent setting. ∎

In general, when a signal splits into several ones, which pass through different paths, and then reconverge together later at the same gate, the signal is called a *reconvergent fanout signal*. It is seen generally that the stuck-at fault of a reconvergent fanout signal cannot be detected by the single-path sensitization approach because it is hard or impossible to find a consistent combination of inputs. The approach to overcoming this difficulty is by the use of a multiple-path sensitization approach to sensitize all paths related to the signal at the same time.

16.2.4.1   D-Algorithm   The essential idea of the D-algorithm is based on the multiple-path sensitization technique. However, its operations are quite complicated

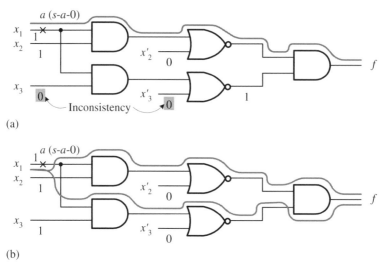

(a)

(b)

FIGURE 16.17 Comparison between (a) single and (b) multiple sensitized paths in the path sensitization approach

[3, 15]. Hence, in the following we only present a simplified version of this. Before describing the simplified D-algorithm, it is necessary to define some basic operators: *primitive D-cube of fault* (or failure) (pdcf), *propagation of D-cube* (pdc) and *singular cover* (sc). All of these for a specified gate are the different views of the truth table of the gate.

The pdcf is used to model faults in a logic circuit and can model any stuck-at fault, bridge fault or any change in logic gate function. For example, the pdcf of the AND gate when its output is stuck-at-1 is "1 0 D'", where D' represents an unknown value. Both the ϕ and D symbols represent a value of 0 or 1. However, all ϕ symbols are set to an arbitrary value independently but all D symbols need to be set to the same value. The pdcfs of the five most commonly used gates are shown in Figure 16.18.

The D-cube is a collapsed truth table entry that can be used to characterize an arbitrary logic gate or block. The pdc is used to propagate a D-cube through the gate under consideration. As we have introduced in Section 2.1.1, any basic gate can be considered as a controlled gate whose output receives the data from one of its inputs and is enabled by the other input of the gate. For example, in order to propagate a D-cube through an AND gate being entered into one input of the AND, it is necessary to set the other input to 1, and the output of the AND gate is still D. If a NAND gate is considered instead, the output is D' because the input D is inverted. The pdcs of the five most commonly used gates are shown in Figure 16.19.

The singular cover (sc) of a logic gate is the input assignments required for a specific output value, including 0 and 1. It is indeed a compressed form of the truth table of the logic gate. For example, for an AND gate, the output is 0 whenever one of its two inputs is 0 and the output is 1 only when both of its two inputs are 1. Hence, the singular cover of AND gate has three rows, as shown in Figure 16.20. In either one of the first two rows, one input is 0 and the other input is ϕ (either 0 or 1), and in the third

x	y	z	Faults covered
0	0	D'	$z/1$
0	1	D'	$x/1, z/1$
1	0	D'	$y/1, z/1$
1	1	D	$x/0, y/0, z/0$

(a)

x	y	z	Faults covered
0	0	D'	$x/1, y/1, z/1$
0	1	D	$x/0, z/0$
1	0	D	$y/0, z/0$
1	1	D	$z/0$

(b)

x	y	z	Faults covered
0	0	D	$z/0$
0	1	D	$x/1, z/0$
1	0	D	$y/1, z/0$
1	1	D'	$x/0, y/0, z/1$

(c)

x	y	z	Faults covered
0	0	D	$x/1, y/1, z/0$
0	1	D'	$x/0, z/1$
1	0	D'	$y/0, z/1$
1	1	D'	$z/1$

(d)

z	Faults covered
0	D $x/1, z/0$
1	D' $x/0, z/1$

(e)

FIGURE 16.18 The primitive D-cube of fault (pdcf) of the five most commonly used gates:
(a) AND gate; (b) OR gate; (c) NAND gate; (d) NOR gate; (e) NOT gate

row both inputs are 1. The singular cover (sc) of the five most commonly used gates
are shown in Figure 16.20.

After defining the required operators, the basic operations of the D-algorithm can
then be described as follows.

Algorithm: A Simplified D Algorithm

1. *Fault sensitization.* Select a pdcf for the fault for which a test pattern is to be
   generated.
2. *D-drive.* Use pdcs to propagate the D signal to at least one primary output of
   the circuit.
3. *Consistency operations.* Use the sc of each logic module to perform the
   consistency operation.

The following example explains how the D-algorithm is used to find test vectors
for a specified fault.

AND gate			OR gate			NAND gate			NOR gate			NOT gate	
x	y	z	x	y	z	x	y	z	x	y	z	x	z
D'	1	D'	D'	0	D'	D'	1	D	D	0	D'	D'	D
1	D'	D'	0	D'	D'	1	D'	D	0	D	D'		
D	1	D	D	0	D	D	1	D'	D	0	D'	D	D'
1	D	D	0	D	D	1	D	D'	0	D	D'		

FIGURE 16.19 The
propagation of D-cube
(pdc) of the five most
commonly used gates

AND gate			OR gate			NAND gate			NOR gate			NOT gate	
x	y	z	x	y	z	x	y	z	x	y	z	x	z
0	ϕ	0	1	ϕ	1	0	ϕ	1	1	ϕ	0	1	0
ϕ	0	0	ϕ	1	1	ϕ	0	1	ϕ	1	0	0	1
1	1	1	0	0	0	1	1	0	0	0	1		

FIGURE 16.20 The singular cover (sc) of the five most commonly used gates

EXAMPLE 16.12   *An Example of the Simplified D-Algorithm*

The first step is fault sensitization. Consider the stuck-at-0 at net β in Figure 16.14 and from Figure 16.18 we obtain the pdcf of the AND gate where one input stuck-at-0 is "1 1 D", which is shown in Figure 16.21 as the first row. The second step is D-drive. In this step, we need to propagate the D cube to the output through an OR gate. Hence, the pdc of the OR gate is employed, which is "0 D D", as shown in the second row in Figure 16.21. The final step is consistency operations. Here, three gates are needed to find consistent assignments by using their scs. After applying the scs to the corresponding gates, the result test vectors for the stuck-at-0 at net β is $\{(1, \phi, 0, \phi),(\phi, 1, 0, \phi) \}$, which is the same as that obtained from the Boolean difference approach. ∎

Like the Boolean difference approach, using the D-algorithm to find test vectors, it is necessary to find the test vectors for each fault, and then build a fault detection table to find the test set of the circuit under consideration.

Review Questions

Q16.21 Describe the basic operations of the path-sensitization approach.

Q16.22 What is the major drawback of the single-path sensitization approach?

Q16.23 What are the basic differences between the single-path and multiple-path sensitization approaches?

Q16.24 Describe the basic operations of the D-algorithm.

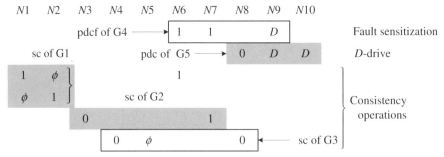

FIGURE 16.21 An illustration of the operations of the D-algorithm

16.3 TESTABLE CIRCUIT DESIGN

As we have discussed, the controllability and observability of a net determine whether a fault at that net can be detected or not. In practice, a logic circuit is not always an irredundant one. Therefore, not every fault of the circuit is detectable. Moreover, it is very difficult to test a sequential circuit because the state diagram of the sequential circuit is not guaranteed to be strongly connected; namely, it is hard, even impossible, to transfer the circuit to a specified state.

As a result, the best way to facilitate the testability of a circuit is to add some extra circuits or modify the original circuit in order to increase the controllability and observability of the circuit. Such an approach, by adding extra circuits so as to reduce the test difficulty, is known as a *testable circuit design* or called *design for testability* (DFT).

The most widely used DFT approaches can be roughly cast into the following three types: *ad hoc approach*, *scan-path method* and *built-in self-test* (BIST). The ad hoc approach uses some means to increase both controllability and observability of the circuit. The scan-path method provides the controllability and observability at each register within the circuit by adding a 2-to-1 multiplexer that may configure all registers of the same scan path into a shift register which can then be accessed externally. The built-in self-test (BIST) relies on augmenting circuits to allow them to perform operations so as to prove the correct operations of the circuit.

16.3.1 Ad hoc Approach

The basic principles behind the ad hoc approach are to increase both the controllability and observability of the circuits by using some heuristic rules. The general guidelines are listed as follows:

1. *Providing more control and test points.* Control points are used to set logic values of some selected signals and test points are used to observe the logic values of the selected signals.

2. *Using multiplexers to increase the number of internal control and test points.* Multiplexers are used to apply external stimuli to the circuit and take out the responses from the circuit.

3. *Breaking feedback paths.* Using AND gates or other appropriate gates breaks the feedback paths of sequential circuits so that sequential circuits are transformed into combinational circuits during testing.

4. *Using state registers* to reduce the additional I/O pins required for testing signals.

Moreover, the exhaustive test is feasible when the number of input signals of a logic circuit is not too many. For a circuit with a large number of input signals, the exhaustive test might still be feasible if the circuit can be partitioned into several smaller modules.

The following example demonstrates how it is possible to perform an exhaustive test by partitioning the circuit into two smaller ones and using multiplexers to increase the number of internal control and test points.

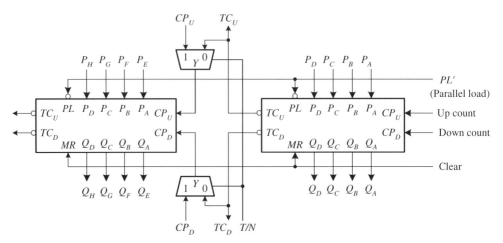

FIGURE 16.22 An example of the exhaustive test

EXAMPLE 16.13 *An Example of the Exhaustive Test Method*

Figure 16.22 shows an 8-bit binary up/down counter consisting of two 4-bit binary up/down counter modules. Due to the symmetry operations of up and down counting, we only consider the up counting operation. By simply using the exhaustive test method, we can continuously apply clock pulses to make it count from 0 to 255, which requires 256 clock cycles in total.

In order to reduce the test time, namely, the required number of clock cycles, it may use two multiplexers to cut the connections between the two 4-bit binary up/down counters, as depicted in Figure 16.22. When the source selection signal T/N (Test/Normal) of the multiplexers is set to 1, the circuit is in test mode. Two 4-bit binary counters operate independently. They can be tested in parallel and require 16 clock cycles. An additional clock cycle is required for testing the carry propagation. When the source selection signal T/N (Test/Normal) of the multiplexers is set to 0, the two 4-bit binary up/down counters are cascaded into an 8-bit binary up/down counter. ∎

Review Questions

Q16.25 Describe the basic rules of the ad hoc approach.

Q16.26 In what conditions can the exhaustive test be applied to test a circuit?

16.3.2 Scan-Path Method

As we have discussed previously, it is very difficult to test a sequential circuit directly without adding any extra circuits. The scan-path method is utilized to alleviate this difficulty by adding a 2-to-1 multiplexer at the data input of each D-type flip-flop to allow access of the D-type flip-flop directly from an external circuit. As a result, both controllability and observability increase to the point where the states of the underlying sequential circuit can be controlled and observed externally.

In the scan-path method, the 2-to-1 multiplexer at the data input of each D-type flip-flop operates in either of two modes: *normal* and *test*. In the normal mode, the data input of each D-type flip-flop connects to the output of its associated combinational logic circuit so as to perform the normal operation of the sequential circuit. In the test mode, all D-type flip-flops are cascaded into a shift register through the use of those 2-to-1 multiplexers associated with the D-type flip-flops. Consequently, all D-type flip-flops can be set to specified values or read out to examine their values externally.

In general, the test approach using the scan-path method for a sequential circuit is as follows:

1. Set the D-type flip-flops into the test mode in order to form a shift register. Shift a specific 0 and 1 sequence into the shift register and then observe whether the shift-out sequence exactly matches the input sequence.

2. Use either of the following two methods to test the circuit:

   (a) Test whether each state transfer of the state diagram is correct or not.

   (b) Test whether any stuck-at fault exists in the combinational logic.

To make this idea clearer, an example is given in the following.

EXAMPLE 16.14   *An Example of the Scan-Path Method*

Figure 16.23 shows a sequential circuit used as a controller for some applications. In the original controller, the 2-to-1 multiplexers at the inputs of the D-type flip-flops do not exist. The purpose of adding 2-to-1 multiplexers is to form a scan path. As the light line shown in the figure, when in test mode all 2-to-1 multiplexers and D-type flip-flops form a 3-bit shift register with *ScanIn* as input and *ScanOut* as output. Therefore, any D-type flip-flop can be set to a specific value by shift operation and the output of any D-type flip-flop can be shifted out as well. That is, each D-type flip-flop has the features of complete controllability and observability. All D-type flip-flops are operated in their original function desired in the sequential circuit when the source selection signal T/N of all 2-to-1 multiplexers is set to 0. ■

Review Questions

Q16.27   Describe the basic principles of the scan-path method.

Q16.28   In what conditions can the scan-path method be applied to test a circuit?

16.3.3   BIST

Two purposes for testing a logic circuit are fault detection and fault location. Fault detection is a process that determines whether any fault occurs in the circuit under test and fault location is a process that determines the location of a fault. In addition, fault coverage is often used to describe how many detectable faults have been detected or located in a test. That is, what fraction of faults can be detected or located from all possible

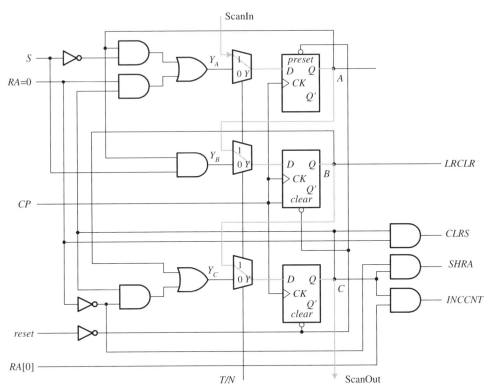

FIGURE 16.23 An example of a scan chain

faults. Commercial CAD tools, such as *Sandia controllability and the observability analysis program* (SCOAP), can be employed to analyze the fault coverage in a circuit.

The BIST principles can be illuminated by using Figure 16.24, which shows the basic concepts of testing a digital system. Figure 16.24(a) shows the scheme using automated test equipment (ATE) to test a digital system. In a typical ATE scheme, the test vectors are stored in memory, then read out and sent to the circuit under test (CUT) as stimuli by a μP. The response signals from the CUT are captured by the ATE and stored in memory, and then compared with the expected results. If they are matched, the test is passed; otherwise, the test fails and hence the circuit is faulty.

Due to the increasing density of current integrated circuits, the complexity of the circuit under test is much higher than before. In order to decrease the test cost, it is necessary to reduce the amount of time for using the ATE or embed the test vector generation circuits into the circuit under test. The approach to embedding the test vector (often called the pattern) generation circuits into the logic circuit is known as the *built-in self-test* (BIST), as shown in Figure 16.24(b). The test vectors required for the test are generated automatically by a circuit and directly applied to the CUT. The response signals from the CUT are compressed into a signature by a response-compression circuit. The signature is then compared with the fault-free signature to determine whether the CUT is faulty or not. Since all of the above test vector generation and response-compression circuits are embedded into the CUT, they have to be simple

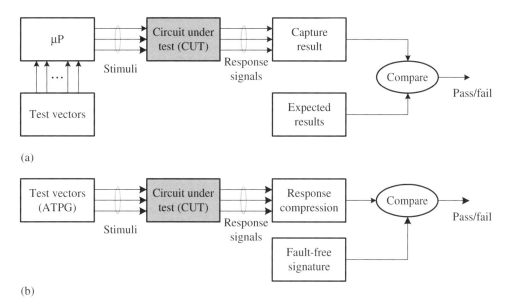

(a)

(b)

FIGURE 16.24 The BIST principles of digital systems: (a) automated test equipment (ATE); (b) built-in self test (BIST)

enough so as not to increase too much cost. The circuit used to generate the test vectors is known as an *automatic test pattern generator* (ATPG) and the response-compression circuit is called a *signature generator*.

16.3.3.1 Random Test and ATPG

For a complicated logic circuit, much time may be needed to generate the test set and it may not be feasible to embed a BIST circuit into the CUT due to too much hardware being required for storing the test vectors. In such a situation, an alternative way, known as the *random test*, may be used instead. The random test only generates enough test vectors randomly.

In the random test, a maximum-length sequence generator, known as an *autonomous linear feedback shift register* (ALFSR) or also called the *PR-sequence generator* (PRSG), is usually used to generate the stimuli. It is not an exhaustive test because it does not generate all combinations of input values and it does not guarantee that the circuit passing the test is completely fault-free. However, the detection rate of detectable faults will approach 100 % when enough test vectors are applied. The two implementations of an ALFSR, accompanied with sample primitive polynomials, have been discussed in Section 9.5.1. The interested reader may refer to that section for details.

The output signals from a PRSG can be an output in one of two modes: *serial* and *parallel*. In the serial output mode, the output signal may be taken from any *D*-type flip-flop; in the parallel output mode, the output signals are taken from all or a portion of all *D*-type flip-flops.

16.3.3.2 Signature Generator/Analysis

As illustrated in Figure 16.24(b), in BIST the response signals have to be compressed into a small amount of data and then compared with the expected result to determine whether the CUT is faulty or not.

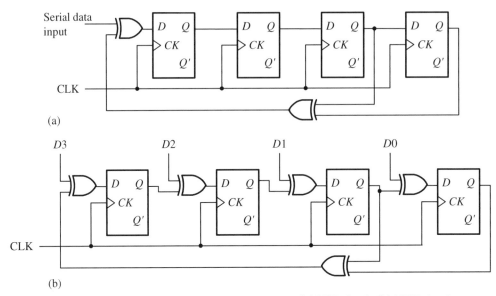

FIGURE 16.25 Two examples of signature generators: (a) SISR circuit; (b) MISR circuit

For the time being, the most widely used response-compression circuit is a circuit using an n-stage linear feedback shift register (LFSR), which is much like the CRC circuit introduced in Section 9.5.2.

Signature generator/analysis can be classified into two types: *serial signature generator/analysis* and *parallel signature generator/analysis*. The n-stage *serial-input signature register* (SISR) is a PRSG with an additional XOR gate at the input for compacting an m-bit message, M, into the standard-form LFSR. The n-bit *multiple-input signature register* (MISR) is a PR-sequence generator with n extra XOR gates at each input of the D-type flip-flop for compacting an m-bit message, M, into the standard-form LFSR. Examples of SISR and MISR are shown in Figures 16.25(a) and (b), respectively. Except for the data inputs, the feedback functions of both circuits are the same as that of the PRSG introduced in Section 9.5.1.

EXAMPLE 16.15 *An Application of Signature Analysis*

An application of signature analysis is shown in Figure 16.26. The inputs x, y and z are generated by an ALFSR. The output response, f, of the circuit is sent to a 4-stage SISR for compressing into a 4-bit signature, as shown in Figure 16.26(a). Figure 16.26(b) lists the outputs and signature values of fault-free, stuck-at-0 at both nets α and β, stuck-at-1 at net α and stuck-at-1 at net β under six input combinations. Consequently, whenever it is needed to test the circuit, the test vectors are generated in sequence and applied to the circuit. Then, the signature of output responses is compared with that of the fault-free circuit to determine whether the circuit is faulty or not. If both signatures are matched, the circuit is fault-free; otherwise, the circuit is faulty. ∎

The basic principle of signature analysis is to compress a longer stream into a short one (i.e. signature). Hence, it is possible to map many streams onto the same

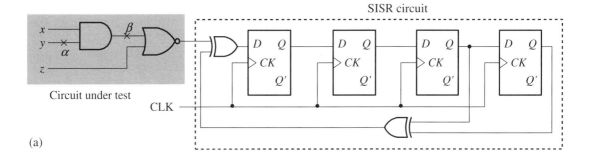

(a)

Input			Fault-free output	Faulty output		
x	y	z	f	$f_{\alpha 0}$ and $f_{\beta 0}$	$f_{\alpha 1}$	$f_{\beta 1}$
0	0	1	0	0	0	0
0	1	0	1	1	1	0
1	0	0	1	1	0	0
1	0	1	0	0	0	0
1	1	0	0	1	0	0
1	1	1	0	0	0	0
			0101	0001	1100	0000

(b)

FIGURE 16.26 An example of the application of digital signature: (a) logic diagram; (b) a numerical example of signature

signature. In general, if the length of the stream is m bits and the signature generator has n stages and hence the signature is n bits, then there are $2^{m-n} - 1$ erroneous streams that will produce the same signature. Since there are a total of $2^m - 1$ possible erroneous streams, the following theorem follows immediately [1].

Theorem 16.3.1 *For an input data stream of length m bits, if all possible error patterns are equally likely, then the probability that an n-bit signature generator will not detect an error is:*

$$P(m) = \frac{2^{m-n} - 1}{2^m - 1} \tag{16.4}$$

where, for m >> n, approaches 2^{-n}.

Consequently, the fault detection capability of an n-stage signature generator will asymptotically approach to be perfect if n is large enough.

16.3.3.3 BILBO As mentioned above, both ALFSR and signature generator need n-stage shift registers and use primitive polynomials. Consequently, both circuits can be combined together and use only one register. The resulting circuit is known as a *built-in logic block observer* (BILBO).

Figure 16.27 shows a typical BILBO circuit. The BILBO has four modes:

1. *Scan mode.* When mode selection signals $M_1 M_0 = 00$, it supports the scan-path method.

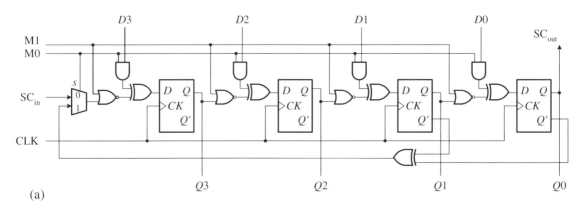

(a)

M1	M0	Function
0	0	Scan mode
0	1	MISR
1	0	Clear register
1	1	Parallel load

(b)

(c)

FIGURE 16.27 An example of a BILBO circuit: (a) logic diagram; (b) function selection; (c) application example

2. *Multiple-input signature register (MISR) mode.* When mode selection signals $M_1 M_0 = 01$, it supports the signature analysis.

3. *Clear mode.* When mode selection signals $M_1 M_0 = 10$, it clears the contents of the register.

4. *Parallel-load mode.* When mode selection signals $M_1 M_0 = 11$, it performs as a register with the capability of parallel load.

The function selection of the circuit shown in Figure 16.27(a) is listed as Figure 16.27(b). A simple application is depicted in Figure 16.27(c). In the normal operation mode, the PRSG at the input of the combinational logic circuit is isolated and removed effectively by a multiplexer. The registers of the signature generator are used as the state registers for the combinational logic circuit so that they combine into a sequential circuit. In the test mode, the inputs of the combinational logic circuit are the random test vectors generated by the PRSG and the responses from the combinational logic circuit are sent to the signature generator to perform the signature analysis.

Comments on DFT When using a scan-chain tool to insert scan-chain circuits into a design, it is often necessary to pay some attention to the following issues [2]:

• *Avoid using gated reset or preset control signal.* A flip-flop is unscannable if its reset is functionally gated in the design.

• *Avoid using gated or generated clocks.* A flip-flop is unscannable if its clock input is functionally gated in the design. When this is necessary, a multiplexer may be used to bypass the gated clock in the scan-mode.

- *Using single edge of the clock in a design.* The same clock edge (positive or negative) is used for the entire design when the design is in the scan-mode.
- *For multiple clock domains.*It is suggested to assign a separate scan-chain for each clock domain because mixing use of clock domains within a scan-chain usually leads to timing problems. Two possible solutions are as follows. *Solution 1* is to group all flip-flops belonging to a common clock domain, and then connect them in series to form a single scan-chain. *Solution 2* is to use a clock multiplexer at the clock source, so that only one clock is used during the scan-mode.

Review Questions

Q16.29 Describe the basic meaning of a random test.

Q16.30 What is an autonomous linear feedback shift register (ALFSR)?

Q16.31 Describe the basic principles of the built-in logic block observer.

Q16.32 What is the probability that an *n*-stage signature generator fails to detect a fault?

Q16.33 Describe the basic components of a BIST.

16.3.4 Boundary-Scan Standard—IEEE 1149.1

The test methods described previously are limited to a single chip or device. It is much more difficult to test the entire system on a printed-circuit board (PCB). In order to decrease the test cost of PCB systems, the *Joint Test Advisory Group* (JTAG) proposed a testable bus specification in 1988 and then this was defined as a standard known as *IEEE 1149.1* by the IEEE [6, 17]. IEEE 1149.1, also called the *boundary scan standard*, has become a standard for almost all integrated circuits that must follow.

The goals of the boundary scan standard are to provide a data transfer standard between the ATE and the devices on a PCB, provide a method of interconnecting devices on a PCB and provide a way of using test bus standard or BIST hardware to find the faulty devices on a PCB as well. An example to show the boundary scan architecture is given in Figure 16.28.

Figure 16.28 shows a PCB-based system which contains four devices facilitated with the boundary scan standard. The boundary scan cells of all four devices are cascaded into a single scan path through connecting the test data output (TDO) of a device into the test data input (TDI) of another device. The resulting system can carry out the following test functions: *interconnection test, normal operation data observation* and *each device test*.

From the above discussion, it can be seen that the essential idea of the boundary scan standard is to extend the scan-path method into the entire PCB system. To reach this, each device has to provide a testable bus interface, as shown in Figure 16.29. The boundary scan standard mainly include five parts: *test access port* (TAP), *data registers, TDO driver, instruction register and decoder* and *TAP controller*.

The boundary scan standard provides the following major modes of operations: *non-invasive mode* and *pin-permission mode*. In the non-invasive mode, the outside world is allowed to serially write in test data or instructions or serially read out test results in an asynchronous way. In the pin-permission mode, the boundary scan standard

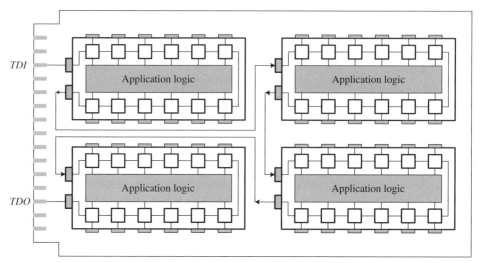

FIGURE 16.28 A boundary scan architecture used to test the entire board-level module

takes control of the IC input/output pins, thus disconnecting the system logic from the outside world. All ICs adhering to the boundary scan standard must be designed to power-up in the non-invasive mode. Moreover, the boundary scan standard allows delivery of BIST (built-in self-test) mode commands (e.g. RUNBIST) through JTAG hardware to the device under test.

16.3.4.1 Test Access Port (TAP) TAP defines four mandatory control signals, *test clock input* (TCK), *test data input* (TDI), *test data output* (TDO) and *test mode select* (TMS), and one optional control signal, *test reset* (TRST).

- TCK is used to clock test data and results into and out of the device, respectively.
- TDI is sampled at the positive edge of the TCK and used to input test data into the device.

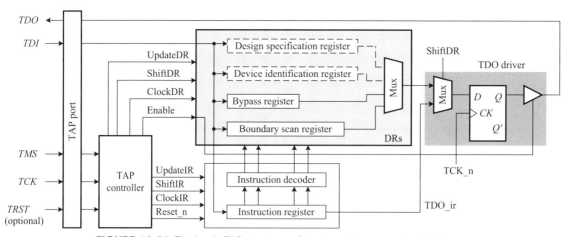

FIGURE 16.29 The basic TAP structure of the boundary scan standard

- TDO is a tristate signal and driven only when the TAP controller is shifting out the data from the device. It outputs test result and status out of the device at the negative edge of the TCK in order to avoid the race with the TDI.

- TMS is sampled at the positive edge of the TCK and decoded by the TAP controller in order to control the test operations.

Besides, one optional reset signal is defined in the following:

- TRST is an optional active-low asynchronous reset signal, used to reset the TAP controller externally.

In the normal mode, TRST and TCK are held low and TMS is held high to disable the boundary scan. To prevent race conditions, inputs are sampled at the positive edge of the TCK and output toggles at the negative edge of the TCK.

16.3.4.2 *Data Registers*   Data registers (DRs) accept the test data input from the TDI. The boundary scan standard defines two mandatory registers, *a bypass register* and a *boundary scan register*, and one optional register, *device identification* (*ID*) register. For some special applications, an optional *design-specification register* can also be used for special requirements. In the following, we will consider both bypass register and boundary scan register in detail.

The goal of a 1-bit bypass register is to bypass the boundary scan cells in a device by way of connecting the *TDO* to the *TDI*. Its structure is shown in Figure 16.30(a). The use of the bypass register in the BYPASS instruction is shown in Figure 16.30(b), which is used in MCMs or PCBs where all devices have a boundary scan chain connected serially but only one device is being tested.

The boundary scan cell is an extension of the scan-path register used in the scan-path method to allow testing of the interconnection among devices on a PCB-based system, test external devices and sample the signals of application logic with the device. A possible boundary scan cell is shown in Figure 16.31(a), which can be utilized as an input cell or an output cell.

When used as an input cell, the data input (*Din*) is connected to an input pad of the device and the data output (*Qout*) is connected to a normal input of the application

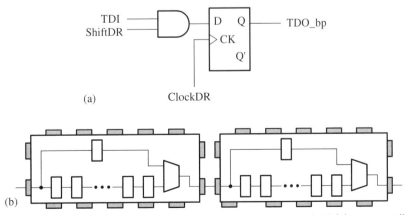

FIGURE 16.30 The bypass register of the boundary scan standard: (a) structure; (b) TDO driver

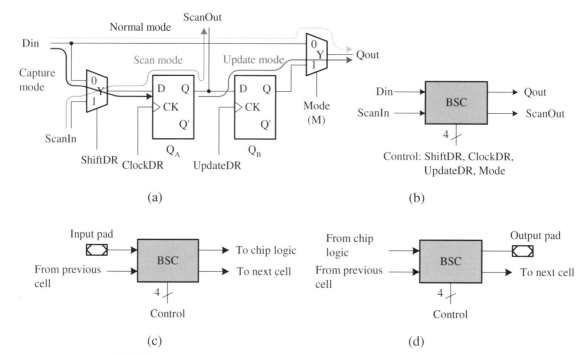

FIGURE 16.31 A possible boundary scan cell: (a) circuit structure; (b) logic symbol; (c) input pad configuration; (d) output pad configuration

logic circuit. When used as an output cell, the data input (Din) is connected to a normal output of the application logic circuit and the data output ($Qout$) is connected to an output pad of the device.

A boundary scan cell can be generally operated in one of the following modes:

- *Normal mode.* In normal mode, the mode selection M is set to 0. The data flow path is directly from the data input (Din) to the data output ($Qout$), as shown in Figure 16.31(a).

- *Scan mode.* In scan mode, all boundary scan cells in a device are cascaded into a shift register, as shown in Figure 16.31(a). The data path is from scan data input ($ScanIn$) to scan data output ($ScanOut$) through the D-type flip-flop Q_A.

- *Capture mode.* In capture mode, the data input (Din) is sampled and stored in the D-type flip-flop Q_A, as shown in Figure 16.31(a). The data output ($Qout$) may be Din or from the output of the D-type flip-flop Q_B, depending on the value of mode selection M.

- *Update mode.* In this mode, the data stored in the D-type flip-flop Q_A may be transferred into the D-type flip-flop Q_B and the data output ($Qout$), as shown in Figure 16.31(a).

The boundary scan cell can be represented as a logic symbol, as shown in Figure 16.31(b). Using this logic symbol, both input and output pad configurations can then be configured as Figures 16.31(c) and (d), respectively. Each of these two simple pad configurations only need one boundary scan cell.

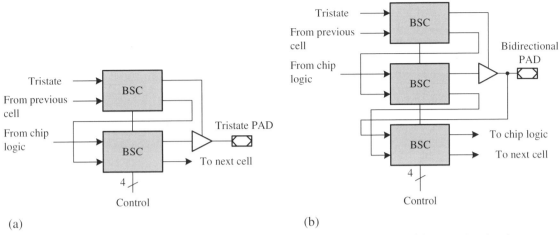

FIGURE 16.32 The boundary scan cells used for (a) tristate and (b) bidirectional pads

The boundary scan cell shown in Figure 16.31(a) can also be used to configure the other two useful pads: tristate and bidirectional. As shown in Figure 16.32(a), each tristate pad needs two boundary scan cells, one for the control signal of the tristate buffer and the other for the input of the tristate buffer. As shown in Figure 16.32(b), each bidirectional pad needs three boundary scan cells, the first for the control signal of the tristate buffer, the second for the output signal and the third for the input signal.

16.3.4.3 TDO Driver The TDO driver is used to drive the TDI of another device in a cascaded manner. One possible structure of the TDO driver is as depicted in Figure 16.33. Some other structure may not contain the negative edge-triggered D-type flip-flop. The input data to the TDO register can be from one of the following sources: instruction register (TDO_ir), boundary scan register (TDO_bs) and bypass register (TDO_bp). TMS is sampled at the positive edge of the TCK whereas TDO is sampled at the negative edge of the TCK.

The following example explains how to model the data registers.

EXAMPLE 16.16 *An Example of the Data Register of the Boundary Scan Standard*

In this example, we assume that the application logic is an n-bit adder, which has two n-bit inputs x and y and one $(n+1)$-bit output *sum*. The data registers include a boundary scan register, a bypass register and a TDO driver.

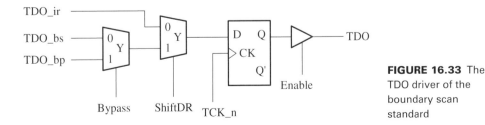

FIGURE 16.33 The TDO driver of the boundary scan standard

```verilog
// data register module
module data_reg(TCK, TDI, TDO_IR, ClockDR, UpdateDR,
  ShiftDR, enable, test_mode, bypass, din_a, din_b,
  fromCore_sum, TDO, sum, toCore_a, toCore_b);
input   TCK, TDI, TDO_IR, ClockDR, UpdateDR, ShiftDR,
        enable;
input   test_mode, bypass;
parameter N = 4;   // define the default size of n-bit
                   // adder
input   [N-1:0] din_a, din_b; // din_a = input a; din_b =
                              // input b
input   [N:0] fromCore_sum;   // sum = fromCore_sum;
output  TDO;
output  [N:0] sum;
output  [N-1:0] toCore_a, toCore_b;
reg     [N-1:0] shift_reg_a, data_reg_a, shift_reg_b,
                data_reg_b;
reg     [N:0] shift_reg_sum, data_reg_sum;
wire    TDO_selected;
reg     TDO_bypass, TDO_delayed;
// boundary scan registers
always @(posedge ClockDR)
   {shift_reg_a, shift_reg_b, shift_reg_sum} <=
        ShiftDR ?
        {TDI, shift_reg_a, shift_reg_b,
          shift_reg_sum[N:1]}:
        {din_a, din_b, fromCore_sum};
always @(posedge UpdateDR)
   {data_reg_a, data_reg_b, data_reg_sum} <=
        {shift_reg_a, shift_reg_b, shift_reg_sum};
assign {toCore_a, toCore_b} = test_mode ?
                  {data_reg_a, data_reg_b} :
                  {din_a, din_b};
assign sum = test_mode ? data_reg_sum : fromCore_sum;
// bypass register
always @(posedge ClockDR)
   TDO_bypass <= TDI & ShiftDR;
// TDO output driver
assign TDO_selected = ShiftDR ?
      (bypass ? TDO_bypass: shift_reg_sum[0]):
      TDO_IR;
always @(negedge TCK)
   TDO_delayed <= TDO_selected;
assign TDO = enable ? TDO_delayed: 1'bz;
endmodule
```

16.3.4.4 Instruction Register and Decoder The instruction register accepts and stores a test command, known as a *test instruction*, from the TDI input at a time. The instruction is then decoded into control signals to select an appropriate data register or perform some specific operations. The structure of the instruction register is similar to that of the boundary scan cell shown in Figure 16.31(a) with one exception that the output 2-to-1 multiplexer is removed, as shown in Figure 16.34. The instruction register has at least two bits in order to provide the following three mandatory instructions: BYPASS, EXTEST and SAMPLE.

- BYPASS (instruction code is all 1s) places the bypass register in the DR chain so that the path from the TDI to TDO only involves one bit register.
- SAMPLE/PRELOAD places boundary scan registers in the DR chain. In the Capture-DR state, it copies the device's I/O values into DRs. New values are shifted into the DRs, but not driven onto the I/O pins.
- EXTEST (instruction code is all 0s) places boundary scan registers in the DR chain and drives the values from the DRs onto the output pads. It allows for the testing off-device circuitry.

The following instructions are also recommended:

- INTEST allows for single-step testing of internal circuitry through the boundary scan registers. It also drives the device core with signals from the DRs rather than from the input pads.
- RUNBIST is used to activate internal self-testing procedures within a device.
- CLAMP causes the state of all signals driven from the output pins of the device to be defined completely by the data held in the boundary-scan register.
- IDCODE is employed to connect the device ID register serially between the TDI and TDO for reading out in the Shift-DR state and for writing into the vendor identification code in the Capture-DR state. It is required when the device ID register is included in the devices. The layout of the device ID register is as follows: bit 0 (LSB) is always set to 1, bits 1 to 11 are manufacturer identity, bits 12 to 27 are part numbers and bits 28 to 31 are version number.
- USERCODE is intended for programmable devices such as FPGAs and Flash memories. The USERCODE instruction allows a user-programmable

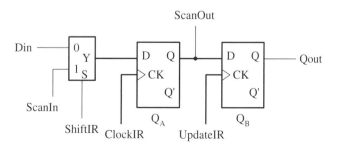

FIGURE 16.34 The instruction register of the boundary scan standard

identification code to be loaded and shifted out for examination and thus allows the programmed function of the component to be determined. It is required when the device ID register is included in the user-programmable devices. Like the IDCODE instruction, the USERCODE instruction connects the device ID register serially between the TDI and TDO for reading out in the Shift-DR state and for writing into the user-programmable identification code in the Capture-DR state.

- HIGHZ causes all output pins of the device to be placed in a high-impedance state.

The following is an example of an instruction register. Here, there are five instructions in total.

EXAMPLE 16.17 *An Example of the Instruction Register of the Boundary Scan Standard*

Due to only five instructions, the instruction register is defined to be 3 bits. In addition to the three mandatory instructions, BYPASS, EXTEST and SAMPLE/PRELOAD, we add two extra instructions, NOP and INTEST.

```
// instruction register module
module instruction_reg (TDI, reset_n, ClockIR, UpdateIR,
   ShiftIR, TDO_IR, test_mode, bypass);
input  TDI, reset_n, ClockIR, UpdateIR, ShiftIR;
output TDO_IR, test_mode, bypass;
reg [2:0] shift_reg, inst_reg;
// define instructions
localparam BYPASS  = 3'b111;
localparam EXTEST  = 3'b000;
localparam INTEST  = 3'b001;
localparam    NOP  = 3'b010;
localparam SAMPLE_PRELOAD  = 3'b011;
// the body of instruction module
always @(posedge ClockIR)
   shift_reg <= ShiftIR ? {TDI, shift_reg[2:1]}: NOP;
always @(posedge UpdateIR or negedge reset_n)
   if (!reset_n) inst_reg <= BYPASS;
   else inst_reg <= shift_reg;
assign TDO_IR   = shift_reg[0];
assign bypass   = (inst_reg == BYPASS);
assign test_mode = (inst_reg == INTEST);
endmodule
```

16.3.4.5 *TAP Controller* The TAP controller is a synchronous 16-state finite-state machine (FSM) with only one signal input TMS and one clock input TCK. The

TMS is used to control the access of the data register (DR) and instruction register (IR). The data register is dispersed into each boundary scan cell, as shown in Figure 16.31(a) and has the purpose of setting the test signals and storing the test results. The instruction register stores the instruction being executed. In the following, we describe each state in detail.

Test-Logic-Reset   The TAP controller enters into this state when power is on or the optional TRST is activated. In addition, no matter what the original state of the controller, it will enter the Test-logic-reset state when the TMS is held high for at least five TCK cycles. When the test-logic is reset, the instruction register is set to the IDCODE instruction if the optional device identification register is provided; otherwise, it is set to the BYPASS instruction.

Run-Test-Idle   The TAP controller will pass through this state regardless of any operations. The controller will stay at this state if the TMS remains 0.

Capture-DR   In this state, the data register selected by the test instruction will load test data in parallel, if any.

Shift-DR   In this state, the test data are shifted in from the TDI at the positive edge of the TCK and the old test data are shifted out from the TDO at the negative edge of the TCK. When the test data need to be shifted a number of positions, it only requires to maintain the TMS at 0 and give the required cycles of the TCK.

Update-DR   In this state, the test data are latched into the data register at the negative edge of the TCK.

Capture-IR   In this state, the shift register contained in the instruction register loads a pattern of fixed logic values at the positive edge of the TCK. The current instruction and its selected test data register remain unchanged.

Shift-IR   In this state, the new test instruction is shifted in from the TDI at the positive edge of the TCK and the old test instruction is removed from the TDO at the negative edge of the TCK. When the test instruction needs to be shifted a number of positions, it only requires to maintain the TMS at 0 and give the required cycles of the TCK. In this state, the current instruction and its selected test data register remain unchanged.

Update-IR   In this state, the test instruction is latched into the instruction register at the negative edge of the TCK. The current instruction is updated to the new instruction.

Pause-DR and Pause-IR   In these states, the shift operations can be temporarily suspended if the TMS remains 0.

The rest states, **Select-DR-scan**, **Select-IR-scan**, **Exit1-DR**, **Exit1-IR**, **Exit2-DR** and **Exit2-IR**, are all temporary control states. They are branch points for changing the control flow of the TAP controller.

The following is an example of a TAP controller. Recall that the TAP controller is a 16-state finite-state machine. Hence, the following example shows how to implement this finite-state machine.

EXAMPLE 16.18 *An Example of a TAP Controller of the Boundary Scan Standard*

All sixteen states are defined as local parameters with each associated a code that can be assigned arbitrarily. The details of the module is described in the state diagram of the TAP controller shown in Figure 16.35.

```
// TAP controller --- IEEE 1149.1 standard
module tap_controller (TCK, TMS, TRST_n, ShiftIR, ShiftDR,
  ClockIR, ClockDR, UpdateIR, UpdateDR, reset_n, enable);
input  TCK, TMS, TRST_n;
output ClockIR, ClockDR, UpdateIR, UpdateDR;
output reg ShiftIR, ShiftDR, reset_n, enable;
reg [3:0] ps, ns;
// TAP controller states --- from IEEE 1149.1 standard
// actually, the following code assignment is arbitrary.
```

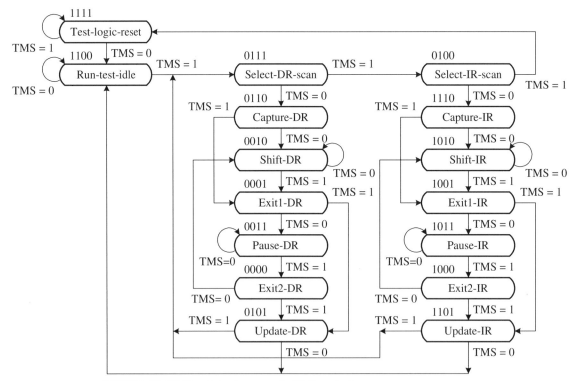

FIGURE 16.35 The state diagram of a TAP controller

```
localparam Test_logic_reset = 4'b1111;
localparam Run_test_idle    = 4'b1100;
localparam Select_DR_scan   = 4'b0111;
localparam Capture_DR       = 4'b0110;
localparam Shift_DR         = 4'b0010;
localparam Exit1_DR         = 4'b0001;
localparam Pause_DR         = 4'b0011;
localparam Exit2_DR         = 4'b0000;
localparam Update_DR        = 4'b0101;
localparam Select_IR_scan   = 4'b0100;
localparam Capture_IR       = 4'b1110;
localparam Shift_IR         = 4'b1010;
localparam Exit1_IR         = 4'b1001;
localparam Pause_IR         = 4'b1011;
localparam Exit2_IR         = 4'b1000;
localparam Update_IR        = 4'b1101;
// update next state
always @(posedge TCK or negedge TRST_n)
   if (!TRST_n) ps <= Test_logic_reset;
   else         ps <= ns;
// compute next state
always @(ps or TMS) case(ps)
   Test_logic_reset : ns = (TMS)? Test_logic_reset: Run_
                           test_idle;
   Run_test_idle    : ns = (TMS)? Select_DR_scan: Run_
                           test_idle;
   Select_DR_scan   : ns = (TMS)? Select_IR_scan:
                              Capture_DR;
   Capture_DR       : ns = (TMS)? Exit1_DR: Shift_DR;
   Shift_DR         : ns = (TMS)? Exit1_DR: Shift_DR;
   Exit1_DR         : ns = (TMS)? Update_DR: Pause_DR;
   Pause_DR         : ns = (TMS)? Exit2_DR: Pause_DR;
   Exit2_DR         : ns = (TMS)? Update_DR: Shift_DR;
   Update_DR        : ns = (TMS)? Select_DR_scan: Run_
                           test_idle;
   Select_IR_scan   : ns = (TMS)? Test_logic_reset:
                              Capture_IR;
   Capture_IR       : ns = (TMS)? Exit1_IR: Shift_IR;
   Shift_IR         : ns = (TMS)? Exit1_IR: Shift_IR;
   Exit1_IR         : ns = (TMS)? Update_IR: Pause_IR;
   Pause_IR         : ns = (TMS)? Exit2_IR: Pause_IR;
   Exit2_IR         : ns = (TMS)? Update_IR: Shift_IR;
   Update_IR        : ns = (TMS)? Select_DR_scan: Run_
                           test_idle;
endcase
```

```
// clock registers on rising edge of TCK at end of state;
// otherwise idle clock high.
assign ClockIR = TCK | ~((ps == Capture_IR)|
                 (ps == Shift_IR));
assign ClockDR = TCK | ~((ps == Capture_DR)|
                 (ps == Shift_DR));
// update registers on falling edge of TCK.
assign UpdateIR = ~TCK & (ps == Update_IR);
assign UpdateDR = ~TCK & (ps == Update_DR);
// change control signals on falling edge of TCK.
always @(negedge TCK or negedge TRST_n)
   if (!TRST_n) begin ShiftIR <= 0; ShiftDR <= 0;
      reset_n <= 0; enable <= 0; end
   else begin
      ShiftIR <= (ps == Shift_IR);
      ShiftDR <= (ps == Shift_DR);
      reset_n <= ~(ps == Test_logic_reset);
      enable  <= (ps == Shift_IR) | (ps == Shift_DR);
   end
endmodule                                                    ■
```

We now illustrate how the boundary scan standard can be used to test an arbitrary logic circuit. Figure 16.36 shows a complete example to integrate a 4-bit adder as an application logic being surrounded by a boundary scan logic. There are thirteen I/O terminals, including two 4-bit inputs x and y, and one 5-bit output *sum*.

The following example integrates the various modules described previously into a top module.

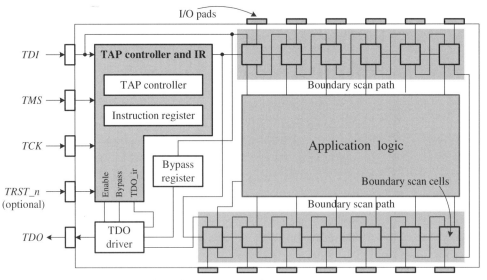

FIGURE 16.36 A complete boundary-scan example

EXAMPLE 16.19 *A Complete Application Example of the IEEE 1149.1 Standard*

This is the top module of the complete example shown in Figure 16.36. Within the module, four modules, the nbit_adder, tap_controller, instruction_reg and data_reg, are instantiated. Moreover, the nbit_adder module is also shown in this example for convenience of illustration.

```
// an example to illustrate how the IEEE 1149.1 coworks
// with core logic.
module ASIC_with_tap(TCK, TMS, TDI, TRST_n, TDO, a, b,
  sum);
input  TCK, TMS, TDI, TRST_n;
output TDO;
parameter N = 4;       // define data width of core logic
input  [N-1:0] a, b;   // primary inputs of n-bit adder
output [N:0] sum;       // primary output of n-bit adder
// define port connection nets
wire [N-1:0] toCore_a, toCore_b;
wire [N:0] fromCore_sum;
wire UpdateIR, ShiftIR, ClockIR;
wire UpdateDR, ShiftDR, ClockDR;
wire reset_n, enable, TDO_IR;
wire test_mode, bypass;
// example core logic instantiation
nbit_adder #(N) example_CoreLogic (toCore_a, toCore_b,
                                   fromCore_sum);
// TAP controller
tap_controller tap_cntl(TCK, TMS, TRST_n, ShiftIR,
                        ShiftDR, ClockIR, ClockDR,
                        UpdateIR, enable);
// instruction register
instruction_reg inst_reg(TDI, reset_n, ClockIR, UpdateIR,
  ShiftIR, TDO_IR, test_mode, bypass);
// test data registers
data_reg #(N) data_reg(TCK, TDI, TDO_IR, ClockDR,
                       UpdateDR, ShiftDR, enable,
                       test_mode,bypass, a, b, from
                       Core_sum, TDO, sum, toCore_a,
                       toCore_b);
endmodule
// the example core logic module
module nbit_adder(x, y, sum);
// I/O port declarations
```

```
parameter N = 4;       // define data width of the
                                        adder
input  [N-1:0] x, y;  // declare as an n-bit array
output [N:0] sum;      // declare as an (n+1)-bit array
// specify the function of an n-bit adder.
assign sum = x + y;
endmodule                                                ■
```

Although the scan-path method is widely used in designing testable systems, it is extremely important to minimize clock skews in order to avoid any hold-time violations in the scan-chain so as to make good use of this approach.

Coding Styles

1. It should order scan-chains appropriately to minimize clock skew.

Review Questions

Q16.34   What is the goal of the boundary scan standard?

Q16.35   What are the ingredients of the boundary scan standard?

Q16.36   What are the operation modes of boundary-scan cells of the boundary scan standard?

Q16.37   What are the four test signals defined in the boundary scan standard?

Q16.38   What instructions can the boundary-scan bus circuit of the boundary scan standard perform?

16.4 SYSTEM-LEVEL TESTING

In this section, we are concerned with system-level testing, including SRAM testing, SRAM BIST, core-based testing and SoC testing. Moreover, the IEEE 1500 standard is introduced.

16.4.1 SRAM BIST and March Test

As we introduced in Section 9.2, SRAM is an important component or module in most digital systems. In practice, SRAM modules with a variety of sizes are often embedded into system designs. Therefore, it is necessary to test them properly; otherwise, they might become the faulty corner of the systems which use them.

At present, there are many schemes used to test memory, such as those reported in Bushnell and Agrawal [3] and Wang *et al.* [17]. Among these, testing the memory at system level with BIST may be the most widely used approach. A simple example of using BIST is shown in Figure 16.37. Here, a 10-bit counter generates the test data and an MISR is used to record the signature. When in test mode, the 10-bit counter takes over the control of the SRAM and generates all data required for testing the SRAM, including both read and write modes. The test data are first written

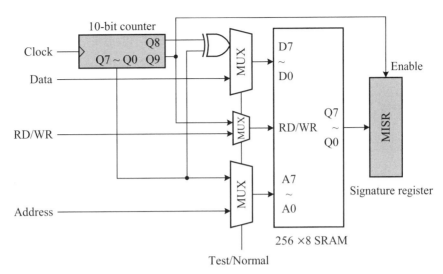

FIGURE 16.37 SRAM BIST with an MISR

into the SRAM and then read out and captured by the MISR to compute the signature. When in normal mode, all BIST hardware are disabled or removed via the use of multiplexers.

Another famous algorithm widely used to test memory with BIST is known as the *March algorithm* or *March test* [3,16]. The required BIST hardware for the March algorithm includes the following (Figure 16.38):

- A BIST controller controls the test procedure.
- An address counter generates the required address during the testing procedure.
- A multiplexer (MUX) circuit is used to feed the memory with the required data during self-test from the controller.

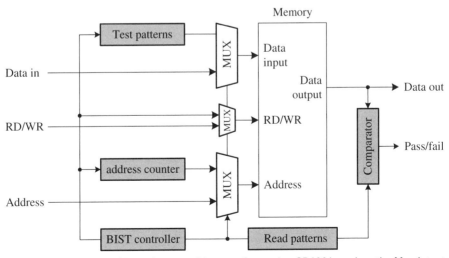

FIGURE 16.38 The BIST hardware architecture for testing SRAM based on the March test

- A comparator checks the response from memory by comparing it with data patterns generated by the read pattern generator controlled by the BIST controller.

- A test pattern generator is used to generate the required test patterns for writing into memory.

- A read pattern generator is used to generate the required test patterns for comparing with the data read from memory.

Notation for the March+ test: $r0$ and $r1$ denote the read a 0 and 1 from the specific memory location, respectively, while $w0$ and $w1$ denote the write a 0 and 1 to the specific memory location, respectively. The arrows are used to the address related operations: $\Uparrow$ and $\Downarrow$ represent that the addressing orders are increasing and decreasing, respectively; $\Updownarrow$ denotes that the addressing order can be either increasing or decreasing.

The March test is applied to each cell in memory before proceeding to the next cell, which means that if a specific pattern is applied to one cell, then it must be applied to all cells. For example, $M0: \Updownarrow(w0); M1: \Uparrow(r0, w1); M2: \Downarrow(r1, w0)$. The detailed operations of the above March test are shown in the following:

Algorithm: MATS+ March test

M0: {March element $\Updownarrow$ (w0)}
    for(i = 0; i <= n - 1; i + +)
        write 0 **to** A[i];
M1: {March element $\Uparrow$ (r0, w1)}
    for (i = 0; i <= n - 1; i + +){
        read A[i]; {expected value = 0}
        write 1 **to** A[i];
    }
M2: {March element $\Downarrow$ (r1, w0)}
    for (i = n - 1; i >= 0; i − −){
        read A[i]; {expected value = 1}
        write 0 **to** A[i];
    }

Even though memory BIST circuitry may be used to test embedded memory blocks, during scan-mode the outputs of these memory blocks are still unknown. To solve this difficulty and increase the fault coverage of the design, a bypass logic is often used, as shown in Figure 16.39. The fault coverage is low due to unknown data appearing at the outputs of the memory when the memory block is not bypassed. However, the fault coverage is increased much more after implementing the bypass logic at the output end of the memory block [2].

Coding Styles

1. It should bypass BIST memory for increasing fault coverage.

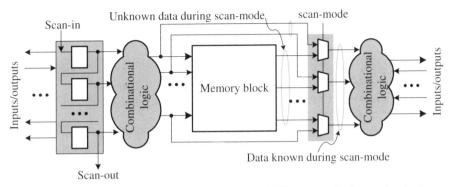

FIGURE 16.39 Illustration of the way to bypass the BIST memory for increasing fault coverage

16.4.2 Core-Based Testing

As we have described previously, a core is an IP of the vendor, which is a predesigned and verified functional block that is included on a chip. In general, the vendor provides tests for the core but the SoC designer must provide the boundary scan to a core embedded on the chip. It will become difficult to test the core and its surrounding logic if the core does not embed the boundary-scan logic.

The most widely used approach to solving the above mentioned difficulty is by surrounding the core with a test logic known as a *test-wrapper* [3], as shown in Figure 16.40. Indeed, this is an application of the concept and technique of the scan-path method.

The test-wrapper includes a cell for each core I/O port. For each input port of the core, the test wrapper has to provide three modes of operations: *normal mode*, *external test mode* and *internal test mode*. The external test mode allows the test-wrapper to observe the core input port for the interconnect test and the internal test mode allows the test-wrapper to test the core function.

For each output port of the core, the test wrapper also has to provide three modes of operations: *normal mode*, *external test mode* and *internal test mode*. The external

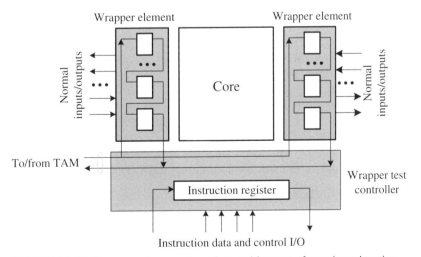

FIGURE 16.40 The general concepts and an architecture of core-based testing

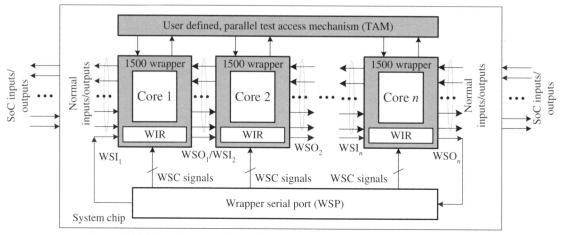

FIGURE 16.41 A system view of the IEEE 1500 standard for testing an SoC

test mode allows the test-wrapper to observe the core output port for the interconnect test and the internal test mode allows the test-wrapper to observe the core output for testing the core function.

16.4.3 SoC Testing

The SoC test is an extension of the core-based test; that is, it uses the technique that is a combination of the test-wrapper concepts and the boundary-scan approach. It also incorporates into the ATPG and signature analysis or other response checking techniques. The overall architecture was defined by the IEEE in 2005 as a standard known as the *IEEE 1500 standard* [7,17]. In the following, we only briefly describe the basic concepts of the standard. The interested reader needs to refer to the IEEE Std 1500-2005 Standard [7] and Wang *et al.* [17] for further details.

A system overview of the IEEE 1500 standard is shown in Figure 16.41, where the system is composed of n cores, with each being wrapped with a 1500 wrapper. The 1500 standard provides both serial and parallel test modes. The serial test mode is facilitated with the *wrapper serial port* (WSP), which is a set of I/O terminals of the wrapper for serial operations. It comprises the *wrapper serial input* (WSI), the *wrapper serial output* (WSO) and several *wrapper serial control* (WSC) signals. Each wrapper has a *wrapper instruction register* (WIR) used to store the instruction to be executed in the corresponding core.

The parallel test mode is achieved by incorporating a user-defined, parallel *test access mechanism* (TAM), as shown in Figure 16.41. Each core can have its own TAM-in and TAM-out ports, being composed of a number of data or control lines for parallel test operations. The test source includes counters, PRSGs or test vectors stored in ROMs and the test sink contains the signature analysis or other response compressions.

Review Questions

Q16.39   How would you test a core that does not have the test access port?

Q16.40   Explain the operations of Figure 16.37.

Q16.41   What is the meaning of a test-wrapper used in core-based testing?

SUMMARY

The testing is an essential step in any PCB-based, cell-based or FPGA-based system, even in a simple digital logic circuit because the only way to ensure that a system or a circuit may function properly is through a careful testing process. The goal of testing is to find the faults in a system or a circuit. The fault is the physical defect in a circuit. When the fault manifests itself in the circuit, it is called a failure. When the fault manifests itself in the signals of a system, it is called an error.

The most common fault models found in CMOS technology include stuck-at faults, bridge faults and stuck-open faults. The stuck-at means that the net will adhere to a logic 0 or 1 permanently. The bridge fault means a shorting between any set of lines (nets) unintentionally. A stuck-open fault is a feature of CMOS circuits and means that some net is broken during manufacturing or after a period of operation. When a circuit is in a stuck-at fault, it still remains a combinational logic but when a circuit is in a stuck-open fault, it will be converted into a sequential circuit.

After the fault models are introduced, we discuss the difficulty of testing sequential circuits. This difficulty motivates the introduction of the scan-path method and the techniques of design for testability, through which test stimuli can be directly applied to the combinational logic portion embedded in between two registers so as to test it easily.

Like the case based on a PCB or bread-board, in order to test a fault in a system or a circuit, we have to input stimuli, observe the responses and analyze the results. The input stimulus special for use in a given system or circuit is often called a test vector. In other words, a test vector is a combination of input values used to detect a fault in the circuit. The collection of test vectors for testing detectable faults in a system or combinational circuit is called a test set. Except for the case of using an exhaustive test, the test set is a sub-set of all possible combinations of input values based on some specific fault models. The four basic types of test vector generations, including fault table, fault simulation, Boolean difference and path sensitization, are described concisely in this chapter. In addition, a simplified D algorithm is introduced and illustrated by an example.

The testability of a circuit is around two basic metrics: controllability and observability. In order to increase these two metric features of a circuit, it needs to add some extra circuits into the design during the design phase. The resulting circuit is known as a design for testability or a testable circuit design. The most widely used approaches can be roughly cast into the following three types: ad hoc approach, scan-path method and built-in self-test (BIST). The ad hoc approach uses some means to increase both the test and observe points of the circuit. The scan-path method provides the capabilities of setting and observing the values of each register within the circuit by incorporating a 2-to-1 multiplexer into the data input of each flip-flop so that all flip-flops associated with the scan-chain can then be configured as a shift register and hence the state of each flip-flop can be set and observed externally. As a consequence, test stimuli can be directly applied to the combinational logic portion. The built-in self-test (BIST) relies on augmenting circuits to allow them to perform operations so as to prove the correct operations of the circuit. The boundary scan standard (namely, IEEE 1149.1) is also introduced.

The final section is concerned with system-level testing based on the scan-path method, including SRAM testing, SRAM BIST, core-based testing and SoC testing. The most widely used approach to testing a core (or IP) is by surrounding the core with a test-wrapper through which test access can be made. The test of SoC is a combination of the core-based test, which uses the test-wrapper concepts and techniques, and the boundary-scan approach. It often also incorporates into the ATPG and signature analysis or other response checking techniques. The overall architecture was defined by the IEEE in 2005 as a standard known as the IEEE 1500 standard.

REFERENCES

1. M. Abramovici, M.A. Breuer and A.D. Friedman, *Digital Systems Testing and Testable Design,* 2nd Edn, IEEE Press, 1996.
2. H. Bhatnagar, *Advanced ASIC Chip Synthesis: Using Synopses Design Compiler and Prime Time,* Kluwer Academic Publishers, Boston, MA, USA, 1999.
3. M.L. Bushnell and V.D. Agrawal, *Essentials of Electronic Testing for Digital, Memory and Mixed-Signal VLSI Circuits,* Kluwer Academic Publishers, Boston, MA, USA, 2000.
4. M.D. Ciletti, *Modeling, Synthesis and Rapid Prototyping with the Verilog HDL,* Prentice-Hall, Upper Saddle River, NJ, USA, 1999.
5. J.P. Hayes, *Introduction to Digital Logic Design,* Reading Massachusettes: Addison-Wesley, Reading, MA, USA, 1993.
6. IEEE Std 1149.1-2001 Standard, *IEEE Standard Test Access Port and Boundary-Scan Architecture,* IEEE Press, New York, NY, USA, 2001.
7. IEEE Std 1500-2005 Standard, *IEEE Standard for Embedded Core Test,* IEEE Press, New York, NY, USA, 2005.
8. B.W. Johnson, *Design and Analysis of Fault Tolerant Digital Systems,* Addison-Wesley, Reading, MA, USA, 1989.
9. Z. Kohavi, *Switching and Finite Automata Theory,* 2nd Edn, McGraw-Hill, New York, NY, USA, 1978.
10. P.K. Lala, *Practical Digital Logic Design and Testing,* Prentice-Hall, Upper Saddle River, NJ, USA, 1996.
11. M.-B Lin, *Digital System Design: Principles, Practices and ASIC Realization,* 3rd Edn, Chuan Hwa Book Company, Taipei, Taiwan, 2002.
12. E.J. McCluskey, *Logic Design Principles,* Prentice-Hall, Englewood Cliffs, NJ, USA, 1986.
13. V.P. Nelson, H.T. Nagle, B.D. Carroll and J.D. Irwin, *Digital Circuit Analysis and Design,* Prentice-Hall, Upper Saddle River, NJ, USA, 1995.
14. S. Palnitkar, *Verilog HDL: A Guide to Digital Design and Synthesis,* 2nd Edn, SunSoft Press, 2003.
15. J.P. Roth, "Diagnosis of automata failures: a calculus and a method", *IBM Journal of Research and Development,* **10**(4), 278–291, 1966.
16. D.S. Suk and S.M. Reddy, "A March test for functional faults in semiconductor random-access memories", *IEEE Transactions on Computers,* **C30**(12), 982–985, 1981.
17. L.-T. Wang, C.-W. Wu and X. Wen, *VLSI Test Principles and Architectures: Design for Testability,* Morgan Kaufmann, New York, NY, USA, 2006.

PROBLEMS

16.1 Supposing that the stuck-at fault model is used, answer the following questions.

    (a) List all equivalent faults of a two-input NAND gate.

    (b) List all equivalent faults of a two-input NOR gate.

16.2 The following problem is related to the 0101-sequence detector introduced in Section 11.1.2.

(**a**) Find all homing sequences from the state diagram.

(**b**) Find all transfer sequences from the state diagram.

(**c**) Is the state diagram a strongly connected graph?

16.3 Consider the logic circuit shown in Figure 16.42.

(**a**) Find the fault table for all possible stuck-at faults.

(**b**) Find the fault detection table.

(**c**) Find a complete test set.

FIGURE 16.42 The logic circuit for Problem 16.3

16.4 Consider the logic circuit shown in Figure 16.43.

(**a**) Find the fault table for all possible stuck-at faults.

(**b**) Find the fault detection table.

(**c**) Find a complete test set.

FIGURE 16.43 The logic circuit for Problem 16.4

16.5 Consider the logic circuit shown in Figure 16.44.

(**a**) Using the Boolean difference approach, find the test vector(s) for the stuck-at-0 fault at net a.

(**b**) Using the Boolean difference approach, find the test vector(s) for the stuck-at-1 fault at net b.

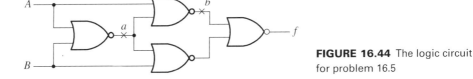

FIGURE 16.44 The logic circuit for problem 16.5

16.6 Consider the logic circuit shown in Figure 16.45.

(**a**) Using the Boolean difference approach, find the test vector(s) for the stuck-at-1 fault at net a.

(**b**) Using the Boolean difference approach, find the test vector(s) for the stuck-at-0 fault at net b.

(c) Using the Boolean difference approach, find the test vector(s) for the stuck-at-0 fault at net c.

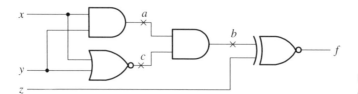

FIGURE 16.45 The logic circuit for Problem 16.6

16.7 Consider the logic circuit shown in Figure 16.44.

(a) Using the single-path sensitization approach, find the test vector(s) for the stuck-at-0 fault at net a.

(b) Using the multiple-path sensitization approach, find the test vector(s) for the stuck-at-0 fault at net a.

(c) Using the path-sensitization approach, find the test vector(s) for the stuck-at-1 fault at net b.

16.8 Consider the logic circuit shown in Figure 16.45.

(a) Using the path sensitization approach, find the test vector(s) for the stuck-at-1 fault at net a.

(b) Using the path sensitization approach, find the test vector(s) for the stuck-at-0 fault at net b.

(c) Using the path-sensitization approach, find the test vector(s) for the stuck-at-0 fault at net c.

16.9 Consider the logic circuit shown in Figure 16.14.

(a) Using the D-algorithm, find the test vector(s) for the stuck-at-0 fault a net α.

(b) Using the D-algorithm, find the test vector(s) for the stuck-at-1 fault at net α.

(c) Using the D-algorithm, find the test vector(s) for the stuck-at-0 fault at net γ.

(d) Using the D-algorithm, find the test vector(s) for the stuck-at-1 fault at net γ.

16.10 The following are problems when using the exhaustive test.

(a) How many test vectors are required for testing an 8-bit ripple-carry adder?

(b) Refer to the 8-bit counter example shown in Figure 16.22, introduced in Section 16.3.1. Divide the 8-bit ripple-carry adder into two 4-bit ripple-carry adders and draw the resulting circuit.

(c) How many test vectors are required for testing the above 8-bit ripple-carry adder constructed by cascading two 4-bit ripple-carry adders?

16.11 Let $f(x) = 1 + x + x^5 + x^6 + x^8$ and design a BILBO to function as follows:

$m_1 m_0 = 00$: ALFSR mode $m_1 m_0 = 10$: Parallel-load mode
$m_1 m_0 = 01$: Scan mode $m_1 m_0 = 11$: MISR mode

16.12 Let $f(x) = 1 + x^3 + x^{10}$ and design a BILBO to function as follows:

$m_1 m_0 = 00$: ALFSR mode $m_1 m_0 = 10$: Parallel-load mode
$m_1 m_0 = 01$: Scan mode $m_1 m_0 = 11$: MISR mode

16.13 Design a parameterized BILBO to function as follows:

$m_1 m_0 = 00$: ALFSR mode $m_1 m_0 = 10$: Parallel-load mode
$m_1 m_0 = 01$: Scan mode $m_1 m_0 = 11$: MISR mode

16.14 Assume that a scan chain would like to be added into the $n \times n$ Booth multiplier introduced in Section 11.2.3. Modify the related Verilog HDL examples given in Section 16.3.4 and write a test bench to verify the functionality of the unsigned array multiplier through the scan-chain logic.

16.15 Assume that a scan chain would like to be added into the $m \times n$ unsigned array multiplier introduced in Section 14.2.1. Modify the related Verilog HDL examples given in Section 16.3.4 and write a test bench to verify the functionality of the unsigned array multiplier through the scan-chain logic.

VERILOG HDL SYNTAX

THIS APPENDIX is reprinted with permission from IEEE Std 1364-2001, (Revision of IEEE Std 1364-1995), IEEE Standard Verilog® Hardware Description Language, Copyright 2001, by IEEE. The IEEE disclaims any responsibility or liability resulting from the placement and use in the described manner.

This appendix contains the formal definition[1] of the Verilog-2001 standard in Backus-Naur Form (BNF). The following conventions are used in describing the syntax:

- All the keywords of Verilog HDL are in lower case and denoted in Courier-New font, such as `always`.

- Courier-New text represents literal words, themselves called terminals, such as `module`, while normal-font text, possibly with underscores, represents syntactic categories, called non terminals, such as for instance, `_identifier`.

- Curly brackets ({ }) surround items that can repeat zero, or more times.

- Square brackets ([]) surround optional items.

- A vertical line (|) separates alternatives.

A.1 KEYWORDS

always	and	assign	automatic
begin	buf	bufif0	bufif1
case	casex	casez	cell
cmos	config	deassign	default
defparam	design	disable	edge

```
else            end             endcase         endconfig
endfunction     endgenerate     endmodule       endprimitive
endspecify      endtable        endtask         event
for             force           forever         fork
function        generate        genvar          highz0
highz1          if              ifnone          incdir
include         initial         inout           input
instance        integer         join            large
liblist         library         localparam      macromodule
medium          module          nand            negedge
nmos            nor             noshowcancelled
not             notif0          notif1          or
output          parameter       pmos            posedge
primitive       pull0           pull1           pulldown
pullup          pulsestyle_                     pulsestyle_
                   onevent                         ondetect
rcmos           real            realtime        reg
release         repeat          rnmos           rpmos
rtran           rtranif0        rtranif1        scalared
showcancelled                   signed          small
specify         specparam       strong0         strong1
supply0         supply1         table           task
time            tran            tranif0         tranif1
tri             tri0            tri1            triand
trior           trireg          unsigned        use
vectored        wait            wand            weak0
weak1           while           wire            wor
xnor            xor
```

A.2 SOURCE SYNTAX

A.2.1 Library Source Text

library_text ::= { library_descriptions }
library_descriptions ::=
    library_declaration
    | include_statement
    | config_declaration
library_declaration ::=
    `library` library_identifier file_path_spec [{ , file_path_spec }]
    [`-incdir` file_path_spec [{ , file_path_spec }];
file_path_spec ::= file_path
include_statement ::= `include` <file_path_spec> ;

A.2.2 Configuration Source Text

config_declaration ::=
      `config` config_identifier ;
      design_statement
      {config_rule_statement}
      `endconfig`
design_statement ::= `design` { [library_identifier.]cell_identifier } ;
config_rule_statement ::=
      default_clause liblist_clause
      | inst_clause liblist_clause
      | inst_clause use_clause
      | cell_clause liblist_clause
      | cell_clause use_clause
default_clause ::= `default`
inst_clause ::= `instance` inst_name
inst_name ::= topmodule_identifier{.instance_identifier}
cell_clause ::= `cell` [library_identifier.]cell_identifier
liblist_clause ::= `liblist` [{library_identifier}]
use_clause ::= `use` [library_identifier.]cell_identifier[: `config`]

A.2.3 Module and Primitive Source Text

source_text ::= { description }
description ::=
      module_declaration
      | udp_declaration
module_declaration ::=
      { attribute_instance } module_keyword module_identifier
          [module_parameter_port_list]
          [list_of_ports] ; { module_item }
          `endmodule`
      | { attribute_instance } module_keyword module_identifier
          [module_parameter_port_list]
          [list_of_port_declarations] ; { non_port_module_item }
          `endmodule`
module_keyword ::= `module` | `macromodule`

A.2.4 Module Parameters and Ports

module_parameter_port_list ::= # (parameter_declaration { , parameter_declaration })
list_of_ports ::= (port { , port })
list_of_port_declarations ::=
      (port_declaration { , port_declaration })
      | ( )

port ::=
    [port_expression]
    | . port_identifier ([port_expression])
port_expression ::=
    port_reference
    | { port_reference { , port_reference } }
port_reference ::=
    port_identifier
    | port_identifier [constant_expression]
    | port_identifier [range_expression]
port_declaration ::=
    {attribute_instance} inout_declaration
    | {attribute_instance} input_declaration
    | {attribute_instance} output_declaration

A.2.5 Module Items

module_item ::=
    module_or_generate_item
    | port_declaration ;
    | { attribute_instance } generated_instantiation
    | { attribute_instance } local_parameter_declaration
    | { attribute_instance } parameter_declaration
    | { attribute_instance } specify_block
    | { attribute_instance } specparam_declaration
module_or_generate_item ::=
    { attribute_instance } module_or_generate_item_declaration
    | { attribute_instance } parameter_override
    | { attribute_instance } continuous_assign
    | { attribute_instance } gate_instantiation
    | { attribute_instance } udp_instantiation
    | { attribute_instance } module_instantiation
    | { attribute_instance } initial_construct
    | { attribute_instance } always_construct
module_or_generate_item_declaration ::=
    net_declaration
    | reg_declaration
    | integer_declaration
    | real_declaration
    | time_declaration
    | realtime_declaration
    | event_declaration
    | genvar_declaration
    | task_declaration
    | function_declaration

non_port_module_item ::=
      { attribute_instance } generated_instantiation
      | { attribute_instance } local_parameter_declaration
      | { attribute_instance } module_or_generate_item
      | { attribute_instance } parameter_declaration
      | { attribute_instance } specify_block
      | { attribute_instance } specparam_declaration
      parameter_override ::= `defparam` list_of_param_assignments ;

A.3 DECLARATIONS

A.3.1 Declaration Types

A.3.1.1 Module Parameter Declarations

local_parameter_declaration ::=
    `localparam` [`signed`] [range] list_of_ param_assignments ;
    | `localparam` `integer` list_of_param_assignments ;
    | `localparam` `real` list_of_param_assignments ;
    | `localparam` `realtime` list_of_param_assignments ;
    | `localparam` `time` list_of_param_assignments ;
parameter_declaration ::=
    `parameter` [`signed`] [range] list_of_param_assignments ;
    | `parameter` `integer` list_of_param_assignments ;
    | `parameter` `real` list_of_param_assignments ;
    | `parameter` `realtime` list_of_param_assignments ;
    | `parameter` `time` list_of_param_assignments ;
specparam_declaration ::= `specparam` [range] list_of_specparam_assignments ;

A.3.1.2 Port Declarations

inout_declaration ::= `inout` [net_type] [`signed`] [range] list_of_port_identifiers
input_declaration ::= `input` [net_type] [`signed`] [range]
    list_of_port_identifiers
output_declaration ::=
    `output` [net_type] [`signed`] [range]
    list_of_port_identifiers
    | `output` [`reg`] [`signed`] [range]
    list_of_port_identifiers
    | `output` `reg` [`signed`] [range]
    list_of_variable_port_identifiers
    | `output` [output_variable_type]
    list_of_port_identifiers
    | `output` output_variable_type
    list_of_variable_port_identifiers

A.3.1.3　Type Declarations

event_declaration ::= `event` list_of_event_identi- fiers ;
genvar_declaration ::= `genvar` list_of_genvar_identifiers ;
integer_declaration ::= `integer` list_of_variable_identifiers ;
net_declaration ::=
  　net_type [`signed`]
  　[delay3] list_of_net_identifiers ;
  　| net_type [drive_strength] [`signed`]
  　[delay3] list_of_net_decl_assignments ;
  　| net_type [`vectored` | `scalared`] [`signed`]
  　range [delay3] list_of_net_identifiers ;
  　| net_type [drive_strength] [`vectored` | `scalared`] [`signed`]
  　range [delay3] list_of_net_decl_assignments ;
  　| `trireg` [charge_strength] [`signed`]
  　[delay3] list_of_net_identifiers ;
  　| `trireg` [drive_strength] [`signed`]
  　[delay3] list_of_net_decl_assignments ;
  　| `trireg` [charge_strength] [`vectored` | `scalared`] [`signed`]
  　range [delay3] list_of_net_identifiers ;
  　| `trireg` [drive_strength] [`vectored` | `scalared`] [`signed`]
  　range [delay3] list_of_net_decl_assignments ;
real_declaration ::= `real` list_of_real_identifiers ;
realtime_declaration ::= `realtime` list_of_real_identifiers ;
reg_declaration ::= `reg` [`signed`] [range]
  　list_of_variable_identifiers ;
time_declaration ::= `time` list_of_variable_identifiers ;

A.3.2　Declaration Data Types

A.3.2.1　Net and Variable Types

net_type ::=
  　`supply0` | `supply1`
  　| `tri` | `triand` | `trior` | `tri0` | `tri1`
  　| `wire` | `wand` | `wor`
output_variable_type ::= `integer` | `time`
real_type ::=
  　real_identifier [= constant_expression]
  　| real_identifier dimension { dimension }
variable_type ::=
  　variable_identifier [= constant_expression]
  　| variable_identifier dimension { dimension }

A.3.2.2　Strengths

drive_strength ::=
  　(strength0 , strength1)
  　| (strength1 , strength0)

      | (strength0 , `highz1`)
      | (strength1 , `highz0`)
      | (`highz0` , strength1)
      | (`highz1` , strength0)
strength0 ::= `supply0` | `strong0` | `pull0` | `weak0`
strength1 ::= `supply1` | `strong1` | `pull1` | `weak1`
charge_strength ::= (`small`) | (`medium`) | (`large`)

A.3.2.3 *Delays*
delay3 ::= # delay_value | # (delay_value [, delay_value [, delay_value]])
delay2 ::= # delay_value | # (delay_value [, delay_value])
delay_value ::=
    unsigned_number
    | parameter_identifier
    | specparam_identifier
    | mintypmax_expression

A.3.3 Declaration Lists
list_of_event_identifiers ::= event_identifier [dimension { dimension }]
    { , event_identifier [dimension { dimension }] }
list_of_genvar_identifiers ::= genvar_identifier { , genvar_identifier }
list_of_net_decl_assignments ::= net_decl_assignment { , net_decl_assignment }
list_of_net_identifiers ::= net_identifier [dimension { dimension }]
    { , net_identifier [dimension { dimension }] }
list_of_param_assignments ::= param_assignment { , param_assignment }
list_of_port_identifiers ::= port_identifier { , port_identifier }
list_of_real_identifiers ::= real_type { , real_type }
list_of_specparam_assignments ::= specparam_assignment { , specparam_assignment }
list_of_variable_identifiers ::= variable_type { , variable_type }
list_of_variable_port_identifiers ::= port_identifier [= constant_expression]
    { , port_identifier [= constant_expression] }

A.3.4 Declaration Assignments
net_decl_assignment ::= net_identifier = expression
param_assignment ::= parameter_identifier = constant_expression
specparam_assignment ::=
    specparam_identifier = constant_mintypmax_expression
    | pulse_control_specparam
pulse_control_specparam ::=
    `PATHPULSE$` = (reject_limit_value [, error_limit_value]) ;
    | `PATHPULSE$`specify_input_terminal_descriptor`$`specify_output_terminal_
        descriptor
        = (reject_limit_value [, error_limit_value]) ;

error_limit_value ::= limit_value
reject_limit_value ::= limit_value
limit_value ::= constant_mintypmax_expression

A.3.5 Declaration Ranges

dimension ::= [dimension_constant_expression : dimension_constant_expression]
range ::= [msb_constant_expression : lsb_constant_expression]

A.3.6 Function Declarations

function_declaration ::=
    `function` [`automatic`] [`signed`] [range_or_type] function_identifier ;
    function_item_declaration { function_item_declaration }
    function_statement
    `endfunction`
    | `function` [`automatic`] [`signed`] [range_or_type] function_identifier
        (function_port_list) ;
    block_item_declaration { block_item_declaration }
    function_statement
    `endfunction`
function_item_declaration ::=
    block_item_declaration
    | tf_input_declaration ;
function_port_list ::= { attribute_instance } tf_input_declaration
        { , { attribute_instance } tf_input_declaration }
range_or_type ::= range | `integer` | `real` | `realtime` | `time`

A.3.7 Task Declarations

task_declaration ::=
    `task` [`automatic`] task_identifier ;
    { task_item_declaration } statement
    `endtask`
    | `task` [`automatic`] task_identifier (task_port_list) ;
    { block_item_declaration }
    statement
    `endtask`
task_item_declaration ::=
    block_item_declaration
    | { attribute_instance } tf_input_declaration ;
    | { attribute_instance } tf_output_declaration ;
    | { attribute_instance } tf_inout_declaration ;
task_port_list ::= task_port_item { , task_port_item }
task_port_item ::=

```
      { attribute_instance } tf_input_declaration
      | { attribute_instance } tf_output_declaration
      | { attribute_instance } tf_inout_declaration
tf_input_declaration ::=
      input [ reg ] [ signed ] [ range ] list_of_port_identifiers
      | input [ task_port_type ] list_of_port_identifiers
tf_output_declaration ::=
      output [ reg ] [ signed ] [ range ] list_of_port_identifiers
      | output [ task_port_type ] list_of_port_identifiers
tf_inout_declaration ::=
      inout [ reg ] [ signed ] [ range ] list_of_port_identifiers
      | inout [ task_port_type ] list_of_port_identifiers
task_port_type ::=
      time | real | realtime | integer
```

A.3.8 Block Item Declarations

```
block_item_declaration ::=
      { attribute_instance } block_reg_declaration
      | { attribute_instance } event_declaration
      | { attribute_instance } integer_declaration
      | { attribute_instance } local_parameter_declaration
      | { attribute_instance } parameter_declaration
      | { attribute_instance } real_declaration
      | { attribute_instance } realtime_declaration
      | { attribute_instance } time_declaration
block_reg_declaration ::= reg [ signed ] [ range ]
      list_of_block_variable_identifiers ;
list_of_block_variable_identifiers ::=
      block_variable_type { , block_variable_type }
block_variable_type ::=
      variable_identifier
      | variable_identifier dimension { dimension }
```

A.4 PRIMITIVE INSTANCES

A.4.1 Primitive Instantiation and Instances

```
gate_instantiation ::=
      cmos_switchtype [delay3]
            cmos_switch_instance { , cmos_switch_instance } ;
      | enable_gatetype [drive_strength] [delay3]
            enable_gate_instance { , enable_gate_instance } ;
      | mos_switchtype [delay3]
            mos_switch_instance { , mos_switch_instance } ;
```

```
          | n_input_gatetype [drive_strength] [delay2]
                  n_input_gate_instance { , n_input_gate_instance } ;
          | n_output_gatetype [drive_strength] [delay2]
                  n_output_gate_instance { , n_output_gate_instance } ;
          | pass_en_switchtype [delay2]
                  pass_enable_switch_instance { , pass_enable_switch_instance } ;
          | pass_switchtype
                  pass_switch_instance { , pass_switch_instance } ;
          | pulldown [pulldown_strength]
                  pull_gate_instance { , pull_gate_instance } ;
          | pullup [pullup_strength]
                  pull_gate_instance { , pull_gate_instance } ;
cmos_switch_instance ::= [ name_of_gate_instance ] ( output_terminal ,
          input_terminal , ncontrol_terminal , pcontrol_terminal )
enable_gate_instance ::= [ name_of_gate_instance ] ( output_terminal ,
          input_terminal , enable_terminal )
mos_switch_instance ::= [ name_of_gate_instance ] ( output_terminal ,
          input_terminal , enable_terminal )
n_input_gate_instance ::= [ name_of_gate_instance ] ( output_terminal ,
          input_terminal { , input_terminal } )
n_output_gate_instance ::= [ name_of_gate_instance ] ( output_terminal
          { , output_terminal } , input_terminal )
pass_switch_instance ::= [ name_of_gate_instance ] ( inout_terminal , inout_terminal )
pass_enable_switch_instance ::= [ name_of_gate_instance ] ( inout_terminal ,
          inout_terminal , enable_terminal )
pull_gate_instance ::= [ name_of_gate_instance ] ( output_terminal )
name_of_gate_instance ::= gate_instance_identifier [ range ]
```

A.4.2 Primitive Strengths

```
pulldown_strength ::=
      ( strength0 , strength1 )
      | ( strength1 , strength0 )
      | ( strength0 )
pullup_strength ::=
      ( strength0 , strength1 )
      | ( strength1 , strength0 )
      | ( strength1 )
```

A.4.3 Primitive Terminals

```
enable_terminal ::= expression
inout_terminal ::= net_lvalue
input_terminal ::= expression
ncontrol_terminal ::= expression
```

output_terminal ::= net_lvalue
pcontrol_terminal ::= expression

A.4.4 Primitive Gate and Switch Types

cmos_switchtype ::= `cmos` | `rcmos`
enable_gatetype ::= `bufif0` | `bufif1` | `notif0` | `notif1`
mos_switchtype ::= `nmos` | `pmos` | `rnmos` | `rpmos`
n_input_gatetype ::= `and` | `nand` | `or` | `nor` | `xor` | `xnor`
n_output_gatetype ::= `buf` | `not`
pass_en_switchtype ::= `tranif0` | `tranif1` | `rtranif1` | `rtranif0`
pass_switchtype ::= `tran` | `rtran`

A.5 MODULE AND GENERATED INSTANTIATION

A.5.1 Module Instantiation

module_instantiation ::=
    module_identifier [parameter_value_assignment]
        module_instance { , module_instance } ;
parameter_value_assignment ::= # (list_of_parameter_assignments)
list_of_parameter_assignments ::=
    ordered_parameter_assignment { , ordered_parameter_assignment } |
    named_parameter_assignment { , named_parameter_assignment }
ordered_parameter_assignment ::= expression
named_parameter_assignment ::= . parameter_identifier ([expression])
module_instance ::= name_of_instance ([list_of_port_connections])
name_of_instance ::= module_instance_identifier [range]
list_of_port_connections ::=
    ordered_port_connection { , ordered_port_connection }
    | named_port_connection { , named_port_connection }
ordered_port_connection ::= { attribute_instance } [expression]
named_port_connection ::= { attribute_instance } .port_identifier ([expression])

A.5.2 Generated Instantiation

generated_instantiation ::= `generate` { generate_item } `endgenerate`
generate_item_or_null ::= generate_item | ;
generate_item ::=
    generate_conditional_statement
    | generate_case_statement
    | generate_loop_statement
    | generate_block
    | module_or_generate_item

generate_conditional_statement ::=
　　　if (constant_expression) generate_item_or_null [else generate_item_or_null]
generate_case_statement ::= case (constant_expression)
　　　genvar_case_item { genvar_case_item } endcase
genvar_case_item ::= constant_expression { , constant_expression } :
　　　generate_item_or_null | default [:] generate_item_or_null
generate_loop_statement ::= for (genvar_assignment ; constant_expression ;
　　　　　genvar_assignment)
　　　begin : generate_block_identifier { generate_item } end
genvar_assignment ::= genvar_identifier = constant_expression
generate_block ::= begin [: generate_block_identifier] { generate_item } end

A.6 UDP DECLARATION AND INSTANTIATION

A.6.1 UDP Declaration

udp_declaration ::=
　　　{ attribute_instance } primitive udp_identifier (udp_port_list) ;
　　　udp_port_declaration { udp_port_declaration }
　　　udp_body
　　　endprimitive
　　　| { attribute_instance } primitive udp_identifier (udp_declaration_port_list) ;
　　　udp_body
　　　endprimitive

A.6.2 UDP Ports

udp_port_list ::= output_port_identifier , input_port_identifier
　　　　　{ , input_port_identifier }
udp_declaration_port_list ::=
　　　udp_output_declaration , udp_input_declaration { , udp_input_declaration }
udp_port_declaration ::=
　　　udp_output_declaration ;
　　　| udp_input_declaration ;
　　　| udp_reg_declaration ;
udp_output_declaration ::=
　　　{ attribute_instance } output port_identifier
　　　| { attribute_instance } output reg port_identifier [= constant_expression]
udp_input_declaration ::= { attribute_instance } input list_of_port_identifiers
udp_reg_declaration ::= attribute_instance reg variable_identifier

A.6.3 UDP Body

udp_body ::= combinational_body | sequential_body
combinational_body ::= table combinational_entry
　　　　　{ combinational_entry } endtable

combinational_entry ::= level_input_list : output_symbol ;
sequential_body ::= [udp_initial_statement] `table` sequential_entry
        { sequential_entry } `endtable`
udp_initial_statement ::= initial output_port_identifier = init_val ;
init_val ::= 1'b0 | 1'b1 | 1'bx | 1'bX | 1'B0 | 1'B1 | 1'Bx | 1'BX | 1 | 0
sequential_entry ::= seq_input_list : current_state : next_state ;
seq_input_list ::= level_input_list | edge_input_list
level_input_list ::= level_symbol { level_symbol }
edge_input_list ::= { level_symbol } edge_indicator { level_symbol }
edge_indicator ::= (level_symbol level_symbol) | edge_symbol
current_state ::= level_symbol
next_state ::= output_symbol | –
output_symbol ::= 0 | 1 | x | X
level_symbol ::= 0 | 1 | x | X | ? | b | B
edge_symbol ::= r | R | f | F | p | P | n | N | *

A.6.4 UDP Instantiation

udp_instantiation ::= udp_identifier [drive_strength] [delay2] udp_instance
        { , udp_instance } ;
udp_instance ::= [name_of_udp_instance] (output_terminal , input_terminal
        { , input_terminal })
name_of_udp_instance ::= udp_instance_identifier [range]

A.7 BEHAVIORAL STATEMENTS

A.7.1 Continuous Assignment Statements

continuous_assign ::= `assign` [drive_strength] [delay3] list_of_net_assignments ;
list_of_net_assignments ::= net_assignment { , net_assignment }
net_assignment ::= net_lvalue = expression

A.7.2 Procedural Blocks and Assignments

initial_construct ::= `initial` statement
always_construct ::= `always` statement
blocking_assignment ::= variable_lvalue = [delay_or_event_control] expression
nonblocking_assignment ::= variable_lvalue <= [delay_or_event_control] expression
procedural_continuous_assignments ::=
     `assign` variable_assignment
     | `deassign` variable_lvalue
     | `force` variable_assignment
     | `force` net_assignment
     | `release` variable_lvalue
     | `release` net_lvalue

function_blocking_assignment ::= variable_lvalue = expression
function_statement_or_null ::=
    function_statement
    | { attribute_instance } ;

A.7.3 Parallel and Sequential Blocks

function_seq_block ::= begin [: block_identifier
    { block_item_declaration }] { function_statement } end
variable_assignment ::= variable_lvalue = expression
par_block ::= fork [: block_identifier
    { block_item_declaration }] { statement } join
seq_block ::= begin [: block_identifier
    { block_item_declaration }] { statement } end

A.7.4 Statements

statement ::=
    { attribute_instance } blocking_assignment ;
    | { attribute_instance } case_statement
    | { attribute_instance } conditional_statement
    | { attribute_instance } disable_statement
    | { attribute_instance } event_trigger
    | { attribute_instance } loop_statement
    | { attribute_instance } nonblocking_assignment ;
    | { attribute_instance } par_block
    | { attribute_instance } procedural_continuous_assignments ;
    | { attribute_instance } procedural_timing_control_statement
    | { attribute_instance } seq_block
    | { attribute_instance } system_task_enable
    | { attribute_instance } task_enable
    | { attribute_instance } wait_statement
statement_or_null ::=
    statement
    | { attribute_instance } ;
function_statement ::=
    { attribute_instance } function_blocking_assignment ;
    | { attribute_instance } function_case_statement
    | { attribute_instance } function_conditional_statement
    | { attribute_instance } function_loop_statement
    | { attribute_instance } function_seq_block
    | { attribute_instance } disable_statement
    | { attribute_instance } system_task_enable

A.7.5 Timing Control Statements

delay_control ::=
      # delay_value
      | # (mintypmax_expression)
delay_or_event_control ::=
      delay_control
      | event_control
      | `repeat` (expression) event_control
disable_statement ::=
      `disable` hierarchical_task_identifier ;
      | `disable` hierarchical_block_identifier ;
event_control ::=
      @ event_identifier
      | @ (event_expression)
      | @*
      | @ (*)
event_trigger ::=
      -> hierarchical_event_identifier ;
event_expression ::=
      expression
      | hierarchical_identifier
      | `posedge` expression
      | `negedge` expression
      | event_expression `or` event_expression
      | event_expression , event_expression
procedural_timing_control_statement ::=
      delay_or_event_control statement_or_null
wait_statement ::=
      `wait` (expression) statement_or_null

A.7.6 Conditional Statements

conditional_statement ::=
      `if` (expression)
          statement_or_null [`else` statement_or_null]
      | if_else_if_statement
if_else_if_statement ::=
      `if` (expression) statement_or_null
      { `else` `if` (expression) statement_or_null }
      [`else` statement_or_null]
function_conditional_statement ::=
      `if` (expression) function_statement_or_null
          [`else` function_statement_or_null]
      | function_if_else_if_statement
function_if_else_if_statement ::=

`if` (expression) function_statement_or_null
{ `else` if (expression) function_statement_or_null }
[`else` function_statement_or_null]

A.7.7 Case Statements

case_statement ::=
    `case` (expression)
        case_item { case_item } `endcase`
    | `casez` (expression)
        case_item { case_item } `endcase`
    | `casex` (expression)
        case_item { case_item } `endcase`
case_item ::=
    expression { , expression } : statement_or_null
    | `default` [:] statement_or_null
function_case_statement ::=
    `case` (expression)
        function_case_item { function_case_item } `endcase`
    | `casez` (expression)
        function_case_item { function_case_item } `endcase`
    | `casex` (expression)
        function_case_item { function_case_item } `endcase`
function_case_item ::=
    expression { , expression } : function_statement_or_null
    | `default` [:] function_statement_or_null

A.7.8 Looping Statements

function_loop_statement ::=
    `forever` function_statement
    | `repeat` (expression) function_statement
    | `while` (expression) function_statement
    | `for` (variable_assignment ; expression ; variable_assignment)
        function_statement
loop_statement ::=
    `forever` statement
    | `repeat` (expression) statement
    | `while` (expression) statement
    | `for` (variable_assignment ; expression ; variable_assignment) statement

A.7.9 Task Enable Statements

system_task_enable ::= system_task_identifier [(expression { , expression })] ;
task_enable ::= hierarchical_task_identifier [(expression { , expression })] ;

A.8 SPECIFY SECTION

A.8.1 Specify Block Declaration

specify_block ::= `specify` { specify_item } `endspecify`
specify_item ::=
     specparam_declaration
     | pulsestyle_declaration
     | showcancelled_declaration
     | path_declaration
     | system_timing_check
pulsestyle_declaration ::=
     `pulsestyle_onevent` list_of_path_outputs ;
     | `pulsestyle_ondetect` list_of_path_outputs ;
showcancelled_declaration ::=
     `showcancelled` list_of_path_outputs ;
     | `noshowcancelled` list_of_path_outputs ;

A.8.2 Specify Path Declarations

path_declaration ::=
     simple_path_declaration ;
     | edge_sensitive_path_declaration ;
     | state_dependent_path_declaration ;
simple_path_declaration ::=
     parallel_path_description = path_delay_value
     | full_path_description = path_delay_value
parallel_path_description ::=
     (specify_input_terminal_descriptor [polarity_operator] =>
         specify_output_terminal_descriptor)
full_path_description ::=
     (list_of_path_inputs [polarity_operator] *> list_of_path_outputs)
list_of_path_inputs ::=
     specify_input_terminal_descriptor { , specify_input_terminal_descriptor }
list_of_path_outputs ::=
     specify_output_terminal_descriptor { , specify_output_terminal_descriptor }

A.8.3 Specify Block Terminals

specify_input_terminal_descriptor ::=
     input_identifier
     | input_identifier [constant_expression]
     | input_identifier [range_expression]
specify_output_terminal_descriptor ::=
     output_identifier
     | output_identifier [constant_expression]

| output_identifier [range_expression]
input_identifier ::= input_port_identifier | inout_port_identifier
output_identifier ::= output_port_identifier | inout_port_identifier

A.8.4 Specify Path Delays

path_delay_value ::=
    list_of_path_delay_expressions
    | (list_of_path_delay_expressions)
list_of_path_delay_expressions ::=
    t_path_delay_expression
    | trise_path_delay_expression , tfall_path_delay_expression
    | trise_path_delay_expression , tfall_path_delay_expression ,
        tz_path_delay_expression
    | t01_path_delay_expression , t10_path_delay_expression ,
        t0z_path_delay_expression,
      tz1_path_delay_expression , t1z_path_delay_expression ,
      tz0_path_delay_expression
    | t01_path_delay_expression , t10_path_delay_expression ,
        t0z_path_delay_expression,
      tz1_path_delay_expression , t1z_path_delay_expression ,
      tz0_path_delay_expression
      t0x_path_delay_expression , tx1_path_delay_expression ,
      t1x_path_delay_expression,
      tx0_path_delay_expression , txz_path_delay_expression ,
      tzx_path_delay_expression
    t_path_delay_expression ::= path_delay_expression
    trise_path_delay_expression ::= path_delay_expression
    tfall_path_delay_expression ::= path_delay_expression
    tz_path_delay_expression ::= path_delay_expression
    t01_path_delay_expression ::= path_delay_expression
    t10_path_delay_expression ::= path_delay_expression
    t0z_path_delay_expression ::= path_delay_expression
    tz1_path_delay_expression ::= path_delay_expression
    t1z_path_delay_expression ::= path_delay_expression
    tz0_path_delay_expression ::= path_delay_expression
    t0x_path_delay_expression ::= path_delay_expression
    tx1_path_delay_expression ::= path_delay_expression
    t1x_path_delay_expression ::= path_delay_expression
    tx0_path_delay_expression ::= path_delay_expression
    txz_path_delay_expression ::= path_delay_expression
    tzx_path_delay_expression ::= path_delay_expression
    path_delay_expression ::= constant_mintypmax_expression
    edge_sensitive_path_declaration ::=
        parallel_edge_sensitive_path_description = path_delay_value
        | full_edge_sensitive_path_description = path_delay_value

parallel_edge_sensitive_path_description ::=
    ([edge_identifier] specify_input_terminal_descriptor =>
    specify_output_terminal_descriptor [polarity_operator] : data_source_expression)
full_edge_sensitive_path_description ::=
    ([edge_identifier] list_of_path_inputs *>
    list_of_path_outputs [polarity_operator] : data_source_expression)
data_source_expression ::= expression
edge_identifier ::= posedge | negedge
state_dependent_path_declaration ::=
    if (module_path_expression) simple_path_declaration
    | if (module_path_expression) edge_sensitive_path_declaration
    | ifnone simple_path_declaration
polarity_operator ::= + | –

A.8.5 System Timing Checks

A.8.5.1 *System Timing Check Commands*
system_timing_check ::=
    $setup_timing_check
    | $hold _timing_check
    | $setuphold_timing_check
    | $recovery_timing_check
    | $removal_timing_check
    | $recrem_timing_check
    | $skew_timing_check
    | $timeskew_timing_check
    | $fullskew_timing_check
    | $period_timing_check
    | $width_timing_check
    | $nochange_timing_check
$setup_timing_check ::=
    $setup (data_event , reference_event , timing_check_limit [, [notify_reg]]) ;
$hold_timing_check ::=
    $hold (reference_event , data_event , timing_check_limit [, [notify_reg]]) ;
$setuphold_timing_check ::=
    $setuphold (reference_event , data_event , timing_check_limit ,
      timing_check_ limit
        [, [notify_reg] [, [stamptime_condition] [, [checktime_condition]
        [, [delayed_reference] [, [delayed_data]]]]]]) ;
$recovery_timing_check ::=
    $recovery (reference_event , data_event , timing_check_limit
      [, [notify_reg]]) ;
$removal_timing_check ::=
    $removal (reference_event , data_event , timing_check_limit
      [, [notify_reg]]) ;
$recrem_timing_check ::=

$recrem (reference_event , data_event , timing_check_limit , timing_check_limit
   [, [notify_reg]] [, [stamptime_condition] [, [checktime_condition]
   [, [delayed_reference] [, [delayed_data]]]]]]) ;
$skew_timing_check ::=
   $skew (reference_event , data_event , timing_check_limit [, [notify_reg]]) ;
$timeskew_timing_check ::=
   $timeskew (reference_event , data_event , timing_check_limit
      [, [notify_reg]] [, [event_based_flag] [, [remain_active_flag]]]]) ;
$fullskew_timing_check ::=
   $fullskew (reference_event , data_event , timing_check_limit ,
      timing_check_ limit
         [, [notify_reg]] [, [event_based_flag] [, [remain_active_flag]]]]) ;
$period_timing_check ::=
   $period (controlled_reference_event , timing_check_limit
      [, [notify_reg]]) ;
$width_timing_check ::=
   $width (controlled_reference_event , timing_check_limit ,
      threshold [, [notify_reg]]) ;
$nochange_timing_check ::=
   $nochange (reference_event , data_event , start_edge_offset ,
      end_edge_offset [, [notify_reg]]) ;

A.8.5.2   System Timing Check Command Arguments

checktime_condition ::= mintypmax_expression
controlled_reference_event ::= controlled_timing_check_event
data_event ::= timing_check_event
delayed_data ::=
      terminal_identifier
      | terminal_identifier [constant_mintypmax_expression]
delayed_reference ::=
      terminal_identifier
      | terminal_identifier [constant_mintypmax_expression]
end_edge_offset ::= mintypmax_expression
event_based_flag ::= constant_expression
notify_reg ::= variable_identifier
reference_event ::= timing_check_event
remain_active_flag ::= constant_mintypmax_expression
stamptime_condition ::= mintypmax_expression
start_edge_offset ::= mintypmax_expression
threshold ::=constant_expression
timing_check_limit ::= expression

A.8.5.3   System Timing Check Event Definitions

timing_check_event ::=
      [timing_check_event_control] specify_terminal_descriptor

[&&& timing_check_condition]
controlled_timing_check_event ::=
    timing_check_event_control specify_terminal_descriptor
        [&&& timing_check_condition]
timing_check_event_control ::=
    posedge
    | negedge
    | edge_control_specifier
specify_terminal_descriptor ::=
    specify_input_terminal_descriptor
    | specify_output_terminal_descriptor
edge_control_specifier ::= edge [edge_descriptor [, edge_descriptor]]
edge_descriptor1 ::=
    01 | 10
    | z_or_x zero_or_one
    | zero_or_one z_or_x
zero_or_one ::= 0 | 1
z_or_x ::= x | X | z | Z
timing_check_condition ::=
    scalar_timing_check_condition
    | (scalar_timing_check_condition)
scalar_timing_check_condition ::=
    expression | ~ expression
    | expression == scalar_constant
    | expression === scalar_constant
    | expression ! = scalar_constant
    | expression ! == scalar_constant
scalar_constant ::= 1'b0 | 1'b1 | 1'B0 | 1'B1 | 'b0 | 'b1 | 'B0 | 'B1 | 1 | 0

A.9 EXPRESSIONS

A.9.1 Concatenations

concatenation ::= { expression { , expression } }
constant_concatenation ::= { constant_expression { , constant_expression } }
constant_multiple_concatenation ::= { constant_expression constant_concatenation }
module_path_concatenation ::= { module_path_expression
        { , module_path_expression } }
module_path_multiple_concatenation ::= { constant_expression
        module_path_concatenation }
multiple_concatenation ::= { constant_expression concatenation }
net_concatenation ::= { net_concatenation_value { , net_concatenation_value } }
net_concatenation_value ::=
    hierarchical_net_identifier
    | hierarchical_net_identifier [expression] { [expression] }

| hierarchical_net_identifier [expression] { [expression] } [range_expression]
| hierarchical_net_identifier [range_expression]
| net_concatenation
variable_concatenation ::= { variable_concatenation_value
    { , variable_concatenation_value } }
variable_concatenation_value ::=
    hierarchical_variable_identifier
    | hierarchical_variable_identifier [expression] { [expression] }
    | hierarchical_variable_identifier [expression] { [expression] }
       [range_expression]
    | hierarchical_variable_identifier [range_expression]
    | variable_concatenation

A.9.2 Function Calls

constant_function_call ::= function_identifier { attribute_instance }
    (constant_expression { , constant_expression })
function_call ::= hierarchical_function_identifier{ attribute_instance }
    (expression { , expression })
genvar_function_call ::= genvar_function_identifier { attribute_instance }
    (constant_expression { , constant_expression })
system_function_call ::= system_function_identifier
    [(expression { , expression })]

A.9.3 Expressions

base_expression ::= expression
conditional_expression ::= expression1 ? { attribute_instance } expression2
    : expression3
constant_base_expression ::= constant_expression
constant_expression ::=
    constant_primary
    | unary_operator { attribute_instance } constant_primary
    | constant_expression binary_operator { attribute_instance } constant_expression
    | constant_expression ? { attribute_instance } constant_expression :
       constant_expression
    | string
constant_mintypmax_expression ::=
    constant_expression
    | constant_expression : constant_expression : constant_expression
constant_range_expression ::=
    constant_expression
    | msb_constant_expression : lsb_constant_expression
    | constant_base_expression + : width_constant_expression
    | constant_base_expression – : width_constant_expression

dimension_constant_expression ::= constant_expression
expression1 ::= expression
expression2 ::= expression
expression3 ::= expression
expression ::=
      primary
      | unary_operator { attribute_instance } primary
      | expression binary_operator { attribute_instance } expression
      | conditional_expression
      | string
lsb_constant_expression ::= constant_expression
mintypmax_expression ::=
      expression
      | expression : expression : expression
module_path_conditional_expression ::= module_path_expression ?
            { attribute_instance } module_path_expression : module_path_expression
module_path_expression ::=
      module_path_primary
      | unary_module_path_operator { attribute_instance } module_path_primary
      | module_path_expression binary_module_path_operator { attribute_instance }
            module_path_expression | module_path_conditional_expression
module_path_mintypmax_expression ::=
      module_path_expression
      | module_path_expression : module_path_expression : module_path_expression
msb_constant_expression ::= constant_expression
range_expression ::=
      expression
      | msb_constant_expression : lsb_constant_expression
      | base_expression + : width_constant_expression
      | base_expression – : width_constant_expression
width_constant_expression ::= constant_expression

A.9.4 Primaries

constant_primary ::=
      constant_concatenation
      | constant_function_call
      | (constant_mintypmax_expression)
      | constant_multiple_concatenation
      | genvar_identifier
      | number
      | parameter_identifier
      | specparam_identifier
module_path_primary ::=
      number
      | identifier

    | module_path_concatenation
    | module_path_multiple_concatenation
    | function_call
    | system_function_call
    | constant_function_call
    | (module_path_mintypmax_expression)
primary ::=
    number
    | hierarchical_identifier
    | hierarchical_identifier [expression] { [expression] }
    | hierarchical_identifier [expression] { [expression] } [range_expression]
    | hierarchical_identifier [range_expression]
    | concatenation
    | multiple_concatenation
    | function_call
    | system_function_call
    | constant_function_call
    | (mintypmax_expression)

A.9.5 Expression Left-Side Values

net_lvalue ::=
    hierarchical_net_identifier
    | hierarchical_net_identifier [constant_expression] { [constant_expression] }
    | hierarchical_net_identifier [constant_expression] { [constant_expression] }
        [constant_range_expression]
    | hierarchical_net_identifier [constant_range_expression]
    | net_concatenation
variable_lvalue ::=
    hierarchical_variable_identifier
    | hierarchical_variable_identifier [expression] { [expression] }
    | hierarchical_variable_identifier [expression] { [expression] }
        [range_expression]
    | hierarchical_variable_identifier [range_expression]
    | variable_concatenation

A.9.6 Operators

unary_operator ::=
    + | − | ! | ~ | & | ~& | | | ~| | ^ | ~^ | ^~
binary_operator ::=
    + | − | * | / | % | == | != | === | !== | && | || | **
    | < | <= | > | >= | & | | | ^ | ^~ | ~^ | >> | << | >>> | <<<
unary_module_path_operator ::=
    ! | ~ | & | ~& | | | ~| | ^ | ~^ | ^~

binary_module_path_operator ::=
        == | ! = | && | | | | & | | | ^ | ^ ~ | ~ ^

A.9.7 Numbers

number ::=
        decimal_number
        | octal_number
        | binary_number
        | hex_number
        | real_number
real_number[1] ::=
        unsigned_number . unsigned_number
        | unsigned_number [. unsigned_number] exp [sign] unsigned_number
exp ::= e | E
decimal_number ::=
        unsigned_number
        | [size] decimal_base unsigned_number
        | [size] decimal_base x_digit { _ }
        | [size] decimal_base z_digit { _ }
binary_number ::= [size] binary_base binary_value
octal_number ::= [size] octal_base octal_value
hex_number ::= [size] hex_base hex_value
sign ::= +|−
size ::= non_zero_unsigned_number
non_zero_unsigned_number[1] ::= non_zero_decimal_digit { _| decimal_digit}
unsigned_number[1] ::= decimal_digit { _| decimal_digit }
binary_value[1] ::= binary_digit { _| binary_digit }
octal_value[1] ::= octal_digit { _| octal_digit }
hex_value[1] ::= hex_digit { _| hex_digit }
decimal_base[1] ::= '[s | S]d | '[s | S]D
binary_base1 ::= '[s | S]b | '[s | S]B
octal_base1 ::= '[s | S]o | '[s | S]O
hex_base1 ::= '[s | S]h | '[s | S]H
non_zero_decimal_digit ::= 1 | 2 | 3 | 4 | 5 | 6 | 7 | 8 | 9
decimal_digit ::= 0 |1 | 2 | 3 | 4 | 5 | 6 | 7 | 8 | 9
binary_digit ::= x_digit | z_digit | 0 |1
octal_digit ::= x_digit | z_digit 0 |1 | 2 | 3 | 4 | 5 | 6 | 7
hex_digit ::=
        x_digit | z_digit | 0 |1 | 2 | 3 | 4 | 5 | 6 | 7 | 8 | 9
        | a | b | c | d | e | f | A | B | C | D | E | F
x_digit ::= x | X
z_digit ::= z | Z | ?

A.9.8 Strings

string ::= "{ Any_ASCII_Characters_except_new_line } "

A.10   GENERAL

A.10.1   Attributes

attribute_instance ::= (* attr_spec { , attr_spec } *)
attr_spec ::= attr_name = constant_expression
      | attr_name
attr_name ::= identifier

A.10.2   Comments

comment ::=
      one_line_comment
      | block_comment
one_line_comment ::= // comment_text \n
block_comment ::= /* comment_text */
comment_text ::= { Any_ASCII_character }

A.10.3   Identifiers

arrayed_identifier ::=
      simple_arrayed_identifier
      | escaped_arrayed_identifier
block_identifier ::= identifier
cell_identifier ::= identifier
config_identifier ::= identifier
escaped_arrayed_identifier ::= escaped_identifier [range]
escaped_hierarchical_identifier[4] ::=
      escaped_hierarchical_branch
      { .simple_hierarchical_branch | .escaped_hierarchical_branch }
escaped_identifier ::= \ { Any_ASCII_character_except_white_space } white_space
event_identifier ::= identifier
function_identifier ::= identifier
gate_instance_identifier ::= arrayed_identifier
generate_block_identifier ::= identifier
genvar_function_identifier ::= identifier /* Hierarchy disallowed */
genvar_identifier ::= identifier
hierarchical_block_identifier ::= hierarchical_identifier
hierarchical_event_identifier ::= hierarchical_identifier
hierarchical_function_identifier ::= hierarchical_identifier
hierarchical_identifier ::=
      simple_hierarchical_identifier

| escaped_hierarchical_identifier
hierarchical_net_identifier ::= hierarchical_identifier
hierarchical_variable_identifier ::= hierarchical_identifier
hierarchical_task_identifier ::= hierarchical_identifier
identifier ::=
  simple_identifier
  | escaped_identifier
inout_port_identifier ::= identifier
input_port_identifier ::= identifier
instance_identifier ::= identifier
library_identifier ::= identifier
memory_identifier ::= identifier
module_identifier ::= identifier
module_instance_identifier ::= arrayed_identifier
net_identifier ::= identifier
output_port_identifier ::= identifier
parameter_identifier ::= identifier
port_identifier ::= identifier
real_identifier ::= identifier
simple_arrayed_identifier ::= simple_identifier [range]
simple_hierarchical_identifier[3] ::=
  simple_hierarchical_branch [.escaped_identifier]
simple_identifier[2] ::= [a-zA-Z_] { [a-zA-Z0-9_$] }
specparam_identifier ::= identifier
system_function_identifier[5] ::= $[a-zA-Z0-9_$]{ [a-zA-Z0-9_$] }
system_task_identifier[5] ::= $[a-zA-Z0-9_$]{ [a-zA-Z0-9_$] }
task_identifier ::= identifier
terminal_identifier ::= identifier
text_macro_identifier ::= simple_identifier
topmodule_identifier ::= identifier
udp_identifier ::= identifier
udp_instance_identifier ::= arrayed_identifier
variable_identifier ::= identifier

A.10.4 Identifier Branches

simple_hierarchical_branch[3] ::=
  simple_identifier [[unsigned_number]]
  [{ .simple_identifier [[unsigned_number]] }]
escaped_hierarchical_branch[4] ::=
  escaped_identifier [[unsigned_number]]
  [{ .escaped_identifier [[unsigned_number]] }]

A.10.5 White Space

white_space ::= space | tab | newline | eof[6]

NOTES

1. Embedded spaces are illegal.

2. A simple_identifier and arrayed_reference shall start with an alpha or underscore (_) character, shall have at least one character and shall not have any spaces.

3. The period (.) in simple_hierarchical_identifier and simple_hierarchical_ branch shall not be preceded or followed by white_space.

4. The period in escaped_hierarchical_identifier and escaped_hierarchical_ branch shall be preceded by white_space, but shall not be followed by white_space.

5. The $ character in a system_function_identifier or system_task_identifier shall not be followed by white_space. A system_function_identifier or system_task_identifier shall not be escaped.

6. End of file.

INDEX

Digital System Designs and Practices Ming-Bo Lin
© 2008 John Wiley & Sons (Asia) Pte Ltd